Cambridge A Level Qualification Chemistry 9701

Azhar ul Haque Sario

Published by Azhar ul Haque Sario, 2023.

CAMBRIDGE A LEVEL QUALIFICATION CHEMISTRY 9701

First edition. September 10, 2023.

Copyright © 2023 Azhar ul Haque Sario.

ISBN: 979-8223182399

Written by Azhar ul Haque Sario.

Table of Contents

....1

Atoms and Ions 3

Isotopes 6

Atomic Theory 8

Ionization Energy in Periodic Table 12

Measurement Units for Particles 16

The Avogadro Constant and Mole 17

Formulas, Equations, and Compounds 18

The Mole Concept and Stoichiometry in Chemistry 21

Electronegativity: Properties and Applications 24

Ionic Bonding 26

Metallic Bonding 27

Examples of Covalent Bonding Molecules 28

Covalent Bonds and Properties 30

Molecule Shapes and Bond Angles with VSEPR Theory 33

Intermolecular Forces 35

Dot-and-Cross Diagrams 39

Pressure in Gases, Ideal Gases, and the Ideal Gas Equation 40

Structure and Bonding 42

Enthalpy Change, Activation Energy, and Standard Conditions in Chemical Reactions 45

Hess's Law 48

The Role of Oxidation-Reduction Reactions 51

Chemical Equilibria 54

Brønsted–Lowry theory of acids and bases 59

Understanding the Factors and Calculating the Rate of Chemical Reactions 63

The Role of Temperature and Activation Energy in Chemical Reactions 65

Catalysts 67

Properties and Variations of Atoms and Ions 69

Period 3 Chemical Periodicity 71

Elements' Periodic Trends 76

Properties of Group 2 elements 78

Group 17 characteristics 81

Properties of Halogens and Hydrogen Halides 83

Halide Ions' Reactions 85

Chlorine Reactions 87

Sulfur and Nitrogen 88

Organic Compound Basics 91

Organic Reaction Types 92

Organic Reactions: Characteristics 94

Shapes of Organic Molecules 96

Isomerism 98

Alkanes 100

Alkenes..103

Halogenoalkanes 107

Alcohols 110

Carboxylic acids 116

Esters 118

Primary Amines 119

Nitriles and Hydroxynitriles 120

Polymerization by Addition 122

Organic Synthesis 124

Infrared Spectroscopy 126

Mass Spectrometry 127

Chemical Energetics 131

Electrochemistry 161

Equilibria 188

Reaction Kinetics 212

Homogeneous and heterogeneous catalysts 231

Group 2 238

Chemistry of Transition Elements 241

Hydrocarbons 304

Halogen Compounds 323

Hydroxy Compounds 327

Carboxylic Acids and Derivatives 348

Nitrogen Compounds 376

Polymerisation 416

Organic Synthesis 434

Analytical Techniques 440

Cambridge A Level Qualification Chemistry 9701

Atoms and Ions

Imagine an atom as a tiny, yet fascinating world! It consists of a central core called the nucleus, which is made up of positively charged particles called protons and neutrally charged particles called neutrons. The nucleus is so compact that it makes up almost the entire mass of an atom. But the real magic happens in the vast empty space surrounding the nucleus, where we find tiny negatively charged particles known as electrons whizzing around in predictable orbits called shells. These shells are like imaginary orbits that surround the nucleus. So, essentially, atoms are made up of mostly empty space, albeit with a small, but dense, positively charged nucleus at the center.

The electrons are found in their shells within the empty space surrounding the nucleus. This unique arrangement allows atoms to interact with each other in fascinating ways and gives rise to everything around us! Let me describe the three magical particles that make up the world we all know and love: protons, neutrons, and electrons. Firstly, protons are positively charged particles that reside in the nucleus of an atom. They have a relative mass of 1, and their charge is equal and opposite to that of the electron. This means that while the electron is negatively charged, the proton is positively charged, and they cancel each other out, resulting in a neutral atom. Next up are neutrons, which are neutral particles that also exist in the nucleus of an atom. They have a relative mass of approximately 1, which is slightly larger than that of the proton. Neutrons are important in maintaining the stability of the nucleus and taking up space, but they do not participate in chemical reactions. Last but not least, electrons are negatively charged particles that orbit the nucleus of an atom. Electrons have a tiny relative mass compared to protons and neutrons, and their negative charge is equal and opposite to that of the proton. The arrangement of electrons in an atom's energy levels determines the chemical properties of the element.

The atomic number of an atom refers to the total number of protons that are present within the atom's nucleus. It is also called the proton number. Each element on the periodic table has a unique atomic number, as the number of protons found in an atom determines what the element is. The mass number, also known as the nucleon number, refers to the total number of particles found in an atom's nucleus. This includes both protons and neutrons. The mass number is always a whole number, and it can be used to determine the number of neutrons in an atom by subtracting the atomic number from the mass number. To elaborate a little more, the atomic number is crucial in determining an element's properties. It is the defining characteristic of an element, as each element has a unique number of protons in its nucleus. For example, all carbon atoms have six protons in their nuclei, giving them an atomic number of 6. This also means that all carbon atoms will behave chemically in a similar way, as they all have the same number of protons. The mass number, on the other hand, is more of an indicator of an atom's weight or size. Since the mass number takes into account the total number of particles in an atom's nucleus, it is affected by both protons and neutrons. For example, an atom of carbon with six protons and six neutrons will have a mass number of 12. However, an atom of carbon with six protons and seven neutrons will have a mass number of 13.

When it comes to the distribution of mass within an atom, most of it is concentrated in the nucleus, which is the central region of the atom where protons and neutrons are located. The nucleus has a positive charge overall, due to the presence of the positively charged protons. The electrons, which are negatively charged particles that orbit the nucleus, reside in specific energy levels or shells that are located outside the nucleus. Compared to the nucleus, electrons are very lightweight. The protons and electrons in an atom are of equal magnitude, which means the positive charge of the protons exactly balances out the negative charge of the electrons, resulting in a net neutral charge for the whole atom. The number of electrons in an atom is what determines an atom's chemical properties, as they are the particles that are involved in chemical bonding. The overall size of an atom is largely determined by the distribution of the electrons, which comprise the majority of an atom's volume. The nucleus, which contains the protons and neutrons, makes up a very small amount of the total volume of the atom. The energy level or distance between the electrons and the nucleus is what determines an atom's size, with larger energy levels indicating larger atoms.

When a beam of charged particles like electrons, protons, or ions is placed in an electric field, it experiences a force due to the electric field. The behavior of the beam depends on the properties of the particles being accelerated by the electric field. Since protons and electrons are charged particles, they experience similar forces in the electric field, while neutrons, which are neutral, do not experience any force in the electric field. If a beam of protons and electrons is accelerated by an electric field, both particles experience an electric force in the same direction. However, since protons are positively charged and electrons are negatively charged, they will experience opposite directions of force. This causes the protons and electrons to move in opposite directions in the electric field. If the beam of protons and electrons is moving at the same velocity, the electrons, being lighter, will move more quickly than the protons. As for neutrons, they are electrically neutral and do not experience any force in the electric field. They will move along with the protons and electrons but will not be deflected or accelerated by the electric field. The behavior of a beam of charged particles in an electric field can be observed in many real-life scenarios. For example, in a particle accelerator, a beam of charged particles like electrons or protons is accelerated by electric fields to very high speeds before being directed into a target. Medical imaging technology such as MRI also uses the principles of charged particles in an electric field to visualize internal structures. By using a magnetic field to temporarily polarize the nuclei in an object's atoms and then releasing them, these protons will return to their original alignment while releasing energy, which is detected by magnetic sensors.

To start, let's first understand what each of these terms means: Atomic number or proton number: The number of protons in an atom's nucleus that determines its identity as an element. The mass number or nucleon number is the sum of the protons and the neutrons found in an atom's nucleus. Charge: The electrical property of an atom or ion, which can be positive, negative, or neutral. Now, let's consider the case of an atom. If we know the atomic number and mass number of an atom, we can easily find the number of protons and neutrons present in the nucleus. For example, a carbon atom with an atomic number of 6 and a mass number of 12 consists of 6 protons and 6 neutrons. If we want to determine the number of electrons present in a neutral atom, we can simply use the atomic number to know the number of electrons present in the atom. In the case of carbon, it has 6 electrons surrounding the nucleus. Moving on to ions, the number of protons in an ion remains the same as in the corresponding atom, but the number of electrons changes according to the ion's charge. For example, if a carbon atom loses two electrons, it becomes a positively charged C^{2+} ion. However, the number of protons remains the same at 6. If instead, the carbon atom gains 2 electrons, it becomes a negatively charged C^{2-} ion. Again, the number of protons remains 6, but the number of electrons is now 8. Finally, to find the number of neutrons present in ions, we can subtract the number of protons from the mass number. For example, if we have a C^{2+} ion with 6 protons and a mass number of 12, then it must have 6 neutrons.

Atomic radius refers to the distance between the nucleus and the outermost electron shell of an atom. On the other hand, ionic radius is the distance between the nucleus and the outermost electron shell of an ion. The size of the atomic and ionic radius can vary across a period and down a group in the periodic table. Across a period, as we move from left to right, the atomic and ionic radii generally decrease. This is because, although the number of protons in the nucleus increases, the electrons added to the same level of the outermost shell do not shield this from the nucleus. As a result, the attractive force of the protons in the nucleus increased, causing the electrons to be pulled closer to the nucleus. This effect causes the atomic and ionic radii to be smaller for elements on the right than the elements on the left. Down a group, as we move from top to bottom, there is generally an increase in both atomic and ionic radii. This can be explained by the fact that as we move down a group, the number of energy levels occupied by the electrons increases. Further from the nucleus, the electrons experience reduced attractive forces and shielding effects by the inner electrons. Thus, the electrons can exist farther away from the nucleus, making the atomic and ionic radii larger. In addition, for ions with the same number of electrons, ions at the bottom of the group would be larger due to addition of extra energy levels. It is essential to note that the size of the atomic and ionic radius might be affected by various environmental factors as well, including the type of material being measured. The size of atoms and ions can also vary due to changes in electron distribution and charge, which can affect the ionization energy and electronegativity of the element.

Isotopes

An isotope can be defined as one of multiple variations of an element, each with a different number of neutrons in its nucleus. Elements are composed of atoms, which in turn consist of protons, neutrons, and electrons. While the number of protons in an atom (known as its atomic number) determines the element that it represents, the sum of the protons and neutrons in the nucleus determines the isotope of that element. Essentially, isotopes are different versions of the same element, differentiated by the number of neutrons they contain. Some isotopes of an element can be stable, while others can be radioactive and decay over time. The atomic mass, or weight, of an element listed on the periodic table is actually the average atomic mass of all its stable isotopes. In simpler terms, imagine you have a fruit bowl with several apples in it. You can think of each apple as an atom of a particular element. Now, imagine some of the apples are slightly larger or smaller than the others. These variations in size represent different isotopes of the same element. So, just as each apple in the bowl is still an apple, each isotope is still the same element, just with a different number of neutrons.

The notation xyA is used to represent a specific isotope of an element. Here, 'x' represents the mass number, and 'y' represents the atomic number or proton number. The mass number, x, is the sum of the number of protons and the number of neutrons present in the nucleus of an atom. The atomic number, y, is the number of protons in the nucleus, which determines the identity of the element. So, imagine an atom of an element with an atomic number of y. This atom may have different isotopes, each with a different number of neutrons in the nucleus. Let's say we have two isotopes of this element, one with 24 neutrons and the other with 25 neutrons. The first isotope would have a mass number of 24 + y, and the second isotope would have a mass number of 25 + y. Using this notation, we can easily identify the isotope of an element by looking at its mass number and atomic number. For example, the element carbon has three naturally occurring isotopes. Carbon-12 has a mass number of 12 and an atomic number of 6, carbon-13 has a mass number of 13 and an atomic number of 6, and carbon-14 has an atom that has 6 protons in its nucleus (atomic number) and has a total of 14 particles including protons and neutrons (mass number). In other words, the xyA notation is an easy and convenient way to represent isotopes and distinguish between different versions of an element. It helps us to understand the composition of the nuclei of atoms and helps us to study the properties of elements and their isotopes.

In chemistry, the behavior of an element is determined by the arrangement of its electrons in the outermost shell, or valence shell. This arrangement determines how an element will react with other substances to form compounds. Since isotopes of an element have the same number of protons, they have identical electron configurations, meaning they will have the same chemical properties as each other. For instance, let's consider the element carbon, which has three naturally occurring isotopes: carbon-12, carbon-13, and carbon-14. While these isotopes have different numbers of neutrons in their nuclei, they all have the same number of electrons in their respective shells. This means that they will all behave in the same way chemically, and they can all form the same types of compounds. So, while isotopes have different masses and may undergo different types of radioactive decay, their chemical reactivity is not affected since chemical reactions primarily involve the electrons in the outermost valence shell. This is why isotopes of the same element are useful in many areas such as medicine and industry, where they can be used interchangeably to fulfill the same roles in various chemical processes without any difference in outcome.

Isotopes are different forms of the same element that vary in their atomic mass due to differing numbers of neutrons in the nucleus while retaining the same atomic number. Since physical properties, such as mass and density, arise due to the arrangement of atoms in a substance, isotopes of the same element can have different physical properties. The mass of an atom is determined by the sum of the masses of its protons, neutrons, and electrons. Although isotopes of the same element have the same number of protons, they can differ in their number of neutrons. As a result, isotopes differ in their atomic mass. The isotopes with more neutrons would have a greater mass resulting in different masses of atoms. In general, the higher the mass of an isotope, the more energy it takes to move or change its motion, which affects its physical properties. This difference in atomic mass directly affects the density of isotopes. The definition of density refers to how much a particular substance weighs in relation to how much space it takes up or its volume, in a given or measured unit. The mass of an isotope ultimately affects the density of the element. As a result, isotopes with higher mass have higher densities than their counterparts. For example, the isotopes of neon are neon-20, neon-21, and neon-22, which vary only in their number of neutrons. Ne-22 is the heaviest isotope of neon and has the highest density. Ne-20 is the lightest with the lowest density while Ne-21 is in between. This difference in mass and density affects the physical properties of materials made from isotopes. For example, the isotopes of uranium, like uranium-235 and uranium-238, have significantly different masses and densities, which affect their ability to undergo nuclear fission reactions.

Atomic Theory

Shells, sub-shells, and orbitals are fundamental terms in atomic theory used to describe the arrangement of electrons around a nucleus. Shells refer to the energy levels of an atom, with the first, innermost shell being the lowest energy level and the outermost shell being the highest. Each shell can hold a specific number of electrons, with the first shell accommodating up to two electrons, the second shell accommodating up to eight electrons, and so on. Sub-shells, on the other hand, are smaller energy divisions within a shell and are labeled using letters. The first shell contains only one sub-shell, labeled as s. The second shell contains two sub-shells, labeled as s and p. The third shell has three sub-shells, labeled as s, p, and d. The fourth shell also has four sub-shells, labeled as s, p, d, and f. Lastly, orbitals refer to the regions within a sub-shell where an electron is most likely to be found. Each orbital can accommodate only two electrons with opposite spins. The s sub-shell contains one orbital, the p sub-shell contains three orbitals, the d sub-shell contains five orbitals, and the f sub-shell contains seven orbitals.

The principal quantum number (n) is a term used in atomic theory that describes the energy level of an electron within an atom. In simpler terms, it refers to the distance of the electron from the nucleus. The n value can be a positive integer starting from 1 and goes up to infinity. It defines the size of the electron cloud and determines the energy of the electron. The higher the value of n, the further away from the nucleus the electron is and the greater amount of energy the electron has. Each principal quantum number is associated with a specific shell in an atom. For instance, an atom with n = 1 would belong to the first shell (also known as the K shell) and can hold up to two electrons. An atom with n = 2 would belong to the second shell (or L shell) and can hold up to eight electrons. The maximum number of electrons an atom can hold is $2n^2$. The principal quantum number plays a crucial role in determining the electronic configuration of an atom, which in turn affects its chemical and physical properties. It also affects the atomic size, which grows without bounds as n increases, showing that an atom with higher n is bigger than the atom with lower n.

The ground state is an essential concept in atomic theory that refers to the lowest possible energy level of an atom's electrons. It is a term that is primarily used in discussing electronic configurations. When an electron is in its ground state, it means that it is in the lowest possible energy level around the atomic nucleus. In this state, it is most stable, and it does not gain or lose any energy. The configuration of electrons in the ground state is known as the atomic structure of an element. The ground state is significant because it plays a crucial role in the chemical behavior of an atom. When an atom is in its ground state, it is considered to have its neutral, unaffected state, and would not bond readily with other atoms. It is the foundation level of the electronic configuration of an atom and determines its properties and behavior. For example, an atom of hydrogen in its ground state has one electron in its electron shell with the configuration $1s^1$. The first shell can hold only two electrons. So, the atom's electronic configuration is complete with one electron in its ground state. However, when an atom absorbs energy, it becomes excited and moves to a higher energy state known as the excited state. Electrons jump to a higher energy level when they receive energy from the surroundings. An atom in this state is in an unstable condition, and it doesn't follow the rules of the ground state configuration.

In atomic theory, sub-shells refer to the smaller energy divisions within a shell, and are labeled using letters. The three most common sub-shells are s, p, and d. Each sub-shell contains a different number of orbitals and can hold a specific number of electrons. The s sub-shell contains only one orbital, which is spherical in shape, and can hold a maximum of two electrons. It is the first sub-shell to be filled in an element's electronic configuration, and its electrons have the lowest energy level. To fill the s sub-shell, the atom needs at least one electron, and an additional electron will pair up with the first. When filled with two electrons it is considered as the s2 sub-shell configuration. The p sub-shell has three orbitals, each of different orientation, and can hold up to six electrons. Its shape looks like a dumbbell, with two nodes on opposite sides that indicate the probability of finding the electron in specific regions surrounding the nucleus. The p sub-shell follows the s sub-shell and is labeled as 2p.

To fill the p sub-shell, the atom needs at least three electrons. When six electrons fill the p sub-shell, it is considered as p6 sub-shell configuration. Lastly, the d sub-shell has five orbitals, each with a different complex shape, and can hold up to ten electrons. In the d sub-shell orbitals, the probability of finding an electron is in more than one direction. The d sub-shell follows the p sub-shell, and it is labeled as 3d. To fill the d sub-shell, the atom needs at least five electrons, and ten electrons in total if a full d sub-shell is required, which is the d10 sub-shell configuration.

In atomic theory, the principle of increasing energy levels of sub-shells is essential in determining the electronic configurations of elements. Within the first three shells, the sub-shells increase in energy in a regular pattern. The first shell (n = 1) contains an s sub-shell, which has the lowest energy level. The second shell (n = 2) has an s sub-shell and a higher energy level p sub-shell. The p sub-shell is higher in energy than the s, as it contains more electrons with higher energy levels. Therefore, the pattern in the second shell is s < p. The third shell (n = 3) follows the pattern established in the first two shells, containing s, p, and d sub-shells in ascending order of energy. The s sub-shell is once again at the lowest energy level, followed by the higher energy level p and d sub-shells. The order of increasing energy levels within the third shell is s < p < d. The 4s and the 4p sub-shells are in the fourth shell (n = 4) and follow a different trend. The 4s sub-shell has a lower energy level than the 3d sub-shell, which belongs to the higher energy third shell. So, in the fourth shell, the order of the sub-shells in increasing energy is 4s < 3d. The 4p sub-shell follows the same pattern as before, higher in energy than the 4s, with the order of increasing energy being 4s < 3d < 4p.

Electronic configuration is an essential concept in atomic theory that describes the arrangement of electrons in an atom. It is represented as a series of numbers and letters that indicate the number of electrons in each sub-shell of the atom. To write the electronic configuration of an atom, we start with the lower energy levels and move upwards, filling each sub-shell before moving to the next. The number before each letter represents the principal quantum number, which is equal to the number of shells. The letter after the number represents the sub-shell, and the superscript represents the number of electrons in that sub-shell. For example, the electronic configuration of hydrogen, which has one electron, is denoted as $1s^1$. This indicates that it has one electron in the first shell (n=1) and in the s sub-shell. The superscript indicates the number of electrons, which in this case is one. The electronic configuration of helium, which has two electrons, is denoted as $1s^2$. The first number (1) denotes the number of the shell, the letter (s) denotes the sub-shell, and the superscript denotes the number of the electrons (2), indicating that helium has a total of two electrons with both electrons occupying the s sub-shell. For atoms with more than two electrons, the electronic configuration becomes more complex. An atom like nitrogen with seven electrons will have the electronic configuration $1s^2\,2s^2\,2p^3$. The lower energy levels are filled first, starting with the 1s sub-shell, followed by the 2s and then the 2p sub-shells. The superscripts represent the total number of electrons in each sub-shell, with the 2p sub-shell having three electrons occupying its three orbitals.

Electronic configuration plays an essential role in determining the properties and behavior of atoms and molecules. It explains how electrons are arranged within an atom's energy levels and how they occupy orbitals in a way that minimizes the inter-electron repulsion. Electrons are negatively charged particles that occupy energy levels surrounding the atomic nucleus. The energy levels are divided into sub-shells that correspond to the shape of the electron cloud. Each sub-shell can hold a specific number of electrons, and the electrons in a given sub-shell have the same energy level. Electronic configurations follow the Aufbau principle and the Pauli exclusion principle, which help in determining the energy levels of electrons and their inter-electron repulsion. According to the Aufbau principle, electrons fill the lower energy sub-shells before moving to the higher ones.

The Pauli exclusion principle states that each electron in an atom has a unique set of quantum numbers, including spin and orientation, and no two electrons in an atom can have the same set of quantum numbers. The inter-electron repulsion arises from the negative charges of the electrons within an atom. Electrons in the same orbital or sub-shell experience higher repulsion due to their closer proximity, making these configurations less stable. To minimize electron-electron repulsion, atoms and molecules arrange electrons in orbitals within sub-shells according to Hund's rule. Hund's rule states that, when electrons are placed in degenerate orbitals (orbitals of equal energy), they prefer to occupy different orbitals with parallel spins. This allows them to maintain the maximum distance between themselves and reduce the overall repulsion energy between electrons. For example, in the electronic configuration of carbon, 1s2 2s2 2p2, it means the first shell has two electrons filling the s orbital, while the second shell has an s orbital filled with two electrons and a p orbital with two electrons, following Hund's rule.

Electronic configuration refers to the arrangement of electrons in an atom or ion within its energy levels or shells and sub-shells. The electronic configuration of an atom or ion may be determined using the atomic number or proton number and the charge on the ion. The electronic configuration may be written in two conventions, the full electronic configuration and the shorthand configuration. The full electronic configuration details every shell, sub-shell, and orbital in which electrons are present. The shorthand configuration employs noble gas notation where the atomic symbol of a noble gas is used to represent the electronic configuration of the preceding elements. Square brackets indicate the noble gas configuration, and the remaining shell orbitals are listed afterward. For example, let us determine the electronic configuration of Fe (Iron), which has an atomic number of 26. Using the full electronic configuration, we start by filling the shells and sub-shells in order of increasing energy levels, following the Aufbau principle and Pauli's exclusion principle. $1s^2, 2s^2, 2p^6, 3s^2, 3p^6, 4s^2, 3d^6$ The complete electron configuration of Fe is $1s^2 2s^2 2p^6 3s^2 3p^6 3d^6 4s^2$. Alternatively, using the shorthand configuration in noble gas notation, we find the noble gas that precedes Fe, which is Argon (Ar). The noble gas is written in brackets $[Ar] 4s^2 3d^6$ This notation represents the Argon core structure, followed by the configuration of the additional electrons found in Fe's outermost shell. Another example is the determination of the electronic configuration of O^{2-} using the atomic number of oxygen, which is 8, and the charge of the ion, which is -2. Oxygen loses two electrons to form the negative ion, which means it has ten electrons. $1s^2, 2s^2, 2p^4$ The complete configuration is written as $1s^2 2s^2 2p^4$. For the shorthand configuration, we find the noble gas closest to O, which is Helium (He). $[He] 2s^2 2p^4$ This notation represents the Helium core structure followed by the configuration of the additional electrons found in oxygen's outermost shell.

The electrons in boxes notation is a way to represent the distribution of electrons in an atom's energy levels or shells. It is a system that is used to show which sublevels are filled with electrons in an atom. This notation system is often used to represent elements in their ground state or the state where they are most stable. For example, let's consider Iron (Fe). The symbol for Iron is "Fe" and it has an atomic number of 26. The electrons of Iron are distributed in different energy levels and sublevels. Electrons fill shells from the innermost shell to outermost shell. Therefore, the first two electrons are located in the first shell (1s sublevel) and the remaining 24 electrons are distributed in the second and third energy levels (2s, 2p, 3s, 3p, 3d sublevels). The electron configuration of Iron can be abbreviated using the electron-in-boxes notation. In this notation, the different energy levels are represented by horizontal rows, and electrons are represented by boxes placed within those rows in the sub-levels corresponding to the atom. The boxes are always filled from left to right, and each sub level can hold only a certain number of electrons. For Fe, the electron configuration has 2 electrons in the 1s level, 2 electrons in the 2s level, 6 electrons in the 2p level, 2 electrons in the 3s level, 6 electrons in the 3p level, and 6 electrons in the 3d level. The electron configuration in notation [Ar] represents that the electron configuration of Iron can be represented as the noble gas Argon (Ar) with all its completed energy sublevels plus the remaining 4 electrons in the Iron's outmost shell (3d and 4s).

The shapes of orbitals are essential to understanding the behavior of electrons in atoms. There are four types of orbitals - s, p, d, and f - each with their unique shape and orientation. In this explanation, I will be focusing on the first two types of orbitals, s and p. The S orbital is spherical and symmetrical. It is represented by a single sphere in three-dimensional space and is centered around the atomic nucleus. This shape can be thought of as a simple electron cloud or probability density function. The probability of finding an electron is high in the region near the nucleus and decreases as you go further away from it. The s-orbital has nodes, regions of zero probability, and is always spherically symmetric. On the other hand, the P orbital is dumbbell-shaped with two lobes that are symmetrical about the nucleus. These two lobes represent the regions where there is a high probability of finding an electron. The P orbital has two values for the magnetic quantum number, $m = -1$ and $m = +1$. These two values correspond to the two lobes of electron density in the x, y, or z direction in three-dimensional space. Therefore, the shape of each P orbital can differ depending on the applied direction of the magnetic field. Additionally, each P orbital has a planar node, a region where the probability of finding an electron is zero, that passes through the nucleus perpendicular to the axis of rotation, and dividing the orbital into two identical regions. The P orbital also has a compression node, where the probability of finding an electron is zero, that forms between the two regions of electron density around the nucleus (the lobes of the P orbital). It is also important to note that both s and p orbitals are non-directional, meaning they do not have a preferred direction in space. This feature of the shape of the orbitals reflects the fact that electrons have wave-particle duality and can behave like waves instead of individual particles.

A free radical is a molecule, atom, or a chemical species that contains one or more unpaired electrons in its outermost energy level. In chemistry, all atoms and molecules need to have a paired electrons in their outermost shells to remain stable and to avoid reactivity. However, in the case of free radicals, there are unpaired electrons that make them highly reactive and unstable. These unpaired electrons seek to pair up with an electron from another atom or molecule to satisfy the octet rule of chemical bonding, making them very reactive and highly reactive. One of the primary sources of free radicals is exposure to ultraviolet radiation from the sun or radiation from other sources. When these types of radiation hit atoms or molecules in the body, they can cause the molecules to lose an electron in their outermost shell, creating a free radical. Other sources of free radicals include chemical byproducts from metabolic processes in the body, as well as environmental pollutants, smoking, and some medications. Free radicals act as oxidizing agents as they seek to form covalent bonds with other atoms or molecules to fulfill their need for paired electrons. When free radicals react with other molecules, they can cause damage to cells and tissues. This damage leads to a condition known as oxidative stress that can lead to a range of diseases, including cancer, diabetes, cardiovascular diseases, and age-related degenerative diseases. However, free radicals can also play a critical role in the process of cell differentiation, signaling, and immune response. They are involved in the regulation of cell growth and signaling pathways. Hence, the body has developed several antioxidant mechanisms that work to neutralize free radicals by donating or receiving electrons to stabilize the unpaired electron.

Ionization Energy in Periodic Table

First ionization energy, or IE, is the amount of energy required to remove one electron from an isolated neutral atom. It is a fundamental concept in chemistry and plays a crucial role in understanding the behavior of atoms and their interactions with other atoms. Think of it like this: an electron is like a shy little creature that's been living inside an atom. Removing it from its comfortable home requires some serious coaxing and energy, hence the need for ionization energy. This concept is incredibly important because it helps us understand the reactivity of different elements and how they react with other elements. The harder it is to remove an electron from an atom, the less reactive that element will be. Elements with low ionization energy are more likely to form chemical bonds, while elements with high ionization energy are more stable. Overall, the first ionization energy is a fascinating and critical concept that enables us to understand the properties of the elements and the behavior of matter itself.

When we talk about ionization energies, there are actually a few different types we might refer to. The most common are the first, second, and subsequent ionization energies. In each case, we're referring to the amount of energy required to remove an electron from an atom or ion. The first ionization energy (IE1) is defined as the energy required to remove the first electron from a neutral atom or molecule. In equation form, this might look like: $A(g) \rightarrow A^+(g) + e^-$ (where A represents an atom) This equation shows that a neutral atom (A) loses an electron (e^-) to become a positively charged ion (A^+). The energy required to make this happen is the first ionization energy. The second ionization energy (IE2) is defined as the energy required to remove a second electron from an ion. This equation will look similar to the equation for the first ionization energy, but with an extra positive charge on the ion: $A^+(g) \rightarrow A^{2+}(g) + e^-$ Here, we're removing a second electron from a positively charged ion (A^+), resulting in an even more positively charged ion (A^{2+}). Subsequent ionization energies refer to the energy required to remove further electrons from an ion. So for example, the third ionization energy (IE3) would be the energy required to remove a third electron from an ion, and so on. Each subsequent ionization energy will require more energy than the one before it, as it's progressively harder to remove electrons from an ion.

The Periodic Table is a fundamental tool in chemistry that organizes the elements based on their atomic structure and properties, including ionization energy. The ionization energy of an element is a measure of how much energy is required to remove an electron from an atom, forming a positively charged ion. When we look at the ionization energies of elements across a period (horizontally), we see that they generally increase from left to right. This is because as we go across a period, the atomic number of the elements increases, meaning there are more protons in the nucleus. This results in a stronger pull on the electrons, making it harder to remove them, and hence requiring more energy. The increase in ionization energy across a period is not constant, but rather shows a gradual increase until a peak, followed by a sudden increase at the beginning of the next period. On the other hand, the ionization energies of elements down a group (vertically) tend to decrease. This is due to the fact that as we move down a group, the number of electron shells and shielding effect also increases, resulting in a lower effective attraction between the nucleus and valence electrons, which are farther away. Therefore, these electrons are more easily removed, and require less energy. The decrease in ionization energy down a group is generally steady and consistent. There are, however, some exceptions to these general trends. For example, elements that have completely filled subshells (such as helium) are especially stable and have higher ionization energies due to their fully occupied energy levels. Understanding the trends in ionization energy across the Periodic Table allows us to predict the behavior of different elements and their interactions with other substances. These trends also provide valuable insights into the mechanisms of chemical reactions, how elements bond together, and the properties of matter in general.

Successive ionization energies are the energies required to remove successive electrons from an atom or ion. As we proceed to remove each electron, the ion becomes more positively charged. This increase in charge causes the remaining electrons to experience a stronger electrostatic force due to the attraction to the increased positive nucleus, making them harder to remove. When we look at the successive ionization energies of an element, we see that the amount of energy required to remove the first electron is usually the lowest, followed by an increase in energy required for the next successive ionization energies. This trend continues until a maximal value is reached when there are only a few electrons left. For example, the ionization energy of magnesium increases significantly after the second ionization energy. Successive ionization energies are an important tool for understanding the electron configuration and chemical behavior of different elements. By analyzing the variation of ionization energies, we can determine the number of electrons in an element's outermost shell, which informs their reactivity and chemical behavior. For example, elements with a low ionization energy (like alkali metals) are more reactive because they have a strong tendency to lose electrons to form cations. Conversely, elements that have high ionization energies (like noble gases) are less reactive and resist forming chemical bonds. The increase in successive ionization energies can also provide insight into the electron configuration of an atom or ion. For example, when there is a significant jump in ionization energy, this indicates that a new energy level is being reached, rather than another electron from the same energy level being removed. These trends can be used to identify the elements or ions present in a particular chemical substance.

Ionization energy refers to the energy needed to dislodge an electron from an atom or molecule. This energy is necessary because it takes a certain amount of force to overcome the attractive pull between the positively charged nucleus and negatively charged electrons that orbit around it. The atomic nucleus contains the positively charged protons and uncharged neutrons, tightly bound together at the center of the atom. Orbiting around the nucleus are negatively charged electrons, which are held in place by the electrostatic force of attraction between their negatively charged electrons and the positively charged nucleus. The outermost electrons, also known as valence electrons, are the ones most easily removed and determine the chemical behavior of an element. The amount of energy required to remove an electron is influenced by the strength of the electrostatic attraction between the nucleus and the electron. The strength of this attraction depends on two factors: the distance between the nucleus and the electron, and the number of protons in the nucleus. As we move across a period of the Periodic Table, the number of protons in the nucleus increase, meaning the positive charge of the nucleus is increasing. This results in a higher electrostatic attraction between the nucleus and outer electron making it tougher to remove the electron. On the other hand, as we move down a group in the Periodic Table, the number of energy levels and the shielding of inner shells between the nucleus and valence electrons increase. This weaker electrostatic force required less energy to remove electrons, hence the ionization energy decreases.

The ionization energy of an element is influenced by a range of factors. These factors can be broadly categorized into four main areas: nuclear charge, atomic/ionic radius, shielding by inner shells and sub-shells, and spin-pair repulsion. Effective nuclear charge or nuclear charge is the extent of the positive charge that's felt by the outermost electrons in an atom. If this charge is higher, then the outer electrons are drawn more strongly to the nucleus, making it more challenging to remove them, which consequently needs more energy. For example, elements on the right side of the Periodic Table, which have high nuclear charges, exhibit high ionization energies. Atomic or ionic radius is another important factor affecting ionization energy. As the atomic or ionic radius of an element increases, the distance between the outer electron and nucleus also increases. Therefore, it becomes easier to remove the outer electron and the ionization energy decreases. For example, the alkali metals, which have larger atomic radii, have lower ionization energies. Shielding by the inner electrons is also important in determining ionization energies. As we move down a group or column, the number of inner shells and sub-shells increases, and electrons in these shells partially shield the outer electrons from the electrostatic attraction of the nucleus. This reduces the effective nuclear charge experienced by the valence electrons, making them easier to remove and requiring less energy to do so. Finally, spin-pair repulsion is a factor that influences the ionization energy of elements. In the case of atoms with more than one electron in an orbital, each electron has spin, either up or down. When two electrons share the same orbital in a pair, they will repulse, exerting an extra force. This makes it harder to remove an electron from an orbital with paired electrons. For example, the half-filled and filled subshells in elements such as nitrogen or oxygen, have repulsions that make it harder to remove electrons, requiring a relatively higher amount of energy.

One of the many ways to determine an element's electronic configuration is by analyzing its successive ionization energy data. This can help us to identify the number of electrons for each energy shell and subshell, and how they interact with other atoms and molecules. Ionization energies follow a general trend and increase as each electron is removed from an atom or molecule. This means that the first ionization energy is typically the lowest, while each subsequent ionization energy is higher than the previous ones. The amount of energy required to remove each electron depends on the electron's level of bonding and the shielding of the inner shells and subshells that reduce the charge effect felt by the outer electrons. By analyzing the trend in the successive ionization energies, we can determine how many electrons the element has in each energy shell and subshell. For example, an element that exhibits a significant jump in ionization energy after losing two electrons indicates that the third electron is located in a new energy level, meaning there must be two filled shells. Similarly, if an element shows minimal or no change in ionization energy after a certain point, it means that the next electron to be removed is from a fully-filled subshell. Successive ionization energy data also provides insight into the elements' electron configurations during chemical reactions. By looking at the specific ionization energy sequences, we can predict how elements may bond together to form molecules, and how they can participate in chemical reactions.

The organization of the elements in the Periodic Table is no simple task. In fact, it is one of the greatest achievements in chemistry. It allows us to understand the properties of the elements and to predict their behavior under different conditions. One way to determine the position of an element in the table is by analyzing its successive ionization energy data. Successive ionization energy is the energy required to remove an electron from an atom or ion. If we plot this data for all the elements, we can see that there are certain jumps in energy levels. These jumps occur when we remove an electron from a filled shell and start removing electrons from the next shell, which requires more energy. By analyzing the pattern of these energy jumps, we can deduce the position of an element in the Periodic Table. For example, if an element has a low ionization energy, it will be located in the left-hand side of the table, in the alkali metal group. These elements have a single electron in their outermost shell, which is easily removed to form a positively charged ion. On the other hand, an element with a high ionization energy will be located in the right-hand side of the table, in the noble gas group. These elements have a full outer shell and are therefore highly stable and unreactive. Furthermore, the trend of increasing ionization energy moving from left to right across a period can be observed in all elements. The reason for this is that as we move across a period, the number of protons in the nucleus increases, which means that the electrons are held more tightly and require more energy to be removed.

Measurement Units for Particles

Atoms are tiny, incredibly small particles that make up everything around us. Because they are so small, scientists needed a way to measure their mass in a precise and consistent manner. That's where the concept of the unified atomic mass unit came in. The unified atomic mass unit (or UAMU for short) is a unit of measurement that scientists use to measure the mass of atoms, molecules, and other tiny particles. Specifically, it is defined as one twelfth of the mass of a carbon-12 atom. Now, you might be wondering, why carbon-12? Well, scientists chose carbon-12 as the standard because it is a relatively stable and common isotope of carbon, the element that makes up the basis of all life on Earth. By using carbon-12 as the standard, scientists were able to create a consistent and universally accepted unit of mass for all particles. So, to recap, the UAMU is a measurement unit that scientists use to determine the mass of tiny particles like atoms and molecules. It is defined as one twelfth of the mass of a carbon-12 atom, which was chosen as the standard because of its stability and prevalence.

Relative atomic mass (Ar) is a measurement of the mass of an atom in relation to the mass of one atom of carbon-12. Essentially, it is the average mass of all the isotopes of an element, taking into account the abundance of each isotope. This means that the Ar of an element takes into consideration the different isotopes that may exist for that element. For example, the Ar of carbon is 12.01, which means that the average mass of carbon atoms is slightly more than 12 times the mass of a carbon-12 atom. Relative isotopic mass is similar to Ar, but it specifically refers to the mass of a particular isotope of an element relative to the mass of one atom of carbon-12.

So, for example, the relative isotopic mass of carbon-14 would be 14.003 as it is 14 times heavier than a carbon-12 atom. Relative molecular mass (Mr) is used to measure the mass of a molecule relative to the mass of one atom of carbon-12. It is calculated by adding up the relative atomic masses of all the atoms in the molecule. For example, the Mr of water (H_2O) would be the sum of the relative atomic masses of two hydrogen atoms and one oxygen atom. So, MR of water would be (2x 1.008) + 16.00 = 18.02 UAMU. Relative formula mass (RFM) is used to measure the mass of a formula unit of a compound relative to the mass of one atom of carbon-12. It is calculated by adding up the relative atomic masses of all the atoms in one formula unit of the compound. For example, the RFM of sodium chloride (NaCl) would be the sum of the relative atomic masses of one sodium atom and one chlorine atom. So, RFM of NaCl would be 22.99 + 35.45 = 58.44 UAMU.

The Avogadro Constant and Mole

The mole is a unit of measurement that is used to quantify the amount of a substance. It is one of the most important concepts in chemistry, as it helps scientists to understand the relationship between the mass of an object and the number of particles it contains. Now, you might be wondering, what is the Avogadro constant? Well, it is defined as the number of particles (atoms, molecules, ions) in one mole of a substance. Specifically, the Avogadro constant is equal to 6.022 x 10^23 particles per mole. So, when we say that a substance has a certain number of moles, we are essentially expressing the number of particles it contains in terms of the Avogadro constant. For example, if we had one mole of water (H2O), we would know that it contains 6.022 x 10^23 molecules of water. Why is this important? Well, the mole is essential for performing calculations and measurements in chemistry. For instance, it allows us to convert between mass and number of particles. By knowing the mass and molar mass of a substance, we can calculate the number of moles in that substance and use the Avogadro constant to find the number of particles in it. And vice versa, we can also calculate mass if the number of particles and molar mass of a substance are known.

Formulas, Equations, and Compounds

To write a formula for an ionic compound, you need to know the charges of each ion that makes up the compound. The sum of the charges in the compound must always add up to zero, which means that the positive and negative charges must balance each other out. Here are some examples:

1. To predict the ionic charge of an element based on its position on the periodic table, you need to know the element's group number. Group 1 elements (such as sodium and potassium) tend to have a 1+ charge, while Group 2 elements (such as calcium and magnesium) tend to have a 2+ charge. Group 17 elements (such as chlorine and fluorine) tend to have a 1- charge, while Group 16 elements (such as oxygen and sulfur) tend to have a 2- charge.

2. Some common ions and their names and formulae are:
- Nitrate ion: NO_3^-
- Carbonate ion: CO_3^{2-}
- Sulfate ion: SO_4^{2-}
- Hydroxide ion: OH^-
- Ammonium ion: NH_4^+
- Zinc ion: Zn^{2+}
- Silver ion: Ag^+
- Bicarbonate ion: HCO_3^-
- Phosphate ion: PO_4^{3-}

Using the charges of these ions, you can write the formulas for ionic compounds. For example:
- Sodium nitrate: $NaNO_3$
- Calcium carbonate: $CaCO_3$
- Magnesium sulfate: $MgSO_4$
- Potassium hydroxide: KOH
- Ammonium chloride: NH_4Cl
- Zinc oxide: ZnO
- Silver bromide: $AgBr$
- Sodium bicarbonate: $NaHCO_3$
- Calcium phosphate: $Ca_3(PO_4)_2$

These examples demonstrate how to use the charges of the ions to write the formulae for different ionic compounds. By knowing the charges of the ions and how they combine, you can write the correct formula for any ionic compound that you encounter.

(a) To write the formula for an ionic compound, you need to determine the charges on the ions that make up the compound. The charge on a cation can be predicted by looking at the element's position in the periodic table, while the charge on an anion is indicated by the oxidation number in Roman numerals.

Here's a basic formula for an ionic compound:

cation (with charge) + anion (with charge) = ionic compound

For example, the ionic compound formed between calcium and chlorine would be written as Ca2+ + Cl- = CaCl2.

(b)

- NO3−: Nitrate ion; Formula: NO3−
- CO32−: Carbonate ion; Formula: CO32−
- SO42−: Sulfate ion; Formula: SO42−
- OH−: Hydroxide ion; Formula: OH−
- NH4+: Ammonium ion; Formula: NH4+
- Zn2+: Zinc ion; Formula: Zn2+
- Ag+: Silver ion; Formula: Ag+
- HCO3−: Hydrogen carbonate (bicarbonate) ion; Formula: HCO3−
- PO43−: Phosphate ion; Formula: PO43−

(a) To write a balanced equation, you must have the same number of atoms of each element on both sides of the equation. An ionic equation only includes the ions involved in the reaction, and not the spectator ions. Here's an example of a balanced ionic equation for the reaction between lead(II) nitrate and potassium iodide:

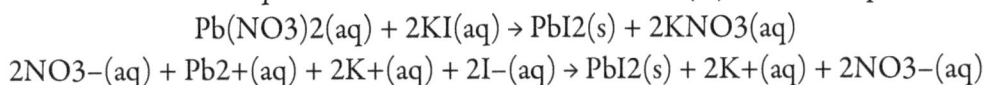

$$Pb(NO3)2(aq) + 2KI(aq) \rightarrow PbI2(s) + 2KNO3(aq)$$

$$2NO3−(aq) + Pb2+(aq) + 2K+(aq) + 2I−(aq) \rightarrow PbI2(s) + 2K+(aq) + 2NO3−(aq)$$

(b) State symbols (aq) for aqueous, (s) for solid, (l) for liquid and (g) for gas are used in equations where necessary. For example, in the equation above, (aq) is used to indicate that the substances are dissolved in water, while (s) is used to indicate that PbI2 is a solid precipitate.

Empirical and molecular formulas are essential concepts in chemistry that are used to represent the chemical composition of substances. These formulas provide chemists with a standardized method for conveying information about the elemental composition of a compound. Empirical formula refers to the simplest whole-number ratio of atoms in a compound. It gives us information about the relative number of each element in a compound. The empirical formula is calculated by dividing the subscripts in a molecular formula by their greatest common factor. This allows for determination of the simplest ratio of atoms, which is what the empirical formula conveys. Molecular formula, on the other hand, refers to the actual number of atoms of each element in a compound. It represents the exact composition of the compound. The molecular formula can be determined by using experimental data such as the molar mass of the compound. To understand the difference between these two types of formulas, consider the example of glucose. The molecular formula for glucose is C6H12O6, which tells us that there are 6 carbon atoms, 12 hydrogen atoms, and 6 oxygen atoms in a single molecule of glucose. Its empirical formula, however, is CH2O, which tells us that there is one carbon atom, two hydrogen atoms, and one oxygen atom for every repeat unit of glucose.

Anhydrous, hydrated, and water of crystallization are terms that are used to describe the physical and chemical properties of some compounds. Anhydrous refers to a substance that does not contain water or any other crystallization liquid in its structure. This means that it is completely free from water molecules. These substances are generally more stable and inert than their hydrated forms. Many substances are anhydrous in their pure form but can absorb water from the environment and become hydrated. Hydrated refers to a substance that contains a specific number of water molecules within its structure. When a compound binds water molecule into its crystal structure, it forms hydrated compounds. These hydrated compounds can vary in their water content, indicating different degrees of hydration. For example, copper sulfate pentahydrate has five water molecules per formula unit of copper sulfate. Water of crystallization refers to the water molecules that are present as part of the crystal lattice structure of a compound. They are considered part of the compound and are not physically adsorbed on its surface. These water molecules are essential parts of the crystal structure that give the compound unique physical and chemical properties. One remarkable characteristic of hydrated compounds and their water of crystallization is that they are generally unstable in anhydrous environments when they lose their water molecules. This is why they are often used as drying agents. When water molecules are removed from the crystal structure, they can absorb water from their surroundings or react with other compounds to form new products. This phenomenon is sometimes called "dehydration."

Calculating empirical and molecular formulae is an essential part of chemistry that helps us understand the chemical composition of a compound. Empirical and molecular formulae are represented by symbols and subscripts that convey information about the relative and absolute amounts of atoms of different elements present in a compound. To calculate the empirical formula of a compound, we need the mass (in grams or moles) of each element in the compound. Once we have this information, we can find the mole ratio between the different elements in the compound. Then, we divide the number of moles of each element by the smallest number of moles to obtain whole-number mole ratios. These whole-number ratios of atoms give us the empirical formula. The molecular formula can be derived from the empirical formula by determining the ratio of the empirical formula mass to the actual molecular mass of the compound. Once we have the empirical formula and the molecular weight of the compound, we can determine the number of empirical units in the molecule and calculate the molecular formula. For example, let's take the compound succinic acid, whose empirical formula is CH_2O_2. To determine the molecular formula, we need to know its molecular weight - which is 118.09 g/mol - and divide it by the empirical formula weight, which is 46.03 g/mol. This gives us the factor by which we need to multiply the empirical formula to obtain the molecular formula - which is 2.56. Therefore, the molecular formula of succinic acid is $C_4H_4O_4$.

The Mole Concept and Stoichiometry in Chemistry

Have you ever wondered how chemists calculate the amount of a particular substance needed for a chemical reaction? Well, they use a unit of measurement called a mole to do so. A mole is a unit that represents the amount of a substance in a given quantity. For example, one mole of carbon atoms contains 6.02×10^{23} carbon atoms.

Now, let's say we have a chemical equation that shows the reaction of hydrogen gas (H_2) with oxygen gas (O_2) to form water (H_2O). The equation would be:

$$2H_2 + O_2 \rightarrow 2H_2O$$

From this equation, we can see that 2 moles of hydrogen react with 1 mole of oxygen to form 2 moles of water. This is where the mole concept comes in handy.

To perform calculations related to reacting masses, we first need to know the molecular formula of each substance involved in the reaction. For example, the molecular formula of hydrogen gas is H_2 and the molecular formula of oxygen gas is O_2.

Let's say we want to know how much water can be formed if we react 4 moles of hydrogen with 2 moles of oxygen. To do this, we need to use the mole ratios from the chemical equation.

According to the equation, 2 moles of hydrogen react with 1 mole of oxygen to produce 2 moles of water. So, if we have 4 moles of hydrogen and 2 moles of oxygen, we have enough reactants to form 4 moles of water.

Percentage yield calculations are also commonly used to determine the efficiency of a chemical reaction. This involves comparing the actual yield of a reaction (the amount of product obtained from the reaction) to the theoretical yield (the amount of product that should be obtained according to calculations).

Let me explain how the mole concept is used to perform calculations involving volumes of gases in chemistry, particularly in the burning of hydrocarbons.

Firstly, we need to understand the relationship between the volume of a gas and the number of moles of that gas. According to Avogadro's Law, equal volumes of all gases at the same temperature and pressure contain the same number of particles, or moles. This means that one mole of any gas will occupy the same volume as one mole of any other gas under identical conditions.

Let's consider the combustion of a hydrocarbon such as methane (CH_4) in oxygen gas (O_2) to form carbon dioxide (CO_2) and water vapor (H_2O):

$$CH_4 + 2O_2 \rightarrow CO_2 + 2H_2O$$

To calculate the amount of gases involved in this reaction, we can use the mole concept to determine the number of moles of each gas. For example, if we know the volume of a gas (such as oxygen) at a given temperature and pressure, we can use Avogadro's Law to calculate the number of moles of that gas:

$$n = V/VM$$

Where n is the number of moles, V is the volume in liters, and VM is the molar volume (which is equal to 22.4 liters/mole at standard temperature and pressure).

Using this equation, we can calculate the number of moles of oxygen needed to react with one mole of methane. In this case, we would need 2 moles of oxygen for each mole of methane.

Additionally, we can use the Ideal Gas Law (PV = nRT) to relate the volume of a gas to its pressure, temperature, and number of moles. This equation can be used to calculate the volume of a gas at a given temperature and pressure when the number of moles is known.

Let me explain how the mole concept is used to perform calculations involving volumes and concentrations of solutions in chemistry.

The mole concept is a key principle in chemistry that is used to convert between the amount of a substance in moles and the mass or volume of that substance. When it comes to solutions, the mole concept can be used to calculate the number of moles of solute present in a given volume of solution as well as the concentration of the solution.

Firstly, let's define some key terms related to solutions. The solute is the substance that is dissolved in a solution, while the solvent is the substance that dissolves the solute to form the solution. The concentration of a solution refers to the quantity of the dissolved substance, also known as the solute, contained within a specific volume of the solution.

For example, let's say we have a solution containing 0.5 grams of sodium chloride (NaCl) dissolved in 50 milliliters of water. To determine the number of moles of NaCl in this solution, we need to use the mole concept.

The molecular weight of NaCl is 58.44 grams/mol, which means that one mole of NaCl weighs 58.44 grams. Using this information, we can calculate the number of moles of NaCl in the solution:

$$n = m/MW$$

Where n is the number of moles, m is the mass of the solute in grams, and MW is the molecular weight of the solute.

In this case, we have 0.5 grams of NaCl, so:

$$n = 0.5 \text{ g} / 58.44 \text{ g/mol} = 0.00855 \text{ moles}$$

To calculate the concentration of the solution, we need to divide the number of moles of solute by the volume of the solution in liters:

$$C = n/V$$

Where C is the concentration in moles per liter, n is the number of moles of solute, and V is the volume of the solution in liters.

In this case, we have 50 milliliters of solution, which is equal to 0.05 liters. Therefore, the concentration of the solution is:

$$C = 0.00855 \text{ moles} / 0.05 \text{ L} = 0.171 \text{ M}$$

Let me explain how the mole concept is used to perform calculations involving limiting reagents and excess reagents in chemistry.

In a chemical reaction, the reactants involved are not always present in the exact stoichiometric amounts required to fully react. The limiting reagent is the reactant that is completely consumed in the reaction, while the excess reagent is the reactant that remains after the reaction is complete.

To determine the limiting and excess reagents, we need to use the mole concept to calculate the number of moles of each reactant and compare them to the stoichiometric coefficients in the balanced chemical equation.

Take, for instance, the chemical reaction that occurs between sodium hydroxide (NaOH) and hydrochloric acid (HCl) which results in the formation of sodium chloride (NaCl) and water (H2O).

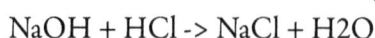

$$NaOH + HCl \rightarrow NaCl + H2O$$

Suppose we have 5.0 grams of NaOH and 3.0 grams of HCl. To determine which reactant is the limiting reagent and which is the excess reagent, we need to use the mole concept to calculate the number of moles of each reactant:

$$n = m/MW$$

Where n is the number of moles, m is the mass of the reactant in grams, and MW is the molecular weight of the reactant.

For NaOH, we have:

$$n = 5.0 \text{ g} / 40.00 \text{ g/mol} = 0.125 \text{ moles}$$

For HCl, we have:

$$n = 3.0 \text{ g} / 36.46 \text{ g/mol} = 0.0822 \text{ moles}$$

Using the balanced chemical equation, we can see that the stoichiometric coefficients are 1:1 for NaOH and HCl. This means that for every one mole of NaOH, one mole of HCl is required for complete reaction. Based on the number of moles calculated, we can see that NaOH is present in excess, while HCl is the limiting reagent. This is because we have 0.0822 moles of HCl, which is less than the 0.125 moles of NaOH.

To calculate the amount of product produced and the amount of excess reactant left over, we need to use the limiting reagent to determine the theoretical yield of the reaction. In this case, since HCl is the limiting reagent, we can calculate the amount of NaCl produced based on the stoichiometry of the balanced chemical equation. With the balanced chemical equation, we know that 1 mole of HCl reacts with 1 mole of NaOH to produce 1 mole of NaCl. Therefore, since we have only 0.0822 moles of HCl, that means we can produce only 0.0822 moles of NaCl.

The field of stoichiometry deals with delineating the interrelation between the amount of reactants and the products generated in a chemical reaction. It involves calculating the amount of substance used or produced in a reaction, based on the given number of moles or volume or mass of the reactants and products. The mole concept is essential to understand stoichiometry, as it allows for the calculation of reactants and products involved in a chemical reaction. For example, to deduce stoichiometric relationships using calculations, we must use the mole ratios from the balanced chemical equations to calculate the number of moles of reactants and products involved in a reaction. Using this information, we can determine the limiting reagent and calculate the theoretical yield of a reaction. Additionally, we can calculate the excess reactant involved and determine the percentage yield of the reaction. Stoichiometry is a critical concept in chemistry as it helps chemists predict and optimize the amounts of reagents needed to carry out a chemical reaction to obtain the desired product. By studying stoichiometry, we can understand the underlying principles of reactions and apply this knowledge to solve complex problems in chemical reactions.

Electronegativity: Properties and Applications

Electronegativity is a term used to describe an atom's ability to draw electrons towards itself. Essentially, it refers to an atom's eagerness to attract and hold on to electrons. This attraction arises because of the positive charge in the nucleus of the atom, which exerts a pull on negatively charged electrons in the outermost shell. The higher an atom's electronegativity value, the more likely it is to attract electrons. This can create polarized bonds between atoms, where one atom gains a slightly negative charge and the other gains a slightly positive charge. Being able to comprehend electronegativity holds immense importance in the sense that it plays a crucial role in defining the character of chemical bonds and the activity of elements in chemical reactions. In short, electronegativity is the measure of how much an atom wants to hold onto its electrons, and how strong its pull is felt on nearby electrons.

Electronegativity is a fundamental property of an atom that measures its ability or tendency to attract electrons towards itself when it participates in a chemical bond with other atoms. The electronegativity of an element is influenced by several factors, including nuclear charge, atomic radius, and shielding by inner shells and sub-shells. The nuclear charge, which is the number of protons in the nucleus of an atom, has a significant impact on its electronegativity. An atom with a higher nuclear charge has a stronger positive charge in its nucleus, which attracts the negatively charged electrons towards itself with greater strength, thereby increasing its electronegativity. Therefore, elements with a higher atomic number and more protons in their nucleus tend to have higher electronegativities. The size of an atom, or its atomic radius, is another significant factor that affects electronegativity. As the size of an atom increases, its outer electrons are farther away from the positively charged nucleus, reducing their attraction towards it. Therefore, larger atoms tend to have lower electronegativities compared to smaller ones. Another factor that influences electronegativity is the shielding effect of inner shells and sub-shells of electrons. Electrons in the inner shells and sub-shells shield the outer valence electrons from the positively charged nucleus, reducing their attraction towards it. As a result, atoms with more inner shells and sub-shells tend to have lower electronegativities than those with fewer inner shells and sub-shells.

Electronegativity refers to the tendency of an atom to attract electrons towards itself when forming a covalent bond with another atom. The Periodic Table enables us to examine the trends in electronegativity across a period and down a group. Periodic trends in electronegativity illustrate that electronegativity increases from left to right across a period. The tendency of elements to have higher electronegativity increases as we move across the periodic table from the left-hand side to the right-hand side. This trend results from the increasing nuclear charge in the elements across a period. As the atomic number increases, the number of protons in the nucleus increases, resulting in a stronger pull on the electron cloud. Be that as it may, the electronegativity of noble gases such as helium, neon, and argon is, by and large, unresponsive because of their stable electronic configuration. The trend in electronegativity down the group illustrates that electronegativity decreases as we move down a group. This means that the electronegativity of the elements tends to decrease from the top of a group to the bottom of it. The reduction happens though the atomic radius increases down the group, which reduces the attraction of the positive nucleus to the outer electrons. This means that an atom with a larger atomic radius has a diminished attraction for the shared electron pair due to the increased distance between the positive nucleus and the electrons. Therefore, the electrons in the outermost shells will be less tightly bound to the nucleus, reducing the overall electronegativity of the element.

Pauling electronegativity values can be used to predict the type of chemical bonding that occurs between two atoms in a molecule. Electronegativity measures the tendency of an atom to attract electrons towards itself in a covalent bond. The higher the difference in electronegativity between the two atoms, the more likely it is that they will form an ionic bond, and the lower the difference, the more likely they will form a covalent bond. When the difference in electronegativity between the two atoms is relatively high, greater than 1.7, the bond between them is usually ionic. In ionic bonding, the atoms lose or gain electrons to form ions with opposite charges. The ionic bond is then formed between the positively charged cation and the negatively charged anion. Examples of molecules that form ionic bonds include NaCl, where the electronegativities of Na and Cl are 0.93 and 3.16, respectively. The difference in electronegativity between Na and Cl is 2.23, which is greater than 1.7, indicating that the bond between them is predominantly ionic. When the difference in electronegativity between two atoms is relatively low, less than 1.7, the bond between them is generally covalent. In covalent bonding, the atoms share electrons to form a stable bond. The degree of sharing of electrons is affected by the electronegativity of each atom. The closer their electronegativities are, the more equal the sharing of their electrons in the bond. For instance, HCl, where the electronegativities of H and Cl are 2.20 and 3.16, respectively, has an electronegativity difference of 0.96, which means the bond between H and Cl is predominantly covalent. However, there are cases where the electronegativity difference is not clear cut, and the bond between atoms is a mixture of ionic and covalent bonds. This type of bond is known as polar covalent bonding. A polar covalent bond occurs when the electron pair is shared unequally between the atoms because their electronegativity difference is less than 1.7 but greater than zero. Water (H_2O) is an example of a molecule with polar covalent bonds, where the electronegativity difference between H and O is 0.98.

Ionic Bonding

At the heart of chemistry lies the phenomenon of ionic bonding, an incredibly powerful force that holds together atoms to create the spectacular variety of inorganic materials that make up our world. Ionic bonding is a miraculous process that arises from the mutual attraction between positively charged cations and negatively charged anions, as they seek to achieve a more stable electron configuration. Imagine two oppositely charged individuals being drawn irresistibly towards each other by a force more powerful than themselves. This is precisely what happens in ionic bonding. Cations, with their deficit of electrons, are desperate to give up those extra protons they carry, while anions, with their surplus of electrons, are eager to share their bounty. This mutual attraction gives rise to an electrostatic force that binds them together in a stable, crystalline structure. The beauty of ionic bonding lies in its simplicity and elegance. It is a force that arises out of the fundamental nature of matter, requiring no gimmicks or complexities to explain its workings. And yet, it is a force of immense power and importance, shaping the very foundations of chemistry and driving much of the modern world's technological progress. So the next time you pick up a piece of salt, or gaze in awe at the intricate patterns of a crystal, remember the wonder of ionic bonding, and how it serves as the key to unlocking the secrets of this amazing universe in which we live.

Ionic bonding is a type of chemical bonding that involves the transfer of electrons between atoms, leading to the formation of positively charged cations and negatively charged anions that attract each other through electrostatic forces. This type of bonding is common among inorganic compounds and is responsible for the formation of many familiar substances. One of the most famous examples of ionic bonding is sodium chloride, or table salt. Sodium chloride is formed from positively charged sodium ions ($Na+$) and negatively charged chloride ions ($Cl-$) that bond together in a crystal structure. Sodium atoms transfer an electron to chloride atoms, giving sodium a positive charge and chloride a negative charge. These opposite charges attract each other, leading to the formation of NaCl. Magnesium oxide is another example of ionic bonding, formed from the reaction of positively charged magnesium ions ($Mg2+$) and negatively charged oxygen ions ($O2-$). Magnesium atoms lose two of their electrons to oxygen atoms, creating the $Mg2+$ and $O2-$ ions, respectively. These ions then combine to form magnesium oxide (MgO), which has a crystalline structure. Calcium fluoride is another example of ionic bonding, formed from positively charged calcium ions ($Ca2+$) and negatively charged fluoride ions ($F-$). In this case, the transfer of electrons occurs between calcium and fluoride atoms, leading to the formation of $Ca2+$ and $F-$ ions that attract each other electrostatically to form a crystal structure.

Metallic Bonding

Metallic bonding occurs when positively charged metal ions are attracted to a group of electrons that are not confined to any one atomic nucleus, known as delocalized electrons. These free-roaming electrons are responsible for many of the unique properties of metals, such as high thermal and electrical conductivity, malleability, and ductility. Imagine a bustling city where the metal ions are skyscrapers, towering above the bustling streets below. The delocalized electrons are like the crowds below, weaving in between the buildings, carrying out their daily business. In metallic bonding, the positive ions are dispersed throughout a lattice of shared electrons, forming an interconnected network that allows the metal to conduct heat and electricity with ease. Through this electrostatic attraction, the positive metal ions are able to maintain a rigid structure while also allowing for flexibility, allowing the metal to be hammered into sheets or drawn out into wires. This flexibility also allows for metallic bonds to be reformed after being broken, giving metals their ability to bend and deform under pressure without breaking. So, in essence, metallic bonding is the strong electromagnetic attraction between the positively charged metal ions and the delocalized electrons that surround them, creating a bond that allows metals to take on remarkable physical properties.

..

Covalent bonding is a type of chemical bond that occurs when two atoms share a pair of electrons in a mutual effort to achieve stability. At its core, covalent bonding is all about creating a sense of cooperation between two atoms, as they work together to achieve a more stable state. Through this process, the atoms are able to create a new, more complex substance that possesses unique properties and characteristics. Think of covalent bonding as a dance between two atoms, in which they share electrons with each other in a sort of "give-and-take" rhythm. This process requires a delicate balance of attraction and repulsion, as the atoms work to find the perfect configuration of shared electrons that will lead to stability. Once the bond is formed, the resulting compound is stronger and more stable than the individual atoms were before.

Examples of Covalent Bonding Molecules

Covalent bonding is a fundamental concept of chemistry that explains how atoms stick together to form molecules. In simple terms, covalent bonding is the sharing of electrons between atoms. This type of bonding occurs when two atoms share a pair of electrons in order to fill their outermost valence shell and become more stable.

Let's take a look at three examples of covalent bonding involving the elements hydrogen, oxygen, and nitrogen.

First, let's examine hydrogen, which is the simplest of the three examples. Hydrogen atoms have only one electron in their outermost shell, which means they need one more electron to fill it up. When two hydrogen atoms come together, they can share their electrons and form a covalent bond. This results in the creation of a molecule of hydrogen gas, or H_2.

Next, let's consider oxygen, which is a bit more complex. Oxygen atoms have six electrons in their outermost shells, but they need eight electrons to fill them up completely. When two oxygen atoms come together, they can each share two electrons with the other atom, forming two covalent bonds. This results in the creation of a molecule of oxygen gas, or O_2.

Finally, let's explore nitrogen, which has similar properties to oxygen. Nitrogen atoms also have six electrons in their outermost shells and need eight electrons to fill them up. When two nitrogen atoms come together, they can each share three electrons with the other atom, forming three covalent bonds. This results in the creation of a molecule of nitrogen gas, or N_2.

Let's look at three examples of covalent bonding involving the elements chlorine, hydrogen, carbon, and nitrogen.

Chlorine, represented by Cl_2, is a diatomic molecule that occurs between two chlorine atoms. Each chlorine atom has seven valence electrons, which means they can share a pair of electrons and become more stable. As such, they create a covalent bond, sharing pairs of electrons until they have a full outer electron shell and form Cl_2, which is a gas at room temperature.

Hydrogen chloride, commonly represented as HCl, is a molecule that forms when a hydrogen atom bonds with a chlorine atom. Hydrogen has only one valence electron, which can be shared with chlorine that has seven valence electrons. The resulting compound has a full outer shell of electrons, leaving it more stable. HCl is a highly corrosive and pungent gas that is extensively used in the chemical industry.

Carbon dioxide (CO_2) is a molecule formed when carbon bonds to two oxygen atoms. Carbon has four valence electrons, while oxygen has six valence electrons. To fill its outer shell, carbon shares two electrons, one with each oxygen atom, resulting in the formation of two covalent bonds. CO_2 is a key greenhouse gas, contributing to the warming of our planet.

Last, ammonia, NH_3, is a molecule that consists of one nitrogen atom sharing three electrons with three hydrogen atoms. Nitrogen has five valence electrons and each hydrogen atom has only one valence electron. Nitrogen shares its electrons effectively with three hydrogen atoms, creating a stable compound which has multiple applications in industry and agriculture.

First, let's examine methane, which is a molecule composed of one carbon atom and four hydrogen atoms, represented by the chemical formula CH_4. Carbon has four valence electrons and hydrogen has one, so carbon shares one electron with each of the four hydrogen atoms, effectively filling the outer shell of each atom. This results in the formation of a covalent bond between the carbon and hydrogen atoms, creating a stable molecule. Next, ethane is a molecule composed of two carbon atoms and six hydrogen atoms, with the chemical formula of C_2H_6. The carbon atoms are bonded to each other by a single covalent bond, while each carbon is bonded to three hydrogen atoms. This is because each carbon atom has four valence electrons and by sharing one with the other carbon, they each have three remaining valence electrons that can be shared with hydrogen.

Finally, ethylene is a molecule composed of two carbon atoms and four hydrogen atoms, with the chemical formula of C_2H_4. The carbon atoms are bonded to each other by a double covalent bond, while each carbon atom is also bonded to one hydrogen atom. Each carbon atom has four valence electrons, but in ethylene they only share three electrons each with the other carbon atom, leaving one remaining valence electron to bond with a hydrogen atom.

Covalent Bonds and Properties

You see, the octet rule is a fundamental principle in chemistry that states that atoms tend to combine in such a way that they have eight electrons in their outermost shell. This makes them more stable and less reactive. However, there are some exceptions to this rule, especially when it comes to elements in the third period of the periodic table. In compounds such as sulfur dioxide (SO_2), phosphorus pentachloride (PCl_5), and sulfur hexafluoride (SF_6), the elements in period 3 can actually expand their octet beyond eight electrons. This happens because of a phenomenon known as "electron promotion," where electrons from the lower energy levels are excited to higher energy levels so that they can participate in bonding with other atoms. So, what does this mean? Well, it means that the elements in period 3 have the ability to form more complex and stable compounds than those in the previous periods. This is a fascinating discovery in the world of chemistry that has contributed greatly to our understanding of chemical bonding and the behavior of different elements.

One of the most interesting topics in this area is the concept of coordinate or dative covalent bonding. In this type of bond, one atom donates a pair of electrons to another atom, which accepts the pair as part of a covalent bond. A great example of this type of bonding can be seen in the reaction between ammonia (NH_3) and hydrogen chloride (HCl) gases to form the ammonium ion (NH_4^+). During this reaction, the nitrogen atom in ammonia donates a pair of electrons to the hydrogen atom in hydrogen chloride, forming a covalent bond. As a result, the hydrogen atom becomes positively charged and the nitrogen atom takes on a formal negative charge, creating the ammonium ion. Another example of coordinate bonding can be seen in the molecule Al_2Cl_6. Here, the aluminum atoms donate pairs of electrons to the chlorine atoms, creating six individual covalent bonds. However, because the electrons are being donated from only one atom (the aluminum), this creates a coordinate bond. This concept of coordinate bonding is an important one in chemistry because it helps us understand how different molecules and ions can form and interact with one another. It is a fascinating way to look at the world of chemical reactions, and it opens up many possibilities for exploration and innovation in the field.

Are you curious about the concept of covalent bonds and how they work? When two atoms come together to form a molecule, they often share electrons to create a covalent bond. In this type of bond, each atom contributes one or more electrons to the bond, and they are held together by a shared electron pair. This results in a stable molecule with a net zero charge. The covalent bond is formed by the overlap of the atomic orbitals of the bonding atoms. This overlap can be direct or indirect. In the case of a σ bond, the atomic orbitals overlap directly between the bonding atoms creating a head-on interaction, which results in a high electron density region between the two nuclei. This region is called a sigma (σ) bond and is responsible for the strength of the bond between the atoms. On the other hand, there is another type of covalent bond called the π bond. This bond is formed by the sideways overlap of two atomic orbitals. This type of bonding is weaker than the σ bond, and has less overlap between the electrons. The π bond is typically formed by p orbitals, which are shaped like dumbbells. The σ bond and the π bond both play an important role in the formation and properties of molecules. They help govern the shape, reactivity, and stability of many compounds in the world around us.

Covalent bonds are a type of chemical bond formed between two atoms that share electrons. These bonds are formed by the overlapping of atomic orbitals, which are the areas around the nucleus where electrons are most likely to be found. Specifically, covalent bonds are formed through the overlapping of orbitals between two atoms. The overlapping of atomic orbitals occurs in two ways, resulting in two types of covalent bonds called sigma (σ) and pi (π) bonds. The sigma bond is formed through the overlapping of atomic orbitals end-to-end along the bond axis. On the other hand, the pi bond is formed through the sideways overlapping of adjacent p orbitals above and below the sigma bond axis. The pi bond results from the side-by-side interaction of the p orbitals, which have a cylindrical shape. This type of bond is weaker than the sigma bond since there is less overlap between the atoms involved. The pi bond is also able to rotate around the bond axis, which allows molecules to have a variety of shapes.

Chemical bonds are formed between atoms through the sharing or transfer of electrons. There are two kinds of covalent bonds that are formed through the overlapping of atomic orbitals between two atoms, sigma (σ) and pi (π) bonds. In a molecule of H_2, two hydrogen atoms share an electron through the formation of a covalent bond. The sigma bond is formed between the two hydrogen atoms from the overlapping of their s-orbitals. Both atoms' s-orbitals overlap along the bond axis, creating a sigma bond. The H_2 molecule does not have any pi bonds since hydrogen atoms only have s-orbitals. In a molecule of C_2H_6, two carbon atoms and six hydrogen atoms share electrons through the formation of covalent bonds. The two carbon atoms share a sigma bond by overlapping their s-orbitals and a pi bond by overlapping their p-orbitals. Each of the two carbon atoms also shares three sigma bonds with three hydrogen atoms through the overlapping of their s-orbitals. In a molecule of C_2H_4, two carbon atoms and four hydrogen atoms form a covalent bond. The pi bond is formed by the sideways overlapping of adjacent p-orbitals above and below the sigma bond axis. The two carbon atoms share a sigma bond from the overlapping of their s-orbitals and two pi bonds from the overlapping of their p-orbitals, resulting in a double bond between the two carbon atoms. In a molecule of HCN, one hydrogen atom, one carbon atom, and one nitrogen atom share electrons through the formation of covalent bonds. The carbon and nitrogen atoms share a triple bond, consisting of one sigma and two pi bonds. The sigma bond is formed from the overlapping of their s-orbitals, while the two pi bonds are formed from the overlapping of their p-orbitals. The hydrogen atom shares a single sigma bond with the carbon atom. In a molecule of N_2, two nitrogen atoms share electrons through the formation of a covalent bond. The two nitrogen atoms share a triple bond consisting of one sigma and two pi bonds. The sigma bond is formed from the overlapping of their s-orbitals, while the two pi bonds are formed from the overlapping of their p-orbitals.

Hybridisation is a concept in chemistry that explains how orbitals mix to form new, hybrid orbitals suitable for bonding. The hybridisation of an atom can be determined by the nature of the molecule and the number of valence electrons in the atom. In sp hybridisation, the s-orbital and one of the three p-orbitals combine to form two sp hybrid orbitals. These sp orbitals are linear and form sigma bonds with other atoms. Examples of molecules that have sp hybridisation include $BeCl_2$ and HCN. In sp2 hybridisation, the s-orbital and two of the three p-orbitals combine to form three sp2 hybrid orbitals. These sp2 orbitals are trigonal planar in shape and form sigma bonds with other atoms. The remaining p-orbital remains unhybridized, forming a pi bond with another atom. Examples of molecules that have sp2 hybridisation include CH_4 and CO_2. In sp3 hybridisation, the s-orbital and all three p-orbitals combine to form four sp3 hybrid orbitals. These sp3 orbitals are tetrahedral in shape and form sigma bonds with other atoms. Examples of molecules that have sp3 hybridisation include CH_3OH and NH_3.

Bond energy is a term used in chemistry to refer to the amount of energy required to break one mole of a particular covalent bond in the gaseous state. It is also known as the bond dissociation energy, and it is a measure of the strength of the covalent bond between two atoms. When two atoms form a covalent bond, their electrons are shared between them. In order to break this bond, energy must be supplied to the system. The bond energy is the amount of energy needed to break the bond, usually measured in kilojoules per mole (kJ/mol). Bond energy values vary depending on several factors, including the types of atoms involved, the length of the bond, and the presence of any surrounding molecules or compounds that affect the bond strength. For example, the bond energy for a carbon-hydrogen bond (C-H) is generally lower than the bond energy for a carbon-carbon bond (C-C), due to differences in the electronegativity of the atoms involved. The bond energy is an important parameter in many chemical reactions, as it determines the energy required to break and form new bonds. Reactions that require a net input of energy to break bonds are endothermic, while those that release energy when bonds are formed are exothermic.

Bond length is a term used in chemistry to describe the distance between two covalently bonded atoms. It is the internuclear distance, or the distance between the nuclei of the two atoms involved in the covalent bond. When two or more atoms form a covalent bond, they share their electrons to attain a more stable electronic configuration. The bond length is determined by a number of factors, including the size of the atoms involved and the strength of the bond between them. Generally, smaller atoms form stronger bonds, leading to shorter bond lengths. This is because smaller atoms have a stronger nuclear charge, which attracts electrons more strongly and causes a tighter bond between the atoms. Conversely, larger atoms form weaker bonds, leading to longer bond lengths. Measuring bond length is an important part of understanding the properties of molecules and predicting their behavior in chemical reactions. By adjusting the bond length between two atoms, chemists can tailor the properties of a molecule for specific applications, such as increased strength, stability, or reactivity.

The reactivity of covalent molecules can be compared using the concepts of bond energy values and bond length. In general, lower bond energies and shorter bond lengths indicate greater reactivity, as these factors make it easier for molecules to participate in chemical reactions. Bond energy values indicate the strength of the covalent bond between two atoms. Stronger bonds require more energy to break, indicating greater stability in the molecule. Weak bonds, on the other hand, require less energy to break, indicating greater reactivity. For example, bonds between elements in the halogen group have low bond energies, making them highly reactive and prone to forming compounds with other elements. Similarly, bond length can also affect the reactivity of a covalent molecule. Shorter bond lengths indicate a stronger bond between atoms, which can make it more difficult for the molecule to break apart or react with other molecules. Longer bond lengths, on the other hand, indicate a weaker bond, making it easier for the molecule to participate in chemical reactions. For example, consider the reaction between methane (CH_4) and oxygen (O_2) to form carbon dioxide (CO_2) and water (H_2O). Methane has shorter bond lengths and higher bond energies compared to oxygen gas. The stronger bonds between the carbon and hydrogen atoms in methane make it less likely to react with oxygen compared to the weaker bonds between the oxygen molecules. Thus, oxygen is more reactive and has a greater tendency to form compounds with other elements.

Molecule Shapes and Bond Angles with VSEPR Theory

VSEPR theory is an important concept in chemistry that helps us in understanding the shape of molecules and the bond angles between their atoms. This theory states that the electron pairs surrounding a central atom in a molecule will repel each other in such a way as to maximise the distance between them. As a result, the molecule will adopt a certain shape and bond angles between atoms.

Let's take some simple examples to understand this theory in more detail. Boron trifluoride (BF_3) is a molecule that has a trigonal planar shape and a bond angle of 120 degrees. This means that the three fluorine atoms surrounding the boron atom are arranged in a flat triangular shape, with each bond angle being 120 degrees. The reason for this shape is the repulsion between the electrons in the bonding pairs and the electrons in the non-bonding pairs.

Another example is carbon dioxide (CO_2), which has a linear shape and a bond angle of 180 degrees. In this molecule, the two oxygen atoms are arranged in a straight line on either side of the central carbon atom. The reason for this shape and bond angle is that the electrons in the bonding pairs are arranged to be as far apart as possible, resulting in a linear shape.

Lastly, methane (CH_4) is a molecule that has a tetrahedral shape and bond angle of 109.5 degrees. In this molecule, the four hydrogen atoms surround the central carbon atom at equal distances from each other, giving rise to a shape that resembles a pyramid with a triangular base. This shape is due to the fact that the electrons in the bonding pairs repel each other in such a way that they are as far apart as possible, leading to the tetrahedral shape.

Ammonia (NH_3) is a molecule that has a pyramid shape and a bond angle of 107 degrees. This means that the three hydrogen atoms surround the central nitrogen atom in a triangular base pyramid shape. Additionally, the lone pair of electrons on nitrogen atom occupies a position that repels the other three bonding pairs and pushes them towards the three hydrogen atoms. This pyramid shape and bond angle are due to the repulsion between the bonding pairs and the lone pair electrons.

Water (H_2O) is a molecule that has a non-linear shape and a bond angle of 104.5 degrees. The two hydrogen atoms are at an angle to each other, and both lie on the same side of the central oxygen atom. This shape and bond angle arise from the four electron pairs of the oxygen atom, two of which are non-bonding pairs. The non-bonding pairs occupy greater amounts of space, causing the two hydrogen atoms to be further apart than expected.

Sulfur hexafluoride (SF_6) is a molecule that has an octahedral shape and bond angle of 90 degrees. This molecule contains six fluorine atoms surrounding a central sulfur atom. The six bonding pairs of electrons in the molecule are arranged in such a way as to be as far apart from each other as possible, resulting in an octahedral shape. The bond angles between each fluorine and sulfur atom are 90 degrees, as a result of the repulsion between each pair of electrons.

Phosphorus pentafluoride (PF_5) is a molecule that has a trigonal bipyramidal shape and bond angles of 120 and 90 degrees. This molecule contains five fluorine atoms and a central phosphorus atom. The three electron pairs are arranged in a trigonal plane around the central phosphorus atom, while the two remaining electron pairs occupy positions in a vertical plane at right angles to the trigonal base. The molecule assumes the shape of a trigonal bipyramid, and the bond angles between the five fluorine atoms and the phosphorus atom are 120 and 90 degrees.

⁂

Molecule Shapes and Bond Angles with VSEPR Theory

⁂

Let's delve into some examples to better understand this theory. First, consider the diatomic molecule of oxygen gas (O2). Since oxygen has six valence electrons, and the molecule has a double bond, there are two electron pairs, both of which are bonding pairs, surrounding the central oxygen atom. This arrangement results in a linear shape with a bond angle of 180 degrees.

Next, let's look at the carbonate ion (CO32-). The central carbon atom has four bonding electron pairs and no lone pairs, while each oxygen atom surrounding the central carbon atom has one bonding electron pair and two lone pairs. This arrangement leads to a trigonal planar shape with a bond angle of 120 degrees.

Another example is nitrate ion (NO3-). The central nitrogen atom has three bonding electron pairs and one lone pair. Each oxygen atom surrounding the central nitrogen atom has one bonding pair and two lone pairs. This leads to a trigonal planar shape with a bond angle of 120 degrees.

Furthermore, let's consider the tetrahedral-shaped molecule known as phosphine (PH3). The central phosphorus atom has three bonding electron pairs and one lone pair. The three hydrogen atoms surrounding the central phosphorus atom are arranged in a triangular base pyramid shape. This arrangement leads to a tetrahedral shape with a bond angle of 109.5 degrees.

Lastly, consider the octahedral-shaped molecule of sulfur hexafluoride (SF6). The central sulfur atom is surrounded by six fluorine atoms, each with a single bonding electron pair. This results in the six fluorine atoms forming a symmetrical octahedral shape around the central sulfur atom, with bond angles of 90 degrees.

In summary, by utilizing the VSEPR theory, we can determine the shapes and bond angles of a wide range of molecules and ions by examining their electron pair configurations. This theory is a fundamental concept in chemistry and is essential for understanding chemical reactions and properties of compounds.

Intermolecular Forces

Hydrogen bonding is a unique intermolecular force that occurs between molecules that contain N–H and O–H groups. This bonding occurs when a hydrogen atom that is covalently bonded to a nitrogen or oxygen atom is attracted to another nitrogen or oxygen atom in a neighboring molecule. The attractive force between these atoms is called a hydrogen bond, and it results from the polarity difference between the electronegative nitrogen or oxygen atom and the positively charged hydrogen atom. Hydrogen bonding is especially significant in biological and chemical systems because it controls the physical and chemical properties of many substances. For example, the bonding structure of water is determined by the hydrogen bonding that occurs between its molecules. In water, each molecule can form up to four hydrogen bonds with neighboring molecules, which causes water to have a higher boiling point and a solid form that is less dense than its liquid form. Ammonia is another simple example of hydrogen bonding, as its molecules also contain N–H groups. In ammonia, each molecule can form up to three hydrogen bonds with neighboring molecules, which causes ammonia to have a higher boiling point than expected based on its molecular weight.

When we think of water, we typically view it as a simple "H2O" molecular compound, but in reality, it's much more complex than that. One of the key reasons for this complexity is the concept of hydrogen bonding, which plays a fundamental role in explaining many of water's unusual properties. Hydrogen bonding occurs when the positively charged hydrogen atoms of one water molecule are attracted to the negatively charged oxygen atoms of another water molecule, creating a weak bond between them. Although these hydrogen bonds are relatively weak compared to the covalent bonds that hold water molecules together, they are still strong enough to have a significant impact on the physical properties of water. One of the most obvious ways that hydrogen bonding affects water is through its relatively high melting and boiling points. Typically, molecular compounds with similar molecular weights and structures to water will have much lower melting and boiling points. However, because of the strong hydrogen bonding between water molecules, it takes much more energy to break those bonds and transition water from a solid to a liquid or gas state. Another significant property of water that is affected by hydrogen bonding is its high surface tension. This means that water molecules at the surface of a body of water are more strongly attracted to each other than they are to the air around them. This results in the formation of a "skin" or surface layer on top of the water, which makes it appear as if it has a sort of "elastic" or "stretchy" surface. The density of ice compared with liquid water is also due to hydrogen bonding. When water freezes, it forms a crystalline structure where each molecule is spaced apart from the others in a rigid lattice structure. The hydrogen bonding between water molecules in this lattice structure makes it more open and less dense than liquid water. So, when ice melts, the hydrogen bonds are broken causing the molecules to move closer together and as a result, the density of liquid water is higher than the density of ice.

Electronegativity, put simply, refers to the tendency of an atom to attract a shared pair of electrons towards itself when it is chemically bonded with another atom. When two atoms of different electronegativity come together to form a chemical bond, one atom tends to attract the electrons more strongly than the other. As a result, the electron density in the bond is not shared equally between the two atoms, creating bond polarity. A molecule is polar when it has a positive and negative end, known as a dipole. This happens when there is an unequal sharing of electrons between the atoms within the molecule, leading to an overall imbalance of electron density. The greater the difference in electronegativity between the atoms in the molecule, the more polar the bond and the greater the dipole moment. For example, consider a molecule of water (H_2O). The oxygen atom has a greater ability to attract electrons than hydrogen atom in O-H bonds which causes the electrons to be more strongly associated with the oxygen atom. This results in hydrogen atoms having a partially positive charge and the oxygen atom having a partially negative charge. This creates a dipole moment within the molecule, with the oxygen end being more negative and the hydrogen end being more positive.

Van der Waals' forces are the molecular interactions that occur between neighboring molecules due to their proximity, without any chemical bonding taking place. These interactions are responsible for many of the physical and chemical properties of substances such as melting and boiling points, density, viscosity, and surface tension. Van der Waals' forces are a type of intermolecular force, which is a force that exists between molecules and other neutral particles. These forces arise due to the interaction between the electrons and nuclei of neighboring particles. The strength of these forces depends on various factors such as the molecular shape, size, and polarity. Van der Waals' forces can be divided into three categories: London dispersion forces, dipole-dipole interactions, and hydrogen bonding. London dispersion forces are the weakest of the three and occur due to the temporary fluctuations in electron density that occur in molecules. Even though these fluctuations are temporary, they can induce similar fluctuations in neighboring molecules, leading to a weak attraction between them. Dipole-dipole interactions, on the other hand, occur between polar molecules. These forces arise due to the attraction between the positive and negative ends of neighboring molecules. For example, in a molecule like hydrogen chloride (HCl), the hydrogen and chlorine atoms have a significant difference in electronegativity, which makes the molecule polar and capable of dipole-dipole interactions. Lastly, hydrogen bonding is a specific type of dipole-dipole interaction that occurs between molecules containing hydrogen atoms bonded to highly electronegative atoms like nitrogen, oxygen, or fluorine. The electronegative atom creates a strong dipole that is capable of forming a hydrogen bond with neighboring molecules.

One of the types of van der Waals' forces is the instantaneous dipole-induced dipole force, also known as London dispersion forces. These forces occur between nonpolar molecules and have a weak but significant effect on intermolecular interactions. London dispersion forces arise due to the enormous fluctuations in electron density and distribution that exist in molecules. At any given moment, the electrons in a molecule may be distributed in such a way that a temporary dipole is created. This temporary dipole induces a polarization in a neighboring molecule, creating a second temporary dipole. These two temporary dipoles then attract each other, creating an interaction between the two molecules. The polarizability of molecules determines the strength of London dispersion forces. Larger molecules with more electrons are more polarizable, and thus have stronger London dispersion forces than smaller molecules. For example, the molecules of iodine are larger and more polarizable than those of chlorine, even though they belong to the same halogen family. As a result, iodine has stronger London dispersion forces than chlorine. Innovatively, think of London dispersion forces as a dance.

Imagine two people engaging in a dance-off. Each of them is a separate molecule, but as they dance, their movements cause temporary dipoles. The movements of one "induce" a dipole in the other, much like the dance moves of one person can inspire their competitor. They are still separate dancers, much like the molecules are still separate entities, but there is an attraction between them that wouldn't have existed without the dance-off. This attraction between the dancing molecules is the London dispersion force.

Another type of van der Waals' force is the permanent dipole-permanent dipole force, also known as the PD-PD force. This force arises between polar molecules that have a permanent separation of charge, meaning one end of the molecule is positively charged, while the other end is negatively charged. Permanent dipoles occur due to differences in the electronegativity of the atoms that make up a molecule. For example, in a molecule of water (H_2O), oxygen is more electronegative than hydrogen, which creates a permanent dipole. The negative end of the molecule is the oxygen atom, while the positive end is the hydrogen atoms. This type of van der Waals' force is significantly stronger than London dispersion forces due to the permanent separation of charge in the polar molecules. The strength of the force depends on the magnitude and orientation of the dipoles in the molecules.

Additionally, hydrogen bonding is a specific type of permanent dipole-permanent dipole force that occurs between molecules containing hydrogen atoms bonded to electronegative atoms like nitrogen, oxygen, or fluorine. Innovatively, consider permanent dipole-permanent dipole forces as a dance between two people who have great chemistry. Unlike the dance-off between nonpolar molecules that results in temporary dipoles, this dance is a bit more permanent because each person has a clearly defined role. Imagine the two dancers in perfect synchronization with each other, their movements identical but opposite in nature. There's a permanence to their dance because they're always moving together, much like the permanent dipoles within polar molecules.

Hydrogen bonding is a type of permanent dipole-permanent dipole force that occurs between molecules containing hydrogen atoms bonded to highly electronegative atoms like nitrogen, oxygen, or fluorine. Hydrogen bonding is a special case of permanent dipole-permanent dipole forces. When a hydrogen atom is covalently bonded to a highly electronegative atom, like oxygen or nitrogen, the electron density in the bond is skewed towards the electronegative atom, creating a partially positive hydrogen end and a partially negative electronegative end. The resulting polarity of the bond creates a permanent dipole, which can then interact with other dipoles via intermolecular forces. The strength of hydrogen bonds depends on several factors, the most important being the electronegativity of the atom to which hydrogen is bonded. The greater the electronegativity, the stronger the hydrogen bond. Additionally, the size and shape of the molecules involved in the bonding also play a role in the strength of the bond. Innovatively, think of hydrogen bonding as a group of people who are bound together by shared interests and passions. Much like the hydrogen atom bonded to an electronegative atom, each person in the group has a unique role to play, but they're all bound together by a common goal. It's this bond that permits the group to interact with other groups and create a more significant community.

When compared to intermolecular forces, the bonds formed through ionic, covalent, and metallic bonding are relatively stronger. This is because ionic, covalent, and metallic bonds involve the sharing, transfer, or sharing of electrons in a much stronger and more permanent way than intermolecular forces. Ionic bonds occur between positively and negatively charged ions in a crystal lattice. They occur when a metal transfers an electron to a non-metal to form a highly stable, oppositely charged cation-anion pair. Ionic bonds are much stronger than intermolecular forces because they involve the complete transfer of electrons, leading to a stable electronic configuration. Covalent bonds are formed when two atoms share electrons to fill their outer valence shells, creating a molecule. Covalent bonding is stronger than intermolecular forces, as each molecule is a discrete entity with a defined structure and shape. Metallic bonding, on the other hand, occurs between metal atoms, which are held together in a "sea" of shared electrons. This bond is unique in the sense that the electrons are free to move throughout the metal lattice, creating a strong electron "glue." The strength of metallic bonding results in the formation of metal alloys, which often have superior properties like increased strength, ductility, and electrical conductivity. Innovatively, think of ionic, covalent, and metallic bonds as the kind of friendships that last a lifetime. These bonds are much stronger than the fleeting connections formed by intermolecular forces.

Think of an ionic bond as a pair of best friends who know each other inside and out and are perfectly complementary. They're deeply committed to each other, and their bond is unique and unbreakable. A covalent bond is like a friendship where both parties share everything and are equally invested in the relationship. The bond between the two is strong and permanent. Finally, the metallic bond can be compared to a group of close friends who all share a common goal - they work together as one, supporting each other as needed, to create a stronger whole.

Dot-and-Cross Diagrams

Dot-and-cross diagrams are used to represent the bonding between atoms in a molecule or compound. These diagrams show how electrons are shared or transferred between atoms to form chemical bonds. There are three types of bonding illustrated by dot-and-cross diagrams: ionic, covalent, and coordinate.

Ionic bonding occurs between a metal and a non-metal. In this type of bonding, electrons are transferred from the metal atom to the non-metal atom to form ions. The metal atom loses one or more electrons to become a positively charged ion, while the non-metal atom gains one or more electrons to become a negatively charged ion. These oppositely charged ions attract each other to form an ionic bond. Dot-and-cross diagrams for ionic compounds show the transfer of electrons and the resulting arrangement of ions.

Covalent bonding occurs between non-metals. In covalent bonding, electrons are shared between atoms to form a molecule. Atoms can form covalent bonds by sharing one or more electrons with each other. This results in a shared pair of electrons between the bonded atoms. Dot-and-cross diagrams for covalent compounds show the sharing of electrons and the resulting molecular structure.

Coordinate bonding is a special type of covalent bonding in which one atom donates both electrons to the shared pair. This type of bonding occurs when an atom has a lone pair of electrons that it donates to another atom to form a covalent bond. The atom that donates the electrons is called the donor, while the atom that accepts them is called the acceptor. Dot-and-cross diagrams for compounds with coordinate bonds show the donor atom donating both electrons and the resulting molecular structure.

Pressure in Gases, Ideal Gases, and the Ideal Gas Equation

Pressure in a gas is a measure of the force exerted per unit area of the container wall by the molecules of the gas. The origin of this pressure is the collisions between gas molecules and the wall of the container. The gas molecules are in a constant state of motion, moving in random directions and at different speeds. As the molecules collide with each other, they transfer energy between them, changing direction and speed. When a molecule collides with the wall of the container, it exerts a force on the wall due to its momentum. This force is perpendicular to the wall and is equal to the rate of change of momentum of the molecule. When multiple molecules collide with the wall of the container, the cumulative force exerted on the wall increases and results in pressure. These collisions occur thousands of times per second, which is why gas molecules exert pressure on the walls of their container. The pressure of a gas is affected by several factors, including temperature, volume, and the number of molecules present in the container. For example, increasing the temperature of a gas will cause the molecules to move faster and collide with the container walls more frequently, resulting in higher pressure. Similarly, reducing the volume of the container will result in more frequent collisions between molecules and the wall, leading to higher pressure.

An ideal gas is a theoretical gas that is characterized by its unique properties and behavior but does not actually exist in the real world. One of the fundamental assumptions made in the study of ideal gases is that they have zero particle volume and no intermolecular forces of attraction. The assumption of zero particle volume implies that the gas particles are infinitesimal points with no size or mass. This is because the particles are assumed to be so small that they do not occupy any space, and the volume they occupy is negligible compared to the volume of the container. In reality, gas particles have a finite size and mass, but the assumption of zero particle volume makes the calculations and predictions of their behavior simpler. The assumption of no intermolecular forces of attraction means that ideal gas particles do not interact with each other in any way. There are no attractive or repulsive forces between the particles, and they move independently of each other. In reality, gas particles do have intermolecular forces of attraction, which can affect their behavior. For example, the attractive forces between particles can cause them to condense into a liquid or solid state under certain conditions. Ideal gases are often used as a model to study the behavior of real gases, which often exhibit different properties due to their molecular interactions. While ideal gases do not exist in reality, they provide a useful tool for understanding and predicting the behavior of gases in various conditions.

The ideal gas equation, also known as the general gas equation, is a fundamental equation that describes the relationship between the pressure (P), volume (V), temperature (T), and amount (n) of gas present in a container. This equation is represented as $pV = nRT$, where R is the universal gas constant.

In this equation, pressure is measured in Pascals, volume is measured in cubic meters, temperature is measured in Kelvin, and the amount of gas present is measured in moles. The value of the universal gas constant (R) is determined by the units used for pressure, volume, and temperature.

This equation can be used to calculate various properties of a gas, such as its mass, molar mass, and density. For example, if we wish to determine the molar mass (Mr) of a gas, we can rearrange the equation to $Mr = (mRT)/(PV)$, where m is the mass of the gas.

Let's say we have a sample of gas with a volume of 2.5 L, a pressure of 101.3 kPa, and a temperature of 25 °C. Our objective is to calculate the molar weight of the gas substance. Using the ideal gas equation, we can convert the temperature to Kelvin (25+273=298K) and plug in the values as follows:

$$(101.3 \text{ kPa} \times 2.5 \text{ L}) = n \times (8.31 \text{ J/K mol} \times 298 \text{ K})$$

$$252.25 = n \times 2473.38$$

$$n = 0.102 \text{ mol}$$

Next, we can calculate the mass of the gas by multiplying the number of moles by its molar mass. Let's assume that the gas is carbon dioxide (CO_2), which has a molar mass of 44.01 g/mol.

$$\text{Mass} = n \times Mr$$

$$\text{Mass} = 0.102 \text{ mol} \times 44.01 \text{ g/mol}$$

$$\text{Mass} = 4.49 \text{ g}$$

Therefore, the molar mass of the gas is 44.01 g/mol, which corresponds to carbon dioxide, and the mass of the sample is 4.49 g.

Structure and Bonding

Let's talk about the lattice structure of giant ionic crystalline solids like sodium chloride and magnesium oxide. First off, what does "giant ionic" mean? Well, it refers to a type of bond between atoms called an ionic bond. This bond is formed when one atom (in this case, a metal) donates an electron to another atom (in this case, a nonmetal), resulting in a positively charged metal ion and a negatively charged nonmetal ion. These ions then attract each other due to their opposite charges and form a lattice structure. Okay, now onto the lattice structure itself. You can think of it like a three-dimensional grid, where each point in the grid represents an ion. In the case of sodium chloride, sodium ions (positive) and chloride ions (negative) alternate in a repeating pattern throughout the crystal lattice. This creates a very strong and stable structure, which is why sodium chloride (also known as table salt!) is such a common and useful compound. The same goes for magnesium oxide – magnesium ions (positive) and oxygen ions (negative) alternate in a repeating pattern throughout the crystal lattice. This lattice structure is what gives magnesium oxide its properties, like its high melting point and hardness.

Let's chat about the lattice structure of simple molecular crystalline solids like iodine, buckminsterfullerene C60, and ice. So, to start with, what does "simple molecular" mean? Well, it refers to a type of structure in which molecules (made up of two or more atoms) are held together by weak intermolecular forces such as van der Waals forces or hydrogen bonding. Unlike giant ionic solids, the intermolecular forces in simple molecular solids are relatively weak. Now, let's look at the three examples you mentioned. Iodine and buckminsterfullerene C60 both have similar lattice structures. They both form a repeating pattern of individual molecules, with the weak intermolecular forces holding those molecules together in a larger, crystal-like structure. In the case of iodine, for example, one iodine molecule will bond to four other iodine molecules to create a square, almost-planar lattice structure. Ice is, of course, a little different. It's still a simple molecular solid, but instead of individual molecules, it's made up of repeating units of water molecules. The hydrogen bonds between those water molecules create a lattice structure in which each water molecule is surrounded by four others in a kind of tetrahedral shape.

Let's talk about giant molecular solids, which have some of the most fascinating and beautiful lattice structures in all of chemistry. We'll be focusing on three giant molecular solids: silicon(IV) oxide, graphite, and diamond. Before we dive into the structures themselves, what exactly is a giant molecular solid? Well, it's a type of solid in which the individual atoms are covalently bonded to each other, resulting in a large, three-dimensional network of atoms. This kind of bond is incredibly strong, which is what gives giant molecular solids their unique properties. So first up is silicon(IV) oxide (also known as quartz). This mineral forms a beautiful crystal lattice structure in which each silicon atom is bonded to four oxygen atoms and each oxygen atom is bonded to two silicon atoms. The resulting structure is a three-dimensional network of tetrahedrons that extend infinitely in all directions. Next is graphite, which is perhaps best known as the "lead" in pencils. Graphite has a unique lattice structure made up of layers of carbon atoms arranged in a two-dimensional hexagonal pattern. Within each layer, each carbon atom is bonded to three neighboring carbon atoms. These layers are then held together by weak intermolecular forces, which is what allows them to slide past each other and give graphite its unique properties. Last but not least is diamond, one of the hardest and most beautiful naturally-occurring substances on Earth. At the atomic level, diamond is made up of a lattice structure of carbon atoms, each of which is covalently bonded to four other carbon atoms in a kind of tetrahedral shape. This results in a very strong, rigid structure that makes diamond incredibly hard and resistant to scratching or chipping.

Let's take a look at the lattice structure of giant metallic solids like copper. Giant metallic solids are a type of crystalline solid in which the atoms are held together by metallic bonds, rather than covalent or ionic bonds. Metallic bonds occur when the outermost electrons of metal atoms are delocalized, meaning they can move freely throughout the metal lattice and are not associated with any one particular atom. This creates a strong electrostatic attraction between the positively charged metal ions and the negatively charged electrons, which holds the structure together. In the case of copper, the lattice structure is made up of a repeating pattern of copper atoms, each of which is bonded to its neighboring atoms through metallic bonds. This creates what's known as a face-centered cubic lattice structure, in which each corner of a cube contains a copper atom and each face of the cube contains an additional copper atom. The resulting lattice structure is incredibly strong and durable, which is why copper has been used for thousands of years in everything from tools and weapons to electrical wiring and cooling systems. The metallic bonds between copper atoms also give this metal its characteristic properties, like its high electrical conductivity and malleability.

So let's dive in and talk about the effect that different types of structures and bonding have on the physical properties of substances, including melting point, boiling point, electrical conductivity, and solubility. First, let's start with melting point and boiling point. The strength of the bonds between the atoms or molecules in a substance has a direct effect on its melting and boiling points. The stronger the bonds, the higher the melting and boiling points will be. For example, giant ionic and giant molecular solids have relatively high melting and boiling points because they have strong covalent or ionic bonds holding them together. Simple molecular solids, on the other hand, have lower melting and boiling points because they only have weak intermolecular forces holding them together. Additionally, the symmetry and complexity of the molecular structure can also affect the melting and boiling points of a substance. Now, let's talk about electrical conductivity. Metals and certain solutions are good conductors of electricity because they have free-moving electrons that can carry an electrical current throughout the solid/liquid. This is mainly due to the nature of metallic and/or ionic bonding, where there is a consistent flow of electrons in the material. On the other hand, covalently bonded nonmetals typically do not conduct electricity well because they have no free-moving electrons to carry a current. These molecules may still dissolve in water or other solvents, but they do not dissociate to form charged particles that can move and conduct electricity. Lastly, let's talk about solubility. The nature of bonding also affects a substance's solubility in different types of solvents. For example, polar solvents like water will dissolve polar substances and ionic compounds. On the other hand, non-polar substances will dissolve in non-polar solvents. This is because like dissolves like. Giant ionic compounds and some covalent compounds tend to be soluble in polar solvents due to their polar nature.

Determining the type of structure and bonding present in a substance is an important skill in many fields of science, from chemistry to materials science and beyond. Here are some ways to deduce the type of structure and bonding based on given information.

1. Look at the substance's physical properties. As we discussed earlier, the type of bonding in a substance can affect its melting point, boiling point, and electrical conductivity, among other properties. If you know the substance's physical properties, you can make some educated guesses about the type of bonding present. For example, if a substance has a high melting point and does not conduct electricity, it may have an ionic or covalent network structure.

2. Analyze the substance's constituent atoms or molecules. The types and arrangements of atoms or molecules in a substance can tell you a lot about the type of bonding present. For example, if a substance is made up of atoms from the left side of the periodic table (metals), it may have metallic bonding. If it is made up of a combination of metals and nonmetals, it may have ionic bonding. Covalent bonding is often present in molecules made up of nonmetal atoms.

3. Consider the substance's chemical formula. The chemical formula of a substance can also provide clues about its structure and bonding. For example, if the formula contains a metal and a nonmetal (like NaCl), it is likely an ionic compound. If it contains only nonmetal atoms, it may be a simple molecular compound. However, it is important to note that some substances can have more complex or mixed bonding, so the formula alone may not provide a definitive answer.

4. Use spectroscopic techniques to analyze the substance's electronic behavior. Techniques like X-ray diffraction, infrared spectroscopy, and electron microscopy can give detailed information about the arrangement of atoms or molecules in a substance, as well as the types of bonding present. These methods can provide more detailed information that can help confirm or refine your initial deductions based on the substance's physical properties, chemical formula, and constituent components.

Enthalpy Change, Activation Energy, and Standard Conditions in Chemical Reactions

Chemical reactions are not just mere transformations of substances, they are more than that. When substances undergo a chemical reaction, they release or absorb energy, which can be observed in the form of heat, light, or even sound. This energy exchange between the reactants and products is known as enthalpy change, and it is a crucial aspect of chemical reactions. Enthalpy changes can be either exothermic or endothermic. Exothermic enthalpy changes refer to reactions that release heat to their surroundings, which causes the surrounding area to become hotter. These types of reactions typically have a negative ΔH value, indicating that energy is being released from the system. On the other hand, endothermic reactions absorb energy from their surroundings, making the surrounding area cooler. These types of reactions have a positive ΔH value, indicating that energy is being absorbed by the system. Understanding enthalpy changes is vital in several fields, from basic chemistry to engineering, and in everyday life, such as when cooking food or burning fuels. Knowing whether a reaction is exothermic or endothermic can provide insights into how it works, how it will behave under specific conditions, and how to control it.

When chemical reactions occur, the reactants must overcome an activation energy barrier before they can transform into products. This energy barrier is a crucial factor in determining the speed of the reaction and the amount of energy that is transferred. A reaction pathway diagram is a graphical representation of the reactions that occur during the transition from reactants to products. These diagrams are beneficial in visualizing how a reaction occurs and estimating the activation energy that is required. The reaction pathway diagrams use arrows to represent the transition from reactants to products. The horizontal axis represents time, while the vertical axis represents the enthalpy change of the reaction. The enthalpy change is determined by calculating the difference between the energy of the initial substances and the energy of the final substances. In other words, it is the amount of heat that is absorbed or released during the reaction. The diagram's shape can give an idea about how exothermic or endothermic the reaction is. If the final product has a lower energy than the initial reactant, then the reaction is exothermic, and the enthalpy change is negative. If the final product has a higher energy than the initial reactants, then the reaction is endothermic, and the enthalpy change is positive. The activation energy is represented by the vertical height between the reactants and the transition state. If the activation energy increases, the speed of the reaction reduces. The activation energy is the minimum amount of energy that the reactants must possess to transform into products. So, if the activation energy is high, fewer molecules will have enough energy to go over the energy barrier.

When chemists work with chemical reactions, they need to have a standard set of conditions to ensure that their experiments can be accurately compared and contrasted with those of other researchers. The standard conditions are conditions of temperature and pressure, and they are denoted by the symbol ⬦. The standard conditions for a chemical reaction are typically 298 Kelvin (25 degrees Celsius) and 101 kilopascals (1 atmosphere). These values are considered the norm for most laboratory experiments since they provide a stable environment that is easy to replicate. The temperature represents the room temperature at which most experiments are conducted, and the pressure is the typical atmospheric pressure at sea level. Using the symbol ⬦

in chemical reactions indicates that the reaction is taking place under standard conditions. This means that the reaction is being carried out at a specific temperature and pressure, and the resulting products can be expressed with a specific enthalpy change or a standard molar energy change. The use of standard conditions is essential in chemistry since they guarantee that different experiments have occurred under the same conditions, allowing for accurate comparisons, and preventing confusion. Standard conditions provide a reference point for the enthalpy change of reaction, standard enthalpy of formation, enthalpy of combustion, enthalpy of reaction, and other thermodynamic properties.

Enthalpy change is the amount of energy gained or lost by a system at a constant pressure. It is often used in chemistry to describe the energy exchange that occurs during a chemical reaction. In particular, there are several types of enthalpy changes that are frequently used to describe chemical reactions. The first of these is the enthalpy change of reaction (ΔHr), which is the amount of energy absorbed or released during a chemical reaction. This value can be positive for endothermic reactions (absorbing heat from its surroundings) or negative for exothermic reactions (releasing heat to its surroundings). The enthalpy change of formation (ΔHf) refers to the amount of energy released or absorbed during the formation of one mole of a substance from its constituent elements under standard conditions (at 298K and 1atm). In other words, it is the energy required to form a substance from its elements at the temperature and pressure of the system. This value is useful in determining the stability of a compound or molecule and can be found in tables of thermodynamic data. The enthalpy change of combustion (ΔHc) is the amount of energy released or absorbed during the combustion of a substance. In other words, it is the energy released when a substance is burned in the presence of oxygen. This value is useful in determining the energy content of fuels and can be used to compare the energy output of different types of fuels. The enthalpy change of neutralization ($\Delta Hneut$) is the amount of energy released or absorbed when an acid and a base react to form a salt and water. This value can be negative or positive, depending on the strength of the acid and base used in the reaction. The stronger the acid and base, the greater the enthalpy change of neutralization.

When chemical reactions occur, energy is always exchanged between the reactants and products. This exchange of energy is due to the breaking and formation of chemical bonds. Chemical bonds are the forces that hold atoms together in a molecule, and when they are broken or formed, energy is released or absorbed. Breaking bonds requires energy, which is usually gained from the surrounding environment. This energy breaks the bond between atoms, making them available for bonding with other atoms. On the other hand, when new bonds are formed between atoms, energy is released, which can be absorbed by surrounding atoms and molecules, causing the temperature rise in that environment. The amount of energy released or absorbed during a chemical reaction is determined by the strength of the chemical bonds involved. Stronger bonds require more energy to break and form, and so the resulting energy release or absorption is more significant. For example, breaking the bonds in a molecule with triple bonds requires more energy than breaking double or single bonds. The breaking and making of chemical bonds during chemical reactions can be exothermic or endothermic. In exothermic reactions, more energy is released than absorbed. The occurrence of this phenomenon is frequently noticed during combustion reactions. Combustion reactions cause the chemical bonds of the reactant molecules to break, and new bonds are formed between the substances and the oxygen present in the environment. This leads to the release of energy in the form of light and heat. Endothermic reactions absorb more energy than they release. These reactions are non-spontaneous and require an external energy input to proceed. For instance, photosynthesis is an endothermic reaction. In this process, plants convert sunlight energy into chemical energy by making new bonds between carbon dioxide and water molecules to form glucose and releasing oxygen. Bond energies are the amount of energy required to break a chemical bond between two atoms. This energy is usually endothermic, i.e., it is absorbed from the surrounding environment. The knowledge of bond energies is instrumental in calculating the overall enthalpy change of a chemical reaction, which is defined as the difference in energy between the products and reactants. To determine the enthalpy change of a reaction using bond energies, one needs to know the bond energies of each bond that breaks or forms during the reaction. On a molecular level, breaking or forming a bond requires a certain amount of energy, which is reflected by the bond energies. If a bond is broken, energy will be absorbed, while the formation of a bond releases energy. The enthalpy change of the reaction, ΔHr, can be calculated using the bond energies as follows, ΔHr(Bonds Broken) - ΔHr(Bonds Formed) = ΔHr The first step is to identify which bonds are breaking and which are forming in the reaction. Once these bonds are identified, their bond energies can be determined from a bond energy table. For example, the energy required to break a C-H bond is approximately 416 kJ/mol. The next step is to multiply the number of bonds that are breaking by their bond energies and the number of bonds that are forming by their respective bond energies this shall give you the bond energies of the reaction. Once the bond energies are calculated, they can be used in the equation above to determine the enthalpy change of the reaction. It's important to note that the bond energies used in this calculation are average values since the actual bond energy for the particular molecule may vary depending on factors such as chemical environments, bond length, and neighboring atoms.

Bond energies refer to the amount of energy required to break a bond between two atoms. The amount of energy required to break a bond can vary depending on the type of bond and the specific atoms involved. Because of this variability, bond energies are often expressed as average values, though some bond energies are exact. Exact bond energies are values that are known precisely for a particular molecule or bond. For example, the bond energy of the nitrogen–nitrogen (N≡N) bond in a dinitrogen molecule (N2) is precisely known, and therefore considered an exact value. These precise values are obtained by the measurement of a bond dissociation energy or by spectroscopic analysis of the molecule. On the other hand, average bond energies are values that are obtained by averaging the bond energies of many molecules or bonds with similar characteristics. The average bond energy values can be found in bond energy tables and are typically used to predict the enthalpy change of reactions. The values in bond energy tables represent an average value because different molecules and environments can affect the bond energies in a specific molecule. It's important to note that the values of average bond energies are only used for predictions, and do not necessarily reflect the exact bond energies present in a particular molecule under specific circumstances. In some cases, it may be necessary to use an exact bond energy value for a specific molecule or bond junction. Using bond energies, whether exact or average, are useful in predicting the enthalpy change of a chemical reaction, which is a crucial concept in chemistry. These predictions can help scientists better understand the behavior and stability of molecules and reactions, facilitating the design of more efficient and effective chemical processes and materials.

Enthalpy is a term that refers to the amount of heat energy released or absorbed during a chemical reaction. This term is often used to describe the relationship between temperature and energy changes. If we want to calculate the enthalpy change of a reaction, we can use the equation $q = mc\Delta T$, where q represents the amount of heat energy released or absorbed, m represents the mass of the substance, c represents the specific heat capacity of the substance, and ΔT represents the temperature change. Another equation that can be used to calculate the enthalpy change is $\Delta H = -mc\Delta T/n$, where ΔH represents the enthalpy change, m represents the mass of the substance, c represents the specific heat capacity of the substance, ΔT represents the temperature change, and n represents the number of moles of the substance involved in the reaction. This equation represents the relationship between the enthalpy change and the temperature change. To calculate the enthalpy change using these equations, we need to perform experiments using the substances involved in the chemical reaction. We need to measure the mass of the substances, the specific heat capacities, and the temperature changes. By putting these values into the equations above, we can calculate the enthalpy change of the reaction.

Hess's Law

Hess's Law is a concept used in chemistry that states that the total enthalpy of a chemical reaction depends only on the difference in enthalpies between the reactants and the products. This means that you can actually add together the enthalpy changes from multiple reactions to get the overall enthalpy change for a reaction that you're interested in. To construct a simple energy cycle using Hess's Law, you can start with a known reaction (maybe from a textbook or lab experiment) and then figure out how to get to the reaction you're interested in by adding or subtracting other known reactions. This can be useful when you don't have enough information about a particular reaction or if it is difficult to measure the enthalpy change directly. You can think of energy cycles as a way to visualize this process. Each reaction is represented as a box with its enthalpy change (ΔH) written inside. The reactants are on the left side of the box and the products are on the right side. To add reactions together, you can match up the products of one reaction with the reactants of another, cancel out any common species, and then add up the enthalpy changes. You can continue to do this until you get to the reaction you're interested in. It's important to remember that when applying Hess's Law, the enthalpy change is considered to be a state function. This means that the enthalpy change value only depends on the initial and final states of the system, and not on the specific path or sequence of reactions that took place between the states. This allows you to construct energy cycles that might not even be possible in real life, since the actual path of the reaction doesn't matter as long as the initial and final states are the same.

energy cycles can be used to calculate enthalpy changes of reactions that cannot be found directly through experiment. These cycles use a series of known reactions to determine the overall enthalpy change of a reaction that is unknown or difficult to measure. Energy cycles are particularly useful when directly measuring the enthalpy change of a reaction is not feasible.

To carry out calculations using cycles and relevant energy terms, you first need to understand the terms involved in energy cycles. In chemistry, important terms include the enthalpy change (ΔH), which measures the amount of heat absorbed or released in a chemical reaction, and the standard enthalpy of formation (ΔHf). This is the enthalpy change that occurs when one mole of a compound is formed from its constituent elements in their standard states - this means that standard enthalpy of formation is a way to measure the energy changes that take place when compounds are formed.

To determine the enthalpy changes that cannot be found by direct experiment, you can start with a series of known reactions and use these reactions to construct an energy cycle. This cycle should include a reaction with the same products and reactants as the reaction you are interested in, but with a known enthalpy change. You can then use Hess's Law to calculate the enthalpy change of the reaction you are interested in.

Here's an example to illustrate this process: let's say you want to determine the enthalpy change of the following reaction, which cannot be measured directly:

$$N2(g) + 3H2(g) \rightarrow 2NH3(g)$$

To do this, you could use the following known reactions and their enthalpy changes:

1. $N2(g) + 3/2 O2(g) \rightarrow N2O3(g)$ ΔH = -76.43 kJ/mol

2. $N_2O_3(g) + H_2O(l) \rightarrow 2HNO_2(g)$ ΔH = -135.6 kJ/mol

3. $2HNO_2(g) \rightarrow N_2O(g) + H_2O(l)$ ΔH = -138.2 kJ/mol

4. $N_2O(g) + NO(g) \rightarrow N_2(g) + O_2(g)$ ΔH = 43.2 kJ/mol

5. $2NO(g) + O_2(g) \rightarrow 2NO_2(g)$ ΔH = -114.2 kJ/mol

6. $4NH_3(g) + 3O_2(g) \rightarrow 2N_2O(g) + 6H_2O(g)$ ΔH = -1782 kJ/mol

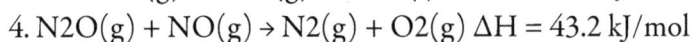

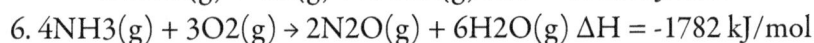

You can then use these known reactions to construct an energy cycle, as follows:

- Start with the reaction you're interested in: $N_2(g) + 3H_2(g) \rightarrow 2NH_3(g)$

- Add reaction 4 in reverse to get $NO(g)$

- Add reaction 5 in reverse to get $N_2O(g)$

- Add reaction 3 to get $2HNO_2(g)$

- Add reaction 2 in reverse to get $N_2O_3(g)$

- Finally, add reaction 1 to return to the starting point: $N_2(g) + 3/2O_2(g) \rightarrow N_2O_3(g)$

By summing up the enthalpy changes of each reaction, with the appropriate sign, you can calculate the enthalpy change of the reaction you're interested in:

ΔH = (-76.43) + (-43.2) + (-138.2) + (135.6) + (114.2) + (-1782) = -1790 kJ/mol

Therefore, the enthalpy change of the reaction is -1790 kJ/mol, which cannot be found directly through experiment.

Bond energy represents the amount of energy required to break one mole of a specific bond in a molecule, and it is measured in kilojoules per mole (kJ/mol).

To carry out calculations using cycles and bond energy data, you would start with the chemical equation of the reaction you're interested in, and then break down the equation into its constituent bonds. You would then calculate the total bond energy in the reactants and the total bond energy in the products. The difference between the two values will give you the total enthalpy change of the reaction.

Here's an example to illustrate this process: Let's say you want to calculate the enthalpy change of the following reaction using bond energy data:

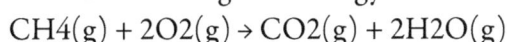

$CH_4(g) + 2O_2(g) \rightarrow CO_2(g) + 2H_2O(g)$

To do this, you would first break down the reactants and products into their constituent bonds:

Reactants:

- 4 C-H bonds in CH_4

- 2 O=O double bonds in O_2

Products:

- 1 C=O double bond and 2 C-O single bonds in CO_2

- 4 O-H bonds in H_2O

Next, you would use bond energy data, which can be found in a reference book or online, to calculate the total bond energy of the reactants and products. Bond energy data varies for different bonds, and it is typically given as the average bond energy across different molecules and materials.

Here are some example bond energy values that will be used for our calculation:

- C-H bond energy = 413 kJ/mol

- O=O bond energy = 498 kJ/mol

- C=O bond energy = 744 kJ/mol

- C-O bond energy = 360 kJ/mol

- O-H bond energy = 463 kJ/mol

Using this data, you can calculate the total bond energy for the reactants and products as follows:

Reactants:

- 4 C-H bonds × 413 kJ/mol = 1652 kJ/mol

- 2 O=O double bonds × 498 kJ/mol = 996 kJ/mol

Total = 2648 kJ/mol

Products:

- 1 C=O double bond × 744 kJ/mol = 744 kJ/mol

- 2 C-O single bonds × 360 kJ/mol = 720 kJ/mol

- 4 O-H bonds × 463 kJ/mol = 1852 kJ/mol

Total = 3316 kJ/mol

The enthalpy change of the reaction is the difference between the total bond energy of the products and the total bond energy of the reactants:

ΔH = 3316 kJ/mol - 2648 kJ/mol = 668 kJ/mol

Therefore, the enthalpy change of the reaction is 668 kJ/mol, which represents the heat evolved or absorbed by the reaction.

The Role of Oxidation-Reduction Reactions

Oxidation numbers play a significant role in understanding chemical reactions. Simply put, the oxidation number of an element is a measure of how many electrons it has gained or lost during a chemical reaction. It is a way to keep track of how the charge of an element changes as it participates in the reaction. To calculate oxidation numbers, we start by assigning a zero value to the atoms in their uncombined state. Then, we assign electrons to the atoms based on a set of rules. For example, Group 1 elements always have an oxidation number of +1, while Group 2 elements always have an oxidation number of +2. Elements in a compound can have different oxidation numbers, depending on the overall charge of the compound. For example, in water (H_2O), hydrogen has an oxidation number of +1, while oxygen has an oxidation number of -2. This means that the overall charge of the compound is zero. When it comes to ions, the oxidation number of an ion is simply equal to the charge of the ion. For example, the oxidation number of the sodium ion ($Na+$) is +1, while the oxidation number of the chloride ion ($Cl-$) is -1.

Balancing chemical equations is an important step in understanding chemical reactions. It ensures that the reactants and products are in the correct ratio to each other. One way to balance chemical equations is by using changes in oxidation numbers.

To use changes in oxidation numbers to balance a chemical equation, we first determine the oxidation state of each element on both the reactant and product side of the equation. We then identify which elements have changed oxidation state and by how much.

Once we have identified the elements that have changed oxidation state, we can use this information to balance the equation. We can add coefficients to the reactants and products to make sure that the number of atoms and charges balance on both sides of the equation.

For example, let's balance the equation:

Fe + HCl -> FeCl3 + H2

We start by identifying the oxidation state of each element:

Fe: 0 -> +3

H: +1 -> 0

Cl: -1 -> -1

We can see that the oxidation state of Fe has increased from 0 to +3, while the oxidation state of H has decreased from +1 to 0. This suggests that a reduction-oxidation (redox) reaction has occurred.

To balance the equation, we first balance the atoms:

Fe + 2HCl -> FeCl3 + H2

Now we can balance the charges by adding electrons to the side that is most deficient:

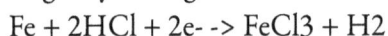

Fe + 2HCl + 2e- -> FeCl3 + H2

Finally, we can cancel out the electrons:

2Fe + 6HCl -> 2FeCl3 + 3H2

And the equation is balanced.

Using changes in oxidation numbers to balance chemical equations can take practice, but it is a valuable skill for any chemist. It allows us to gain a deeper understanding of chemical reactions and ensures that our calculations are accurate.

Redox reactions, or reduction-oxidation reactions, involve the transfer of electrons between reactants. These reactions can be explained in terms of oxidation, reduction, and disproportionation.

Oxidation is the process by which an element loses electrons. It results in an increase in the oxidation number of that element. For example, when iron (Fe) reacts with oxygen (O2) to form iron oxide (Fe2O3), the iron atom loses electrons and undergoes oxidation, while the oxygen atoms gain electrons and undergo reduction:

$$4Fe + 3O2 -> 2Fe2O3$$

Reduction is the opposite of oxidation. It is the process by which an element gains electrons, leading to a decrease in its oxidation number. For example, in the reaction above, oxygen undergoes reduction by gaining electrons from iron.

Disproportionation occurs when a single element undergoes both oxidation and reduction in the same reaction. This leads to the formation of two different compounds with different oxidation states. For example, when chlorine gas (Cl2) reacts with water (H2O), it can form both hydrochloric acid (HCl), where chlorine has gained electrons and undergone reduction, and chlorine gas (Cl2), where chlorine has lost electrons and undergone oxidation:

$$Cl2 + H2O -> HCl + HOCl$$

Redox reactions are important in many areas of chemistry, including energy storage, metabolism, and corrosion. Understanding the concepts of oxidation, reduction, and disproportionation can help us predict the outcome of a chemical reaction and design new chemical processes.

In a redox reaction, one reactant gains electrons (reduction) while another reactant loses electrons (oxidation). The substance that undergoes oxidation is known as the reducing agent while the substance that undergoes reduction is called the oxidizing agent. Reducing agents are substances that tend to donate electrons to other substances. This causes a decrease in their own oxidation state or an increase in the oxidation state of the other substance. In other words, reducing agents are electron donors that promote reduction in the reaction. For example, hydrogen gas (H2) is a powerful reducing agent that is commonly used in a variety of reduction reactions, including in the production of amines, alcohols, and other organic compounds. Oxidizing agents, on the other hand, tend to accept electrons from other substances. This results in an increase in their own oxidation state or a decrease in the oxidation state of the other substance. Oxidizing agents are electron acceptors that promote oxidation in the reaction. Common examples of oxidizing agents include chlorine gas (Cl2), hydrogen peroxide (H2O2), and potassium permanganate (KMnO4). Oxidizing and reducing agents are important in many chemical reactions, including combustion, corrosion, and biological processes like respiration and photosynthesis. In combustion, for example, oxygen gas (O2) acts as an oxidizing agent that reacts with a fuel source to release energy in the form of heat and light.

Roman numerals can be used as a notation to indicate the magnitude of the oxidation number of an element. The Roman numeral is placed after the symbol of the element in parentheses and represents the numerical value of the oxidation number. The use of Roman numerals is especially important when an element has more than one possible oxidation number. For example, the element iron (Fe) can have an oxidation number of +2 or +3. In the compound FeCl2, the oxidation number of iron is +2, while in the compound FeCl3, the oxidation number of iron is +3. By including the Roman numeral in the chemical formula, it indicates the specific oxidation state of the element in a compound. This information is useful for understanding the properties and behavior of the compound. It also helps us to determine the electron transfer in a chemical reaction. To determine the oxidation state of an element, there are several rules and guidelines that should be followed.

However, if the compound does not have a Roman numeral notation for an element, it can be difficult to determine the oxidation state.

Chemical Equilibria

Have you ever played with a toy that can transform, like a transformer or a rubix cube? Those toys can change and transform back and forth, right? That's kind of like what a reversible reaction is. Basically, a reversible reaction is a chemical reaction that can go back and forth between reactants and products. It's like a constant state of transformation - the reactants turn into products, but then the products can also turn back into reactants. It's like a never-ending cycle of transformation. Imagine you're at a party and you walk into a room with a dance floor. The people on the dance floor are like the reactants, and the people sitting on the sidelines are like the products. As soon as someone gets up from the sidelines and joins the dance floor, they become a reactant, and the people they left behind on the sidelines become products. But then, if someone gets tired and sits back down, they become a product again - and the person who took their place on the dance floor becomes a reactant. So a reversible reaction is kind of like a dance floor - always moving and changing, with the reactants and products in a constant state of transformation. It's pretty cool, isn't it?

Dynamic equilibrium is a concept that describes a state in which the rate of the forward reaction is equal to the rate of the reverse reaction, resulting in a stable concentration of both reactants and products. This state of balance is the result of a delicate dance between the opposing forces of the chemical reaction. As one molecule of reactant is converted into product, another molecule of product reverts back to reactant, effectively canceling out any net change in concentration. In simpler terms, dynamic equilibrium is like a game of tug-of-war between two equal and opposite teams. Neither side is able to gain an advantage over the other, resulting in a constant equilibrium between the two teams. Similarly, in a chemical reaction, the forward and reverse reactions are like two teams working against each other. While the forward reaction tries to convert all the reactants into products, the reverse reaction attempts to turn all the products back into reactants. Ultimately, the equilibrium between these two opposing forces results in a stable and unchanging concentration of both reactants and products. Ultimately, dynamic equilibrium is a fascinating concept that describes the complex interplay between chemical reactions and the concentrations of reactants and products. Understanding this phenomenon is vital in fields ranging from biochemistry to chemical engineering and beyond.

The establishment of dynamic equilibrium requires a closed system where the reactants and products are in a closed container without any exchange with the surroundings. This is because in an open system, the equilibrium cannot be established as the reactants and products can escape from the system and can be replaced by new reactants or products from the surroundings. A closed system ensures that the total amount of matter involved in the reaction is constant, which is essential to establish the equilibrium between the forward and reverse reactions. In a closed system, the reactants and products reach a state of balance as they continue to interact. Since the system is closed, the concentration of both reactants and products are conserved, even as they shift between each other in what is known as a reversible reaction. The forward reaction of reactants converting into products is balanced by the reverse reaction of products turning back into reactants, all the while leaving the overall concentration of both reactants and products unchanged. This creates a stable equilibrium where the rates of the forward and reverse reactions are equal. To better understand this concept, imagine a seesaw where two equally weighted individuals are seated on both sides. This seesaw represents the closed system during a reversible chemical reaction. Like the individuals on this seesaw, the reactants and products are constantly moving back and forth, but at the same rate so that neither side can gain an advantage over the other. Their equal and opposing actions keep them at equilibrium, allowing both sides to remain balanced, which is essential for the establishment of dynamic equilibrium. Ultimately, a closed system is essential for the establishment of dynamic equilibrium as it helps to maintain a fixed concentration of reactants and products, thereby allowing the forward and reverse reactions to balance each other. This concept has important implications for various fields like chemistry, biochemistry and chemical engineering and provides a foundation for understanding many chemical processes.

Le Chatelier's principle is a fundamental concept in chemistry that explains how a system at dynamic equilibrium responds to changes in its conditions. At its core, Le Chatelier's principle states that if a chemical system experiences a change, the position of equilibrium will shift in such a way as to minimize the impact of that change on the overall system. In other words, Le Chatelier's principle helps to predict how a chemical reaction will react to changes in factors that affect it, such as temperature, pressure or concentration. If a change is made to one of these factors, the reaction will adjust to counteract the change and remain in equilibrium. For example, if the concentration of a reactant is increased, the reaction will favor the consumption of that reactant and the formation of more products in order to rebalance the system. Similarly, if the temperature of a reaction is increased, the system may react by favoring the endothermic direction of the chemical reaction. An innovative way to think of Le Chatelier's principle is to consider it like a balancing act. When a system is in equilibrium, it's as if a balance has been struck between the various reactants and products, and they are all in harmony with one another. Any change to the system can be thought of as someone adding or removing weight from one side of the balance. This change in weight causes the balance to move in an attempt to restore equilibrium. In a comparable approach, the Le Chatelier's principle envisages that if a chemical system undergoes any alteration, it will adjust in a manner to balance and restore the equilibrium of the system.

Le Chatelier's principle is a significant idea in chemistry that helps to anticipate how an equilibrium system will respond to adjustments in concentration, temperature, pressure, or the availability of a catalyst. By understanding how these changes impact the equilibrium position, we can better control the outcome of chemical reactions. When it comes to changes in temperature, the application of Le Chatelier's principle suggests that increasing the temperature of an endothermic reaction will shift the equilibrium position towards the products, and decreasing the temperature will shift it towards the reactants. Conversely, for an exothermic reaction, increasing the temperature will shift the equilibrium position towards the reactants, and decreasing the temperature will shift it towards the products. In other words, temperature changes affect the equilibrium position by favoring the reaction that either produces or absorbs heat. Changes in concentration or pressure, on the other hand, depend on which side of the equation has more moles. An increase in concentration of reactants will shift the equilibrium position towards the products to counteract the increase in concentration, while an increase in concentration of products will shift the equilibrium position towards reactants. Similarly, an increase in pressure will favor the side of the equation with fewer moles of gas, while a decrease in pressure will favor the side with more moles of gas. The presence of a catalyst is another factor that can affect the equilibrium position. A catalyst increases the rate of a reaction by providing an alternate pathway with lower activation energy, but it does not affect the equilibrium position. This is because it increases the rate of forward and reverse reactions equally but at the same time it does not change the balance between the reactants and products. An innovative way to think about Le Chatelier's principle is by considering it like a dance between the reactants and products. Imagine two basketball teams dancing on a dance floor, representing the forward and backward reactions. A change in temperature can be depicted by turning up the music, which makes the teams dance faster and the equilibrium position shift. An increase in the concentration of reactants can be likened to adding more members to the forward team and the backward team compensating by removing some of its members. Overall, changes in chemical reactions can be viewed as a dynamic dance, constantly trying to balance the reactions and products in equilibrium.

Equilibrium constants, written as Kc, are important mathematical expressions used to describe the extent of chemical reactions at equilibrium. Kc refers to a ratio of equilibrium concentrations, with the numerator representing the product concentrations and the denominator representing the reactant concentrations. The value of Kc is constant only for a specific temperature and it can help predict the direction of a reaction.

To deduce expressions for Kc in terms of concentrations, we must first understand how we can relate the concentration of reactants and products to the equilibrium constant. For a general chemical reaction of the form A + B <—> C + D, the expression for Kc can be written as:

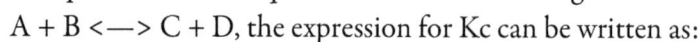

$$Kc = [C]^c[D]^d/[A]^a[B]^b$$

Where [A], [B], [C], and [D] represent the molar concentrations of the respective species at equilibrium and a, b, c and d represent the stoichiometric coefficients of reactants and products in the balanced chemical equation. These coefficients indicate how many molecules of each species are involved in the reaction.

An innovative way to think of Kc is like a recipe of a meal. For example, think of a recipe for baking a cake. The recipe requires a certain amount of flour, sugar, eggs, and other ingredients. Kc is similar in that it tells us how much of each ingredient we need to create a cake with a desirable texture and flavor, in this case the state of equilibrium. The concentration of each reactant or product in the expression for Kc tells us how much of each "ingredient" is used to form the desired product.

Mole fraction and partial pressure are important concepts used in chemistry to describe the properties of gases in mixtures. They both help to determine the behavior of gas mixtures by relating the composition of the mixture to the physical quantities like pressure and volume.

Mole fraction is a measure of the relative proportions of different gases in a mixture. The mole fraction of a gas in a mixture is equal to the number of moles of that gas divided by the total number of moles in the mixture. This value is used to describe which gases are present and in what proportion they are mixed. It's essential to take into account the mole fraction when discussing the properties of gases. They help us to understand how the components dissolve to form the mixture.

Partial pressure is a phrase that pertains to the force that a solitary gas exerts within a combination, as if it was in an independent container. For instance, consider a mixture of two gases, A and B, in a container. The total pressure of the mixture is equal to the sum of the partial pressures of gas A and gas B. We can represent partial pressure mathematically as $p\_A = X\_A * P\_total$, where p_A is the partial pressure of gas A, X_A is its mole fraction, and P_total is the total pressure of the mixture.

An innovative way to think of mole fraction and partial pressure is comparing them to a group of students in a classroom. Think of a classroom with a fixed number of students, where different students represent different gases in the mixture. Mole fraction equates to the number of students belonging to a particular group, and their fraction of the total number of students. The partial pressure is like the noise level made by the students of one particular group; this group's gas's partial pressure reflects this noise and corresponds to the pressure the gas would exert if it were the only gas in the mixture.

Equilibrium constants, represented by K, are important in chemistry as they relate to the concentration or partial pressure of species at the point of equilibrium. The value of the equilibrium constant can be obtained using several formulas, depending on the nature of the chemical reaction. The most common formula used is Kc when expressing the equilibrium constant in terms of concentration. Kp, on the other hand, is another formula that expresses the equilibrium constant in terms of partial pressures and it assumes a reaction between gaseous species.

To deduce expressions for equilibrium constants in terms of partial pressures, we need to consider the relationship between the total pressure of a gaseous reaction mixture and the partial pressures of its constituent gases. Consider the general reaction $aA + bB \leftrightarrow cC + dD$, where A, B, C and D are gaseous reactants/products and 'a', 'b', 'c' and 'd' are stoichiometric coefficients. Kp, which describes the extent of the reaction based on partial pressures, can be expressed as follows:

$$Kp = (P\_C)^c (P\_D)^d / (P\_A)^a (P\_B)^b$$

Where P_A, P_B, P_C, and P_D are the partial pressure of gases A, B, C, and D, respectively, at equilibrium, and a, b, c, and d are the stoichiometric coefficients of the balanced chemical equation. Essentially, Kp relates the activity of the gaseous reactants or products at equilibrium to their original partial pressures.

An innovative way to think about Kp is to imagine a group of people dancing together. The number of people in the group represents the total pressure of the reaction mixture, and each person represents a specific gas. The partial pressures of each gas would correspond to the space they occupy on the dance floor. The greater the amount of space occupied by a person, the greater their contribution to the overall pressure in the area.

Using Kc and Kp expressions in calculations involves determining the equilibrium constant values by either concentration in moles per liter or partial pressure values. These expressions provide a mathematical framework for predicting the direction and extent of chemical reactions that reach a state of equilibrium. Unlike other equations that require solving quadratic equations, Kc and Kp expressions do not necessitate complex computations, making them an efficient tool for analyzing and understanding chemical reactions. To perform calculations using Kc and Kp expressions, it is important to understand the underlying principles that govern chemical equilibrium. Equilibrium constants are determined based on the ratio of the concentration or partial pressure of products to reactants at equilibrium. These values provide insight into the relative stability of the reactants and products and allow scientists to predict how the reaction will proceed under different conditions.

When using Kc expression, concentration values of all species involved in the chemical reaction are used to calculate the equilibrium constant. The value of Kc can be used to determine the concentration of the products and reactants in a system at equilibrium. On the other hand, Kp expression uses partial pressure values of gases involved in the reaction to calculate the equilibrium constant. Although the values of Kc and Kp are different, they are related to each other by the ideal gas equation.

The process of calculating quantities present at equilibrium involves determining the concentrations or partial pressures of the products and reactants at the point of chemical equilibrium. This requires the use of equilibrium constants, which provide a mathematical framework for predicting the direction and extent of chemical reactions. To calculate quantities at equilibrium, we need appropriate data such as the initial concentrations or pressures of reactants and products, and the equilibrium constant value for the chemical reaction under consideration. Using this data, we can determine the concentrations or partial pressures of the species involved in the reaction, which will be present once the reaction reaches equilibrium. The equilibrium constant, represented by Kc or Kp, is a ratio of the concentration of the products to reactants or the partial pressures of the products to reactants, respectively. Therefore, once we know the value of the equilibrium constant, we can determine the concentration or partial pressure of each component at equilibrium using simple algebraic calculations. It's important to note that changes to the initial conditions can affect the equilibrium concentrations and, therefore, affect the outcome of the reaction. For example, changes to temperature, pressure, or concentration can shift the position of the equilibrium, modifying the concentrations or partial pressures of the reactants and products.

The equilibrium constant is a fundamental concept in chemical reactions that describes the relationship between the concentrations of reactants and products when the reaction reaches equilibrium. The equilibrium constant value is determined based on the relative amounts of reactants and products in a reaction mixture. It is a constant value at a specific temperature, concentration, and pressure under a given set of conditions. However, the values of temperature, concentration, pressure, and the presence of a catalyst can cause changes in the value of the equilibrium constant. Temperature is one of the factors that can affect the value of the equilibrium constant. When the temperature of the system is altered, this can shift the position of the equilibrium. This movement may result in a change in the relative amounts of reactants and products and, ultimately, can cause the equilibrium constant to change. A reaction that is exothermic will shift to the right when the temperature decreases, whereas it will shift towards the left when the temperature increases. Conversely, for an endothermic reaction, the opposite is true. Concentration is another critical factor that can impact the equilibrium constant.

Alterations to the concentration of any of the reactants or products at equilibrium can cause shifts in the direction of chemical reactions so as to maintain the equilibrium constant. Therefore, if the concentration of any of the species present in the equilibrium is changed, the equilibrium will shift to restore equilibrium by adjusting its concentrations, eventually changing the value of Kc or Kp. Pressure also causes changes in the equilibrium constant, particularly in gas-phase reactions. An increase in pressure causes the equilibrium to shift towards the side of the reaction involving fewer moles of gas, whereas the decrease in pressure will shift the equilibrium towards the side of the reaction involving more moles of gas. A catalyst's existence has no effect on the equilibrium constant's magnitude. A catalyst can increase the rate of the forward and reverse reactions to the same extent; hence, the position of the equilibrium remains the same.

The Haber process and the Contact process are two chemical industrial processes that rely on understanding dynamic equilibrium and the application of Le Chatelier's principle. The Haber process is a chemical process utilized to create ammonia from nitrogen gas and hydrogen gas. On the other hand, the Contact process involves the production of sulfuric acid from sulfur dioxide gas and oxygen gas. Both of these industrial processes operate under specific conditions to achieve an optimal outcome. In the Haber process, the necessary conditions for achieving equilibrium are a pressure of around 200 atmospheres, a temperature of 450 degrees Celsius, and the presence of a catalyst, typically iron. The high pressure of the system pushes the reactive gases into a state of equilibrium that produces ammonia, with the help of the catalyst. Therefore, once the equilibrium is reached, the forward and reverse reactions continue at the same rate to maintain equilibrium concentrations. Ammonia is a crucial component in fertilizers, polymers production, and even explosives. The Contact process requires an understanding of dynamic equilibrium and Le Chatelier's principle. The process takes place at temperatures of approximately 450 to 500 degrees Celsius and a pressure of about 1 to 2 atmospheres. The process employs a catalyst vanadium oxide to oxidize sulfur dioxide gas to produce sulfur trioxide gas which reacts with water to form sulfuric acid. As the reaction between sulfur dioxide and oxygen is exothermic, increasing the temperature of the system causes the equilibrium to shift to the left, reducing the amount of sulfur trioxide produced. However, Le Chatelier's principle states that by reducing the temperature, the equilibrium will shift to the right towards the formation of sulfur trioxide, increasing the yield of sulfuric acid.

Brønsted–Lowry theory of acids and bases

Acids are compounds with a pH level lower than 7 which means they give out H+ ions when dissolved in water. Some common acids that you may come across include hydrochloric acid, sulfuric acid, nitric acid, and ethanoic acid. Let's dive a bit deeper into these acids and see what makes them unique. Starting with hydrochloric acid or HCl, it is a clear, colorless, and pungent-smelling solution. It is composed of hydrogen and chlorine and has a chemical formula of HCl. It is one of the strongest acids available, with a pH close to zero. This acid is produced in the stomach to break down food and aid in digestion. Next up is sulfuric acid or H2SO4. It is a powerful and highly reactive acid, with a molecular formula of H2SO4. This acid is commonly used in the production of fertilizers, dyes, and detergents. It is also used in laboratory experiments and research. The third acid on our list is nitric acid or HNO3. Similar to sulfuric acid, it is a powerful oxidizing agent with a molecular formula of HNO3. It is used in the production of fertilizers, explosives, and dyes. Last but not least, we have ethanoic acid or CH3COOH. It is also known as acetic acid and commonly found in vinegar, giving it that acidic and sour taste that we all know and love. It has a chemical formula of CH3COOH, and it is used in the production of plastics, dyes, and textiles.

Alkalis are basic substances that have a pH level higher than 7 and they release hydroxide ions when dissolved in water. Here are the names and formulas of three common alkalis. The first alkali on our list is sodium hydroxide or NaOH. It is commonly known as caustic soda and is a highly corrosive compound. It is chemically made up of one atom of sodium, one atom of hydrogen, and one atom of oxygen. Sodium hydroxide is used to break down fats and oils in soap making, medicine production, and various industrial applications. The second alkali is potassium hydroxide, also known as KOH. It is a strong base with a chemical formula of KOH and is often used in the production of soaps, detergents, and other cleaning products. It is also used as a substitute for sodium hydroxide in some instances. The third and final alkali on our list is ammonia, represented by the chemical formula NH3. Ammonia is a colorless gas with a pungent odor that can be harmful when inhaled. It is commonly used in the production of agricultural fertilizers, cleaning products, and refrigerants.

The Brønsted–Lowry theory is one of the most widely accepted theories of acids and bases. Johannes Brønsted, who was a Danish chemist, and Thomas Lowry, who was an English chemist, came up with this hypothesis in 1923. According to their theory, an acid is a substance that donates a proton (H+) whereas a base is a substance that accepts a proton (H+). This means that in a reaction, there must always be an acid and a base present. The acid donates a proton or hydrogen ion to the base, thereby forming a conjugate base and acid. For example, let's consider the reaction of hydrochloric acid (HCl) and water (H2O). Hydrochloric acid is an acid because it donates a hydrogen ion to water, which is a base, resulting in the formation of hydronium ions (H3O+) and chloride ions (Cl-). This concept applies to other acid-base reactions as well. For instance, in the reaction between ammonia (NH3) and hydrochloric acid (HCl), ammonia is the base because it accepts a proton from hydrochloric acid, which is the acid. This leads to the formation of ammonium ions (NH4+) and chloride ions (Cl−). The Brønsted–Lowry theory therefore emphasizes the importance of proton transfer in acid-base reactions, and it is a useful guide for understanding not only the properties of acids and bases but also the behavior of reactions that involve them.

When a strong acid is dissolved in an aqueous solution, it completely breaks down, or dissociates, into its constituent ions. For example, if hydrochloric acid (HCl) is dissolved in water, all of the HCl molecules will be converted into free protons (H+) and chloride anions (Cl-). Basically, the entire molecule falls apart into its constituent ions. Other examples of strong acids include sulfuric acid (H2SO4) and nitric acid (HNO3). Strong acids are highly reactive and can quickly cause chemical reactions. Similarly, when a strong base is dissolved in water, it dissociates completely into hydroxide ions (OH-) and cations. For example, if sodium hydroxide (NaOH) is dissolved in water, all of the NaOH molecules will be converted into Na+ ions and OH- ions. Potassium hydroxide (KOH) and calcium hydroxide (Ca(OH)2) are amongst the few strong bases available. On the other hand, weak acids and bases only partially dissociate or ionize in aqueous solution, forming reversible equilibria. This means that only some of the molecules are converted into ions when dissolved in water. For instance, acetic acid (CH3COOH) is a weak acid that is only partially dissociated in water, meaning that not all of the acid molecules will break apart to form H+ and acetate ions. Some other examples of weak acids include citric acid and carbonic acid. Similarly, weak bases only partially dissociate in aqueous solution to form hydroxide ions (OH-). A common example of a weak base is ammonia (NH3), which forms ammonium ions (NH4+) and hydroxide ions (OH-) in water.

The pH scale is a method that is used to determine the acidic or basic nature of a solution. It ranges from zero to 14, with seven representing a neutral solution. A solution with a pH of less than 7 is considered acidic, while a solution with a pH greater than 7 is considered basic, or alkaline. Water, which is a neutral substance, has a pH of exactly 7. This means that the concentration of hydrogen ions (H+) and hydroxide ions (OH-) in water is balanced, and the solution is neutral. However, when an acid is added to water, the concentration of hydrogen ions increases and the pH decreases below 7. On the other hand, when a base is added, the concentration of hydroxide ions increases and the pH increases above 7. Acidic solutions have a pH of less than 7, and the lower the pH, the stronger the acid. Strong acids such as hydrochloric acid and sulfuric acid have a pH that is close to zero, while weak acids such as citric acid and acetic acid have a pH closer to 7 (but still lower than 7). Acidic solutions can be harmful and corrosive to living tissues and materials, and can cause chemical reactions to speed up. Alkaline solutions, also referred to as basic solutions, have a pH greater than 7. Unlike acidic solutions, the higher the pH, the stronger the basic solution. Strong bases such as sodium hydroxide and potassium hydroxide have a pH close to 14, while weak bases such as ammonia have a pH closer to 7 but still above 7. Alkaline solutions are often used in cleaning products and can also be harmful to living tissues.

Strong and weak acids have different properties and react differently with other substances, which can be qualitatively observed through various methods, such as with a reactive metal or through measuring pH values using a pH meter, universal indicator, or conductivity.

Reactivity with Reactive Metals:

Strong acids readily react with metals, especially reactive metals like magnesium (Mg) and zinc (Zn), generally producing hydrogen gas (H2) in the process. For example, hydrochloric acid (HCl) reacts with magnesium (Mg) to produce magnesium chloride (MgCl2), and hydrogen gas (H2).

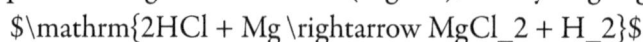

$$\mathrm{2HCl + Mg \rightarrow MgCl_2 + H_2}$$

Weak acids, on the other hand, do not react in the same way with reactive metals like magnesium or zinc, as their acidic values are much closer to neutral and they do not have as many free protons to donate in an acid-base reaction. As such, a weak acid like acetic acid (CH3COOH) will have no reaction with metal, or a much slower rate of reaction than strong acids like hydrochloric or sulfuric acid.

Differences in pH Values:

The pH value of a solution is a measure of its acidity or basicity. Strong acids have a very low pH value, usually between 0 and 3, due to their complete dissociation in aqueous solution, while weak acids have a pH value closer to 7 but still less than 7. For example, hydrochloric acid has a pH of around 1, while acetic acid has a pH of around 4.6 at a concentration of 1M.

When measuring pH, a pH meter gives accurate readings of the solution's pH level by measuring the electric potential difference between two electrodes. A universal indicator is a liquid mixture of several different pH indicators that gives a color change depending on the pH of the solution. Conductivity is also a useful approach to explain the difference in behaviour between strong and weak acids, as strong acids are more conductive because they produce a high concentration of H+ ions in solution, while weak acids produce fewer ions in solution and hence less conductive.

In chemistry, neutralisation reactions occur when an acid and a base react to form a salt and a neutral compound, usually water (H_2O). These reactions result in the transfer of protons (H+) between the acid and the base, which ultimately neutralises the solution.

The concept of neutralisation is based on the Brønsted–Lowry theory, which states that an acid donates a proton (H+) to a base, which accepts it. When we add an acidic solution to a basic solution, there is a transfer of H+ ions from the acidic solution to the basic solution. This causes a rise in OH- ions in the solution, and as the amount of OH- ions increase to meet the concentration of H+, they react and form water molecules.

For example, when hydrochloric acid (HCl) reacts with sodium hydroxide (NaOH), the H+ ion from hydrochloric acid reacts with the OH- ion from sodium hydroxide to form water (H_2O), leaving behind a salt, sodium chloride (NaCl).

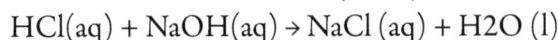

$$HCl(aq) + NaOH(aq) \rightarrow NaCl\ (aq) + H_2O\ (l)$$

Here, the hydrogen ion (H+) from hydrochloric acid combines with the hydroxide ion (OH-) from sodium hydroxide to form water, which is a neutral compound, with no acidic or basic properties. The solution becomes neutral after all the acid has been neutralised by the base.

This process is a fundamental concept in chemistry and has many practical applications, such as in the production of medicines, food, and fertilizers. It is also an essential concept to understand in areas such as environmental science, where we can use neutralisation reactions to neutralise acids and bases in order to protect our environment.

Neutralisation reactions usually result in the formation of a salt, which is a chemical compound that is formed from the positive ion of a base and the negative ion of an acid. When an acid is neutralised with a base, the two oppositely charged ions react to form a neutral compound or salt.

The formation of a salt in a neutralisation reaction is essential, as it means that the reaction has reached completion, and the acid and base have effectively cancelled out each other's properties. Salts are usually solid, crystalline compounds that have high melting and boiling points and are often soluble in water.

For example, when hydrochloric acid is neutralised by sodium hydroxide, sodium chloride (NaCl) is formed. This process can be represented using a chemical equation:

$$HCl + NaOH \rightarrow NaCl + H_2O$$

Here, the hydrogen ion (H+) from hydrochloric acid (HCl) combines with the hydroxide ion (OH-) from sodium hydroxide (NaOH) to form water, and sodium ion (Na+) from the base combines with the chloride ion (Cl-) from the acid to form sodium chloride (NaCl), which is a salt.

Similarly, when sulfuric acid (H_2SO_4) is neutralised by calcium hydroxide ($Ca(OH)_2$), calcium sulfate ($CaSO_4$) is formed, along with water:

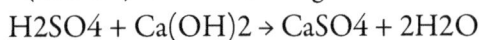

$$H_2SO_4 + Ca(OH)_2 \rightarrow CaSO_4 + 2H_2O$$

Here, the hydrogen ion (H+) from sulfuric acid combines with the hydroxide ion (OH-) from calcium hydroxide to form water, and the calcium ion (Ca2+) from the base combines with the sulfate ion (SO4 2-) from the acid to form calcium sulfate (CaSO4), which is again a salt.

When it comes to titrations, there are many different combinations of acids and alkalis that can be used. Each of these combinations will have a unique pH titration curve, which will give you information about how the pH of the solution changes as you add more acid or alkali.

One important thing to note is that the type of acid and alkali used in the titration will have a big impact on the shape of the pH titration curve. If you are using a strong acid and a strong alkali, for example, the curve will look very different from if you were using a weak acid and a strong alkali.

In general, the pH titration curve will start out low (on the acidic side) and gradually rise as more alkali is added. At some point, it will reach its endpoint, where the pH suddenly jumps up to a high value. This endpoint is usually when all of the acid has been neutralized by the alkali.

If you are using a strong acid and a strong alkali, the pH titration curve will be very steep and rapid, since the strong acid will be completely neutralized by the strong alkali. As a result, the endpoint will occur very quickly and sharply.

If you are using a weak acid and a strong alkali, on the other hand, the pH titration curve will be more gradual and less steep. This is because the weak acid will not be completely neutralized by the strong alkali, since it does not dissociate as readily. As a result, the pH will rise more slowly until it eventually reaches the endpoint.

Similarly, if you are using a strong acid and a weak alkali, the pH titration curve will be more gradual and less steep. This is because the weak alkali will not be able to completely neutralize the strong acid, so the pH will rise more slowly until the endpoint is reached.

Finally, if you are using a weak acid and a weak alkali, the pH titration curve will be the most gradual and least steep of all. This is because both the weak acid and the weak alkali will be slow to dissociate and neutralize, so the pH will rise very slowly until it eventually reaches the endpoint.

Indicators are essential for acid-alkali titrations, as they help determine the endpoint of the reaction, where the acid and alkali are completely neutralized. Choosing the right indicator for a titration is very important, as it has a major impact on the accuracy and precision of the final result.

When selecting an indicator for an acid-alkali titration, the primary consideration is the range of pH at which the indicator changes color. Ideally, the indicator should change color very close to the endpoint of the reaction, which means that its pH range should be very close to the pH at which the end point of the titration occurs.

One common indicator used in acid-alkali titrations is phenolphthalein. Phenolphthalein is a weak organic acid with a pH range of approximately 8.2-10, meaning that it changes color at around pH 8.2-10. This makes it a suitable choice for titrations where the endpoint occurs at a pH near to 8.2-10. In such cases, phenolphthalein will change from colorless to pink as the solution becomes increasingly alkaline, indicating the endpoint of the titration.

Another commonly used indicator is methyl orange. Like phenolphthalein, methyl orange is an organic compound, but it has a pH range that is much lower at around 3.1-4.4. This makes it a suitable choice for titrations where the endpoint occurs at a pH close to 3.1-4.4. In such cases, methyl orange will change from red to yellow as the solution becomes increasingly basic, indicating the endpoint of the titration.

In addition to phenolphthalein and methyl orange, there are many other indicators that can be used for acid-alkali titrations. Some of these include bromothymol blue, thymol blue, and litmus. Each of these indicators has its own unique pH range where it changes color, so the choice of indicator will depend largely on the specific titration being performed.

Understanding the Factors and Calculating the Rate of Chemical Reactions

Essentially, the rate of reaction refers to how quickly a chemical reaction takes place. It is often measured in terms of the amount of product produced per unit time or the amount of reactant used up per unit time. But what affects the rate of reaction? One major factor is the frequency of collisions. When reactant particles collide, there's a chance that a reaction will take place. The more often they collide, the higher the likelihood of a successful reaction. However, not all collisions are created equal. Effective collisions are those which result in a reaction. This occurs when particles collide with enough energy to break bonds and form new ones. Non-effective collisions, on the other hand, lack sufficient energy to cause a reaction. Think of it like a game of pool. The frequency of collisions is like the number of balls hitting each other on the table. But just like how not every ball hitting another will result in a pocketed shot, non-effective collisions do not lead to a reaction. Effective collisions, meanwhile, are like a ball hitting a pocket, resulting in a successful reaction.

To understand the effect of concentration and pressure changes on the rate of a chemical reaction! When two or more reactant molecules come together and collide, it is called a collision. Not all collisions, however, are effective in producing a reaction. The frequency of effective collisions between reactants plays a vital role in determining the rate of a chemical reaction. Now, let's talk about how concentration affects the rate of a reaction. Concentration refers to the amount of a substance that is dissolved in a given solution. When the concentration of a reactant increases, it means that there are more particles of that substance in the solution per unit volume. This increases the likelihood of collisions between particles, resulting in more effective collisions, which in turn can increase the rate of a reaction. So, if the concentration of a reactant increases, the rate of a reaction also increases. Next, let's discuss how pressure changes affect the rate of a reaction. Pressure changes are typically relevant in gas-phase reactions. Increasing the pressure of a gas system can increase the frequency of collisions between gas molecules, which can increase the rate of the reaction. This is because an increase in pressure means that there are more gas molecules in the same volume, which results in more collisions between the gas molecules, increasing the number of effective collisions.

how to calculate the rate of a reaction using experimental data! When we talk about the rate of a reaction, we're referring to how quickly a reaction occurs over time. To calculate the rate of a reaction, we need to use experimental data, specifically, how the concentrations of the reactants and products change over time. Typically, when determining the rate of a reaction, we use the initial rates method, which involves monitoring how the concentration of a particular reactant or product changes at the start of a reaction. This is done because the rate of the reaction is usually highest at the beginning before any significant amount of reactants have been consumed. Once we have determined the initial concentration of a particular reactant or product, we can measure how the concentration changes over time. This is usually done by taking a series of readings at specific intervals, then plotting the results on a graph. The slope of the straight line that fits through these points is known as the rate of the reaction. However, the rate of a reaction is not always constant over time. In many reactions, the rate will decrease as reactants are used up and the reaction progresses towards equilibrium. In these cases, we may need to use more advanced mathematical models to calculate the rate, such as the integrated rate law. This involves knowing the orders of the reactants and/or products and using appropriate equations to calculate the rate of the reaction at any point in time. Finally, it's important to note that the rate of a reaction can be influenced by many factors, including temperature, concentration, and the presence of catalysts or inhibitors. By understanding the factors that affect the rate of a reaction, we can develop strategies to modify or control the rate to optimize the chemical reactions for various applications.

The Role of Temperature and Activation Energy in Chemical Reactions

Activation energy, also commonly known by its abbreviation EA, refers to the minimum amount of energy that is required to make a collision between atoms, ions, or molecules effective. In other words, the EA represents the threshold needed for a chemical reaction to occur. It's like a key that opens the door to a reaction. Without the correct amount of EA, the collision will fail, and the reaction will not happen. To understand this concept, imagine two cars colliding at high speed. If the cars do not have a minimum amount of speed or momentum, they will not create any significant impact or damage. Similarly, in chemistry, for atoms and molecules to react, they must meet a minimum level of energy. This energy determines the fate of the chemical reactions, whether it ends up producing a useful product, a harmful substance, or nothing at all. Activation energy plays a critical role in various scientific fields such as thermodynamics, biochemistry, and physics. It's a fundamental concept that helps us understand how different molecules interact and transform under specific conditions. Think of it as a unique fingerprint that each reaction possesses, which determines its outcome. In summary, Activation Energy is the kickstart that sets the ball rolling in a chemical reaction and is crucial for the understanding of chemical reactions and the engineering of new substances.

The Boltzmann distribution is a crucial concept in thermodynamics that explains the distribution of energies in a system composed of particles. It plays a significant role in understanding the significance of activation energy in chemical reactions. To explain this, let's consider an example of a reaction that occurs when two gases collide. In the absence of activation energy, not all collisions between the two gases will result in a reaction. Instead, only those with enough energy to overcome the activation energy barrier will be successful in causing a chemical reaction. The Boltzmann distribution explains this phenomenon by showing that particles in a system have different energy levels, and only those with higher energy levels will have the potential to surpass the activation energy barrier. The Boltzmann distribution graph can illustrate this point effectively. The graph shows a bell curve with energy levels displayed on the x-axis and the frequency of molecules with that energy on the y-axis. As temperature increases, the peak of this curve shifts towards the right. The area under the curve represents the total energy of the system and is constant at all temperatures. Now to apply this to activation energy, imagine that the top of the activation energy barrier sits at the point where the curve meets the y-axis. This means that only the tiny number of molecules on the tail end of the distribution's higher energy side have enough energy to reach the activation energy barrier. Thus, the likelihood of a reaction occurring increases as the temperature rises, as more molecules have a higher energy level, giving them a better chance of overcoming the activation energy barrier. At temperatures too low, even if some molecules have an energy level equal to or higher than the activation energy, they still cannot pass through the activation energy barrier.

Temperature plays a crucial role in chemical reactions by affecting the rate of reaction. By understanding how temperature changes affect the Boltzmann distribution and the frequency of effective collisions, we can explain how temperature changes impact the reaction rate. The Boltzmann distribution can illustrate the energy distribution of particles in the system and how the distribution changes with a temperature change. As temperature is increased, the energy distribution of molecules becomes wider. This means that a larger portion of molecules has energies that are greater than the activation energy. As a result, more collisions will result in effective collisions, leading to a higher reaction rate. On the other hand, at lower temperatures, the Boltzmann distribution is narrower, meaning a smaller proportion of molecules have energies higher than the activation energy. This leads to fewer effective collisions and consequently, to a slower reaction rate. Apart from the Boltzmann distribution, the frequency of effective collisions is another factor that affects the reaction rate. Effective collisions are those that cause a chemical change to occur. As temperature increases, the frequency of these effective collisions also increases, as more molecules possess the minimum required energy to collide successfully. Effective collisions are mostly a function of the speed or velocity of molecules rather than the energy carried by them. As the temperature rises, the speed of molecules increases, and they collide with a higher frequency, increasing the number of effective collisions. Together, the Boltzmann distribution and the frequency of collisions give us a good qualitative understanding of the effect of temperature change on the rate of reaction. As the temperature rises, the system acquires energy, which results in a wider distribution of energies in the system, and higher-energy collisions that increase the probability of an effective collision occurring, thereby increasing the rate of the reaction.

Catalysts

Essentially, when we talk about a reaction having a different mechanism in the presence of a catalyst, what we mean is that the introduction of a catalyst fundamentally changes the way that the reaction occurs. A catalyst, in essence, provides an easier route for the reactants to become transformed into the products of the reaction. This easier route can be thought of as a different mechanism because, without the catalyst present, the reaction would not occur in the same way. The reason that a catalyst is so effective at changing the mechanism of a reaction has to do with the effect it has on the activation energy of the reaction. Activation energy refers to the energy that is required for the reactants to overcome the energy barrier that separates them from becoming products. Essentially, the higher the activation energy of a reaction, the less likely it is to occur. However, in the presence of a catalyst, the activation energy of a reaction is lowered. This means that the energy barrier that the reactants must overcome in order to become products is significantly reduced. As a result, the reaction is much more likely to occur, and it occurs in a different way than it would without the catalyst present.

To understand how a catalyst lowers the activation energy of a reaction and enables it to occur more easily, we can look at the Boltzmann distribution. The Boltzmann distribution is a statistical distribution that describes the distribution of energy levels of particles in a system at a given temperature. When we add a catalyst to a reaction, what is essentially happening is that new reaction pathways become available to the reactants. These new pathways have different activation energies than the original pathway, and typically have lower activation energies. The Boltzmann distribution tells us that molecules exist at different energy levels, with some having more energy than others. However, at any given temperature, the vast majority of molecules will be at lower energy levels. In the absence of a catalyst, the majority of reactant molecules do not possess the necessary energy to undergo the reaction, because the activation energy is too high. With the introduction of a catalyst, however, the activation energy of the reaction is lowered. As a result, more of the reactant molecules can now overcome the energy barrier and engage in the reaction, as they now possess enough energy to do so. The Boltzmann distribution therefore shifts to include a larger number of molecules that now have sufficient energy to react. This increase in the number of molecules with sufficient energy drives the reaction forward, as the rate at which the reaction occurs is directly proportional to the number of reacting molecules.

A reaction pathway diagram is essentially a visual representation of the steps that occur during a chemical reaction. The diagram presents the series of chemical species that are involved in the reaction, along with the intermediates and transition states that occur during the course of the reaction. It is a helpful tool for scientists to understand the mechanism of the reaction and determine how best to optimize the reaction conditions. Let's suppose we are considering a simple chemical reaction between reactant A and reactant B, which produces product C. Without a catalyst, the reaction pathway diagram might look something like this:

A + B -> Intermediate 1 -> Intermediate 2 -> C

Here, the reaction proceeds through a series of intermediates with high activation energies that are required to be overcome in order for the reaction to proceed. This process is slow and highly energy-demanding, which makes the reaction difficult to proceed.

Now, let's consider the same reaction in the presence of an effective catalyst. The catalyst provides a lower energy pathway for the reaction to occur, which significantly lowers the activation energy and allows the reaction to proceed more easily. This results in a different reaction pathway diagram with the catalyst:

A + B -> Intermediate 1 -> Catalyst -> Intermediate 2 -> C

In this case, the reaction now proceeds through a catalyst intermediate. The catalyst interacts with one or both reactants to generate an activated complex of lower energy that is then able to proceed through a different pathway with intermediate 2. This pathway requires less energy than the original pathway, and the catalyst enables the reaction to proceed more easily by lowering the activation energy barrier for the reaction.

Properties and Variations of Atoms and Ions

Atomic radius, ionic radius, melting point, and electrical conductivity are all important properties that define the behavior of an element. Understanding these properties is crucial in analyzing the behavior of atoms in a chemical and or physical reaction. In general, these properties vary within a group or a period of the periodic table.

Atomic radius refers to the distance between the nucleus of an atom and its outermost electron. As one moves horizontally across a period from left to right, the atomic radius decreases due to the increased attraction of electrons to the nucleus. In contrast, as one moves vertically down a group, the atomic radius increases because of the addition of a new shell of electrons to the atom.

Ionic radius, on the other hand, is the distance between the nucleus of an ion and its outermost electron. The periodicity of ionic radius follows a similar pattern with atomic radius. As one moves from left to right across a period, the ionic radius decreases due to the increase in the number of protons, which attracts the negative charges, thereby making the ion smaller. Conversely, as one moves down a group, the ionic radius increases due to the addition of new shells to the ion.

Melting point, which is the temperature at which a solid turns to a liquid, also varies across periods and groups. Generally, the melting point increases as one moves from left to right in a period. This is because metallic bonds, which hold metals together, increase in strength as the number of electrons increases. Similarly, the melting point decreases as one moves down a group, this is attributed to the increase in atomic size which weakens the metallic bond.

Lastly, electrical conductivity refers to the ability of a substance to conduct electricity. It varies in patterns that are similar to those of the other properties. Electrical conductivity decreases from left to right along a period because fewer electrons become delocalized as the metallic bond strengthens. Conversely, electrical conductivity increases down a group since more delocalized electrons are available to move and conduct electricity.

Melting point and electrical conductivity are two important properties that reflect the structure and bonding of elements. These properties vary across elements and can be affected by the types of bonding that occur between atoms and the structure of the lattice they form.

The melting point of an element refers to the temperature at which it changes state from a solid to a liquid. Generally, elements that have strong metallic bonds have higher melting points. Metallic bonds occur when a sea of delocalized electrons surround positively charged metal atoms. These delocalized electrons form a network that binds the atoms together in a lattice structure. The lattice structure is held together by strong electrostatic attractions, and therefore greater energy is required to break these bonds, leading to higher melting points.

Electrical conductivity, on the other hand, refers to the ability of an element to conduct electricity. It is determined by the extent to which electrons are free to move within the lattice structure. Elements with strong metallic bonds tend to be good conductors of electricity. This is because the sea of delocalized electrons that surround the metal atoms can move freely throughout the structure, and they can facilitate an electrical current.

Ionic compounds, which consist of charged ions, also display some variation in their melting points and electrical conductivity. An ionic bond is formed when an element donates electrons to another, resulting in two oppositely charged atoms. In a solid ionic compound, these ions pack together in a lattice structure, which is held together by strong electrostatic forces, similar to metallic bonds. Salts such as NaCl have high melting points because of the strong ionic bonds between the positively charged Na+ and the negatively charged Cl-ions.

However, ionic compounds do not conduct electricity in their solid state because the ions are held in place and cannot move. When these compounds are melted or dissolve in water, the ions become free to move and can transfer electric charges, facilitating electrical conductivity.

Period 3 Chemical Periodicity

When elements react with oxygen, they undergo a process known as oxidation. This occurs because oxygen is a very electronegative element which attracts electrons strongly, and so it tends to strip electrons from other elements that it combines with. The resulting compounds are called metal oxides, and each one has a unique set of properties based on the nature of the element it is derived from.

For example, sodium (Na) reacts with oxygen (O_2) to form sodium oxide (Na_2O). The reaction can be written as:

$$4\,Na + O_2 \rightarrow 2\,Na_2O$$

As another example, magnesium (Mg) reacts with oxygen to form magnesium oxide (MgO):

$$2\,Mg + O_2 \rightarrow 2\,MgO$$

Aluminum (Al) reacts with oxygen to form aluminum oxide (Al_2O_3):

$$4\,Al + 3\,O_2 \rightarrow 2\,Al_2O_3$$

The reaction of phosphorus with oxygen generates phosphorus pentoxide (P_4O_{10}):

$$P_4 + 5\,O_2 \rightarrow P_4O_{10}$$

The combination of sulfur with oxygen results in the creation of sulfur dioxide (SO_2):

$$S + O_2 \rightarrow SO_2$$

Moving on to how elements react with chlorine, this is known as a halogenation reaction. In this case, the elements undergo a process known as reduction because chlorine is very electronegative, it attracts the electrons from other elements. This reaction results in the formation of a salt, and again each one has unique properties based on the element it is derived from.

For example, sodium (Na) reacts with chlorine (Cl_2) to form sodium chloride (NaCl). This reaction can be written as:

$$2\,Na + Cl_2 \rightarrow 2\,NaCl$$

Magnesium (Mg) reacts with chlorine to form magnesium chloride ($MgCl_2$):

$$Mg + Cl_2 \rightarrow MgCl_2$$

Aluminum (Al) reacts with chlorine to form aluminum trichloride ($AlCl_3$):

$$2\,Al + 3\,Cl_2 \rightarrow 2\,AlCl_3$$

Silicon (Si) reacts with chlorine to form silicon tetrachloride ($SiCl_4$):

$$Si + 2\,Cl_2 \rightarrow SiCl_4$$

Phosphorus (P) reacts with chlorine to form phosphorus pentachloride (PCl_5):

$$P_4 + 10\,Cl_2 \rightarrow 4\,PCl_5$$

When sodium and magnesium react with water, they undergo a process known as hydrolysis, which results in the formation of a hydroxide and hydrogen gas. This reaction can be described as:

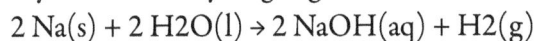

$$2\,Na(s) + 2\,H_2O(l) \rightarrow 2\,NaOH(aq) + H_2(g)$$
$$Mg(s) + 2\,H_2O(l) \rightarrow Mg(OH)_2(aq) + H_2(g)$$

To explain the variation in the oxidation number of the oxides and chlorides based on the valence shells of their outer electrons. First, let's define what we mean by oxidation number. An oxidation number is the charge that an element would have if all of its bonds were ionic or if it were a monatomic ion. In other words, it's the number of electrons an atom gains or loses when it forms a compound. Now, focusing on the oxides, we can see that the oxidation numbers vary based on the number of valence electrons in the outer shell of each element. Sodium (Na), for example, has one valence electron, and it loses that electron to become Na^+. Magnesium (Mg) has two valence electrons, and it loses those electrons to become Mg^{2+}. Aluminum (Al) has three valence electrons, but it loses only one electron to become Al^{3+}. Phosphorus (P) has five valence electrons, and it gains three electrons to become P^{5+}. Sulfur (S) has six valence electrons, and it gains two or three electrons to become SO_2 or SO_3, respectively. Therefore, the variation in oxidation number among oxides is due to the number of valence electrons present in the outer shell of each element. Now, let's turn our attention to the chlorides. The oxidation numbers of the chlorides also depend on the number of valence electrons in the outer shell of the element. Sodium (Na) has one valence electron, and it loses that electron to become Na^+. Magnesium (Mg) has two valence electrons, and it loses both electrons to become Mg^{2+}. Aluminum (Al) has three valence electrons, but it gains one electron to become $AlCl_3$. Silicon (Si) has four valence electrons, and it gains four electrons to become $SiCl_4$. Phosphorus (P) has five valence electrons, and it gains five electrons to become PCl_5. Therefore, the variation in oxidation number among chlorides is also due to the number of valence electrons present in the outer shell of each element.

To describe the reactions of the oxides Na_2O, MgO, Al_2O_3, SiO_2, P_4O_{10}, SO_2, and SO_3 with water and provide equations for each. I can also give you an idea of the likely pH levels of the solutions produced.

First, let's start with Na_2O and water. Na_2O is an oxide of sodium with a neutral pH. When Na_2O reacts with water, it forms sodium hydroxide, $NaOH$. This reaction is exothermic which means that heat is released. The equation for this reaction is:

$$Na_2O + H_2O \rightarrow 2NaOH$$

The resulting solution will have a pH of around 14, making it highly basic.

Next is MgO and water. MgO is an oxide of magnesium, which is also neutral. When MgO reacts with water, it forms magnesium hydroxide, $Mg(OH)_2$. This reaction is also highly exothermic and the equation for this reaction is:

$$MgO + H_2O \rightarrow Mg(OH)_2$$

The resulting solution will be slightly basic, with a pH of around 9.5.

Now let's explore the reaction between aluminum oxide, Al_2O_3, and water. Al_2O_3 is acidic in nature. When it reacts with water, it forms aluminum hydroxide, $Al(OH)_3$. The equation for this reaction is:

$$Al_2O_3 + 3H_2O \rightarrow 2Al(OH)_3$$

The resulting solution will be slightly acidic, with a pH of around 4.5.

Silicon dioxide, SiO_2, is a neutral oxide. When it reacts with water, it forms silicic acid, H_4SiO_4. The equation for this reaction is:

$$SiO_2 + 2H_2O \rightarrow H_4SiO_4$$

The resulting solution will be slightly acidic, with a pH around 7.

The reaction between P_4O_{10} and water is highly acidic. P_4O_{10} is an oxide of phosphorus with a pH of around 0. On reacting with water, it produces phosphoric acid, H_3PO_4. The equation for this reaction is:

$$P_4O_{10} + 6H_2O \rightarrow 4H_3PO_4$$

The resulting solution will be highly acidic with a pH of around 2.

Finally, let's look at the reactions of sulfur dioxide, SO2, and sulfur trioxide, SO3, with water. SO2 is slightly acidic, while SO3 is highly acidic. Sulfur dioxide reacts with water, and from this reaction, sulfurous acid (H2SO3) is produced. The equation for this reaction is:

$$SO_2 + H_2O \rightarrow H_2SO_3$$

The resulting solution will be acidic, with a pH around 4. The reaction between sulfur trioxide and water produces sulfuric acid (H2SO4). The equation for this reaction is:

$$SO_3 + H_2O \rightarrow H_2SO_4$$

The resulting solution will be highly acidic, with a pH of around 1.

To understand the acid/base behavior of the oxides and hydroxides you mentioned, explain how they can act as acids or bases in certain circumstances, and write equations to illustrate the chemical reactions that occur.

Let's start with the oxides. When these oxides dissolve in water, they can be classified as acids or bases. Oxides that have a higher electronegativity tend to act as non-metal oxides and as acidic oxides, while the opposite is true for oxides with lower electronegativity. For example, sodium oxide, Na2O, is a basic oxide, and magnesium oxide, MgO, is also basic. Aluminum oxide, Al2O3, is amphoteric, meaning it can act as both an acid and a base. Phosphorus pentoxide, P4O10, is acidic because it reacts with water to form a corresponding acid, i.e., H3PO4. Sulfur dioxide, SO2, and sulfur trioxide, SO3, are acidic oxides.

So let's look at the chemical reactions from specific oxides and hydroxides.

First, let's start with Na2O. When Na2O dissolves in water, it produces sodium hydroxide (NaOH), which is a strong base and dissociates completely in water, as shown below:

$$Na_2O + H_2O \rightarrow 2NaOH$$

Similarly, when MgO dissolves in water, it produces magnesium hydroxide (Mg(OH)2), which is a weak base and does not dissociate completely in water:

$$MgO + H_2O \rightarrow Mg(OH)_2$$

Next up, let's examine Al2O3. It can act as both an acid and a base because it has both acidic and basic properties. When it reacts with a strong acid, such as hydrochloric acid, it will act as a base and neutralize the acid, forming AlCl3:

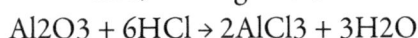

$$Al_2O_3 + 6HCl \rightarrow 2AlCl_3 + 3H_2O$$

When it reacts with a strong base, such as sodium hydroxide, it will act as an acid and neutralize the base, forming sodium aluminate:

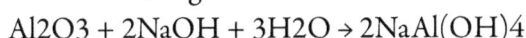

$$Al_2O_3 + 2NaOH + 3H_2O \rightarrow 2NaAl(OH)_4$$

Now let's look at P4O10. When P4O10 reacts with water, it forms phosphoric acid, H3PO4, which is a weak acid:

$$P_4O_{10} + 6H_2O \rightarrow 4H_3PO_4$$

Lastly, let's examine SO2 and SO3. These both dissolve in water to form sulfuric acid (H2SO4), which is a strong acid:

$$SO_2 + H_2O \rightarrow H_2SO_3$$

$$SO_3 + H_2O \rightarrow H_2SO_4$$

To understand the reactions of the chlorides NaCl, MgCl2, AlCl3, SiCl4, and PCl5 with water, and the pH levels of the resulting solutions.

Let's start with NaCl. When NaCl dissolves in water, it forms an aqueous solution of sodium ions (Na+) and chloride ions (Cl-). Sodium chloride is an ionic compound that dissociates completely in water, so the reaction is:

$$NaCl + H_2O \rightarrow Na^+ + Cl^- + H_2O$$

The resulting solution is neutral, with a pH around 7.

Next, let's look at MgCl2. Like NaCl, MgCl2 is an ionic compound that easily dissociates in water to form magnesium ions (Mg2+) and chloride ions (Cl-):

$$MgCl2 + H2O \rightarrow Mg2+ + 2Cl- + H2O$$

The resulting solution is slightly acidic, with a pH around 6.

When aluminum chloride, AlCl3, dissolves in water, it undergoes hydrolysis, i.e., reacts with water to produce H+ ions and aluminate ions (Al(OH)4-):

$$AlCl3 + 6H2O \rightarrow Al(OH)4- + 3H3O+$$

The resulting solution is acidic, with a pH around 3.

The reaction of SiCl4 with water yields hydrogen chloride (HCl) and silicic acid (H4SiO4):

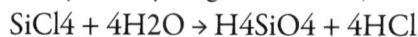

$$SiCl4 + 4H2O \rightarrow H4SiO4 + 4HCl$$

The resulting solution is acidic, with a pH of around 3.

Finally, when PCl5 reacts with water, it forms phosphoric acid (H3PO4) and hydrogen chloride (HCl):

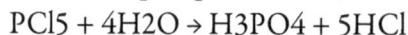

$$PCl5 + 4H2O \rightarrow H3PO4 + 5HCl$$

The resulting solution is acidic, with a pH around 1.

The variation and trends in the oxidation numbers of oxides and chlorides can be explained through the bonding and electronegativity of each element. Electronegativity is a measure of an element's attraction for electrons in a bond, and bond polarity is determined by the difference in electronegativity between the atoms involved. An element with higher electronegativity will tend to pull electrons towards itself, resulting in a polar bond.

In the case of oxides, the electronegativity of each element in the compound plays a crucial role. If the oxide is formed between a metal and a non-metal, the metal usually loses electrons, becoming positively charged, while the non-metal gains electrons, becoming negatively charged. This is because metals generally have low electronegativity, while non-metals have high electronegativity. Therefore, the number of valence electrons in the outer shell of each element determines the oxidation number of the oxide.

Regarding chlorides, similar rules apply. Chlorides are formed between a metal and a non-metal, and the metal usually loses electrons, becoming positively charged, while the non-metal gains electrons, becoming negatively charged. The number of valence electrons in the outer shell of each element determines the oxidation number of the chloride.

The acidity or basicity of oxides and chlorides can also be explained by bond polarity. If an oxide or chloride has ionic bonding, it will tend to be basic. On the other hand, if it has covalent bonding, it can be either acidic or basic, depending on the electronegativity of the atoms involved. An oxide or chloride with polar covalent bonding will typically be acidic because the more electronegative element tends to pull electrons towards itself, resulting in a hydrogen ion (H+) donor in water and thus can act as an acid. In contrast, an oxide or chloride with non-polar covalent bonding will typically be basic.

Overall, the variation and trends in the oxidation numbers, acidity or basicity of oxides and chlorides, can be attributed to the electronegativity and bonding nature of the elements involved. The more electronegative the element is, the more likely it is to form polar covalent bonds, making it acidic. On the other hand, non-metal oxides and metal chlorides typically have ionic bonding and are basic. These concepts can be illustrated with various examples and equations, allowing for a better understanding of the chemical properties of these compounds.

The types of chemical bonding present in oxides and chlorides can be identified by observing their physical and chemical properties. Oxides and chlorides can have either ionic or covalent bonding, which is determined by the electronegativity of the elements involved in the compound.

Ionic bonding is a type of chemical bonding in which electrons are transferred from one atom to another atom to achieve a stable electron configuration. In oxides and chlorides, ionic bonding is observed when the compound contains a metal and a non-metal element. Ionic compounds typically have high melting and boiling points, are hard and brittle, and are good conductors of electricity and heat when in the molten or dissolved state.

On the other hand, covalent bonding is a type of chemical bonding in which electrons are shared between two atoms to complete their outer shells. In oxides and chlorides, covalent bonding is observed when the compound contains two or more non-metal elements. Covalent compounds typically have low melting and boiling points, are soft and brittle, and are poor conductors of electricity and heat.

For example, sodium chloride (NaCl) is an ionic compound formed from the reaction between a metal, sodium, and a non-metal, chlorine. It has a high melting and boiling point, is hard and brittle, and is a good conductor of electricity in the molten state. Therefore, sodium chloride exhibits ionic bonding.

In contrast, silicon dioxide (SiO2) is a covalent compound formed from the reaction between two non-metal elements, silicon, and oxygen. It has a low melting and boiling point, is soft and brittle, and is a poor conductor of electricity and heat. As a result, silicon dioxide exhibits covalent bonding.

Aluminum chloride (AlCl3) is an ionic compound that undergoes hydrolysis in water, indicating its ionic nature. Hydrolysis occurs when a compound reacts with water to form a base or an acid. Aluminum chloride is a Lewis acid, meaning it tends to accept a pair of electrons to form a coordinate bond. This further indicates its ionic nature.

In summary, the types of chemical bonding present in oxides and chlorides can be suggested based on their physical and chemical properties. Ionic bonding is observed in compounds containing a metal and a non-metal element, while covalent bonding is observed in compounds containing two or more non-metal elements. Observations such as melting and boiling points, conductivity, and reactivity with water can help to identify the type of bonding present.

Elements' Periodic Trends

Chemical periodicity refers to the regular and predictable patterns in the properties of elements based on their position on the periodic table. In a given group, elements have the same number of valence electrons in their outermost shell. This leads to similarities in their physical and chemical properties.

For instance, elements in Group 1 also known as the Alkali Metals have a single valence electron, making them highly reactive and highly reactive. They are soft, low melting, and boiling metals with low densities. They readily react with electrons to form ions with a charge of +1.

Similarly, elements in Group 2 also known as the Alkaline Earth Metals have two valence electrons. They are also highly reactive, but less so than Group 1 elements. Compared to alkali metals, these elements are denser, harder, and have higher boiling and melting points. They also form ions with a charge of +2.

The transition metals in Groups 3 to 12 are characterized by their wide range of properties, including high melting and boiling points, varying colors, and variable valence states. They often form colorful coordination complexes with ligands.

Elements in Group 17 or the Halogens all have seven electrons in their outermost shell, making them highly reactive non-metals. They are highly electronegative, easily gaining an electron to form an ion with a charge of -1. Halogens can also form diatomic molecules such as Cl_2, Br_2, and I_2.

Finally, Group 18 elements, also known as noble gases, are characterized by their stable electronic configurations. They have a full outermost shell and are unreactive, hence called inert gases. They are odourless, colourless, monatomic gases with low boiling and melting points and are mostly used in lighting.

One of the most fundamental concepts in chemistry is the ability to identify unknown elements using physical and chemical properties. To deduce the identity and possible position of an unknown element on the periodic table, one must rely on known trends in chemical properties.

The first step is to examine any available physical properties of the element, such as its color, texture, and density. These properties can provide clues about the element's identity and possible position on the periodic table. For example, if the element has a low density and is a shiny silvery solid, it could be a Group 1 metal such as sodium or potassium.

Next, one must examine the chemical properties of the unknown element. In particular, the number of valence electrons and electronegativity can provide vital clues.

If the unknown element readily loses electrons to form cations, it is likely a metal. Knowing this, we can deduce its position on the periodic table by determining the number of valence electrons and comparing it to other known elements. For instance, if the element has only one or two valence electrons and low electronegativity, it could belong to Groups 1 or 2 respectively. If it has three valence electrons, it could belong to the Group 13 or IIIA, and so on.

On the other hand, if the unknown element readily gains electrons to form anions, it is likely a non-metal. Again, we can deduce its possible position on the periodic table based on the number of valence electrons. For example, if it has seven valence electrons, it could belong to Group 17 or VIIA, the halogens. If it has six valence electrons, it could belong to Group 16 or VIA, the chalcogens.

Properties of Group 2 elements

The reactions of different elements with oxygen, water, dilute hydrochloric acid, and sulfuric acid vary depending on their properties. Let us look at the reactions of a few elements:

1. Reactions of Metals with Oxygen:

Most of the metals react with oxygen to produce metal oxides. The reactions are generally exothermic and are known as combustion reactions. For example:

$$4Na + O_2 \rightarrow 2Na_2O$$
$$2Mg + O_2 \rightarrow 2MgO$$

2. Reactions of Non-metals with Oxygen:

The non-metals combine with oxygen to form different types of oxides depending on their valency. For instance, carbon, which has a valency of 4, forms carbon dioxide on reacting with oxygen.

$$C + O_2 \rightarrow CO_2$$

3. Reactions with Water:

The behavior of elements with water is dependent upon their position on the periodic table. The alkali metals and alkaline earth metals are highly reactive with water, producing hydrogen gas and an alkali solution. For example, Sodium reacts vigorously with water:

$$2Na + 2H_2O \rightarrow 2NaOH + H_2$$

4. Reactions with acids:

Metals react with dilute hydrochloric and sulfuric acids to form their corresponding salts, hydrogen gas, and water. The products of these reactions depend on the metal and acid used. For example, the reaction between Zinc and hydrochloric acid gives Zinc chloride, hydrogen gas, and water:

$$Zn + 2HCl \rightarrow ZnCl_2 + H_2$$

Lastly, non-metals do not react with hydrochloric acid, but some react with sulfuric acid on heating to produce sulfur dioxide gas.

$$Cu + 2H_2SO_4 \rightarrow CuSO_4 + SO_2 + 2H_2O$$

Chemical reactions can occur between different types of compounds, including oxides, hydroxides, and carbonates, and other substances such as water and acids. Let us explore the behavior of these compounds when they react with water, dilute hydrochloric acid, and sulfuric acid:

1. Reactions of Oxides with Water and Acids:

The definition of oxides is that they are chemical compounds made up of one or more oxygen atoms that are chemically bonded to one or more other elements. When oxides react with water, they may form acids or bases, depending on the oxide's nature. Here are some examples:

$$CO_2 + H_2O \rightarrow H_2CO_3$$
$$SO_3 + H_2O \rightarrow H_2SO_4$$

When oxides react with dilute hydrochloric acid, they liberate the corresponding acid. For example, calcium oxide (CaO) reacts with dilute HCl to form calcium chloride and water:

$$CaO + 2HCl \rightarrow CaCl_2 + H_2O$$

2. Reactions of Hydroxides with Water and Acids:

Hydroxides are compounds containing hydroxyl (OH-) groups and a metal cation. When hydroxides are dissolved in water, they form alkaline solutions because the hydroxide ion is a strong base. Here are some examples:

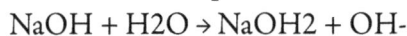

$$NaOH + H_2O \rightarrow NaOH_2 + OH-$$
$$Mg(OH)_2 + 2H_2O \rightarrow Mg_{2+} + 2OH- + 2H_2O$$

When hydroxides react with acids, they form their corresponding salts and water. For example, Sodium hydroxide reacts with hydrochloric acid to form Sodium chloride and water:

$$NaOH + HCl \rightarrow NaCl + H_2O$$

3. Reactions of Carbonates with Water and Acids:

Carbonates are compounds containing carbonate (CO_{32-}) groups and a metal cation. Carbonates react with water and acid to form carbonic acid, which decomposes to carbon dioxide and water, and the metal salt. Here are some examples:

$$CaCO_3 + H_2O + CO_2 \rightarrow Ca(HCO_3)_2$$
$$MgCO_3 + 2HCl \rightarrow MgCl_2 + CO_2 + H_2O$$

The thermal decomposition of nitrates and carbonates is a chemical process in which the compounds undergo a thermal change in their structure leading to the breaking up into simpler substances. The process is characterized by the release of gases such as carbon dioxide, water vapor, and oxides of nitrogen.

1. Thermal Decomposition of Nitrates:

The stability of nitrates increases down the group, that is, nitrates with large ions and lower charge density are more stable and require more heat to decompose. For example:

$$2KNO_3 \rightarrow 2KNO_2 + O_2$$
$$2Ca(NO_3)_2 \rightarrow 2CaO + 4NO_2 + O_2$$

2. Thermal Decomposition of Carbonates:

The trend in thermal stability for carbonates increases down the group. The carbonates of Group 1 and Group 2 metals undergo thermal decomposition readily and form the corresponding oxide and carbon dioxide. For example:

$$CaCO_3 \rightarrow CaO + CO_2$$
$$Na_2CO_3 \rightarrow Na_2O + CO_2$$

The carbonates of the transition metals are usually stable and do not readily decompose on heating.

The thermal decomposition of nitrates and carbonates serves as an important process in the manufacture of substances such as metal oxides, pyrotechnics, and fertilizers. The products of the thermal decomposition of carbonates often serve as important industrial feedstocks for many chemical products.

The solubility of hydroxides and sulfates varies depending on the nature of the metal ion in the compound, as well as the anions present in the compound. The solubility of hydroxides and sulfates is an important factor to consider in various chemical processes, including the production of fertilizers and other chemicals.

1. Solubility of Hydroxides:

The solubility of hydroxides depends on the size and charge density of the metal ion. Generally, the hydroxides of Group 1 and Group 2 metals are soluble in water and form alkaline solutions. As we move down the Group 1 and Group 2, the solubility of the hydroxides increases. For example, the solubility of hydroxides of Group 1 elements (LiOH, NaOH, KOH, etc.) and Group 2 elements (Mg(OH)2, Ca(OH)2, etc.) increase as we move down the group.

2. Solubility of Sulfates:

The solubility of sulfates also varies depending on the nature of the metal ion. The sulfates of Group 1 and Group 2 metals are generally highly soluble in water. However, the solubility of sulfates decreases as we move down the group. For example, the solubility of sulfates of Group 2 elements (e.g., MgSO4, CaSO4) decreases down the group. The sulfates of transition metals also generally have low solubility and often precipitate out of solution.

The solubility of hydroxides and sulfates is an essential consideration in many industries, such as agriculture, mining, and chemical manufacturing. For example, in agriculture, the solubility of certain hydroxides and sulfates is important for the production of fertilizers. In mining, the solubility of hydroxides and sulfates can affect how certain minerals are extracted from ores. In chemical manufacturing, the solubility of hydroxides and sulfates can affect the purity and yield of the products.

Chemical reactions are central to our understanding of the world around us, as they allow us to predict how different elements and compounds will interact with each other. Through the reactions of metals and non-metals with oxygen, water, and acids, we can observe the properties of elements and predict their behavior in various situations.

Metals are generally highly reactive with oxygen, producing metal oxides which are often exothermic combustion reactions. Non-metals, on the other hand, combine with oxygen to form different types of oxides based on their valency. The behavior of elements with water depends on their position on the periodic table, with some being highly reactive and producing hydrogen gas and an alkali solution. Reactions with acids result in the formation of salts, hydrogen gas and water, although the specific products depend on the metal and acid used.

Compounds such as oxides, hydroxides, and carbonates, can also undergo reactions with water, dilute hydrochloric acid and sulfuric acid. Oxides can form acids or bases depending on their nature, with carbon dioxide forming carbonic acid. Hydroxides form alkaline solutions when dissolved in water and react with acids to form the corresponding salts and water. Carbonates react with water and acids to form carbonic acid and a metal salt.

Thermal decomposition of nitrates and carbonates is a process characterized by the release of gases such as carbon dioxide, water vapor, and oxides of nitrogen. It serves as an important process in the manufacture of substances such as metal oxides, pyrotechnics, and fertilizers. The solubility of hydroxides and sulfates varies based on the metal ion and anions present in the compound.

Overall, understanding the trends in physical and chemical properties of different elements and compounds is crucial in predicting the behavior of these substances in various situations.

Group 17 characteristics

Chlorine, Bromine, and Iodine are all halogens found in Group 17 of the periodic table. They are non-metallic elements that tend to exist as diatomic molecules in their gaseous state. One of the distinctive properties of these halogens is their color, which ranges from a yellowish-green for chlorine, a reddish-brown for bromine, to a violet-black for iodine.

The halogens have their highest volatility in their gaseous form, while they exist as crystals when in their solid state. As we move down Group 17 from chlorine to iodine, the halogens' volatility decreases. This trend is a result of the increase in the size and mass of the halogen atoms as we move from chlorine to iodine. The greater the atomic mass, the stronger the intermolecular forces between the atoms (van der Waals forces) which hold the molecules together. While the halogens are all highly reactive, their reactivity decreases down the group. Chlorine is a highly reactive gas that is commonly used as a disinfectant and bleaching agent. When in contact with water, it forms hydrochloric acid, which can cause irritation and corrosion. Bromine, on the other hand, is a reddish-brown liquid that is less reactive than chlorine but still highly toxic. The manufacturing of drugs, insecticides, and several other kinds of chemicals involves the application of it. Finally, Iodine is a grayish-black solid that is rare to find in nature. It is used in medical applications, including the treatment of thyroid disorders and as a disinfectant.

In summary, the halogens' colors vary as we move down the group from chlorine to iodine, ranging from a yellowish-green, reddish-brown to a violet-black. The trend in volatility decreases as we move down the group due to the increase in the size and mass of the halogen atoms. Though highly reactive, their reactivity decreases down the group. Considering the individual properties of halogens and their compounds is crucial when using them in various applications.

The halogens are a group of highly reactive non-metallic elements found in Group 17 of the periodic table. They exist as diatomic molecules in their natural state, with two halogen atoms sharing a covalent bond. The strength of this bond differs between halogens and follows a particular trend as we move down the group.

The halogen molecules' bond strength is related to the size and radius of the halogen atoms. As we move down Group 17, the size of the halogen atoms increases, which results in an increase in the number of electrons and energy levels. These additional electrons and energy levels also increase the distance between the halogen nuclei, resulting in a weaker bond.

In addition to the atom's size, the number of electrons the atoms share also affects the bond strength. The more electrons an atom shares with its neighbour, the stronger the bond. When we move down Group 17, the number of electrons shared between the two halogen atoms increases. The additional electrons contribute to the bond's strength, which partially offsets the effect of larger atomic size in increasing the bond's distance.

As a result, the trend in bond strength of halogen molecules is not entirely straightforward and depends on a balance between increasing atomic size and increasing the number of electrons in the shared bond. Fluorine, the smallest halogen in the group, has the strongest bond of all the halogens. As we move down the group from fluorine to iodine, the bond strength decreases, with iodine having the weakest bond.

In summary, the bond strength of halogen molecules follows a trend that is related to varying size and radius of halogen atoms. The larger the size of the atom, the weaker the bond. However, the sharing of electrons between the two halogen atoms partially offsets the effect of size on bond strength. Overall, understanding the trend in bond strength is essential in predicting the behaviour and reactivity of halogen molecules in various chemical reactions.

The volatility of elements is a measure of their tendency to vaporize. It is related to the strength of the intermolecular forces between atoms or molecules, particularly instantaneous dipole-induced dipole forces. Instantaneous dipole-induced dipole forces are attractive forces that occur between non-polar molecules. These forces are a result of the temporary polarization of electric charges in molecules when the electrons move to one side of the atom, creating a partial negative charge. The instantaneous dipole induces a dipole in an adjacent molecule, and these dipoles momentarily interact, resulting in a weak attractive force.

The volatility of an element depends on the strength of the instantaneous dipole-induced dipole forces. For example, elements with a weaker attractive force and low molecular mass will have a higher volatility than those with a stronger attractive force and higher molecular mass. The weaker attractive force results in a less stable bond between molecules, which makes it easier for the molecules to escape into the gas phase.

In terms of volatility, halogens are particularly interesting.

As we move down the group, the molecular size and mass increase, resulting in increased numbers of electrons. The larger number of electrons leads to a stronger instantaneous dipole-induced dipole force and van der Waals' forces. This trend, along with the simultaneous increase in molecular weight, contributes to decreasing volatility in the group.

Properties of Halogens and Hydrogen Halides

When it comes to the reactivity of Group 17 elements as oxidizing agents, each element varies in its ability to acquire or share electrons. Fluorine, at the top of the group, is the most electronegative element and thus has the highest propensity to attract electrons. It has a great tendency to react and tends to be highly oxidizing. On the other hand, the reactivity of elements decreases moving down the group, with the exception of iodine, which is less reactive than other elements in the group due to its large atomic size. Consequently, in comparison to fluorine, chlorine has less ability to capture an electron and becomes a less powerful oxidizing agent. Similarly, bromine is not as reactive as chlorine, and iodine is the least reactive of the group. This indicates a pattern of reactivity within the group that is related to the trend of the increasing size of the atom.

When elements react with hydrogen, they form compounds known as hydrides. However, the reactivity of the elements with hydrogen varies, depending on their electronegativity, atomic size, and ionization energy. The most reactive elements with hydrogen include the alkali metals and alkaline earth metals, which are on the left-hand side of the periodic table. These elements have a low ionization energy, which means they lose their valence electrons easily and react vigorously with hydrogen, forming ionic hydrides.

Other elements that are reactive with hydrogen include the group 13 and 14 elements, such as aluminum and silicon, respectively. These elements have a moderate reactivity with hydrogen, forming covalent hydrides. On the other hand, the transition metals, which include groups 3-12, are typically not reactive with hydrogen, except for a few exceptions such as palladium and platinum that form metallic hydrides.

The nonmetals, which are located on the right-hand side of the periodic table, are generally not reactive with hydrogen, except for the halogens and chalcogens. These elements form molecular hydrides through covalent bonding with hydrogen. The halogens are the most reactive nonmetals with hydrogen, forming hydrogen halides such as hydrogen chloride and hydrogen fluoride.

The hydrogen halides, which include hydrofluoric acid (HF), hydrogen chloride (HCl), hydrogen bromide (HBr), and hydrogen iodide (HI), exhibit different thermal stabilities depending on their respective bond strengths.

The thermal stability of a compound is a measure of its ability to resist decomposition upon heating. The hydrogen halides are gases at room temperature and can dissociate easily into their constituent hydrogen and halogen molecules upon heating. The bond strengths between the hydrogen and halogen atoms in these molecules determine their relative thermal stabilities.

Hydrofluoric acid, which has the strongest bond strength of all the hydrogen halides, is the most thermally stable. This is because the hydrogen and fluorine atoms form a polar covalent bond, creating a strong dipole moment that holds the molecule together. As a result, hydrofluoric acid requires a higher temperature to dissociate into its constituent atoms.

In contrast, hydrogen iodide, which has the weakest bond strength of all the hydrogen halides, is the least thermally stable. This is because the bond between hydrogen and iodine atoms in the molecule is a weak covalent bond that can be easily broken upon heating. As a result, hydrogen iodide dissociates at a much lower temperature compared to the other hydrogen halides.

Hydrogen chloride and hydrogen bromide have intermediate bond strengths and thermal stabilities. Hydrogen chloride has a stronger bond strength than hydrogen bromide due to the smaller size of chlorine atoms, which leads to a higher electrostatic attraction between the hydrogen and chlorine atoms. Therefore, hydrogen chloride is more thermally stable than hydrogen bromide.

Halide Ions' Reactions

As far as halide ions are concerned, their ability to act as reducing agents depends on their atomic size and electronegativity. Generally speaking, larger halide ions are better reducing agents than smaller ones because they have a weaker hold on their electrons. So, if we take a look at the halogens in order of decreasing atomic size (i.e. F, Cl, Br, I), we can see that the reducing ability of the halide ions increases down the group. This is because the larger ions have more electron shells and therefore a weaker hold on their valence electrons, which makes them more willing to donate them to other atoms or compounds. Similarly, electronegativity plays a role in determining a halide ion's reducing ability. Halogens with lower electronegativity values - such as iodine - have a weaker attraction to electrons, which makes them more willing to give them up and act as reducing agents. It's also worth noting that the solvent used in a reaction can affect the reactivity of halide ions as reducing agents. For example, in non-aqueous solvents like acetone or dimethylformamide, fluoride ions tend to be the best reducing agents due to their high reactivity with polarizability.

When halide ions react with aqueous silver ions followed by aqueous ammonia, a series of fascinating reactions occur, resulting in the formation of different silver-halide complexes. The initial reaction involves the mixing of aqueous silver ions (Ag^+) with halide ions (e.g. Cl^-, Br^-, I^-) to form insoluble silver halide precipitates, such as $AgCl$, $AgBr$, or AgI, depending on which halide ion is used. These precipitates are often solids that form as a result of the interaction between the oppositely charged Ag^+ and halide ions, which produce compounds that are highly insoluble in water. In a second reaction, an aqueous solution of ammonia (NH_3) is then added to the mixture of precipitates. The ammonia binds to the silver ions and forms a series of soluble silver-ammonia complexes, including $[Ag(NH_3)_2]^+$ and $[Ag(NH_3)_4]^{2+}$, among other possible variations. These complexes are soluble in aqueous ammonia due to the complex formation with the NH_3 ligand that results in a weaker bond strength with the silver ion compared to halide ions. When halide ions are present in the mixture of precipitates, they can interact with the ammonia and also form their own complexes with silver, such as $[AgCl(NH_3)_2]^+$, $[AgBr(NH_3)_2]^+$, and $[AgI(NH_3)_2]^+$. These halide complexes are similar to the previously mentioned $[Ag(NH_3)_2]^+$ complex, where the halide ion replaces one of the ammonia ligands. The formation of these silver-halide complexes can be used as a chemical test to identify the presence of halide ions. For example, when silver nitrate is added to an unknown aqueous solution, and a precipitate forms, the type of precipitate can be identified using a series of tests such as adding aqueous ammonia or heating. The type of precipitate can indicate the presence of a particular halide ion in the unknown solution.

When halide ions react with concentrated sulfuric acid (H_2SO_4), several different reactions can occur. Let me break it down for you.

Firstly, it's important to understand that concentrated sulfuric acid is a strong dehydrating agent, meaning it removes water molecules from other compounds. When a halide ion (such as Cl^-, Br^-, or I^-) is introduced to concentrated sulfuric acid, it can undergo different reactions depending on the specific halide ion involved. For example, when chloride ions (Cl^-) react with concentrated sulfuric acid, the following reaction occurs:

$$2H_2SO_4 + 2NaCl \rightarrow Na_2SO_4 + 2HCl + SO_2 + 2H_2O$$

In this reaction, the concentrated sulfuric acid reacts with the chloride ion to produce hydrogen chloride gas (HCl), sulfur dioxide (SO2), and water (H2O). The resulting product is sodium sulfate (Na2SO4), which is a white crystalline solid that is highly soluble in water.

Similarly, when bromide ions (Br-) react with concentrated sulfuric acid, the following reaction occurs:

$$2H2SO4 + 2NaBr \rightarrow Na2SO4 + Br2 + SO2 + 2H2O$$

In this reaction, the bromide ion reacts with concentrated sulfuric acid to produce bromine gas (Br2), sulfur dioxide (SO2), and water (H2O). The resulting product is sodium sulfate (Na2SO4).

Finally, when iodide ions (I-) react with concentrated sulfuric acid, the following reaction occurs:

$$8H2SO4 + 16KI \rightarrow 16KHSO4 + 8H2O + 4I2 + S8$$

This is a particularly interesting reaction as it produces iodine (I2) which sublimes to form purple vapors. The sulfuric acid here acts as both a dehydrating agent and an oxidizing agent. In this reaction, the concentrated sulfuric acid reacts with the iodide ions to produce iodine gas, water, and sulfate ions. Additionally, the sulfuric acid is reduced from a +6 oxidation state to a +4 oxidation state, forming sulfur (S8) as a product.

Chlorine Reactions

When chlorine reacts with sodium hydroxide, it undergoes a process known as disproportionation reaction. This reaction involves simultaneous oxidation and reduction of the atoms in a single compound. In the case of chlorine reacting with cold aqueous sodium hydroxide, the oxidation number of chlorine decreases from 0 to -1, while the oxidation number of oxygen increases from -2 to -1. This reduction and oxidation process results in the formation of chloride ions and hypochlorite ions. On the other hand, the reaction of chlorine with hot aqueous sodium hydroxide involves the oxidation of chlorine from 0 to +1 and the oxidation of oxygen from -2 to -1. This reaction forms chloride ions and chlorate ions. In both of these reactions, the oxidation numbers of chlorine and oxygen are changing. The process of disproportionation, where atoms of the same element undergo simultaneous oxidation and reduction reactions, is occurring. It is important to note that these reactions occur in aqueous solutions of sodium hydroxide which acts as a strong reducing agent. Without it, these reactions may not occur. Understanding the changes in oxidation numbers during these reactions can help to predict the products formed and can also be used to balance these chemical equations.

Water is a vital resource for our existence, and it is important to ensure that it is free of harmful microorganisms. Chlorination is a widely used method to disinfect water and make it safe for human consumption. The use of chlorine in water purification involves the production of two active species: Hypochlorous acid (HOCl) and Hypochlorite ions (ClO-), which are responsible for killing bacteria and other harmful microorganisms present in water.

In the water treatment process, chlorine gas is added to water, which leads to the formation of hypochlorous acid (HOCl) and hypochlorite ions (ClO-). This reaction occurs according to the following chemical equation:

$$Cl_2 + H_2O \rightarrow HOCl + H^+ + Cl^-$$

Once the hypochlorous acid (HOCl) is formed, it reacts with bacteria present in the water, causing damage to their cell walls and ultimately killing them. The hypochlorite ion (ClO-) also plays a vital role in disinfection by killing bacteria by oxidizing their cellular components, disrupting their metabolism and eventually killing the bacteria.

HOCl and ClO- are the active species responsible for the disinfection process. The relative amounts of HOCl and ClO- in water depend on the pH level of the water. At lower pH levels, more HOCl is present, while at higher pH levels, more ClO- is present. The use of chlorine gas in water treatment is ideal because it is inexpensive and easy to use, and it can be adjusted based on the requirements of the specific application. However, it is important to note that chlorine gas can also react with organic matter present in water, producing disinfection by-products such as trihalomethanes (THM). These by-products have been associated with adverse health effects, which is why the concentration of chlorine needs to be carefully controlled during water treatment.

Sulfur and Nitrogen

Nitrogen, an element that makes up 78% of the air we breathe, is known for its unique lack of reactivity. Despite its abundance, nitrogen is a relatively inert gas and does not readily react with other elements or compounds. This can be attributed to the strength of its triple bond and the lack of polarity between nitrogen atoms, which contribute to the element's unreactive nature. The triple bond between two nitrogen atoms is one of the strongest bonds in nature, and requires a significant amount of energy to break. This bond is formed by the sharing of three pairs of electrons between the two nitrogen atoms. This results in a very stable molecule, as the three shared electron pairs hold the atoms very close together, making it difficult for other atoms or molecules to approach the nitrogen. Moreover, the triple bond also contributes to the lack of polarity between nitrogen atoms. Since both nitrogen atoms share equal electronegativity, there is no net difference in charge distribution between them. This means there are no partial charges on either end of the molecule, making it difficult for other charged molecules or ions to interact with it. These factors, combined with the fact that nitrogen is a noble gas and has a full outer electron shell, make it an extremely stable element that is resistant to chemical reactions.

Ammonia is a compound that is known for its basicity. According to the Brønsted–Lowry theory, a substance can be defined as a base if it is capable of donating a pair of electrons to another molecule. In the case of ammonia, its basicity is a result of its ability to donate a pair of electrons to a proton, or hydrogen ion, in a chemical reaction. In the simplest form of such a chemical reaction, ammonia can react with a proton to form the ammonium ion. During the reaction, ammonia donates one of its lone pair of electrons to the proton, creating a new bond between the two atoms. This results in the formation of ammonium, which is a cation with a positive charge. From a chemical perspective, this reaction represents a transfer of a hydrogen ion, or proton, from an acid to ammonia, which functions as a base. Indeed, in this scenario, ammonia acts as a Brønsted–Lowry base, while the proton functions as a Brønsted–Lowry acid. The basicity of ammonia can be explained in terms of the lone pair of electrons present on the nitrogen atom in the molecule. These electrons are able to act as a nucleophile, or electron pair donor, in a chemical reaction. By donating this pair of electrons to a proton from an acid, ammonia is able to form a new covalent bond, this time with a hydrogen atom, in such a way that the resulting ammonium ion is more stable than the initial ammonia molecule due to its complete outer electron shell.

The ammonium ion is a positive ion that is formed when ammonia reacts with a proton from an acid. The resulting ion has a tetrahedral structure that is composed of one nitrogen atom and four hydrogen atoms. An acid-base reaction example is the creation of the ammonium ion. In an acid-base reaction, a substance that can donate a proton, called an acid, reacts with a substance that can accept a proton, called a base. In the case of the formation of ammonium ion, ammonia is the base. It contains a lone pair of electrons that can interact with a proton, which functions as the acid. By accepting the proton, the ammonia molecule is converted to the ammonium ion. The resulting ammonium ion is a positively charged tetrahedral structure that is composed of one nitrogen atom and four hydrogen atoms. The nitrogen atom is located at the center of the tetrahedron, with one hydrogen atom positioned at each of the four vertices. The nitrogen atom also has a positive charge, which is balanced by the four negative charges on the four hydrogen atoms. In terms of chemical bonding, the ammonium ion is a covalent molecule that is formed by the sharing of electrons between the nitrogen and hydrogen atoms. Each of the hydrogen atoms is covalently bonded to the nitrogen atom, with a single electron pair shared between them. Furthermore, the nitrogen atom is also bonded to a fifth electron pair, which is the lone pair of electrons that remains after it accepts the hydrogen ion in the acid-base reaction.

The displacement of ammonia from ammonium salts is a chemical reaction that occurs when an acid is introduced to a solution containing an ammonium salt. In such a reaction, the acid functions as a proton donor, while the ammonium salt acts as a proton acceptor. As a result of this reaction, ammonia is displaced from the ammonium salt and is released as a gas. The reaction can be explained in terms of the Brønsted–Lowry theory of acid and base reactions. According to this theory, an acid is a substance that donates a proton, while a base is a substance that accepts a proton. In the case of the displacement of ammonia from ammonium salts, the acid donates a proton to the ammonium ion, which accepts the proton to become ammonia. In such a reaction, ammonium ions present in the ammonium salt solution act as the base. The ammonium ion has an excess of hydrogen ions, leaving it positively charged. The acid, on the other hand, acts as the donor of the proton, which leads to the displacement of ammonia from the ammonium salt and its subsequent release as a gas. The reaction can be symbolized in a chemical equation where ammonium salts and acid each have their respective chemical formulas. The chemical equation would indicate that the products are ammonia and the conjugate acid of the acid used. This reaction can be employed practically in several applications. For instance, in the process of determining the purity of a sample compound, the amount of ammonia displaced from an ammonium salt by an acid can be quantified. Similarly, in the laboratory, this reaction is commonly used for the production of pure gases, as ammonia can be easily generated from ammonium salts by reacting them with an appropriate acid.

Oxides of nitrogen (NOx), in particular nitric oxide (NO) and nitrogen dioxide (NO2), are a group of air pollutants that are released into the atmosphere from both natural and human-made sources. Natural sources of NOx include lightning strikes, volcanic activity, forest fires, and microbial activity in soil. On the other hand, human-made sources of NOx emissions include transportation, industrial processes, and energy generation. Internal combustion engines, found in automobiles, trucks, and other vehicles, are the main human-made sources of NOx pollutants. This is because high temperatures in the combustion process lead to the oxidation of nitrogen in the air, leading to the formation of NOx. The efficiency of combustion engines also plays a role in NOx emissions, as incomplete combustion creates more NOx. The presence of NOx in the atmosphere contributes to various environmental and health problems. NOx can react with other pollutants in the atmosphere to form smog, which can cause respiratory problems in humans. NOx is also implicated in acid rain formation, which can damage crop yields, forests, and aquatic ecosystems. Various catalytic converters are used to remove NOx from the exhaust gases of internal combustion engines before they are released into the atmosphere. These converters use a chemical reaction to convert NOx into less harmful gases. The most prevalent type of catalytic converter utilized for NOx removal is the three-way catalyst. The three-way catalyst achieves NOx removal by promoting a series of chemical reactions. Firstly, the catalyst converts nitrogen oxides to nitrogen and oxygen by reducing the oxides with carbon monoxide (CO) or hydrocarbons. Secondly, the catalyst reduces nitrogen dioxide (NO2) to nitric oxide (NO) using carbon or other reducing agents. Finally, the nitric oxide formed in the second step is reacted with reducing agents to form nitrogen gas and water. Selective catalytic reduction (SCR) is another method of NOx removal from exhaust gases and is commonly used in large-scale industrial processes. In the SCR process, ammonia is used as a reducing agent to convert NOx into nitrogen gas and water.

Atmospheric oxides of nitrogen, including nitric oxide and nitrogen dioxide, can react with unburned hydrocarbons to form peroxyacetyl nitrate (PAN). PAN is a component of photochemical smog, which is a type of air pollution that is usually formed under strong sunlight and high temperatures, especially in urban and industrial areas.

The formation of PAN in the atmosphere is due to the complex reaction of various pollutants, including nitrogen oxides and hydrocarbons. When nitrogen oxides and hydrocarbons are released into the atmosphere by various human activities, they combine with sunlight and other environmental factors to form photochemical smog.

PAN is formed when nitrogen oxides react with other pollutants, specifically unburned hydrocarbons in the presence of sunlight. This reaction involves the formation of peroxy radicals (ROO*) from hydrocarbons, which then react with nitrogen dioxide to form PAN. The equation for this reaction is generally expressed as follows:

$$ROO^* + NO_2 \rightarrow PAN$$

PAN is a powerful irritant to the eyes and respiratory system, and exposure to high levels of PAN can cause severe health effects. It is also responsible for the characteristic brown color observed in photochemical smog.

Several methods can be used to reduce the formation of PAN in the atmosphere, including the reduction of emissions from vehicles and industrial processes, as well as the use of cleaner-burning fuels and the improvement of air quality management practices. The use of catalytic converters in vehicles can also reduce the formation of PAN by converting NOx into nitrogen gas and water.

Nitric oxide (NO) and nitrogen dioxide (NO_2), collectively known as nitrogen oxides (NOx), play an important role in the formation of acid rain. Acid rain is a type of precipitation in which the pH is lower than normal rainwater due to high levels of sulfuric acid (H_2SO_4) and nitric acid (HNO_3). The presence of NOx is a key contributor to the formation of nitric acid, which can cause acid rain.

NOx can directly contribute to acid rain through the following reactions:

$$NO_2 + H_2O \rightarrow HNO_3$$
$$NO + O_3 \rightarrow NO_2 + O_2$$

In the first reaction, NO_2 reacts with water to form nitric acid (HNO_3), which contributes to acid rain. In the second reaction, NO reacts with ozone (O_3) to form nitrogen dioxide (NO_2), which then participates in the first reaction.

Furthermore, NOx can also play a catalytic role in the formation of acid rain by oxidizing sulfur dioxide (SO_2) to form sulfur trioxide (SO_3), which can then react with water to form sulfuric acid (H_2SO_4). The reaction can be expressed as follows:

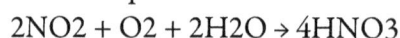

$$2NO_2 + O_2 + 2H_2O \rightarrow 4HNO_3$$
$$SO_2 + 2NO_2 + O_2 + 2H_2O \rightarrow H_2SO_4 + 4HNO_3$$

In the first reaction, NO_2 reacts with oxygen and water to form nitric acid. In the second reaction, NOx mediates the oxidation of SO_2 to form sulfuric acid, in addition to forming nitric acid directly. This reaction is particularly problematic as it accelerates the formation of acid rain caused by sulfuric acid, which can cause damage to aquatic and terrestrial ecosystems, as well as buildings and infrastructure.

The acid rain can also release toxic heavy metals like cadmium, lead and mercury from soil thereby polluting rivers and lakes. Furthermore, acid rain can cause damage to crops and forests, and can harm water quality in bodies of water.

Organic Compound Basics

Hydrocarbon is a type of organic compound that consists of hydrogen and carbon atoms. These atoms connect with each other through covalent bonds, creating a variety of structures and forms. The naturally occurring hydrocarbons play a crucial role in the formation of fossil fuels such as coal, oil, and natural gas. In addition, hydrocarbons are widely used in many industries, including pharmaceuticals, plastics, electronics, and automotive. The diversity of hydrocarbons make them an important building block for numerous products and materials that we use in our daily lives. Overall, hydrocarbons are fascinating and complex molecules whose properties and functions continue to be explored by scientists around the world.

Alkanes are a fundamental group of hydrocarbon molecules that are comprised solely of carbon and hydrogen atoms. Unlike other, more complex hydrocarbon molecules, alkanes do not contain any functional groups. This means that they are the simplest and most basic type of hydrocarbon. An alkane's molecular structure consists of a simple chain of alternating carbon and hydrogen atoms, with each carbon atom bonded to four other atoms (either carbon or hydrogen). Due to the relatively low reactivity and stability of their molecular structure, alkanes are often used as solvents and as fuels for combustion engines. They can also serve as a building block for more complex hydrocarbon molecules, such as alkenes and alkynes. Despite their simple structure, alkanes play a critical role in the world of chemistry and industry, offering a multitude of uses and applications.

I suggest following these steps to deduce the molecular and/or empirical formula of a compound, given its structural, displayed, or skeletal formula:

1. Determine the number of atoms of each element in the molecular formula using the atomic symbols and subscripts in the structural, displayed, or skeletal formula.

2. Convert the molecular formula to the empirical formula by dividing each subscript by the greatest common factor of all the subscripts. If the molecular formula is already the empirical formula, skip this step.

3. If the molar mass of the compound is given, calculate the molecular formula using the empirical formula, the molar mass, and the molar mass of the empirical formula.

For example, let's use the displayed formula of glucose:

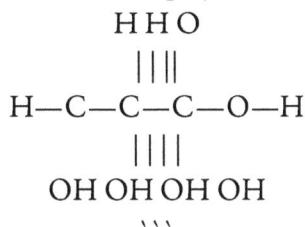

$$
\begin{array}{c}
\text{H H O} \\
| \; | \; | \\
\text{H—C—C—C—O—H} \\
| \; | \; | \; | \\
\text{OH OH OH OH}
\end{array}
$$

1. Count the number of atoms of each element: 6 carbon atoms, 12 hydrogen atoms, and 6 oxygen atoms.

2. Divide each subscript by the greatest common factor (in this case, 6): the empirical formula is CH_2O.

3. The molar mass of glucose is 180 g/mol, and the molar mass of the empirical formula CH_2O is 30 g/mol. The ratio of the molar masses is 180/30 = 6, so the molecular formula is 6 times the empirical formula: $C_6H_{12}O_6$.

Therefore, the molecular formula of glucose is $C_6H_{12}O_6$.

Organic Reaction Types

(a) A homologous series refers to a group of organic compounds that have similar arrangements of atoms and exhibit common properties. Members of a homologous series have a similar general formula, but differ in the number of carbon atoms in their structure, giving them unique physical and chemical characteristics. These compounds share a characteristic functional group, which plays a vital role in their chemical reactions. Understanding homologous series helps us to predict and interpret the properties of different organic compounds, facilitating the creation of innovative approaches for synthesizing new compounds with unique characteristics. So, in simple terms, a homologous series is like a family of compounds that share similar features and exhibit predictable behavior.

(b) In organic chemistry, the terms "saturated" and "unsaturated" are used to describe the types of hydrocarbon molecules. Saturated hydrocarbons are molecules that contain only single bonds between carbon atoms and are "saturated" with hydrogen atoms. On the other hand, unsaturated hydrocarbons have one or more double or triple bonds between carbon atoms, meaning they are capable of forming additional bonds. Due to their structure, unsaturated hydrocarbons are generally more reactive than saturated hydrocarbons.

(c) Homolytic and heterolytic fission are terms used to describe the breaking of chemical bonds during a reaction. In homolytic fission, a bond is broken so that each atom retains one of the two electrons in the bond, creating two free radicals. In contrast, heterolytic fission involves the splitting of a bond between two atoms, with one atom keeping both electrons from the broken bond. This releases an anion and a cation, resulting in the formation of ions.

(d) In organic chemistry, free radical, initiation, propagation, and termination are important terms that describe the reactions of organic compounds. Free radicals are highly reactive molecules that have unpaired electrons, and can be formed during chemical reactions. Initiation involves the formation of free radicals through the breaking of a bond using either heat or light. The newly formed free radical then leads to the propagation stage, where the reactive species will react with other molecules to produce another free radical, continually propagating the reaction. Termination is the final stage of the reaction, where free radicals are removed from the system, and no more reaction takes place. This process often occurs in pairs, with two radicals reacting and removing each other from the system. This process of free radical reactions plays a crucial role in the synthesis of many organic compounds.

(e) Nucleophile and electrophile refer to molecules or ions that participate in chemical reactions. Nucleophiles are species that donate a pair of electrons to an electron-deficient centre of another molecule or ion to form a new bond. Electrophiles, on the other hand, are species that seek to accept a pair of electrons from a nucleophile to form a new bond. The terms nucleophilic and electrophilic are used to describe the reactivity of compounds towards a nucleophile or an electrophile, respectively.

(f) Addition, substitution, elimination, hydrolysis, and condensation refer to different types of organic chemical reactions. A type of chemical reaction called an addition reaction happens when two molecules come together and form a new and larger molecule. Substitution reactions, on the other hand, involve the replacement of an atom or group of atoms within a molecule by another atom or group of atoms. Elimination reactions are reactions where a small molecule is eliminated from a larger molecule upon breaking of a covalent bond. Hydrolysis involves the reaction of water with a compound to break it down into smaller compounds. Finally, condensation reactions involve the combination of two smaller molecules into a larger molecule, which involves the loss of a small molecule, usually water.

(g) Oxidation and reduction reactions involve one species losing electrons while the other gains electrons as a result of a chemical reaction. Oxidation is a process in which a species loses one or more electrons, while reduction is a process in which a species gains one or more electrons. This exchange of electrons, known as redox reactions, plays an essential role in many organic chemical reactions. The reactions can be represented using symbols such as [O], which represents an atom of oxygen from an oxidizing agent, and [H], which represents an atom of hydrogen from a reducing agent. These reactions can be used to synthesise a variety of organic compounds with unique characteristics.

Organic mechanisms refer to the step-by-step processes that happen when chemicals react with one another. In our class, we've been talking about two types of mechanisms: free-radical substitution and electrophilic addition.

Free-radical substitution is a type of organic reaction where a molecule (usually an alkane) is gradually replaced by another molecule due to the presence of free radicals. Free radicals are atoms or molecules with one or more unpaired electrons that make them highly reactive. During the process of free-radical substitution, a free radical attacks a covalent bond in the molecule, causing it to break apart and release more free radicals. These free radicals then attack other molecules, continuing the replacement process until it is complete.

On the other hand, in electrophilic addition, the reaction occurs between an electrophile and a nucleophile. Electrophiles are molecules that promote the addition of electrons, while nucleophiles are molecules that have an affinity for electrons. In this type of reaction, the electrophile is attracted to the pi-bonds of the nucleophile and attacks them, breaking them apart so that the electrons can be shared between the two molecules. The result is a new molecule that is formed by the addition of the electrophile.

Nucleophilic substitution refers to the process where a nucleophile attacks an electrophile that is attached to a molecule, causing the electrophile to be replaced by the nucleophile. This type of mechanism is typically seen in reactions between alkyl halides and nucleophiles such as hydroxide or cyanide ions. The process involves a curved arrow notation, which represents the movement of electron pairs. The arrow begins at the bond between the nucleophile and the molecule being attacked, and ends at the bond that is being broken.

In nucleophilic addition, the nucleophile is added to a molecule with an electrophilic center, such as a carbonyl group. The process involves a curved arrow notation too, which again represents the movement of electron pairs. The arrow begins at the lone pair on the nucleophile and ends at the electrophilic center.

The use of curly arrows is critical to both nucleophilic substitution and nucleophilic addition. These arrows help to show the flow of electrons throughout the reaction, which is essential for understanding how the mechanism works. Mastering the technique of curved arrow notation is an essential part of studying organic chemistry, and it is one of the most critical tools that students need to learn in order to better understand organic mechanisms.

Organic Reactions: Characteristics

Organic chemistry can be a somewhat complicated subject, but I'll do my best to explain the terminology associated with types of organic compounds and reactions.

First off, (a) we have the homologous series. Essentially, a homologous series is a group of organic compounds that have similar chemical properties because they have the same functional group (a specific combination of atoms that give the compound its properties) and the same general molecular formula, but differ in the length of their carbon chains. Each member of the series is called a homologue and they have similar physical and chemical properties.

Next up, (b) we have saturated and unsaturated compounds. When we talk about saturation in organic chemistry, we're not talking about how much vinegar is on your fries. Instead, we're talking about the number of carbon-carbon bonds in a molecule. A saturated compound has only single bonds between its carbon atoms, while an unsaturated compound has at least one double or triple bond between its carbons. The carbon-carbon double or triple bonds create a region of electron density that can react more readily with other molecules, leading to different physical and chemical properties than saturated compounds.

Finally, (c) we have homolytic and heterolytic fission. These are two different types of bond breaking reactions. In homolytic fission, a bond is broken in a way that splits the shared electrons evenly, with one electron going to each atom. This results in two separate and highly reactive entities called radicals. In heterolytic fission, a bond is broken in a way that one of the atoms involved takes both shared electrons, resulting in the formation of two charged species, an anion and a cation.

Starting with (d), we have free radicals, initiation, propagation, and termination. A free radical is a highly reactive molecule or atom that has an unpaired electron. Initiation is the first step of radical reactions where a high energy radical is formed by breaking a bond and forming two free radicals. Propagation is the second step where reactive species react with other monomers and polymers, propagating the reaction to other molecules. Termination is the final step where two free radicals react with one another, thus stopping the chain reaction.

For (e), we have nucleophiles and electrophiles. A nucleophile is an electron-rich species that is attracted to positively charged or electron-poor atoms or molecules. They tend to have a negative charge or a lone pair of electrons. An electrophile, on the other hand, is an electron-poor species that is attracted to negatively charged or electron-rich atoms or molecules. They tend to have a positive charge or a partial positive charge. Nucleophilic reactions involve the attack of a nucleophile on an electrophile, where a bond is formed between the nucleophile and the electrophile. Conversely, electrophilic reactions involve the attack of an electrophile on a nucleophilic centre.

Now, (f) involves five different types of reactions: addition, substitution, elimination, hydrolysis, and condensation. The process of addition reaction involves the combination of multiple molecules leading to the formation of a single, larger molecule. Substitution reaction occurs when a functional group of a molecule is substituted with a new functional group in a chemical reaction, resulting in the formation of another compound. An elimination reaction is where a molecule loses atoms or ions from its structure. Hydrolysis is a reaction where water is used to break a molecule into smaller subunits. In contrast, a condensation reaction is where two or more smaller molecules combine to form a larger molecule, with the elimination of a smaller molecule, usually water, as a byproduct.

Finally, in (g), we have oxidation and reduction. In organic chemistry, oxidation refers to the loss of electrons, whereas reduction refers to the gain of electrons. An oxidizing agent is a substance that removes an electron from another molecule, while a reducing agent is a substance that donates an electron to another molecule. The shorthand symbols [O] and [H] can be used in organic redox reactions to represent oxygen and hydrogen atoms and show the loss or gain of electrons.

Terminology associated with types of organic mechanisms:

(a) Free-radical substitution is a chemical process where a free radical replaces the hydrogen atom in a molecule. In this mechanism, a free radical, often generated from a halogen or other radical initiator, attacks a molecule with a weak C-H bond. This results in the formation of a new molecule where the hydrogen atom has been replaced by the free radical. This mechanism is commonly found in reactions such as halogenation of alkanes, where a hydrogen atom is replaced by a halogen atom.

(b) Electrophilic addition is another type of chemical reaction that occurs between a molecule with a double or triple bond and an electrophile. In this mechanism, electrons of the double or triple bond in the molecule are attracted to an electrophile with a partial positive charge. As the pi-electrons are attracted to the electrophile, the double or triple bond "opens up" and forms a new bond with the electrophile, effectively producing a single bond. This mechanism is commonly found in reactions such as the addition of hydrogen halides to an alkene or the addition of water to an alkyne.

(c) Nucleophilic substitution is a type of chemical reaction where a nucleophile replaces a leaving group in a molecule. In this mechanism, a nucleophile with a negative or partial negative charge attacks a molecule in which a leaving group is present and replaces it, resulting in the formation of a new molecule. This type of reaction is commonly found in organic chemistry in reactions such as the substitution of halogens by hydroxyl groups to give alcohols, or the substitution of amines by alkyl or acyl groups.

(d) Nucleophilic addition is another type of chemical reaction where a nucleophile attacks an electrophilic carbon atom, resulting in the formation of a new bond. In this mechanism, a nucleophile adds to an electrophile in such a way that a new bond is formed between the nucleophile and electrophilic carbon atom in the molecule. This type of reaction is commonly found in organic chemistry in reactions such as the hydration of alkenes or the reaction of aldehydes or ketones with a nucleophile to form an alcohol.

Shapes of Organic Molecules

Organic molecules are the building blocks of all living organisms and play a crucial role in the chemistry of life. These molecules are primarily composed of carbon atoms that bond with other atoms such as hydrogen, oxygen, nitrogen, and sulfur to form different structures. Based on their structure, organic molecules can be classified into three types - straight-chained, branched, or cyclic.

Straight-chained organic molecules are those in which the carbon atoms are arranged in a linear chain, with each carbon atom linked to the next by a single bond. These chains can be short or long, depending on the number of carbon atoms. For instance, hexane, an organic molecule consisting of six carbon atoms and 14 hydrogen atoms, is a straight-chained molecule. Such molecules are usually flexible and can easily bend and twist.

On the other hand, branched organic molecules contain one or more branches that extend from the main carbon chain. These branches can either be short or long and can contain additional carbon and hydrogen atoms. This structure adds complexity to the molecule and can result in unique chemical properties. For example, isopentane, an organic molecule with five carbon atoms, is a branched molecule.

Circular organic molecules are called cyclic molecules. These structures are formed when carbon atoms form a ring, with each carbon atom bonded to two other carbon atoms. The simplest example of a cyclic molecule is cyclopropane, a three-carbon ring with each carbon atom bonded to two hydrogen atoms. These structures can be small or large, and their properties depend on the number of carbon atoms in the ring and the type of atoms bonded to them.

When atoms bond with each other to form molecules, they undergo hybridization to create a new set of orbitals that allow for new bonding possibilities. Hybridization is a process in which the orbitals of two or more atomic orbitals combine to create new orbitals, producing a new set of molecular orbitals. Three types of hybridizations are commonly observed in organic molecules, including sp, sp2, and sp3 hybridization.

An sp hybridized atom is created by mixing one s and one p orbital from the same shell of an atom. This hybridization results in the formation of two hybrid orbitals. The two hybrid orbitals are oriented at an angle of 180^0 to each other, giving the molecule a linear shape. This bonding arrangement leads to the formation of a straight line, where the two atoms attached to the sp hybridized atom are located opposite to each other.

In the sp2 hybridized atoms, one s and two p orbitals from the same shell hybridize to form three hybrid orbitals. These orbitals are arranged in the same plane and are separated by a 120^0 angle from each other. This angle enables the atoms to be spaced evenly apart and bonded to the sp2 hybridized atom in a trigonal planar manner. The trigonal planar shape leads to some interesting chemical properties, including the compound's propensity towards electrophilic and nucleophilic addition reactions.

The sp3 hybridized atoms are a combination of one s orbital and three p orbitals from the same shell. The result of this hybridization gives rise to four hybrid orbitals that are separated by tetrahedral angles of 109.5^0. The tetrahedral angle creates a symmetrical distribution of electrons, which allows atoms attached to the sp3 hybridized atom to be spaced evenly apart. This results in a shape that is roughly spherical, with the four bonds pointing to the corners of a tetrahedron. This shape is responsible for many of the characteristic properties of organic molecules, including their ability to form rings and complex structural shapes.

The arrangement of σ and π bonds in molecules containing sp, sp2, and sp3 hybridized atoms depends on the type of hybridization and the geometry of the molecule. A σ bond is formed by the end-to-end overlap of two atomic orbitals, while a π bond is formed by the side-by-side overlap of two parallel atomic orbitals.

In sp hybridized atoms, there are only two hybrid orbitals, resulting in two σ bonds. The two bonds are oriented opposite to each other along the linear axis of the molecule. No π bonds are formed because there are no p orbitals available for their formation.

In sp2 hybridized atoms, there are three hybrid orbitals, two of which form σ bonds with two other atoms, while the third p orbital forms a π bond with an adjacent atom. The σ bonds are located in a trigonal planar arrangement, while the π bond is located perpendicular to the plane of the σ bonds.

For sp3 hybridized atoms, there are four hybrid orbitals, three of which form σ bonds with three other atoms. One unhybridized p orbital is left in the sp3 hybridized atom, which overlaps with the p orbitals of adjacent atoms to form two π bonds. The arrangement of the four hybrid orbitals and unhybridized p orbitals leads to a tetrahedral geometry for the sp3 hybridized atoms.

The σ and π bonds in organic molecules play an essential role in their reactivity and chemical properties. The presence of multiple π bonds in molecules containing sp2 hybridized atoms makes them more reactive than molecules containing only σ bonds. Additionally, molecules containing π bonds are more planar than those containing only σ bonds, which can affect their overall stability.

The term "planar" is often used in organic chemistry to describe the arrangement of atoms in a molecule. Specifically, a planar molecule is one in which all of the atoms are situated in a single flat plane. This means that the molecule's bonds and orbitals lie in the same plane and have no significant angle between them.

For example, consider the organic molecule ethene, which has the chemical formula C2H4. Ethene consists of two carbon atoms double bonded to each other, and each carbon atom is bonded to two hydrogen atoms. Due to the nature of the carbon-carbon double bond, this molecule has a planar structure.

In ethene, the carbon atoms are situated in the same plane, with the hydrogen atoms oriented perpendicular to the plane. The carbon-carbon double bond is formed by the sideways overlap of the p-orbitals of each carbon atom, leading to the formation of a π bond. The π bond is stronger than a single bond, but weaker than a triple bond.

Several important properties of organic molecules can be attributed to their planarity. For example, many reactive intermediates in organic reactions are planar, due to the need for orbital overlap. Furthermore, the angle between the bonds in a molecule can affect its reactivity, with planar molecules often being more reactive than non-planar molecules.

Isomerism

1. Structural isomerism refers to the phenomenon where molecules with the same molecular formula have different bonding arrangements between atoms. This results in the formation of different structural isomers. There are three main types of structural isomers: chain, positional, and functional group isomerism. Chain isomers are formed when two molecules have the same molecular formula but differ in the arrangement of their carbon chains. Positional isomers are formed when two molecules have the same molecular formula and carbon chain but differ in the position of the functional group. Functional group isomers are formed when two molecules have the same molecular formula but the functional groups are different.

2. Stereoisomerism refers to the phenomenon where two molecules have the same molecular formula and same bonding patterns but differ in their spatial arrangements. There are two types of stereoisomers: geometrical isomerism and optical isomerism. Geometrical isomerism is further divided into cis/trans isomerism. Cis/trans isomers are formed when two molecules have the same molecular formula and bonding pattern, but differ in their spatial arrangements due to restricted rotation around a bond. Optical isomerism is formed when two molecules are mirror images of each other and cannot be superimposed on each other.

3. Geometrical isomerism is a type of stereoisomerism found in alkenes. Alkenes have double bonds between two carbon atoms, which do not allow free rotation around the bond. Due to this restricted rotation, the two substituents around the double bond can be arranged in different spatial arrangements, resulting in the formation of cis/trans isomers. When two substituents are attached to opposite sides of a double bond, it is said to be the trans configuration. Conversely, if they are located on the same side, it is referred to as the cis configuration. This phenomenon arises due to the presence of the pi bond, which restricts rotation around the double bond leading to the formation of different spatial isomers.

4. A chiral centre is a carbon atom that has four different substituent groups attached to it. The presence of such a centre results in the formation of two optical isomers, also known as enantiomers, which are mirror images of each other and cannot be superimposed on each other. This is because the four substituent groups are arranged in a specific sequence, leading to different spatial arrangements of the molecule. Enantiomers have the same physical and chemical properties except for their interaction with plane-polarized light.

5. To identify chiral centres in a molecule, one can look for carbon atoms that have four different substituent groups attached to them. If a carbon atom has two of the same substituent groups attached to it, it is not a chiral centre. In terms of geometrical isomerism, one can look for double bonds in a molecule. If there are two different substituent groups on either side of the double bond, it can exist in the form of cis/trans isomers. This phenomenon also applies to cyclic compounds, where one can find chiral centres and cis/trans isomerism.

6. To deduce the possible isomers for an organic molecule of known molecular formula, one can first check if the molecule has a chiral centre. If it does, then it can give rise to two optical isomers, also known as enantiomers. If the molecule has multiple chiral centres, then the number of possible isomers increases exponentially. On the other hand, if the molecule does not have a chiral centre, it may still have geometrical isomers if there is a double bond present. Furthermore, one can use the molecular formula to determine the

functional groups present in the molecule, which can help in identifying possible isomers. Overall, understanding the properties and characteristics of chiral centres and geometrical isomers can aid in deducing the isomers for an organic molecule with a known molecular formula.

Alkanes

Alkanes are a type of hydrocarbon that are characterized by single bonds between their carbon atoms. Interestingly, there are two common reactions that can be used to produce alkanes. The first reaction is called hydrogenation reaction, which involves adding hydrogen (H_2) to an alkene in the presence of a platinum (Pt) or nickel (Ni) catalyst and heating the mixture. This reaction results in the formation of an alkane and requires the double bond in an alkene to break and be replaced by single bonds.

The second reaction is called cracking and involves breaking down longer chain alkanes into smaller, more useful ones. This reaction is typically carried out by heating the alkane in the presence of a solid catalyst like aluminum oxide (Al_2O_3). This reaction breaks the longer chain into shorter ones and follows a process called homolysis, where the carbon-carbon bonds are broken evenly and lead to the formation of free radicals.

When oxygen is present, alkanes undergo complete combustion, which produces carbon dioxide and water as the end products. This reaction is characterized by the release of energy in the form of heat and light, and is often used as a source of energy in combustion engines and household heating. On the other hand, incomplete combustion occurs when there is a limited supply of oxygen and results in the formation of carbon monoxide, carbon soot, and water vapor as products.

Another important reaction of alkanes is free-radical substitution, in which a hydrogen atom on an alkane is replaced by a halogen atom (chlorine or bromine) through a radical mechanism. This reaction occurs in the presence of ultraviolet light, which provides the energy needed to break the halogen molecule into free radicals. The free radical then attacks the alkane and replaces a hydrogen atom to form a halogenated alkane and a new halogen radical. This reaction is often used to produce halogenated solvents and organic intermediates.

As an example, let's consider the free-radical substitution of ethane with chlorine. In the presence of ultraviolet light, chlorine molecules break apart into chlorine free radicals. These radicals then initiate the reaction by attacking an ethane molecule and replacing one of the hydrogen atoms. The product of this reaction is chloroethane (C_2H_5Cl) and a new ethyl radical. The ethyl radical can then combine with another chlorine molecule to continue the reaction and form 1,2-dichloroethane ($C_2H_4Cl_2$) and a new chlorine radical. This process continues until all the available hydrogens in ethane are replaced by chlorine atoms, resulting in a fully halogenated alkane.

Free-radical substitution is a chemical reaction that involves the replacement of a hydrogen atom in an alkane with a halogen atom (such as chlorine or bromine) in the presence of ultraviolet light. This reaction follows a mechanism consisting of three steps: initiation, propagation, and termination.

In the initiation step, the reaction is started by the formation of free radicals. In the case of the reaction between chlorine and methane, for example, ultraviolet light provides the energy required to break the Cl_2 molecule into two chlorine free radicals: Cl. These free radicals are highly reactive and can start the substitution reaction by attacking a methane molecule, resulting in the formation of a new methyl radical: CH_3.

In the propagation step, the new methyl radical reacts with another chlorine molecule to produce one molecule of chloromethane (CH_3Cl) and another chlorine radical. This chlorine radical can then attack another methane molecule to continue the reaction, producing another molecule of CH_3Cl and another methyl radical. The methyl radical can then react with another chlorine molecule, generating another molecule of CH_3Cl and another chlorine radical. This process continues until all the hydrogen atoms in methane have been replaced with chlorine atoms and the final product, chloromethane, is formed.

In the termination step, the free radicals that are produced in the propagation step are either consumed in an unproductive way or react with each other to form stable molecules. For example, two methyl radicals can react with each other to produce ethane (C2H6), while two chlorine radicals can react with each other to produce Cl2.

Cracking is a key process in the petroleum industry for obtaining useful alkanes and alkenes from heavier crude oil fractions. These heavier fractions typically contain a mixture of long-chain hydrocarbons which are not as useful as their shorter-chain counterparts. By using cracking, the larger hydrocarbons can be broken down into smaller ones, making them more useful as raw materials for other chemical reactions or as fuels.

The process of cracking involves breaking the carbon-carbon bonds in the heavier hydrocarbons through the application of heat and pressure. This is typically carried out in the presence of a solid catalyst like aluminum oxide. As the long-chain hydrocarbons are heated, they vaporize and are then passed over the catalyst, which causes the bonds to break and form smaller fragments.

The products of cracking are smaller hydrocarbons, which can have a range of carbon numbers depending on the specific conditions of the process. These products include alkanes like propane and butane, as well as alkenes like ethene and propene. These smaller hydrocarbons are much more useful as fuels and raw materials for chemical reactions than the larger hydrocarbons they were derived from.

Cracking is typically carried out in large-scale refineries that process crude oil into a range of useful products. By cracking the heavier fractions, these refineries can produce a larger volume of products that are in high demand, such as gasoline, diesel, and other fuels. Additionally, the smaller hydrocarbons produced by cracking can be used as feedstock for other chemical reactions, such as the production of plastics and pharmaceuticals.

The C-H bond is a strong bond compared to other bonds that may form between carbon and other atoms, due to the fact that the carbon and hydrogen atoms have similar electronegativities. This makes it difficult for a polar reagent to break the C-H bond. In addition, since these hydrocarbons only have nonpolar C-H bonds and no polar functional groups, they are not easily or readily attracted to polar solvents.

Furthermore, alkanes are nonpolar molecules, which means they lack an electrical pole difference and have low polarity. This makes it difficult for polar reagents to react with them because chemical reactions generally require reagents that are polar or have opposite polarity. Therefore, since alkanes do not have a polar character, they do not readily react with polar reagents.

In general, the lack of chemical reactivity of alkanes is the result of their high C-H bond strength and nonpolar characteristics, which together make them less reactive to chemical reactions. However, when alkanes are subjected to high temperature, radiation, or other extreme conditions, they can undergo reactions such as combustion or cracking, which break some of the C-H bonds.

The combustion of alkanes in internal combustion engines releases several pollutants into the environment, including carbon monoxide, oxides of nitrogen, and unburnt hydrocarbons. These pollutants can have severe environmental consequences, including air pollution, smog formation, and negative impacts on public health. Carbon monoxide is a colorless, odorless gas that is produced when hydrocarbons are incompletely burned. It is toxic to humans, and high levels of exposure can lead to headaches, dizziness, and even death. Oxides of nitrogen (NOx) are compounds formed by the reaction of nitrogen and oxygen in high-temperature combustion environments. They contribute to the formation of smog and acid rain and can have negative impacts on human health, including respiratory issues and irritation of the eyes and nose.

Unburnt hydrocarbons are hydrocarbons that are not fully combusted in the engine. They contribute to the formation of smog and can also react with nitrogen oxides in the presence of sunlight to produce ground-level ozone. Ground-level ozone is a dangerous pollutant that can cause respiratory problems and contribute to the formation of acid rain.

To mitigate the environmental consequences of these pollutants, catalytic converters have been developed. Catalytic converters are devices that are installed in the exhaust system of a vehicle to reduce emissions of harmful pollutants. Their method of action involves the transformation of carbon monoxide into carbon dioxide, nitrogen oxides into nitrogen and oxygen, as well as unburnt hydrocarbons into carbon dioxide and water.

Alkenes

Alkenes are a class of hydrocarbons with a double bond between two adjacent carbon atoms. There are various methods to produce alkenes, including the elimination of HX from a halogenoalkane, the dehydration of an alcohol, and the cracking of a longer chain alkane.

In the first method, alkene formation involves removing HX from a halogenated hydrocarbon using a solution of ethanolic NaOH and then applying heat. This reaction is known as dehydrohalogenation. Ethanolic NaOH acts as a strong base to abstract the acidic hydrogen from the halogenated hydrocarbon, resulting in the formation of an alkene and a salt.

The second method involves the dehydration of an alcohol to form alkenes. Alcohols, with the removal of one water molecule, can undergo dehydration to yield an alkene product. This process is catalyzed by a heated catalyst, such as Al_2O_3, or a concentrated acid, such as concentrated sulfuric or phosphoric acid.

Lastly, the cracking of a longer chain alkane involves the breaking down of larger hydrocarbons from crude oil into smaller and more useful molecules, such as alkenes. This process requires high temperatures and a catalyst to break the bonds of the larger hydrocarbons and form small molecule products, including alkenes.

Alkenes refer to hydrocarbons which contain a double bond between two carbon atoms and are classified as unsaturated compounds. Due to the presence of this pi bond, alkenes readily undergo electrophilic addition reactions where an electrophile is added to the double bond. Some of the common electrophilic addition reactions of alkenes include hydrogenation, hydration, addition of hydrogen halides, and halogenation.

The first reaction, hydrogenation, is the addition of hydrogen gas to an alkene, which occurs in the presence of a metal catalyst such as platinum or nickel. The double bond of the alkene acts as an electron-rich site that attracts the electrophilic hydrogen molecule, breaking the double bond and forming a single bond with each carbon atom. This reaction is exothermic and releases energy in the form of heat.

The second reaction, hydration, involves the addition of water (H_2O) to the double bond of an alkene. The reaction occurs in the presence of a strong acid catalyst such as sulfuric acid or phosphoric acid. The acid catalyst protonates the water molecule, generating a more electrophilic hydronium ion. The positive charge of the hydronium ion attracts the pi electrons of the double bond, leading to the formation of a carbocation intermediate. A water molecule then attacks the carbocation intermediate, adding an H and OH group to each carbon atom, forming an alcohol.

The third reaction is the addition of hydrogen halides, such as hydrogen chloride (HCl), hydrogen bromide (HBr), and hydrogen iodide (HI), to the double bond of an alkene. The reaction occurs at room temperature and does not require a catalyst. The pi electrons of the double bond attack the electrophilic hydrogen ion of the hydrogen halide, leading to the formation of a carbocation intermediate. The carbocation intermediate then reacts with the halide ion of the hydrogen halide, resulting in the addition of H and X to each carbon atom, forming a halogenoalkane.

The fourth reaction is halogenation, which involves the addition of halogens, such as chlorine (Cl_2), bromine (Br_2), and iodine (I_2), to the double bond of an alkene. This reaction occurs in the presence of a halogen carrier, such as iron or chlorine gas, which enables the halogen molecules to break apart into electrophilic halogen ions. The double bond of the alkene attracts the electrophilic halogen ions, forming a cyclic halonium ion intermediate. The halonium ion is then attacked by a halide ion, causing the cyclic intermediate to open up and form a dihalogenated alkane.

Alkenes are versatile chemical compounds that engage in different types of reactions leading to a wide array of useful products. Some of these reactions include oxidation, polymerization, and addition reactions.

One of the most important reactions of alkenes is oxidation. When alkenes undergo oxidation by cold dilute acidified potassium permanganate (KMnO4), they form diols. The reaction is commonly used to identify the presence of double bonds in organic compounds. Under these conditions, the oxidizing agent breaks the double bond of an alkene and forms a cis-diol derivative. The reaction proceeds rapidly, especially in the presence of a protonating acid, and it results in two neighboring hydroxyl (-OH) functional groups on the same side of the molecule. The cis-diol product is stable and used in various chemical reactions as a starting material for the synthesis of different products.

In contrast, the oxidation of alkenes by hot concentrated acidified KMnO4 leads to the rupture of the carbon-carbon double bond. The reaction proceeds through an oxidative cleavage mechanism, which breaks the carbon-carbon double bond and replaces it with two carbon-oxygen double bonds. The reaction is used to determine the position of the alkene linkages in larger molecules. The product of the reaction is a mixture of two carboxylic acids, each having a carbonyl group at the carbon end of the carbon chain. The identity of the carboxylic acids formed indicates the position of the original double bond in the reactant alkene molecule.

Addition polymerization is another important reaction of alkenes. It involves the addition of monomers to form a polymer chain. In this process, the double bond of an alkene opens up, allowing free-radical initiators to bond with one monomer unit. Subsequently, a chain of monomers forms and a polymer chain grows until a stable product is reached. Ethene and propene are examples of alkenes that undergo addition polymerization. Polyethylene is produced through the addition polymerization of ethene, while polypropylene is produced through the addition polymerization of propene. Addition polymerization of alkenes is crucial for the production of many materials, including plastics, rubber, and adhesives.

Aqueous bromine, also known as the bromine water test, is a common chemical test used to detect the presence of a C=C (carbon-carbon double) bond in organic compounds. This test works through the addition of bromine to the double bond, leading to a visible color change in the solution.

When an organic compound containing a C=C bond is mixed with aqueous bromine, the orange-brown color of the bromine water fades as the bromine molecule is added to the double bond of the C=C bond. The C=C bond is more reactive than single bonds and undergoes electrophilic addition reactions, where an electrophile is added to the C=C bond, breaking the double bond and forming a single bond. In this case, the bromine molecule serves as the electrophile and gets added to the double bond, resulting in the formation of a dibromoalkane.

The color change of the solution is due to the formation of dibromoalkane, which is colorless. The reaction occurs rapidly, providing a useful means of determining the presence of unsaturated compounds with a high degree of sensitivity.

The test is simple, quick, and inexpensive, making it a useful tool in both laboratory and field settings. It is often used to confirm the identity of alkenes and other unsaturated compounds, as well as in the detection of unsaturation in chemical synthesis and analysis. Additionally, the bromine water test can differentiate between different types of unsaturated compounds, such as alkenes and alkynes, as alkynes undergo addition reactions that are more rapid than alkenes.

The mechanism of electrophilic addition in alkenes is a fundamental chemical process that involves the addition of an electrophile (an electron-deficient species) to the double bond of an alkene. This process results in breaking the double bond and the formation of a new single bond between the electrophile and the carbon atoms of the alkene. The electrophilic addition mechanism is critical in organic synthesis and has different types of applications in chemical reactions. The reaction mechanism of electrophilic addition in alkenes can be explained using two examples; the addition of bromine to ethene and the addition of hydrogen bromide to propene.

In the first example, the addition of bromine to ethene involves the bromine molecule acting as the electrophile. The reaction begins by the bromine molecule approaching the pi bond of the ethene molecule. The electron-rich pi bond then serves as the nucleophile, attacking the electrophilic bromine molecule by breaking the double bond. As this happens, the carbon atoms of the C=C bond create a partial positive charge that attracts negatively charged bromine ions to them. This attraction leads to the formation of a cyclic intermediate known as a bromonium ion. The bromine ion is added to one carbon atom, while the other carbon carries the positive charge in the bromonium ion. This intermediate is highly reactive, and it quickly opens to create an active intermediate with a carbocation. Finally, the bromide ion attacks the carbocation, displacing the proton and forming a dihalogenated alkane compound. This reaction mechanism illustrates how electrophiles add to the double bond of an alkene to create a new single bond in alkenes.

The second example involves the addition of hydrogen bromide to propene. In this reaction mechanism, the electrophile is a hydrogen ion, with the hydrogen attached to the halide ion. The halide ion is attracted to the pi bond of the propene molecule by breaking the double bond. As a result, the intermediate produced comprises a positively charged carbocation. This intermediate is highly reactive, and it forms a stable carbon-hydrogen bond. The reaction occurs in two steps; the first step is the addition of the hydrogens to one of the carbon atoms of the double bond, and the second step is the nucleophilic attack of the bromide ion to produce a halogenated alkane.

The inductive effect refers to the influence of nearby atoms or groups on the electron density of a molecule or ion. It is an important concept in organic chemistry that helps explain the stability of primary, secondary, and tertiary cations formed during electrophilic addition reactions.

In electrophilic addition reactions, a cationic intermediate forms when the electrophile attacks the double bond of the alkene. The stability of this intermediate governs the regioselectivity of the reaction, particularly in Markovnikov addition, where the electrophile adds to the carbon atom with fewer alkyl groups. The guiding principle is that the more stable the intermediate, the more likely it is to be formed.

In the case of primary cations, there are no alkyl groups present to 'donate' electrons to the positive charge, hence, the positive charge is on a carbon atom which cannot contribute to the positive charge through the +I effect. Consequently, primary cations are usually less stable than tertiary cations.

On the other hand, tertiary cations are more stable than primary cations. This is because the alkyl groups on carbon atoms adjacent to the positively charged carbon carbenium provide a +I 'push' (a greater number of electrons being less attracted to and thus shielding the positive charge) towards the center of the carbocation, which helps to stabilize its charge. This stabilizing effect is known as hyperconjugation, which is a type of inductive effect. Therefore, the greater the number of alkyl groups in a cationic intermediate, the more stable it is.

Markovnikov addition occurs in electrophilic addition reactions when the electrophile adds to the carbon atom of the alkene that already carries more hydrogen atoms. This regioselectivity can be explained by the inductive effect of alkyl groups, which stabilizes the positive charge on the tertiary carbon of the intermediate. The alkyl groups adjacent to the positively charged carbon carbenium provide a stronger +I 'push' in comparison to the hydrogen atoms, which stabilizes the positive charge on the tertiary carbon more effectively than the primary carbon, thus resulting in the Markovnikov product.

Halogenoalkanes

Halogenoalkanes, also known as alkyl halides, can be produced through various reactions involving different reagents and conditions.

The first way is through a free-radical substitution of alkanes using chlorine or bromine in the presence of ultraviolet light. This type of reaction can be exemplified by the reaction of ethane, where the hydrogen atoms in the molecule are substituted by chlorine or bromine atoms.

Another way is through electrophilic addition where an alkene reacts with a halogen like chlorine or bromine, or a hydrogen halide like hydrochloric acid or hydrobromic acid, at room temperature. This type of reaction results in the addition of the halogen or hydrogen halide to the double bond of the alkene, producing a halogenoalkane.

Substitution of an alcohol is another way to produce halogenoalkanes. This can be done through various reactions like the reaction with hydrogen halides or alkali metal halides like potassium bromide and an acid catalyst like sulfuric acid or phosphoric acid. Additionally, phosphorus trichloride with heat and phosphorus pentachloride or thionyl chloride can also be used. These reactions result in the replacement of the hydroxyl group of the alcohol with a halogen like chlorine or bromine.

Halogenoalkanes are a group of chemicals that contain at least one halogen atom bonded to a carbon atom of an alkane. These halogens include fluorine (F), chlorine (Cl), bromine (Br), and iodine (I). Depending on the position of the halogen atom, halogenoalkanes can be classified as primary, secondary, or tertiary.

A primary halogenoalkane is one in which the halogen atom is attached to a carbon atom that is only bonded to one other carbon atom, meaning that there are no other carbon atoms connected to the carbon atom with the halogen. For example, chloromethane (CH_3Cl) is a primary halogenoalkane because the chlorine atom is bonded to the carbon atom that is only bonded to one other carbon atom.

A secondary halogenoalkane is one in which the halogen atom is attached to a carbon atom that is bonded to two other carbon atoms. In other words, there are two other carbon atoms attached to the carbon atom with the halogen. For example, 2-chloropropane ($CH_3CHClCH_3$) is a secondary halogenoalkane because the chlorine atom is attached to the carbon atom that is bonded to two other carbon atoms.

A tertiary halogenoalkane is one in which the halogen atom is attached to a carbon atom that is bonded to three other carbon atoms. In other words, there are three other carbon atoms attached to the carbon atom with the halogen. For example, 2-chloro-2-methylpropane ($CH_3C(CH_3)_2Cl$) is a tertiary halogenoalkane because the chlorine atom is bonded to the carbon atom that is bonded to three other carbon atoms.

Each type of halogenoalkane has different properties and reactivities due to the different arrangements of carbon atoms and halogen atoms. This classification can be useful in predicting chemical reactions and understanding the behavior of halogenoalkanes in different conditions.

Nucleophilic substitution is a type of chemical reaction where a molecule or ion with an available pair of electrons, referred to as a nucleophile, substitutes a leaving ion which is a functional group that can easily separate. These types of reactions can result in the formation of new functional groups.

In the reaction with NaOH(aq) and heat, a halogenoalkane reacts with sodium hydroxide (NaOH) in water and heat. The halogen is displaced by the hydroxide (OH-) ion of the base and an alcohol is formed. The reaction proceeds through an SN2 (substitution nucleophilic bimolecular) pathway in which the hydroxide ion attacks the carbon atom of the halogenoalkane and displaces the halogen, resulting in the formation of an alcohol and a halogen ion as a leaving group.

When a halogenoalkane is reacted with KCN in ethanol and heat, a nitrile is produced. The cyanide ion (CN-) acts as a nucleophile to replace the halogen, again via an SN2 mechanism, resulting in the formation of nitriles, a functional group consisting of a carbon atom triple-bonded to a nitrogen atom (-C≡N).

Another example of a nucleophilic substitution reaction is the reaction with NH3 in ethanol, heated under pressure, which produces an amine. This reaction follows an SN2 mechanism, with the ammonia molecule acting as a nucleophile to replace the halogen. Amine is a functional group containing nitrogen atoms with hydrogen atoms attached to them.

Lastly, the reaction with aqueous silver nitrate in ethanol is a way to identify the halogen present in a halogenoalkane. This experiment is exemplified by bromoethane. The halogenoalkane is reacted with silver nitrate and ethanol; the silver halide is insoluble in water and forms a precipitate. Depending on the identity of the halide ion, the precipitate will have different colors or properties. In this case, the solution turns brownish-black, indicating the presence of bromine in the bromoethane molecule.

Elimination reactions are chemical reactions in which a molecule is broken down into smaller units by removing atoms, usually resulting in the formation of a double bond. One example of an elimination reaction is when bromoethane reacts with sodium hydroxide (NaOH) in ethanol and heat to produce an alkene. The reaction is carried out in a single step known as E2 (elimination bimolecular) mechanism, in which the hydroxide ion (OH-) acts as a base, accepting a proton from the carbon adjacent to the carbon with the leaving group, the bromine atom. This leads to the formation of a carbanion, which is an unstable intermediate with a negative charge on the carbon atom. The carbanion then attacks the hydrogen of the neighboring carbon and breaks its bond to the bromine atom. This reaction results in the formation of a double bond between the two carbons, and the bromine atom is lost, resulting in the formation of an alkene. Specifically, in the reaction of bromoethane, the bromine atom of the halogenoalkane is replaced by the OH- ion of the base to produce a negatively charged intermediate called the alkoxide ion. The alkoxide ion then loses a molecule of water in a single step, resulting in the formation of an alkene.

Nucleophilic substitution is a fundamental type of organic chemical reaction in which a nucleophile reacts with an alkyl halide, replacing the halogen atom with the nucleophile. There are two mechanisms that can occur in nucleophilic substitution reactions: SN1 and SN2. The SN1 mechanism is a two-step reaction involving a rate-determining step, which is the formation of a carbocation intermediate. In this process, the halogenoalkane molecule separates into two components, leaving a carbocation intermediate with a positively charged carbon atom. The positively charged carbon atom then rapidly reacts with the nucleophile to form a new product. In contrast, the SN2 mechanism is a one-step reaction that involves the simultaneous breaking of a bond between the nucleophile and the halogen atom. This mechanism involves the attack of the nucleophile on the carbon atom to which the halogen is attached, which results in the departure of the halogen and incorporation of the nucleophile. As an alkyl group is added to the carbon atom, the electron density of the carbon atom is affected by the alkyl group in a process known as inductive effect, which can have an impact on the nucleophilic substitution reaction. For example, when an alkyl group is present, it can stabilize the carbocation intermediate in the SN1 mechanism by providing electron density through the inductive effect. This stabilizing effect of the alkyl group increases as the number of alkyl groups at the carbon increases. Therefore, tertiary carbocations that possess three alkyl substituents are more stable than secondary or primary carbocations. On the other hand, the addition of alkyl groups in the SN2 mechanism can cause steric hindrance, affecting the access of the nucleophile to the carbon atom, which decreases the rate of the reaction. Therefore, tertiary halogenoalkanes are less reactive to nucleophilic substitution using SN2 mechanism, as they have a high degree of steric hindrance, while primary halogenoalkanes are more reactive.

Nucleophilic substitution reactions involving halogenoalkanes occur by two mechanisms: SN1 and SN2. The reaction mechanism that takes place is dependent on the structure of the halogenoalkane and the leaving group. The SN2 mechanism is commonly observed in the reaction of primary halogenoalkanes. In this mechanism, the nucleophile attacks the carbon atom to which the halogen is attached, and the halogen atom departs simultaneously. The reaction is a one-step process or occurs with simultaneous bond-breaking and forming. Since primary halogenoalkanes only have one carbon atom attached to the carbon bearing the halogen atom, there is minimal steric hindrance to nucleophile attack, and this mechanism occurs rapidly.

Tertiary halogenoalkanes, on the other hand, tend to undergo the SN1 mechanism. In this mechanism, the halogen atom departs first, leaving a carbocation intermediate. This carbocation intermediate is then rapidly attacked by a nucleophile to form a new product. Tertiary halogenoalkanes have three alkyl groups attached to the carbon atom bearing the halogen, leading to a greater degree of alkyl stabilization of the carbocation intermediate, thus favoring the SN1 mechanism.

Lastly, secondary halogenoalkanes have a mixture of the two mechanisms depending on their structure. This is because the reaction pathway will depend on the strength of the nucleophile, stability of the carbocation intermediate, as well as the steric hindrance of the intermediate. Secondary halogenoalkanes have two alkyl groups and are somewhere between primary halogenoalkanes and tertiary halogenoalkanes in terms of reactivity.

Halogenoalkanes are organic compounds that contain at least one halogen atom bonded to a carbon atom. The relative reactivity of halogenoalkanes depends on the strength of the carbon-halogen bond (C–X) and its susceptibility to cleavage by nucleophiles. The reaction of halogenoalkanes with aqueous silver nitrate is a common test used to determine the reactivity of these compounds.

In general, tertiary halogenoalkanes are the least reactive because they have the strongest carbon-halogen bond of all the halogenoalkanes due to the hyperconjugation and steric factors of the molecule. These factors contribute to the relative stability of the alpha position carbocation intermediate generated in either SN1 or E1 mechanism's of the halogenoalkane. Additionally, the tetravalent carbon atom with three alkyl groups increases the steric hindrance for the nucleophile during the reaction. Therefore, tertiary halogenoalkanes tend to react via SN1 mechanism or E1 mechanism under certain conditions, and these reactions are usually slower than those of primary and secondary halogenoalkanes.

Primary halogenoalkanes are the most reactive, and they tend to react via SN2 mechanism. SN2 mechanism occurs by backside attack of the nucleophile resulting in instantaneous formation of a bond with the carbon intermediate. Therefore, if the conditions are correct, the reaction will proceed very quickly. This is because the primary halogenoalkane has the highest electronegativity difference between the carbon and the halogen and hence the weakest carbon- halogen bond.

Secondary halogenoalkanes have some intermediate level of reactivity. They react slower than primary halogenoalkanes because of the two alkyl groups that stabilizes the halogen ion leaving group due to increased electron density of the carbon atoms surrounding it. They can either undergo SN1 reaction or SN2 reaction depending on the nucleophilic strength, steric factors, and reaction conditions.

Alcohols

There are different methods you can use to produce alcohol from different starting materials.

The first method is called electrophilic addition of steam to an alkene. In this method, you add steam to an alkene using a H3PO4 catalyst. The addition of steam to the double bond of the alkene creates an unstable intermediate. The H3PO4 catalyst helps to stabilize this intermediate so that water can be added to the molecule to form an alcohol.

The second method involves the reaction of alkenes with cold dilute acidified potassium manganate(VII). This reaction forms a diol, which is a compound that contains two hydroxyl (-OH) groups. The potassium manganate(VII) acts as an oxidizing agent, which helps to add oxygen atoms to the double bond of the alkene. The acid catalyzes the reaction and helps to increase the rate of the reaction.

The third method used to produce alcohols is the substitution of a halogenoalkane using NaOH(aq) and heat. In this method, the halogenoalkane is reacted with sodium hydroxide and heated. The NaOH(aq) acts as a nucleophile and attacks the carbon-halide bond of the halogenoalkane. This substitution reaction results in an alcohol and a salt (NaX).

One of the ways to produce an alcohol is through the reduction of aldehydes or ketones using either NaBH4 or LiAl H4. In this method, we use a reducing agent to add hydrogen atoms to the oxygen atoms present in the aldehyde or ketone, which causes it to become an alcohol. Sodium borohydride (NaBH4) and lithium aluminum hydride (LiAl H4) donate hydride, or H-, ions to the carbonyl group of the aldehyde or ketone which reduces it to an alcohol.

Next, we have the reduction of a carboxylic acid using LiAl H4. This reduction reaction involves the reaction of a carboxylic acid with a strong reducing agent like LiAl H4. The LiAl H4 reduces the COOH group to a primary alcohol by adding hydrogen to the oxygen atom in the carbonyl group. This reaction is not possible with NaBH4.

Lastly, we have hydrolysis of an ester using dilute acid or dilute alkali and heat. This reaction involves breaking apart an ester molecule into its component parts by adding water. Adding an acid catalyst or base catalyst to the reaction mix helps promote the reaction. This method is used to make primary, secondary and tertiary alcohols. Dilute acid or alkali causes the ester to split into an alcohol and a carboxylic acid (alkali hydrolysis) or an alcohol and a carboxylic acid ester(acid hydrolysis).

The different reactions that can take place when substances react with oxygen, halogenoalkanes, and sodium respectively.

First, the reaction with oxygen, also known as combustion. Combustion is a reaction where a substance reacts with oxygen gas (O2) to produce carbon dioxide and water. During this reaction, the molecule will release energy in the form of heat and/or light. This reaction is highly exothermic and has many practical applications such as burning fossil fuels and creating energy.

Now, let's move on to substitution reactions with halogenoalkanes. There are various types of substitution reactions that can be used, all of which replace a halogen atom in the halogenoalkane with a nucleophile. For instance, halogenoalkanes can be reacted with hydrogen halides (such as HX or KBr with H2SO4 or H3PO4), which causes the halogen atom to be substituted by the nucleophile. Additionally, halogenoalkanes can be reacted with phosphorous trichloride (PCl3) and heat, which forms a chlorinated product, or they can be reacted with phosphorous pentachloride (PCl5) or thionyl chloride (SOCl2) which result in a halide substitution.

Lastly, the reaction with sodium. When sodium reacts with water, heat and hydrogen gas (H2) are produced. Sodium atoms are highly reactive and easily lose an electron to form positively charged cations. In the reaction with water, sodium loses an electron to form a sodium ion (Na+), and hydrogen ions in water (H+) react with hydroxide ions (OH-) to make water while releasing H2 gas.

To describe the oxidation reactions of organic compounds using acidified K2Cr2O7 or acidified KMnO4. When acidified K2Cr2O7 or KMnO4 is used to oxidize organic compounds, primary alcohols are oxidized to aldehydes and further oxidation of aldehydes give carboxylic acids. Secondary alcohols are oxidized to ketones while tertiary alcohols cannot be oxidized. In both cases, the oxidizing agent acts as a source of oxygen, which is added to the substrate molecule. The addition of oxygen causes the substrate molecule to be oxidized, while the chromium or manganese ions in the oxidizing agent balance the redox reaction.

When trying to oxidize a primary alcohol into an aldehyde, we distill the reaction mixture immediately to avoid overoxidation of the aldehyde to a carboxylic acid. The resulting aldehyde can be further oxidized to a carboxylic acid if the reaction is allowed to continue for a longer period. In the case of secondary alcohols and tertiary alcohols, they do not undergo overoxidation and the elimination of a carbon molecule results in the production of a ketone.

The oxidation reaction can be carried out using either acidified K2Cr2O7 or acidified KMnO4. In both cases, the acid used is typically dilute sulfuric acid (H2SO4), which provides the acidic environment necessary for the oxidation to occur. Chromic acid (H2CrO4) and permanganate ion (MnO4-) present in K2Cr2O7 and KMnO4 respectively act as oxidizing agents, supplying oxygen to the substrate molecules. The overall oxidation reaction releases heat, and care must be taken as the excess heat generated can build up pressure in a semi closed test tube, which can led to an explosion.

Dehydration is a reaction that involves the removal of a molecule of water from the starting material to form an alkene. This process is typically carried out using a heated catalyst. The most common catalysts used in dehydration reactions are Al2O3 or a concentrated acid.

In the presence of a heated catalyst such as Al2O3, the starting material is dehydrated to form an alkene. The Al2O3 acts as a dehydrating agent, which removes the water molecule from the starting material, creating a double bond between two adjacent carbon atoms. As the reaction is carried out under high temperature conditions, the product may have thermodynamic can isomers due to the reversible nature of the reaction.

Alternatively, concentrated acid can be used as a catalyst in the process of dehydration to an alkene. An acid like sulfuric acid (H2SO4) can protonate the alcohol group to give good leaving group, a molecule of water. Protonation of leaving group generates a carbocation intermediate which then forms a double bond between the nearby carbon atoms. Sulfuric acid is a strong acid and better suited for tertiary alcohols but lesser protonating agents like phosphoric acid might work better for primary alcohols.

Dehydration has many industrial applications, including the production of polymers. The reaction can be used to create alkenes for the production of plastics, synthetic fibers and solvents. In a laboratory setting, dehydration is also used to synthesize alkenes, which can then be used for further reactions.

The formation of esters is a common reaction in organic chemistry that involves the reaction of a carboxylic acid and an alcohol in the presence of a strong acid catalyst, such as H2SO4 or H3PO4. As an example, we can consider the reaction of ethanol with a carboxylic acid to form an ester.

In this reaction, the hydroxyl (-OH) group of the carboxylic acid reacts with the hydroxyl (-OH) group of the alcohol. The reaction is catalyzed by a strong acid, which acts as a proton donor and facilitates the formation of a

---

good leaving group, such as a molecule of water. When the two hydroxyl groups come together, they eliminate water to form an ester linkage.

For example, ethanoic acid reacts with ethanol through esterification in the presence of a catalyst, H2SO4, to form ethyl ethanoate. The synthesis is exothermic, meaning the reaction creates heat. The reaction takes place at ambient temperature and can be an equilibrium reaction.

The combination of a carboxylic acid with an alcohol yields a wide variety of esters. The reaction conditions and catalyst used can affect the reaction rate and the yield of the ester product. Additionally, other factors such as the nature, size, and shape of the substituents present in the alcohol and carboxylic acid can also affect the reaction rate and selectivity of the ester product.

In practice, esterification is used commercially to produce flavorings, fragrances, solvents and plasticizers. It also has many other applications in everyday products like perfumes, nail polish removers & adhesives.

Different types of alcohols can be identified based on the number of alkyl groups they contain attached on the same carbon atom that the hydroxyl (-OH) group is also attached. Primary, secondary, and tertiary alcohols are classifications based on the number of alkyl groups present in the molecule.

Let me explain each type of alcohol with examples.

Primary alcohols are those where the carbon atom that bears the -OH group is only attached to one other carbon. For example, ethanol is a primary alcohol since the -OH group is attached to a carbon atom that is directly bonded to only one other carbon atom. Methanol and propanol are other examples of primary alcohols. Primary alcohols can also have more than one alcohol group, such as diols (glycols) like ethylene glycol and 1,3-propanediol.

Secondary alcohols are those that have a carbon atom that bears the -OH group that is bonded to two other carbon atoms. If we consider the molecule 2-propanol, we can see that the -OH group is attached to a carbon atom that is bonded to two other carbon atoms. Other examples of secondary alcohols include 2-butanol, isobutanol, and cyclohexanol.

Tertiary alcohols are those in which the carbon atom attached to the hydroxyl group is bonded to three other carbon atoms. For example, 2-methyl-2-propanol is a tertiary alcohol as the carbon that bears the -OH group is attached to three other carbon atoms. Other examples of tertiary alcohols include tert-butanol and 2,2-dimethyl-1-propanol.

It is also important to note that alcohols can have more than one -OH group attached to different carbon atoms. For example, ethylene glycol is a diol with two hydroxyl groups attached to different carbon atoms. Propanetriol (glycerol) is a triol with three hydroxyl groups.

One of the unique and defining features of reactions is the way they interact with other substances, which can often be observed through changes in color or other physical properties. One example of this is mild oxidation with acidified K2Cr2O7, which causes a distinct transition in color from orange to green. This seemingly simple reaction can be used to identify specific chemical compounds and assess their overall reactivity, making it a valuable tool for researchers and students alike. By understanding these characteristic distinguishing reactions and how they relate to various chemical processes, we can gain a deeper understanding of the underlying principles governing chemical reactions and unlock new opportunities for scientific discovery and innovation.

Chemical reactions can be used to identify specific functional groups within organic compounds, such as the presence of a CH3CH(OH)- group in an alcohol with the molecular formula CH3CH(OH)-R. One way to do this is by reacting the alcohol with an alkaline I2(aq) solution, leading to the formation of a yellow precipitate of tri-iodomethane and an ion known as RCO2-. This reaction occurs due to the ability of the CH3CH(OH)- group to act as a reducing agent, which forms a complex with the I2(aq) solution, leading to the formation of the yellow precipitate.

The reaction between the alcohol and alkaline I2(aq) solution can be attributed to the oxidation of the CH3CH(OH)- group, which releases electrons that are then used to reduce the I2(aq) solution. As a result, the yellow precipitate serves as a clear indicator of the presence of a CH3CH(OH)- group in the original alcohol molecule, highlighting the power and versatility of chemical reactions in identifying specific functional groups within complex organic compounds.

When discussing the acidity of alcohols compared to water, one must consider the unique chemical properties of each substance. In general, alcohols are considered less acidic than water due to the presence of the hydroxyl functional group (-OH) in their molecular structure, which can weaken the overall acidity of the molecule. This is due to the fact that the oxygen atom in the hydroxyl group is often more electronegative than the carbon atoms in the rest of the molecule, leading to a partial negative charge that can partially neutralize the acidity of the hydrogen atoms.

One way to measure the relative acidity of alcohols and water is to look at their respective pKa values, which are indicators of the strength of the acid. In general, the lower the pKa value of a compound, the stronger the acid, with values closer to zero indicating the greatest acidity. In the case of water, the pKa value is around 15.7, making it a weak acid. On the other hand, the pKa value of alcohols is generally around 16-18, indicating that they are even weaker acids than water.

However, it is important to note that the overall acidity of a given alcohol can vary based on the specific molecular structure, with factors such as branching and the presence of electron-withdrawing substituents potentially impacting the acidity. Additionally, certain types of alcohols, such as phenols, can exhibit greater acidity due to the stabilizing influence of resonance in the phenoxide ion.

understanding the reactions that produce aldehydes and ketones. In simple terms, aldehydes and ketones are organic compounds that consist of a carbonyl group (-C=O) attached to a carbon atom. The production of these compounds involves the oxidation of alcohols, which is a common reaction in organic chemistry. There are two types of alcohols: primary and secondary. Primary alcohols can be oxidized to produce aldehydes, while secondary alcohols can be oxidized to produce ketones. This reaction can be carried out using two common oxidizing agents, namely potassium dichromate (K2Cr2O7) and potassium permanganate (KMnO4). To produce aldehydes, a primary alcohol is mixed with acidified potassium dichromate or potassium permanganate, and the mixture is then heated and distilled. During the distillation process, the produced aldehyde is collected in a receiving flask. On the other hand, to produce ketones, a secondary alcohol is mixed with acidified potassium dichromate or potassium permanganate, and the mixture is heated and distilled. The produced ketone is similarly collected in a receiving flask. It's important to note that the conditions of the reaction, such as acidity and temperature, play a crucial role in the success of the reaction. Additionally, there are other methods of producing aldehydes and ketones through various other reactions.

To understand the reductions of aldehydes and ketones, as well as the reaction of these compounds with hydrogen cyanide (HCN). The reduction of aldehydes and ketones is an important reaction in organic chemistry. This reaction involves the use of reducing agents like sodium borohydride (NaBH4) or lithium aluminum hydride (LiAl H4) to convert aldehydes and ketones into alcohols. When NaBH4 or LiAl H4 is added to a solution of aldehyde or ketone, they reduce the carbonyl group (-C=O) to form an alcohol (-C-OH) group. The reduction reaction is selective and works specifically on the carbonyl group. This reaction can be carried out under mild conditions, without the need for strongly acidic or basic conditions. On the other hand, when aldehydes or ketones react with hydrogen cyanide (HCN) in the presence of a catalyst like potassium cyanide (KCN) and heat, hydroxynitriles are produced. Hydroxynitriles are organic molecules that consist of a hydroxyl (-OH) group and a cyano (-CN) group attached to the same carbon atom. For example, when ethanal and propanone react with HCN, they form 2-hydroxypropanenitrile and 2-hydroxy-2-methylpropanenitrile, respectively. This reaction is useful in organic synthesis to produce a range of complex molecules containing both hydroxyl and cyano groups in the same molecule. It's important to note that the use of HCN and its derivatives is highly toxic and requires proper safety measures to be taken.

The nucleophilic addition reaction of hydrogen cyanide (HCN) with aldehydes and ketones is an important reaction in organic chemistry. In general, a nucleophilic addition reaction is a type of chemical reaction in which a nucleophile (an electron-rich compound or element) attacks an electrophile (an electron-deficient compound or element) and forms a new chemical bond. In the case of the reaction of HCN with aldehydes and ketones, the nucleophile is HCN, and the electrophile is the carbonyl group (-C=O) of the aldehyde or ketone.

The reaction mechanism of HCN with aldehydes and ketones involves two steps: the addition of the nucleophile and the subsequent loss of a leaving group. First, the HCN molecule attacks the polarized carbonyl group through nucleophilic addition. This forms a new intermediate species known as a cyanohydrin, which is a compound that has a hydroxyl (-OH) group attached to a carbon atom and a cyano (-CN) group attached to another carbon atom. During this step, the carbonyl group becomes protonated, and an oxyanion is formed. Next, the leaving group (-OH) departs, forming a new bond between the C atom and the N atom. This step results in the formation of a stable product, a hydroxynitrile (-C(OH)=CN).

Overall, the reaction mechanism of HCN with aldehydes and ketones involves the nucleophilic attack of the HCN molecule on the carbonyl group of the aldehyde or ketone. This forms a highly unstable intermediate, which then leads to the formation of a stable hydroxynitrile product through the loss of a leaving group. The reaction mechanism of HCN with aldehydes and ketones is a significant reaction in organic chemistry due to its application in synthesis and regulation of biological processes.

2,4-dinitrophenylhydrazine, or simply 2,4-DNPH reagent, is a commonly used chemical compound that can be utilized to detect the presence of carbonyl compounds. When a carbonyl compound, such as a ketone or an aldehyde, reacts with 2,4-DNPH reagent in the presence of an acid catalyst, it forms a distinctive orange-red precipitate. This precipitate can be easily observed and used as a key indicator for the presence of carbonyl compounds in a given sample. The beauty of this test lies in its versatility and simplicity. It can be used to identify a wide range of carbonyl compounds, including those that are found in natural products, pesticides, pharmaceuticals, and so on. Additionally, the test is easy to perform, requiring only a small quantity of sample and a minimal amount of reagent.

If you have an unknown carbonyl compound and you want to determine whether it is an aldehyde or a ketone, there are several simple tests that you can perform to obtain these results. The two most common tests used to determine the nature of the carbonyl compound are Fehling's and Tollens' reagents.

Fehling's tests involve the use of blue copper(II) ions, which are reduced to red copper(I) ions in the presence of aldehydes and ketones. When Fehling's reagent is added to an unknown carbonyl compound and heated, a red precipitate, which is copper(I) oxide, is formed. In this test, aldehydes reduce Fehling's reagent more effectively than ketones due to their ability to be oxidized to carboxylic acids. However, ketones do not reduce Fehling's reagent, so if no red precipitate forms, the unknown carbonyl compound is probably ketone.

For Tollens' test, a silver mirror is formed when aldehydes are oxidized by Tollens' reagent. In this test, the unknown carbonyl compound is mixed with Tollens' reagent, which is basically a solution of ammonia, silver nitrate, and sodium hydroxide. When heated, a silver mirror forms on the inside of the test tube if the unknown carbonyl compound is an aldehyde. This mirrors forms because the aldehydes are easily oxidized by Tollens' reagent to carboxylic acids and metallic silver is deposited on the test tube's internal surface. Ketones do not give the same silver mirror result, so if the silver mirror doesn't form, it is likely that the unknown carbonyl compound is a ketone.

To deduce the presence of a CH3CO- group in an aldehyde or ketone, we can use a simple reaction with alkaline I2(aq) to form a yellow precipitate of tri-iodomethane and an ion, RCO2-. This is also known as the iodoform test.

When a carbonyl compound containing a CH3CO- group is treated with alkaline I2(aq), a yellow precipitate of tri-iodomethane is formed. This indicates the presence of a CH3CO- group in the carbonyl compound. This is because the CH3CO- group is oxidized by iodine and hydroxide ions to form carboxylate ion (RCO2-) and tri-iodomethane. The tri-iodomethane precipitates out of the solution, indicating the presence of the CH3CO- group.

It is important to note that not all carbonyl compounds will give a positive response for this test, as the presence of other functional groups could interfere with the reaction. Substituent groups, such as methoxy or nitro groups, may inhibit the reaction as well.

Carboxylic acids

Let me help you recall the reactions by which carboxylic acids can be produced!

Carboxylic acids are organic compounds that contain a carboxyl group (-COOH) bonded to a hydrocarbon chain. There are several ways in which carboxylic acids can be produced.

One way is through the oxidation of primary alcohols and aldehydes using acidified $K_2Cr_2O_7$ or acidified $KMnO_4$, while refluxing. In this reaction, the aldehyde or primary alcohol is oxidized to a carboxylic acid via the intermediate formation of a carboxylic acid aldehyde. The reaction conditions involve the use of a strong oxidizing agent such as acidified potassium dichromate or acidified potassium permanganate, which is necessary to generate the oxidizing agent CrO_3 or MnO_2. The carboxylic acid formed will have one more carbon than the original aldehyde or alcohol.

Another reaction that can produce carboxylic acids is the hydrolysis of nitriles with dilute acid or dilute alkali followed by acidification. In this reaction, a nitrile undergoes hydrolysis in the presence of an acid or alkali to form an amide. The amide is then hydrolyzed further to produce its corresponding carboxylic acid. The reaction is characterized by a water molecule hydrating to the nitrile and forming a carboxylic acid via the intermediate of an amide.

Finally, another route for carboxylic acid formation involves the hydrolysis of esters with dilute acid or dilute alkali and heat followed by acidification. This reaction involves an ester reacting with a strong acid or a strong base in the presence of heat to produce an alcohol and its corresponding acid or the carboxylic acid. This reaction is an example of nucleophilic acyl substitution, where a nucleophile (the hydroxide or hydronium ion) adds to the carbonyl carbon of the ester, forming a tetrahedral intermediate which eventually forms a carboxylic acid.

(a) The reaction of a reactive metal with an acid produces a salt and hydrogen gas. This is a redox reaction where the reactive metal acts as a reducing agent, causing the acid to partially or fully reduce.

For example, when magnesium reacts with hydrochloric acid, the magnesium atoms lose electrons, reducing the hydrogen ions (H+) from the acid to hydrogen gas (H_2), while the magnesium ions (Mg^{2+}) join with the chloride ions (Cl-) from the acid to form magnesium chloride ($MgCl_2$). The overall equation for this reaction is

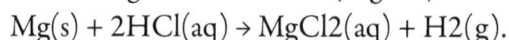

$$Mg(s) + 2HCl(aq) \rightarrow MgCl_2(aq) + H_2(g).$$

This reaction can also be used to identify an unknown metal by observing the hydrogen gas that is produced. If the unknown metal reacts with acid to produce hydrogen gas, it is most likely a reactive metal such as magnesium, zinc, or aluminum.

(b) The reaction of an acid with an alkali produces a salt and water. This type of reaction is called a neutralization reaction because the acid and alkali cancel out each other's effects, resulting in a neutral product.

For example, if hydrochloric acid (HCl) is mixed with sodium hydroxide (NaOH), the hydrogen ions (H+) from the acid combine with the hydroxide ions (OH-) from the alkali to form water (H_2O), while the sodium ions (Na+) from the alkali join with the chloride ions (Cl-) from the acid to form sodium chloride (NaCl). The overall equation for this reaction is $HCl(aq) + NaOH(aq) \rightarrow NaCl(aq) + H_2O(l)$.

Neutralization reactions are very useful, as they can be used to prepare salts and are widely used in industries where quality control is vital, like the food industry and pharmaceutical industry.

(c) The acid-base reaction between an acid and a carbonate produces a salt, water, and carbon dioxide. This type of reaction occurs between a metal carbonate, such as calcium carbonate ($CaCO_3$), and an acid like sulfuric acid (H_2SO_4). The carbonate reacts with the acid, resulting in the formation of a salt, such as calcium sulfate ($CaSO_4$), water, and carbon dioxide (CO_2). The equation for this reaction is:

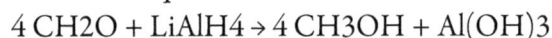

$$H_2SO_4 + CaCO_3 \rightarrow CaSO_4 + H_2O + CO_2$$

Carbon dioxide is a gas that is released during the reaction and could be removed via ventilation or selectively isolated or transformed into other molecules.

(d) Esterification is the process by which an ester is formed from an alcohol and a carboxylic acid. It is typically catalyzed by an acid such as concentrated sulfuric acid (H_2SO_4) or hydrochloric acid (HCl), which acts as a dehydrating agent to remove water, promoting ester formation. During the reaction, the hydroxyl group (-OH) of the carboxylic acid and the hydrogen atom (-H) of the alcohol combine, and a water molecule is removed, resulting in the formation of the ester and another water molecule. An example of this esterification reaction is the reaction between ethanol and acetic acid, which produces ethyl acetate and water. The equation for this reaction is:

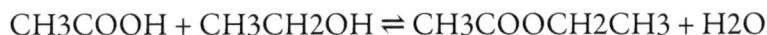

$$CH_3COOH + CH_3CH_2OH \rightleftharpoons CH_3COOCH_2CH_3 + H_2O$$

(e) The reduction of a carbonyl compound, such as an aldehyde or ketone, to a primary alcohol can be accomplished using a reducing agent such as lithium aluminum hydride ($LiAlH_4$). During this reaction, the carbonyl compound reacts with $LiAlH_4$ in anhydrous conditions, resulting in the formation of a primary alcohol. For example, formaldehyde, CH_2O, is reduced to methanol, CH_3OH.

The equation for this reaction is:

$$4\,CH_2O + LiAlH_4 \rightarrow 4\,CH_3OH + Al(OH)_3$$

The $LiAlH_4$ acts as a nucleophile, attacking the carbonyl carbon of the aldehyde or ketone, forming an alkoxide ion that subsequently reacts with H^+ to produce the primary alcohol and aluminum hydroxide as a byproduct.

Esters

Let me recall the reaction by which esters can be produced.

Esters can be produced through a condensation reaction between an alcohol and a carboxylic acid, which requires a catalyst such as concentrated sulfuric acid (H2SO4). During the reaction, the acidic proton in the carboxyl group (-COOH) of the carboxylic acid reacts with the hydroxyl group (-OH) of the alcohol, forming a molecule of water and an ester.

For example, the reaction between ethanol (CH3CH2OH) and acetic acid (CH3COOH) produces ethyl acetate (CH3COOCH2CH3) and water (H2O). This reaction can be represented by the following equation:

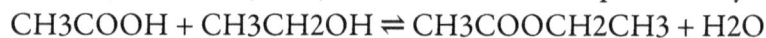

$$CH3COOH + CH3CH2OH \rightleftharpoons CH3COOCH2CH3 + H2O$$

The concentrated sulfuric acid (H2SO4) acts as a catalyst and facilitates the esterification reaction by helping to remove the water that is produced, in order to shift the equilibrium of the reaction towards the formation of the ester. In essence, the H2SO4 helps protonate the carbonyl oxygen in the carboxylic acid, making the carbon more electrophilic and able to attack the nucleophile, which is the alcohol.

Esters are an important group of organic compounds that are commonly used as fragrances in the cosmetics industry, in flavorings for food, and in the production of plastics, solvents and resins. Understanding the synthesis of esters is particularly useful as it allows scientists to manufacture these compounds on a large scale for a multitude of applications.

Esters are commonly found in nature and in human-made substances, such as fragrances, flavors, and plastics. When esters react with either dilute acid or dilute alkali and heat, a hydrolysis reaction occurs.

In the presence of dilute acid, the hydrolysis of esters involves the cleavage of the ester linkage by the addition of a water molecule. This reaction is catalyzed by the acid, which protonates the carbonyl oxygen of the ester and increases the electrophilic nature of the carbonyl carbon. The water molecule then nucleophilically attacks the carbonyl carbon, leading to the formation of a tetrahedral intermediate. The tetrahedral intermediate then collapses and releases an alcohol molecule and carboxylic acid molecule.

On the other hand, in the presence of dilute alkali and heat, the hydrolysis of esters involves the formation of a carboxylate ion and an alcohol molecule. This reaction is catalyzed by the hydroxide ion, which acts as a nucleophile and attacks the carbonyl carbon of the ester. This results in the formation of an unstable tetrahedral intermediate, which then collapses and releases an alcohol molecule and a carboxylate ion.

It is important to note that the reaction conditions for hydrolysis of esters differ depending on whether dilute acid or dilute alkali is used. Dilute acid is used in the presence of water and heat to create a more acidic environment, thereby facilitating the protonation of the carbonyl oxygen. On the other hand, dilute alkali is used in the presence of water and heat to create a more basic environment, which allows for the deprotonation of the carbonyl oxygen.

Primary Amines

Amines are organic compounds that contain a nitrogen atom with a lone pair of electrons. They are commonly used in the production of pharmaceuticals, polymers, and dyes. There are various methods to synthesize amines, one of which is the reaction of a halogenoalkane with ammonia in ethanol under pressure.

In this reaction, a halogenoalkane is reacted with a solution of ammonia in ethanol under pressure at a high temperature. The ethanol acts as a solvent and also helps to promote the formation of the amine product. The halogenoalkane is first converted to a primary or secondary amine due to the nucleophilic substitution of the halogen with the nitrogen atom of the ammonia.

The reaction occurs through a stepwise mechanism, where the halogenoalkane reacts with the ammonia to form an ammonium halide intermediate. This intermediate then undergoes elimination of the halide ion and forms the corresponding primary or secondary amine. The primary amine is formed when the halogenoalkane is reacted with one equivalent of ammonia. Secondary amine is formed when the halogenoalkane is reacted with two equivalents of ammonia.

It is important to note that excess ammonia should be avoided so that it doesn't react with the amine product formed to form a N-substituted amide instead of a primary or secondary amine.

Nitriles and Hydroxynitriles

Nitriles are versatile organic compounds that contain a cyano group (-CN) which is an important functional group that finds applications in various fields including pharmaceuticals, agrochemicals, and polymer production. There are numerous methods to synthesize nitriles, with one of them being the reaction of a halogenoalkane with KCN in ethanol and heating.

The synthesis of nitriles from halogenoalkanes is a multi-step reaction process. In this reaction, a halogenoalkane is reacted with KCN in ethanol under heat to form a nitrile. The reaction mechanism primarily involves the substitution of the halogen atom by the cyanide ion. This can happen via an S_N2 pathway, where the nucleophile (cyanide ion) attacks the carbon that the halogen was bonded to, displacing the halide ion and forming the nitrile group. The reaction takes place in the presence of KCN, which acts as the source of the cyano group, and ethanol as the solvent.

However, the reaction requires heat to proceed efficiently and rapidly. The mechanism of the reaction proceeds via an intermediate that involves the loss of the halide ion from the alkyl halide, followed by the formation of a complex between the cyano ion and the carbocation intermediate. The final product, which is the nitrile, is formed by the hydrolysis of the intermediate complex in the presence of an acid.

Furthermore, it is important to note that the reaction requires careful attention, as cyanides can be toxic and highly reactive. Therefore, this reaction should be carried out in a well-ventilated area with proper safety precautions to minimize exposure to toxic fumes.

Hydroxynitriles, also known as cyanohydrins, are important intermediates in the production of numerous organic compounds such as amino acids, pharmaceuticals, and agrochemicals. There are several methods to synthesize hydroxynitriles, and one of them is the reaction of aldehydes and ketones with hydrogen cyanide (HCN) in the presence of KCN as a catalyst and heating.

The synthesis of hydroxynitriles is an important process that involves the nucleophilic addition reaction of cyanide ion on carbonyl carbon. In this reaction, aldehydes and ketones are made to react with hydrogen cyanide and a catalytic amount of potassium cyanide (KCN) under heating conditions. In the presence of KCN, HCN is activated and readily forms the cyanide ion.

The reaction mechanism is initiated by the attack of the electrophilic carbonyl carbon by the nucleophilic cyanide ion. This results in the formation of a highly reactive intermediate, which then undergoes a proton shift from the hydroxyl group to give the final product, which is a hydroxynitrile. The hydroxynitriles have a characteristic -OH and -CN group on the same carbon atom, which has versatile applications in various fields. The overall reaction requires high temperatures and its optimization is important, as hydrogen cyanide is a toxic substance that requires careful handling. Therefore, it is imperative to carry out the reaction in a well-ventilated area with proper safety measures.

Nitriles are important organic compounds that contain a functional group composed of a carbon triple-bonded to a nitrogen atom (-CN). They are widely used in many chemical, pharmaceutical, and biomedical industries.

When nitriles are hydrolyzed with a dilute acid or dilute alkali followed by acidification, they decompose to form carboxylic acid.

The hydrolysis of nitriles proceeds in two steps. In the first step, the nitrile group (-CN) is hydrolyzed to produce an amide that is subsequently converted to a carboxylic acid in the second step. The first step involves the nucleophilic addition of either a dilute acid or alkali to the nitrile carbon, which results in the formation of a negatively charged imino group. The imino group then reacts with either H^+ or OH^- to give the amide intermediate.

In the second step, the amide intermediate is hydrolyzed to form the corresponding carboxylic acid. The amide intermediate will react with water to form a carboxylic acid and ammonia. In the case of hydrolysis of nitriles in the presence of dilute alkali followed by acidification, the basic solution containing the carboxylate ion is made acidic to give the corresponding carboxylic acid. In the case of hydrolysis of nitriles in the presence of dilute acid, the carboxylic acid is obtained as such.

It is important to note that the reaction conditions vary depending on whether dilute acid or dilute alkali is used. In the case of dilute acid, a weak acid is added to the nitrile along with water. This results in the protonation of the intermediate amide, which is then hydrolyzed to give the carboxylic acid. In the case of dilute alkali, a weak base is added to the nitrile along with water, creating a basic environment where the intermediate amide ion reacts with water to form an anionic carboxylate ion. Acidification then results in the formation of the carboxylic acid.

Polymerization by Addition

Addition polymerization is a process in which monomers are joined together to form a polymer by the addition of unsaturated monomers. This polymerization process is widely used to produce various types of synthetic polymers. Poly(ethene) (commonly known as polyethylene) and Poly(chloroethene) (commonly known as PVC) are two examples of polymers produced through addition polymerization.

In the process of polymerization, monomers containing unsaturated carbons are joined together to form a long chain through the addition of polymerization initiators. These initiators react with the monomers and form reactive intermediate species known as free radicals.

Polyethylene is produced by the polymerization of ethylene molecules. Ethylene molecules contain double bonds which are activated by the addition of a suitable polymerization initiator, such as peroxides. The double bond is then broken to form two reactive intermediates that combine with other ethylene molecules to initiate the polymerization process. The resulting polymer chain contains only carbon-carbon single bonds.

PVC, on the other hand, is produced from vinyl chloride molecules. Polymerization of the vinyl chloride molecule takes place by addition of initiators, leading to the formation of a free radical. This free radical then attacks the double bond between the chloride and carbon atoms in the vinyl chloride molecule. The attack leads to the opening of the double bond, yielding an intermediate molecule with single bonds. This intermediate molecule can then react with another vinyl chloride molecule to produce a chain of poly(chloroethene) polymer.

The property of the final polymer product is determined by the addition of other monomers, addition of comonomers, and variation in the polymerization process. The addition polymerization process is a highly effective method of producing a wide range of products, including plastics, elastomers, and fibers. Furthermore, adjusting the type of initiators, reaction conditions, and addition of comonomers offers the possibility of producing polymers with unique chemical and physical properties.

Addition polymerization involves the formation of a polymer chain through the repeated addition of monomers containing unsaturated double bonds. To deduce the repeat unit of an addition polymer obtained from a given monomer, it is important to consider the structure of the original monomer, the type of unsaturation present in the monomer, and the reaction conditions under which the polymerization process occurred.

The repeat unit of an addition polymer corresponds to the structure of the monomer after it has undergone addition polymerization. For example, in the case of polyethylene, the repeat unit is represented by -CH2- because ethylene contains a double bond that is broken during polymerization to form a polymer chain composed of repeating units of -CH2-. The monomer used in the process of polymerization holds the monomers together.

In the case of PVC, the repeat unit is represented by -CH2-CH(Cl)-. This is because vinyl chloride contains a double bond between the carbon and chlorine atoms, which is broken during the polymerization process. The carbon atom forms a single bond with another carbon atom of another molecule, while the chlorine atom is lost. Hence, the resulting polymer contains repeating units of -CH2-CH(Cl)-.

The repeat unit of an addition polymer obtained from a monomer can be deduced by analyzing the structure of the monomer and determining which bond(s) between atoms have been broken and which new bond(s) have been formed under the given reaction conditions. In general, the repeat unit of an addition polymer is a fragment of the original monomer repeating unit, and it is represented in parentheses.

in understanding the monomer(s) present in a given section of an addition polymer molecule.

Addition polymerization is a process that involves converting a monomer(s) into a polymer by adding monomers in a chain reaction. A monomer is a small molecule that can bond with other monomers to form a polymer. These monomers can be identical or different in structure and have reactive ends that can join together to form a long chain called a polymer.

The monomers present in a given section of an addition polymer molecule can vary depending on the specific polymerization process and the desired properties of the resulting polymer. However, common monomers used in addition polymerization include ethylene, propylene, styrene, and vinyl chloride.

For example, in the production of polyethylene, the monomer ethylene undergoes addition polymerization, where the reactive double bond in ethylene is broken, and the two carbon atoms bond with two other ethylene monomers, forming long chains of polyethylene. Similarly, in the production of polypropylene, the monomer propylene undergoes addition polymerization, where three or more propylene monomers link together, forming a long chain of polypropylene.

Understanding the monomers present and the polymerization process allows for tailoring the properties of the polymer according to the intended application. By controlling the reaction conditions and the type of monomers used, different properties in the polymer, such as strength, chemical resistance, or flexibility, can be achieved.

I'll be glad to help you understand the difficulty in the disposal of poly(alkene)s and the negative impacts they can have on the environment and human health.

Poly(alkene)s, or commonly known as polyethylene and polypropylene, are widely used in the production of various everyday items, including plastic bags, food packaging, and household products. However, these polymers are non-biodegradable, which means that they cannot be broken down and decomposed by natural processes. Consequently, when disposed of, they could persist for hundreds of years in the environment, leading to accumulation and pollution.

One of the most significant challenges with the disposal of poly(alkene)s is their non-biodegradability, as this makes them difficult to remove from the environment and leads to the accumulation in landfills and water systems. This accumulation can have detrimental effects on wildlife, as they can ingest or become entangled in the plastics, leading to suffocation or starvation.

Moreover, the incineration of poly(alkene)s produces toxic and harmful combustion products that can have long-lasting negative effects on human health and the environment. When burned, these polymers release toxic chemicals such as carbon monoxide, nitrogen oxides, and toxic halogenated gases, which contribute to air pollution, increasing the risk of respiratory diseases.

To tackle the problem of poly(alkene) disposal, it is essential to consider alternatives to these non-biodegradable plastics, such as bio-based and biodegradable polymers. Additionally, recycling and reusing poly(alkene)s can help reduce waste accumulation, leading to a cleaner environment and a more sustainable future.

Organic Synthesis

How to identify organic functional groups in a molecule containing several functional groups, as well as how to predict their properties and reactions.

Organic molecules can contain several functional groups, which are specific atoms or groups of atoms that determine their physical and chemical properties and reactivity. Common organic functional groups include alkanes, alkenes, alkynes, alcohols, aldehydes, ketones, carboxylic acids, esters, and amines.

To identify these functional groups, you can use the reactions and reactions mechanisms covered in your syllabus. For example, to identify alkenes, look for double bonds between carbon atoms. Alcohols contain a hydroxyl (-OH) functional group, while ketones contain a carbonyl functional group (-C=O) between two carbon atoms. Amines contain an amino group (-NH2), carboxylic acids contain a carboxyl group (-COOH), and esters contain an ester group (-COO-).

Having identified the functional groups, you can then predict their properties and reactivity. For example, alkanes are generally unreactive due to their strong carbon-carbon and carbon-hydrogen bonds. Alkenes, on the other hand, are more reactive than alkanes and can undergo addition reactions with electrophiles such as halogens or hydrogen halides.

Alcohols are polar molecules and exhibit hydrogen bonding, making them more soluble in water than alkanes or alkenes. Aldehydes and ketones react with nucleophiles such as water and alcohols, form hemiketals and hemiacetals. Carboxylic acids are acidic due to their carboxyl groups and can form salts with bases or undergo esterification reactions with alcohols. Amines demonstrate basicity and can undergo salt formation with acids.

In addition, the physical properties of an organic molecule can be determined by the functional groups. For instance, the presence of polar functional groups enhances intermolecular forces, resulting in higher melting and boiling points. Non-polar functional groups, on the other hand, result in weaker intermolecular forces and lower boiling points.

To devise multi-step synthetic routes for preparing organic molecules using reactions covered in your syllabus! Organic synthesis involves combining simple chemical compounds or reagents in a series of reaction steps, leading to the formation of more complex organic molecules. Multi-step organic synthesis requires careful planning and execution, with an understanding of the different reactions covered in the syllabus.

The first step in devising a synthesis route is to consider the target molecule's desired functional groups and its starting materials. Then, identify the series of reactions required to build the target molecule step-by- step. These reactions may include acid-base reactions, nucleophilic or electrophilic substitution, oxidation, reduction, and elimination.

For example, let's consider the synthesis of 2-methyl-2-pentanol from 2-methyl-2-butene. The synthesis route would begin with the addition of HBr to 2-methyl-2-butene to yield 2-bromo-2-methylbutane. The next step involves converting the bromide group to an alcohol using NaBH4 reduction to produce 2-methyl-2-butanol. The third step involves further oxidation of the alcohol group to the corresponding carbonyl using sodium dichromate to produce the ketone. Lastly, the ketone is reduced to produce the target molecule, 2-methyl-2-pentanol using NaBH4 reduction.

Another example is the synthesis of benzyl alcohol from toluene. The synthesis route would begin with the conversion of toluene to benzyl chloride using Cl_2 and $AlCl_3$ as the catalysts. The next step involves the reduction of the chloride group to an alcohol using $NaBH_4$ reduction to produce benzyl alcohol.

Analyzing a given synthetic route in terms of the type of reaction, reagents, and possible by-products. When analyzing a synthetic route, it is useful to consider the selection of reagents and the type of reaction utilized in each step and their possible outcomes. Many reactions produce by-products that can affect the overall yield of the synthesis and could also produce undesired degradation of the molecules.

For instance, let us consider the synthesis of aspirin by reacting salicylic acid with acetic anhydride. In the first step, the reaction involved the preparation of acetic anhydride from acetic acid and acetic anhydride, by using sulfuric acid as an acid catalyst. The second reaction involves esterification of salicylic acid with acetic anhydride to yield aspirin and acetic acid.

In the first reaction, the use of sulfuric acid as an acid catalyst helps to dehydrate the carboxylic acid functional groups present in acetic acid and forms acetic anhydride by removing the water molecule. The final products of the reaction will be acetic anhydride and acetic acid.

In the second reaction, the acetic anhydride is used as an acetyl donor in the esterification of the salicylic acid. The reaction is driven using a weak base, such as pyridine. The final products would be aspirin and acetic acid. However, there is a possibility of the coupling of two salicylic acid molecules instead of the expected acetylation reaction. This side reaction causes reduced yield and potential by-product, which is salicylic anhydride.

Moreover, in other reactions, the choice of reagents and potential by-products can determine the synthetic route's efficiency and yield. Reactions such as oxidation, reduction, coupling, substitution, and elimination can generate multiple products with varying amounts that can impact the overall synthetic success and purity.

Infrared Spectroscopy

To analyze an infrared spectrum of a simple molecule to identify functional groups.

Infrared spectroscopy is a technique that measures the absorption of infrared radiation by molecules. It is a valuable tool for identifying functional groups in organic molecules as each functional group has a unique absorption frequency in the IR spectrum.

To identify functional groups using an IR spectrum, first, identify the regions of the spectrum that correspond to the absorption of specific functional groups. Common functional groups and their corresponding absorption region include: carboxylic acid (1700-1725 cm^-1), amide (1650-1690 cm^-1), carbonyl (C=O) (1620-1680 cm^-1), alcohol (3200-3600 cm^-1), and alkanes (2800-3000 cm^-1).

Next, locate the absorption peaks in the IR spectrum and compare them to the known absorption region for specific functional groups. For example, the presence of a sharp peak at 1720 cm^-1 indicates a carboxylic acid functional group, while a peak at 1650 cm^-1 indicates an amine functional group.

Let's consider the IR spectrum of a simple molecule, methane (CH_4). The IR spectrum for methane contains only a broad peak centered around 3000 cm^-1, representing the C-H stretching vibration. There are no other functional groups present or indicated by the spectrum.

An example of a molecule with multiple functional groups is propanone (CH_3COCH_3). The IR spectrum of propanone would have a sharp absorption peak at around 1700 cm^-1, which represents the C=O stretching vibration associated with the ketone functional group. In addition, the spectrum would have a broad peak around 3200-3600 cm^-1, indicating the presence of the O-H stretching vibration associated with the alcohol functional group.

Mass Spectrometry

To analyze mass spectra in terms of m/e values and isotopic abundances.

Mass spectrometry is an analytical technique used to identify and characterize molecules based on their mass-to-charge ratio (m/e). A mass spectrometer ionizes a molecule and separates it into charged fragments based on their m/e values. These fragments are then detected, and their relative intensities recorded.

Most molecular ions in a mass spectrum are represented by the base peak, which is the most abundant ion, and the molecular ion peak, which represents the parent molecular ion. The molecular ion peak has a mass-to-charge ratio (m/e) equal to the molecular weight of the compound, and its intensity is related to isotopic abundances.

Isotopic abundances and m/e values are intrinsically related as they both involve the mass of the atoms in the molecule. Isotopes refer to the types of atoms that belong to the same element, but they have different atomic masses due to the varying number of neutrons that they possess.

Therefore, molecules containing isotopes have a higher mass than their non-isotopic counterparts.

The measurement of isotopic abundances in mass spectrometry allows the characterization and identification of the presence and composition of specific isotopes in a molecule. Knowing the relative isotopic abundances of atoms in a molecule allows prediction of the relative intensity of the molecular ion and the fragmentation pattern of the molecule.

For example, let us consider the mass spectrum of methane (CH_4). The molecular ion peak of methane is at m/e 16, which corresponds to the mass of the CH_4 molecular ion. Methane has two stable isotopes, carbon-12 and carbon-13, and one hydrogen isotope, deuterium. The relative abundance of carbon-12 and carbon-13 isotopes in methane is 99% and 1%, respectively, and the abundance of deuterium in methane is around 0.015%. These isotopic abundances contribute to the overall mass of the molecule, and as a result, the molecular ion peak at m/e 16 has two sub-peaks at m/e 15 and m/e 17, corresponding to the different isotopes.

How to calculate the relative atomic mass of an element given the relative abundances of its isotopes or its mass spectrum!

Let me start by giving you a brief overview of what an isotope is. Every element has atoms with the same number of protons in their nucleus, but they may have different numbers of neutrons. These different varieties of an element are called isotopes.

Now, let's move on to the concept of relative atomic mass. The relative atomic mass of an element is the average mass of the atoms of that element in a given sample, compared to the mass of an atom of carbon-12, which is assigned a mass of exactly 12 units. This means that if the average mass of the atoms of an element is 2 times that of carbon-12, then the relative atomic mass of that element would be 24 (2 x 12).

To calculate the relative atomic mass of an element given the relative abundances of its isotopes, you need to understand the concept of weighted average. This is done by multiplying the mass of each isotope by its relative abundance, and then adding up these values for all the isotopes. The result is then divided by the total relative abundance of all the isotopes of that element.

If you are given a mass spectrum of the element, you can still calculate its relative atomic mass. Simply look for the peaks in the spectrum, which represent the different isotopes of the element. The size of the peaks indicates how much of each isotope is present relative to the others. By using the same formula mentioned above, you can calculate the relative atomic mass of the element.

Innovatively, we can compare the concept of relative atomic mass to a pizza. Imagine you have a pizza with different toppings, where the toppings represent the different isotopes of an element. You need to calculate the average mass of the pizza with all its toppings, and then divide it by the total number of toppings on the pizza. This will give you the weighted average mass of the pizza, just like how we calculated the relative atomic mass of an element.

How to deduce the molecular mass of an organic molecule from the molecular ion peak in a mass spectrum! First, let me give you a brief overview of mass spectrometry. It is a scientific technique that helps us determine the mass of molecules in a sample. It works by ionizing and then separating the molecules based on their mass-to-charge ratio (m/z) using a magnetic field.

To deduce the molecular mass of an organic molecule from the molecular ion peak in a mass spectrum, we need to understand the concept of molecular ion peaks. When an organic molecule is ionized in the mass spectrometer, it forms a cation, which is also known as the molecular ion. The molecular ion peak in a mass spectrum represents the cation that has the same mass as the original organic molecule.

To determine the molecular mass of the organic molecule, simply locate the molecular ion peak, and note its m/z value. The molecular mass of the organic molecule is equal to the m/z value of the molecular ion peak, multiplied by the charge of the cation. Most often, the charge of the cation is +1, so the molecular mass is equal to the m/z value of the molecular ion peak.

Innovatively, we can think of the molecular ion peak as a fingerprint of the organic molecule. Just like how our unique fingerprints can identify us, the molecular ion peak with its unique m/z value can identify the organic molecule. By measuring the m/z value of the molecular ion peak, we can deduce the molecular mass of the organic molecule.

It is important to note that the molecular ion peak in a mass spectrum is not always present or can be hard to identify due to other overlapping peaks. In this case, other peaks in the mass spectrum can be used to deduce the molecular mass, such as the isotopic peaks that are caused by the presence of different isotopes in the organic molecule.

How to suggest the identity of molecules formed by simple fragmentation in a given mass spectrum.

In mass spectrometry, fragmentation of an organic molecule often occurs when the molecule is ionized. This results in the formation of smaller fragments that have different masses and can be represented as peaks in the mass spectrum. These peaks are useful in identifying and characterizing the molecular structure of the organic molecule.

To suggest the identity of molecules formed by simple fragmentation in a given mass spectrum, we need to understand the concept of simple fragmentation. Simple fragmentation occurs when weak bonds in the organic molecule are broken, resulting in the formation of smaller, simpler fragments.

To identify the fragments of an organic molecule, it is important to look at the mass spectrum and locate peaks that correspond to smaller fragments that are derived from the original organic molecule. These fragments can be identified by comparing their mass-to-charge ratio (m/z) values to the molecular ion peak.

For example, if we have an organic molecule that has a molecular ion peak at m/z 150, and there is a smaller peak at m/z 75, we can suggest that the molecule has fragmented into two equal parts. Alternatively, if there is a peak at m/z 105, we can suggest that the molecule has lost a group or an atom with a mass of 45 (150-45=105).

Innovatively, we can think of simple fragmentation as similar to breaking down a Lego structure into smaller pieces. Each piece represents a fragment of the original structure, and by analyzing the pieces, we can figure out the original structure. In the same way, by analyzing the small fragments in the mass spectrum, we can deduce the structure of the original organic molecule.

It is important to note that identifying the fragments of an organic molecule in a mass spectrum requires knowledge and understanding of organic chemistry, including the types of bonds and functional groups that make up the molecule. In some cases, multiple fragmentation pathways may produce complex peaks in the mass spectrum, making identification of the fragments more challenging.

How to deduce the number of carbon atoms, n, in a compound using the M +1 peak and the formula n = 100 × abundance of M +1 ion ÷ 1.1 × abundance of M + ion.

In mass spectrometry, the M peak represents the molecular ion peak, which is the peak in the spectrum that represents the intact molecular ion of the compound. The M+1 peak is the peak that is one mass unit higher than the M peak, and it represents the molecular ion that contains one more carbon-13 isotope compared to the molecular ion represented by the M peak.

To use the M+1 peak to deduce the number of carbon atoms, we need to use the formula n = 100 × abundance of M +1 ion ÷ 1.1 × abundance of M + ion. This formula relates the relative abundance of the M+1 and M peaks to the number of carbon atoms in the compound.

The logic behind this formula is that the probability that a compound will contain a carbon-13 isotope is about 1.1% (since carbon-13 accounts for about 1.1% of the carbon atoms on Earth). So, if the M+1 peak is present at an abundance that is 1.1% of the M peak abundance, it is likely that one of the carbons in the compound is a carbon-13 isotope.

For example, if a compound has an M peak at m/z 200, and an M+1 peak at m/z 201 with an abundance of 1%, we can calculate the number of carbon atoms in the compound using the formula: n = 100 × abundance of M +1 ion ÷ 1.1 × abundance of M + ion = 100 × 0.01 ÷ 1.1 × 1 = 0.909. This means that the compound contains approximately 0.909 carbon atoms, which we can round up to 1 carbon atom.

Innovatively, think of the M+1 peak as a "magic number" that can tell us how many carbon-13 isotopes are present in the compound. If we know that the probability of finding a carbon-13 isotope is around 1.1%, then we can use that information, along with the relative abundance of the M and M+1 peaks, to deduce the number of carbon atoms in the compound.

It's important to note that this formula assumes that the compound contains only carbon and hydrogen atoms, and that there are no other elements present that may contribute to the M+1 peak. Additionally, this method may not work for more complex compounds with a higher level of structural variation.

I hope this helps you understand how to deduce the number of carbon atoms in a compound using the M+1 peak and the formula n = 100 × abundance of M +1 ion ÷ 1.1 × abundance of M + ion.

How to deduce the presence of bromine and chlorine atoms in a compound using the M+2 peak.

In mass spectrometry, the M peak represents the molecular ion peak, which is the peak in the spectrum that represents the intact molecular ion of the compound. The M+2 peak is the peak that is two mass units higher than the M peak, and it represents the molecular ion that contains two more heavy isotopes (with an atomic number of 35 or 81) compared to the molecular ion represented by the M peak. These heavy isotopes are ^81Br (boron-81) and ^37Cl (chlorine-37).

To deduce the presence of bromine and chlorine atoms in a compound using the M+2 peak, we need to look for the presence of the M+2 peak in the mass spectrum. If the M+2 peak is present in the mass spectrum, it indicates that the molecular ion has two more heavy isotopes, which means that the compound may contain a bromine or a chlorine atom (or both).

To identify whether the compound contains a bromine or a chlorine atom, there are two methods:

1. Use the isotopic ratio - The isotopic ratio of the M+2 peak relative to the M peak can be used to determine the presence of a specific heavy isotope, as different heavy isotopes have different natural abundances. For example, the natural abundance of ^{81}Br is around 49%, while the natural abundance of ^{37}Cl is around 24%. So, if the M+2 peak is present at an abundance that is close to 49% of the M peak abundance, it is likely that the compound contains a bromine atom. Similarly, if the M+2 peak is present at an abundance that is close to 24% of the M peak abundance, it is likely that the compound contains a chlorine atom.

2. Use the relative position of the M+2 peak - The relative position of the M+2 peak can also help identify the atom present. For example, if the M+2 peak is 2 mass units higher than the M peak (m/z = M + 2), it indicates the presence of ^{81}Br. Alternatively, if the M+2 peak is 1 mass unit higher than the M peak (m/z = M + 1), it indicates the presence of ^{37}Cl.

Innovatively, we can think of the M+2 peak as a "sneaky" peak that can help us identify the presence of bromine or chlorine atoms in a compound. If the M+2 peak is present in the mass spectrum, it suggests that there are two more heavy isotopes in the molecular ion, which means that the compound could contain a bromine or a chlorine atom. The isotopic ratio or relative position of the M+2 peak can then be used to deduce the specific atom present.

It's important to note that other elements could potentially contribute to the M+2 peak, so interpretation of the mass spectrum should be done with caution. Additionally, this method may not work for more complex compounds with a higher level of structural variation.

Chemical Energetics

Part 1: Physical Chemistry

Lattice energy and Born-Haber cycles

Enthalpy change of atomisation, often denoted as ΔHat, is a term used in thermodynamics to describe the amount of energy absorbed when one mole of a given substance is converted into individual atoms in the gas phase. This is typically measured at standard conditions of 1 atmosphere and 25°C. It is an important concept in physical chemistry, particularly in the areas of bonding and thermochemistry.

The term 'enthalpy' refers to the total energy content of a system, including both kinetic and potential energy. 'Change' refers to the difference in this energy before and after a chemical reaction or phase transition. 'Atomisation', in this context, refers to the process of breaking down a substance into its constituent atoms. ΔHat is typically determined experimentally, using techniques such as calorimetry to measure the heat absorbed or released during the atomisation process. The enthalpy change of atomisation can be positive or negative, depending on whether energy is absorbed or released.

In general, the enthalpy change of atomisation is positive, indicating that energy is required to break the bonds between atoms in the molecule. This is because it usually takes energy to overcome the attractive forces that hold atoms together in a molecule. The magnitude of ΔHat can give insights into the strength of these bonds. For example, the enthalpy change of atomisation of a diatomic molecule like hydrogen (H_2) would involve breaking the bond between the two hydrogen atoms to form individual hydrogen atoms. This requires energy, so the ΔHat for this process is positive.

On the other hand, the enthalpy change of atomisation can be negative in some cases, such as when a substance is in a high-energy state and releases energy as it moves to a lower-energy state. This can occur, for example, when a substance is in an excited state and emits light as it transitions to a lower energy state.

The enthalpy change of atomisation is an important parameter in various chemical models and theories. For instance, it plays a crucial role in the Born-Haber cycle, a method used to calculate lattice energies of ionic compounds.

Moreover, ΔHat values are used in the calculation of bond energy, as they provide a measure of the energy required to break a particular type of bond. This information can be used to predict the stability of molecules and the feasibility of chemical reactions.

Lattice Energy

Lattice energy, denoted as $\Delta Hlatt$, is a measure of the energy released when ions in the gas phase come together to form a solid crystal lattice. This term is a fundamental concept in chemistry, particularly in the field of physical chemistry, as it is crucial in understanding the stability, solubility, and volatility of ionic compounds.

More specifically, lattice energy is the enthalpy change of the process by which isolated gaseous ions, under standard conditions, are packed together to form an ionic solid. Lattice energy is determined by the charges of the ions and the distance between the ions in the crystal lattice. The electron affinity is directly related to the multiplication of the charges of the ions and inversely related to the separation between the centers of the ions. As such, the greater the charges of the ions and the smaller the distance between them, the greater the lattice energy.

In the context of ionic compound formation, lattice energy plays a pivotal role. For instance, the formation of sodium chloride (NaCl) from sodium and chlorine atoms involves several steps, including the ionization of the sodium atom and the addition of an electron to the chlorine atom to form ions. These ions then combine to form a sodium chloride crystal, in a process known as lattice formation, which releases a significant amount of energy, known as lattice energy.

In the Born-Haber cycle, a thermochemical cycle that relates various physical and chemical processes to one another, lattice energy is a key component. This cycle allows scientists to analyze the total energy involved in the formation of an ionic compound, as well as the energy required to break it down into its constituent elements or ions. The lattice energy is one of the crucial values in this cycle, indicating the strength of the ionic bonds in the compound.

Let's consider an example of magnesium oxide (MgO). Both magnesium and oxygen are found in nature, although not in their elemental forms due to their high reactivity. They tend to form compounds with other elements, particularly with each other, to form MgO. The formation of this compound is an exothermic process, releasing energy in the form of heat and light. The lattice energy in this case is the energy released when Mg^{2+} and O^{2-} ions in the gaseous state come together to form a solid lattice of MgO.

In a real-world application, understanding lattice energy is essential in the field of materials science, specifically in the design and synthesis of new materials. For instance, high lattice energy often corresponds to high melting points and low solubility in water, properties that are crucial in the manufacture of ceramics and other heat-resistant materials.

The lattice energy is also significant in predicting the solubility of ionic compounds in polar solvents, such as water. Compounds with high lattice energy are generally less soluble in water, as the energy required to break the ionic bonds in the solid and dissolve the compound is much higher than the energy gained from the solvation process.

Furthermore, the concept of lattice energy is pivotal in understanding and predicting the volatility of ionic compounds. Ionic compounds with high lattice energy are generally less volatile, as the high energy required to break the ionic bonds in the solid and convert the compound into a gas is usually not available at normal temperatures.

First Electron Affinity (EA)

The term 'first electron affinity' (EA) refers to the amount of energy released when a neutral atom in its gaseous state accepts an electron to form a negatively charged ion. In the field of physical chemistry, electron affinity is a critical property of an atom that influences its chemical reactivity and bonding patterns. The electron affinity quantifies the level of attraction between the incoming electron and the nucleus of the atom.

The first electron affinity can be understood as the energy change that occurs when one mole of electrons is added to one mole of atoms in the gaseous state. Typically, electron affinity is represented in kilojoules per mole (kJ/mol) units. The higher the electron affinity of an atom, the more readily the atom accepts an electron. As such, atoms with high electron affinities tend to form negative ions more readily.

The concept of first electron affinity is crucial in predicting the likelihood of an atom to gain an electron during a chemical reaction. It provides valuable insights into an atom's ability to form chemical bonds and its reactivity. For instance, non-metals generally have higher first electron affinities compared to metals, explaining why they tend to gain electrons and form anions during chemical reactions.

First electron affinity is a key factor in determining the nature of chemical reactions. It influences the formation of ionic compounds, where one atom transfers one or more electrons to another atom. The atom with a higher electron affinity will typically gain the electron, forming a negative ion, while the atom with a lower electron affinity will lose the electron, forming a positive ion. In the periodic table, there is a general trend in the first electron affinity values. Moving from left to right across a period, the first electron affinity tends to increase. This is mainly due to the increase in the nuclear charge, which attracts the incoming electron more strongly. On the contrary, moving down a group, the first electron affinity tends to decrease because the added electron is further from the nucleus, reducing the attractive force.

However, there are exceptions to these trends. For example, noble gases have low first electron affinities because they have a stable electron configuration. Adding an electron would disrupt this stability, requiring energy rather than releasing it. Similarly, elements in Group 2 (the alkaline earth metals) and Group 15 (the pnictogens) have lower electron affinities than expected due to the added electron occupying a higher energy orbital. While the first electron affinity involves the gain of one electron, atoms can also gain additional electrons. The energy change associated with the addition of the second electron is known as the second electron affinity. However, the second electron affinity is always endothermic, meaning it requires an input of energy. This is because adding a second electron involves overcoming the repulsion from the negatively charged ion formed during the first electron affinity.

Factors Affecting the Electron Affinities of Elements

Electron affinity is influenced by several factors such as atomic size, nuclear charge, electron configuration, and the effective nuclear charge.

The first factor that influences the electron affinity is the atomic size. With increasing atomic size, the electron affinity tends to decrease. This is because the outermost electrons are farther away from the nucleus and hence, the attractive force of the nucleus on these electrons is weaker. Consequently, the energy released when an electron is added to an atom decreases with increasing atomic size.

The second factor that affects the electron affinity is the nuclear charge. The nuclear charge refers to the total charge of all the protons in the nucleus of an atom. An increase in the nuclear charge increases the attractive force of the nucleus on the electrons. This means that the atom can more easily gain an electron, thereby increasing the electron affinity.

The third factor influencing the electron affinity is the electron configuration of the atom. Atoms with full or half-full subshells have lower electron affinities compared to atoms with nearly full or nearly empty subshells. This is because adding an electron to a full or half-full subshell would result in electron-electron repulsion, which would require energy instead of releasing energy.

The fourth factor affecting the electron affinity is the effective nuclear charge. The effective nuclear charge is the total positive charge that an electron experiences within a multi-electron atom. The higher the effective nuclear charge, the higher the electron affinity. This is because a higher effective nuclear charge means a stronger attraction between the nucleus and the electrons, making it easier for the atom to gain an electron.

The fifth factor is the element's position on the periodic table. For example, elements in Group 17 (halogens) have high electron affinity because they have seven valence electrons and need one more to achieve a stable electron configuration. In contrast, noble gases in Group 18 have low electron affinity because their electron shells are already filled.

The sixth factor is the shielding effect. Shielding effect refers to the ability of the innermost electrons to shield the outermost electrons from the full attraction of the nucleus.

A higher shielding effect results in a lower electron affinity because the outermost electrons are less attracted to the nucleus, making it harder for the atom to gain an electron.

The seventh factor is the atomic structure. The atomic structure, particularly the distribution of electrons in different energy levels, also plays a crucial role in determining the electron affinity. Atoms with more balanced distributions of electrons across their energy levels tend to have higher electron affinities.

The eighth factor is the subshell type. Electrons are more likely to be added to the s and p subshells than the d and f subshells. This is because the s and p subshells are closer to the nucleus and are less shielded, resulting in a higher electron affinity.

The ninth factor is the ionization energy. There is an inverse relationship between ionization energy and electron affinity. If an atom has a high ionization energy, it is less likely to accept an additional electron, and thus, its electron affinity is lower.

The tenth factor is the electronegativity of the atom. Electronegativity pertains to an atom's capability to draw electrons towards itself in a chemical bond. An atom with higher electronegativity tends to have a higher electron affinity.

These factors are interrelated, and an understanding of these factors is crucial for predicting the chemical behavior of elements. For instance, knowing that halogens have high electron affinities can help predict their reactivity. Similarly, understanding that large atoms have lower electron affinities can provide insights into the properties of transition metals.

Trends in the electron affinities of Group 16 and Group 17 elements

Group 16 and Group 17 elements, also known as the Chalcogens and Halogens respectively, exhibit interesting trends in their electron affinities. Electron affinity refers to the energy change that occurs when an electron is added to a neutral atom to form a negative ion. It is a measure of an atom's ability to accept an electron.

In the periodic table, electron affinity generally increases across a period from left to right and decreases down a group. This trend is due to two key factors: the effective nuclear charge and the atomic radius. As we move across a period, the number of protons increases, which increases the effective nuclear charge. This enhanced positive charge draws electrons more powerfully, leading to a greater electron affinity. Meanwhile, as we move down a group, the atomic radius increases, which decreases the effective nuclear charge and lowers the electron affinity. Starting with Group 16, the elements include Oxygen (O), Sulfur (S), Selenium (Se), Tellurium (Te), and Polonium (Po). Oxygen has the highest electron affinity in this group, with a value of 141 kJ/mol. This is because Oxygen, being at the top of the group, has the smallest atomic radius and the highest effective nuclear charge. As we move down the group, the atomic radius increases and the electron affinity decreases. For instance, Sulfur has an electron affinity of 200 kJ/mol, Selenium 195 kJ/mol, Tellurium 190 kJ/mol, and Polonium has a significantly lower electron affinity.

The Halogens, which are part of Group 17, consist of Fluorine (F), Chlorine (Cl), Bromine (Br), Iodine (I), and Astatine (At). Fluorine, at the top of the group, has the highest electron affinity, reflecting its small atomic radius and high effective nuclear charge. Fluorine's electron affinity is 328 kJ/mol. Similar to Group 16, as we move down Group 17, the atomic radius increases and the electron affinity decreases. Chlorine has an electron affinity of 349 kJ/mol, Bromine 325 kJ/mol, Iodine 295 kJ/mol, and Astatine has a considerably lower electron affinity.

In comparing the two groups, Group 17 elements have higher electron affinities than Group 16 elements. This is primarily because Group 17 elements have one less electron than they need to achieve a stable electron configuration, so they have a particularly strong attraction for additional electrons. Group 16 elements, on the other hand, require two additional electrons to achieve a full outer electron shell, so they're less eager to accept an additional electron.

These trends of electron affinity are not perfect. For example, although electron affinity generally increases across a period, the electron affinity of Fluorine is actually less than that of Chlorine. This anomaly can be attributed to Fluorine's small atomic size, which results in electron-electron repulsion that reduces its electron affinity.

Furthermore, the electron affinity of Oxygen is also less than that of Sulfur. This is again due to the smaller atomic size of Oxygen, which leads to a greater degree of electron-electron repulsion and therefore a lower electron affinity.

These exceptions highlight the complexity of predicting chemical behavior based on periodic trends. While the overall increase of electron affinity across a period and its decrease down a group provide useful generalizations, the specific behavior of individual elements can be influenced by additional factors such as atomic size and electron-electron repulsion.

Another interesting point to note is the behavior of the heaviest elements in these groups, Polonium and Astatine. These elements have significantly lower electron affinities than their lighter counterparts. This is due to the relativistic effects, which are particularly pronounced in heavy elements. These effects can greatly influence the electron affinities, leading to deviations from the general trend.

Construct and use Born–Haber cycles for ionic solids

The Born-Haber cycle is a thermochemical cycle that relates to the formation of an ionic solid. This cycle is commonly utilized to compute the lattice energy of an ionic solid, which is the energy required to disassemble an ionic solid into its constituent ions. This principle is particularly applicable to +1 and +2 cations, and -1 and -2 anions.

The concept of the Born-Haber cycle was first introduced by Max Born and Fritz Haber in the early 20th century. They proposed this cycle as a method to calculate lattice energies, which are difficult to measure directly. The Born-Haber process relates to Hess's law, which states that the total enthalpy change in a chemical reaction is independent of the path taken. One can begin to understand the Born-Haber cycle by considering the formation of an ionic compound from its constituent elements, such as the formation of NaCl from sodium and chlorine. The first step in this process is the sublimation of the metallic element, which in this case is sodium. Sublimation refers to the phase transition from a solid to a gas, bypassing the liquid state. This step requires energy, referred to as the sublimation energy.

The next step is the ionization of the gaseous metal atoms. Ionization is the process by which an atom or molecule acquires a charge by gaining or losing electrons. This step also requires energy, known as the ionization energy. For NaCl, this involves the removal of an electron from each sodium atom to form Na+ ions. Following this, the non-metal element, in this case, chlorine, undergoes bond dissociation to form individual atoms. This process requires the bond dissociation energy. Now, chlorine atoms can accept an electron from the environment, a process known as electron affinity, to form chloride ions. The final step in the cycle is the formation of the ionic solid from the gaseous ions. This process is exothermic, which means it discharges energy. This energy is what we refer to as the lattice energy. Although this process is exothermic, the overall energy change from the formation of an ionic compound is generally endothermic due to the energy required for sublimation, ionization, and bond dissociation.

The Born-Haber cycle can be represented schematically, with each step of the cycle corresponding to a particular reaction and associated energy change. The sum of all these energy changes gives the overall enthalpy change for the reaction.

The Born-Haber cycle is a valuable tool in understanding the properties of ionic solids. For example, it helps to clarify why some ionic solids have higher melting and boiling points than others. This is because ionic solids with greater lattice energies require more energy to break apart, leading to higher melting and boiling points. Moreover, the Born-Haber cycle can also be applied to other ionic compounds, not just +1 and +2 cations, and -1 and -2 anions. By carefully considering each step in the formation of an ionic compound, one can gain a deeper understanding of the factors that contribute to its stability and reactivity.

Calculations involving Born–Haber cycles

Let's look at an example problem to demonstrate how to use a Born-Haber cycle. Consider the formation of sodium chloride (NaCl), which involves the reaction between sodium (Na) and chlorine (Cl) gases to give NaCl solid:

$$Na(s) + ½ Cl_2(g) \longrightarrow NaCl(s)$$

To calculate the enthalpy of formation (ΔH^f) of NaCl, we need to consider the following steps:

1. Sublimation of sodium metal: $Na(s) \longrightarrow Na(g)$
2. Ionization of sodium atoms: $Na(g) \longrightarrow Na^+(g) + e^-$
3. Dissociation of chlorine gas: $Cl_2(g) \longrightarrow 2\,Cl(g)$
4. Electron affinity of chlorine atoms: $Cl(g) + e^- \longrightarrow Cl^-(g)$
5. Formation of the ionic solid: $Na^+(g) + Cl^-(g) \longrightarrow NaCl(s)$

Now, let's calculate the enthalpy of formation of NaCl using the given data:

Sublimation energy of sodium ($\Delta Hsub$) = +108 kJ/mol
Ionization energy of sodium ($\Delta Hion$) = +496 kJ/mol
Dissociation energy of Cl_2 ($\Delta Hdis$) = +242 kJ/mol
Electron affinity of chlorine (ΔHea) = -349 kJ/mol
Lattice energy of NaCl ($\Delta Hlat$) = -786 kJ/mol

To calculate ΔH^f for NaCl, we sum up the energies for each step: $\Delta H^f = \Delta Hsub + \Delta Hion + \Delta Hdis + \Delta Hea + \Delta Hlat$

$$\Delta H^f = +108 + (+496) + (+242) + (-349) + (-786) = -289 \text{ kJ/mol}$$

Therefore, the enthalpy of formation of NaCl is -289 kJ/mol, indicating an exothermic process.

Now, let's consider another example problem:

Determine the enthalpy change of formation of calcium fluoride (CaF_2) using the following data:

ΔH^f for $CaF_2(s)$ = -1223 kJ/mol
$\Delta Hsub$ for $Ca(s)$ = +178 kJ/mol
$\Delta Hion$ for $Ca(g)$ = +590 kJ/mol
$\Delta Hdis$ for $F_2(g)$ = +159 kJ/mol
ΔHea for $F(g)$ = -328 kJ/mol

Given that the lattice energy ($\Delta Hlat$) is unknown, we can use the given data and the equation:

$$\Delta H^f = \Delta Hsub + \Delta Hion + \Delta Hdis + \Delta Hea + \Delta Hlat$$
$$-1223 = +178 + 590 + 159 + (-328) + \Delta Hlat$$

Simplifying the equation, we find:

$$\Delta Hlat = -14 \text{ kJ/mol}$$

Hence, the lattice energy of CaF_2 is approximately -14 kJ/mol.

How the ionic charge and ionic radius affect the numerical magnitude of a lattice energy?

Ionic compounds are created through the attraction between positively charged ions, known as cations, and negatively charged ions, referred to as anions. This attraction occurs due to the presence of electrostatic forces.

The strength of this attraction, known as the lattice energy, depends on the magnitude of the charges and the sizes of the ions involved.

Effect of Ionic Charge:

The magnitude of the lattice energy is directly influenced by the ionic charges. The greater the magnitude of the charges, the stronger the electrostatic attraction between the ions, resulting in a higher lattice energy. For example, a compound composed of ions with +2 and -2 charges will have a higher lattice energy compared to a compound with +1 and -1 charges.

Sample Problem 1: Determine which compound is expected to have a higher lattice energy: MgO (magnesium oxide) or NaCl (sodium chloride).

Solution: Both compounds consist of +2 and -2 charges. However, Mg has a higher effective nuclear charge compared to Na, leading to a stronger attraction between the Mg^{2+} and O^{2-} ions. Hence, MgO is expected to have a higher lattice energy than NaCl.

Effect of Ionic Radius: The numerical magnitude of the lattice energy is also influenced by the sizes of the ions. Generally, as the size of the ions decreases, the lattice energy increases. This is because smaller ions allow for closer packing in the crystal lattice, resulting in stronger electrostatic attractions.

Sample Problem 2: Compare the lattice energy of NaCl (sodium chloride) with that of KBr (potassium bromide).

Solution: Both compounds consist of +1 and -1 charges. Since K and Cl have larger ionic radii compared to Na and Br, respectively, KBr is expected to have a lower lattice energy compared to NaCl. The larger ionic radii of K^+ and Br^- ions result in weaker electrostatic interactions.

It's important to note that these trends are generalizations and may not always hold in more complex situations. Factors such as crystal structure, polarizability, and hydration energy can also affect the numerical magnitude of the lattice energy.

Sample Problem 3: Calculate the lattice energy of CaO (calcium oxide) using the given data:

$\Delta Hsub$ for Ca = +192 kJ/mol

$\Delta Hion$ for Ca^{2+} = +590 kJ/mol

$\Delta Hsub$ for O = +249 kJ/mol

ΔHea for O^{2-} = -141 kJ/mol

Solution: The lattice energy ($\Delta Hlat$) can be calculated using the formula: $\Delta Hlat = \Delta Hion + \Delta Hea + \Delta Hsub - \Delta Hsub$. Substituting the given values, we have: $\Delta Hlat$ = (+590) + (-141) + (+192) + (+249) = +890 kJ/mol.

Enthalpies of solution and hydration

Enthalpy change: hydration, $\Delta Hhyd$, and solution, $\Delta Hsol$

Enthalpy change refers to the amount of heat energy exchanged during a chemical process or reaction. It is denoted by ΔH, where Δ represents the change and H represents enthalpy. Enthalpy is a thermodynamic property that relates to the heat content of a system.

When we talk about enthalpy changes related to hydration and solution, we specifically refer to $\Delta Hhyd$ and $\Delta Hsol$, respectively.

$\Delta Hhyd$, or the enthalpy change of hydration, is the enthalpy change that occurs when an ionic substance dissolves in water. It represents the amount of heat energy released or absorbed during the process of dissolving. In simple terms, $\Delta Hhyd$ measures the heat involved in breaking the ionic bonds of the solute and the energy released or absorbed when the solute particles interact with water molecules.

138

On the other hand, ΔHsol, or the enthalpy change of solution, refers to the overall enthalpy change when a solute is dissolved in a solvent to form a solution. It takes into account both the enthalpy change of hydration and any additional heat effects caused by mixing or formation of new bonds between the solute and solvent molecules. When a solute dissolve in water, it usually involves a process of breaking bonds within the solute and bonds within the solvent, followed by the formation of new bonds between the solute and solvent molecules. This process can either release or absorb heat, depending on the specific solute-solvent interaction. ΔHhyd and ΔHsol can have positive or negative values. A positive value signifies that the process is endothermic, indicating that it takes in heat from the surroundings. This implies that energy is required to break the bonds and establish new solute-solvent interactions. Conversely, a negative value indicates that the process is exothermic, releasing heat to the surroundings.

Important factors affecting ΔHhyd and ΔHsol include the nature of the solute and solvent, the concentration and temperature of the solution, and the presence of any other substances that may interact with the solute or solvent.

Problem 1: Calculate the enthalpy change of hydration for the dissolution of sodium chloride (NaCl) in water.

Solution: To calculate the enthalpy change of hydration (ΔHhyd) for the given reaction, you need to know the standard enthalpy of formation of NaCl and the standard enthalpy of solution of NaCl. By applying the Hess's law, you can subtract the standard enthalpy of formation of NaCl from the standard enthalpy of solution of NaCl to obtain ΔHhyd.

Problem 2: Determine the enthalpy change of solution (ΔHsol) for dissolving 5 grams of sugar (C12H22O11) in 100 mL of water.

Solution: To find the enthalpy change of solution (ΔHsol) for the given scenario, you need to measure the temperature change during the dissolution process using a calorimeter. By applying the equation q = mcΔT (where q is the heat energy absorbed or released, m is the mass of the solution, c is the specific heat capacity of the solution, and ΔT is the temperature change), you can calculate the amount of heat released or absorbed and determine ΔHsol.

Problem 3: In an experiment, a student determines the enthalpy change of hydration of copper(II) sulfate (CuSO4). However, they mistakenly mix 10 mL of CuSO4 solution with 10 mL of water and measure the temperature change. How can they determine ΔHhyd correctly?

Solution: Since the student mistakenly diluted the CuSO4 solution before measuring the temperature change, the resulting enthalpy change will not reflect the true value of ΔHhyd. To determine ΔHhyd correctly, the student should repeat the experiment with the undiluted CuSO4 solution and measure the temperature change accurately. The proper method includes measuring the initial and final temperatures of the undiluted solution and using those values to calculate the correct enthalpy change.

Problem 4: What is the enthalpy change of hydration when solid calcium chloride (CaCl2) dissolves in water?

Solution: To determine the enthalpy change of hydration (ΔHhyd) for the dissolution of solid calcium chloride (CaCl2) in water, you need to know its standard enthalpy of hydration value. This value can be obtained from reference sources or data tables. It represents the enthalpy change when one mole of CaCl2 dissolves in water to form its hydrated ions. By using the given value, you can calculate ΔHhyd for the dissolution of a specified amount of calcium chloride.

Problem 5: A student is given a solution of hydrochloric acid (HCl) and asked to determine its enthalpy change of solution. However, they don't have access to a calorimeter. How can they measure ΔHsol?

Solution: If a calorimeter is not available, the student can use a method called "adiabatic solution calorimetry" to measure the enthalpy change of solution (ΔHsol) for the hydrochloric acid solution. This method involves adding a known amount of the solution to a known amount of water at a measured initial temperature. The resulting temperature change is then measured to determine the heat exchanged. Using the equation $q = mc\Delta T$, where q represents the heat transferred, m is the mass of the solution, c is the specific heat capacity of the solution, and ΔT represents the temperature change, the student can determine ΔH_{sol} by utilizing the calculated heat transfer.

Ionic dissolution: Energy cycle analysis
Introduction

In the field of chemistry, understanding energy changes is crucial for comprehending various chemical processes. One important concept is the energy cycle involving solution enthalpy change, lattice energy, and hydration enthalpy change. By studying these interrelated processes, we gain insights into the energetics of solution formation, crystal lattice formation, and hydration reactions. This article aims to explain these concepts in a simple yet professional manner.

1. Enthalpy Change of Solution: Enthalpy change of solution refers to the energy change that occurs when a solute dissolves in a solvent to form a solution. This process involves breaking intermolecular forces within the solute and solvent, and forming new solute-solvent interactions. The enthalpy change of solution can be either exothermic (releasing energy) or endothermic (absorbing energy), depending on the strengths of the solute-solvent interactions.

2. Lattice Energy: Lattice energy represents the energy required to completely separate one mole of a solid ionic compound into its gaseous ions. It is a measure of the strength of the ionic bonds within the crystal lattice. Lattice energy is always an exothermic process as energy is released when the ions come together to form a crystal lattice.

3. Enthalpy Change of Hydration: Enthalpy change of hydration is the energy change that occurs when one mole of gaseous ions dissolves in water to form hydrated ions. This process involves the breaking of solute-solute interactions and the formation of solute-water interactions. Similar to the enthalpy change of solution, enthalpy change of hydration can be either exothermic or endothermic.

4. Relationship between Enthalpy Change of Solution and Lattice Energy: The enthalpy change of solution is related to the lattice energy through the Born-Haber cycle. This cycle allows us to calculate the enthalpy change of solution using known values of lattice energy and other relevant energy changes, such as ionization energy and electron affinity.

5. Steps in the Born-Haber Cycle: The Born-Haber cycle involves several steps, including the formation of gas-phase ions from the elements, the creation of a solid ionic compound, and the dissolving of the compound in water. Each step corresponds to a specific energy change, such as ionization energy, electron affinity, lattice energy, and enthalpy change of hydration.

6. Impact of Ionic Bond Strength on Entropy Change of Hydration: The strength of ionic bonds within a crystal lattice affects the enthalpy change of hydration. Generally, compounds with stronger ionic bonds have higher lattice energies, leading to more exothermic enthalpy changes of hydration.

7. Coulomb's Law and Lattice Energy: Coulomb's law, which describes the force between charged particles, plays a major role in determining the lattice energy of an ionic compound. According to Coulomb's law, the lattice energy increases as the charges of the ions increase, and as the distance between the ions decreases.

8. Factors Affecting Enthalpy Change of Solution: The enthalpy change of solution depends on various factors, including the nature of the solute and solvent, temperature, pressure, and concentration. Stronger solute-solvent interactions generally lead to more exothermic enthalpy changes of solution.

9. Applications of Energy Cycles: Energy cycles involving enthalpy change of solution, lattice energy, and enthalpy change of hydration are widely applied in industry and research. They help in understanding and predicting the solubility, stability, and reactivity of substances, as well as designing efficient chemical processes.

10. Significance in Analytical Chemistry: The energy cycle concept is particularly significant in analytical chemistry. By studying the energy changes associated with solution formation, chemists can determine unknown concentrations of substances through techniques such as calorimetry and enthalpy changes.

11. Thermodynamic Considerations: Energy cycles involving enthalpy change of solution, lattice energy, and enthalpy change of hydration are governed by thermodynamic principles. These cycles demonstrate how energy is transferred and exchanged within chemical systems, providing a foundation for understanding chemical equilibrium and reaction spontaneity.

12. Real-world Examples: The energy cycle concept is pervasive in many real-world examples. For instance, the dissolution of salt in water, the hydration of metal ions in aqueous solutions, and the preparation of highly soluble pharmaceutical drugs all involve the interplay of enthalpy change of solution, lattice energy, and enthalpy change of hydration.

13. Relationship to Solubility: The energy cycle concept also sheds light on solubility phenomena. Compounds with more exothermic enthalpy changes of solution and lower lattice energies generally exhibit greater solubility because the energy released during dissolution compensates for the energy required to break the crystal lattice.

14. Limitations and Assumptions: Energy cycles provide a simplified representation of complex energetic processes. They assume idealized conditions and neglect factors such as heat losses, non-ideal interactions, and structural changes. These limitations should be considered when applying energy cycles to real-world scenarios.

15. Experimental Determination: Experimental methods, such as calorimetry, electrochemistry, and spectroscopy, are employed to determine the energy changes involved in enthalpy change of solution, lattice energy, and enthalpy change of hydration, providing valuable data for constructing energy cycles.

16. Energy Cycle Calculation Examples: Calculations involving energy cycles can be complex, but they follow established principles and equations. By utilizing known values and using Hess's Law, one can determine unknown energy changes or validate experimental data.

17. Importance of Systematic Approaches: Building and utilizing energy cycles require a systematic approach. Proper identification of energy changes, careful consideration of directionality, and the application of thermochemical principles ensure accurate predictions and consistent interpretation of experimental results.

18. Practical Implications: Understanding energy cycles has practical implications in various chemical industries, including pharmaceuticals, environmental chemistry, materials science, and energy production. It facilitates the design of efficient processes, the prediction of product stability, and the optimization of reaction conditions.

19. Continued Research and Development: Advancements in computational methods and experimental techniques contribute to ongoing research and development in the field of energy cycles involving enthalpy change of solution, lattice energy, and enthalpy change of hydration. These investigations aim to refine existing models and provide deeper insights into chemical systems.

Problem 1: Explain the energy changes that occur during the process of dissolving an ionic compound in water.

Solution: The process involves the breaking of ionic bonds in the solid compound (lattice energy), the formation of new interactions between the dissolved ions and water molecules (enthalpy change of hydration), and the final recombination of the hydrated ions to form the dissolved ionic compound.

Problem 2: How does the enthalpy change of solution affect the solubility of a substance?

Solution: Substances with greater enthalpy change of solution, indicating more favorable solvation, tend to have higher solubility in the solvent. The energy balance between the breaking of solute-solute and solvent-solvent interactions and the formation of solute-solvent interactions determines solubility.

Problem 3: Differentiate between lattice energy and enthalpy change of hydration.

Solution: Lattice energy represents the energy released when gaseous ions combine to form a solid ionic lattice. Enthalpy change of hydration, on the other hand, refers to the energy change when gaseous ions dissolve in water to form aqueous ions. Lattice energy is an energy input required to break ionic bonds, while enthalpy change of hydration is the energy released when ions are solvated.

Problem 4: How does lattice energy influence the solubility of an ionic compound?

Solution: Compounds with higher lattice energies require more energy to be overcome to break their ionic bonds. This translates to a higher energy barrier for dissolution, resulting in lower solubility in the solvent. Therefore, compounds with lower lattice energies are generally more soluble.

Problem 5: Explain the role of enthalpy change of hydration in determining the solubility of a substance.

Solution: Enthalpy change of hydration represents the interaction between ions and water molecules during the dissolution process. Substances that release a significant amount of heat (exothermic) during the hydration process tend to have more favorable solvation and higher solubility.

Problem 6: How can the energy cycle involving enthalpy change of solution, lattice energy, and enthalpy change of hydration help determine the overall energy changes during a dissolution process?

Solution: The energy cycle provides a systematic approach to understanding the energy changes involved in the dissolution process. It allows students to visualize and calculate the net energy change by considering the respective magnitudes and signs of lattice energy, enthalpy change of hydration, and enthalpy change of solution.

Problem 7: How do the energy changes in the energy cycle help explain the solubility trends of different ionic compounds?

Solution: By considering the relative strengths of lattice energy and enthalpy change of hydration, students can explain why certain compounds have higher or lower solubilities. Compounds with lower lattice energies and/or greater enthalpy change of hydration generally exhibit higher solubilities in water.

Problem 8: Apply the energy cycle to compare the solubility of two ionic compounds given their lattice energies and enthalpy changes of hydration.

Solution: By comparing the magnitudes of lattice energy and enthalpy change of hydration for the two compounds, students can evaluate which compound is expected to be more soluble in water. The compound with lower lattice energy and/or higher enthalpy change of hydration is likely to have greater solubility.

Problem 9: What are the practical implications of understanding the energy cycle involving enthalpy change of solution, lattice energy, and enthalpy change of hydration?

Solution: Understanding these energy changes helps explain the solubility patterns of ionic compounds and can be applied in various fields like pharmaceuticals (drug solubility), materials science (designing new compounds), and environmental engineering (water pollution treatment).

Problem 10: Describe how the energy cycle can be used to predict and explain chemical reactions involving ionic compounds in aqueous solutions.

Solution: The energy cycle aids in predicting and explaining chemical reactions by considering the energy changes during the solvation and subsequent recombination of ions. Students can assess the feasibility of reactions based on the energy differences between the reactants and products within the energy cycle.

Ionic charge impact: Enthalpy change., Ionic radius impact: Enthalpy magnitude

The enthalpy change of hydration is the energy that is either absorbed or released when an ionic compound dissolves in water. In qualitative terms, the ionic charge and ionic radius of the ions in the compound have significant effects on the numerical magnitude of this enthalpy change.

Let's explore these effects in detail.

When it comes to ionic charge, ions with higher charges generally experience stronger attractive forces with water molecules during hydration. This means that more energy is required to overcome these attractions and separate the ions from each other in the crystal lattice. Therefore, the enthalpy change of hydration tends to be more exothermic (releasing more heat energy) for ions with higher charges, as more energy is released when the ions are surrounded by water molecules.

On the other hand, the size of the ions, represented by their ionic radius, also influences the enthalpy change of hydration. Larger ions have more surface area available for water molecules to interact with, leading to a greater number of water molecules surrounding each ion. This hydration shell effectively shields the charges on the ions from each other, thereby reducing the overall attractive forces between the ions. Consequently, the enthalpy change of hydration becomes less exothermic for larger ions compared to smaller ions.

In terms of specific examples, consider the halide ions in Group 17 of the periodic table. As we move from fluoride to chloride, bromide, and iodide, the ionic radius increases, while the charge remains the same. As a result, the enthalpy change of hydration becomes less exothermic (less negative) for these ions. This trend can be attributed to the larger size of the ions, which reduces the attraction between the ions and water molecules.

Another example can be observed with the alkali metal cations in Group 1 of the periodic table. As we move from lithium to sodium, potassium, and finally cesium, the ionic radius increases while the charge remains the same. In this case, the enthalpy change of hydration becomes more exothermic (more negative), indicating stronger interactions between the cations and water molecules. The larger cations have greater charge densities, leading to stronger attractive forces between the cations and the partially negative oxygen atoms of water.

It is important to note that the effect of ionic charge and ionic radius on the enthalpy change of hydration depends not only on the magnitude of these factors but also on the specific properties of the ions involved. Other factors, such as the nature and strength of intermolecular forces, can also influence the enthalpy change.

Problem 1: How does the ionic charge of an ion affect the enthalpy change of hydration?

Solution: The ionic charge of an ion influences the strength of the attractive forces between the ions and water molecules during hydration. Higher charges result in stronger attractions, making the enthalpy change of hydration more exothermic. Remember that enthalpy change is a measure of heat energy released or absorbed.

Problem 2: Explain the trend in enthalpy change of hydration across the halide ions in Group 17 of the periodic table.

Solution: As we move from fluoride to chloride, bromide, and iodide, the size of the ions (ionic radius) increases while the charge remains the same. This increase in size weakens the attractions between the ions and water molecules, resulting in less exothermic enthalpy change of hydration.

Problem 3: How does the ionic radius of an ion affect the enthalpy change of hydration?

Solution: The ionic radius of an ion influences the surface area available for water molecules to interact with. Larger ions have more surface area, which leads to a greater number of water molecules surrounding each ion. This increases the shielding effect and reduces the attractive forces between the ions, resulting in a less exothermic enthalpy change.

Problem 4: Compare and contrast the enthalpy change of hydration for the alkali metal cations in Group 1 of the periodic table.

Solution: The enthalpy change of hydration becomes more exothermic (more negative) as we move from lithium to sodium, potassium, and finally cesium. This trend can be attributed to the increasing ionic radius, which strengthens the attractive forces between the cations and water molecules due to their higher charge densities.

Problem 5: Explain the factors other than ionic charge and ionic radius that can affect the enthalpy change of hydration.

Solution: The nature and strength of intermolecular forces, such as hydrogen bonding or dipole-dipole interactions, can also influence the enthalpy change of hydration. Additionally, any other chemical reactions occurring between the ions and water molecules need to be considered.

Problem 6: Justify why the enthalpy change of hydration for a particular ion is more exothermic compared to another ion with the same charge but different ionic radius.

Solution: The larger ionic radius distributes the charge of the ion over a larger area, reducing the attraction between the ion and water molecules. As a result, the enthalpy change of hydration becomes less exothermic for ions with larger radii compared to ions with smaller radii but the same charge.

Problem 7: Explain the effect of ionic charge and ionic radius on the solubility of an ionic compound in water.

Solution: While the enthalpy change of hydration influences the solubility of an ionic compound, other factors such as the lattice energy and the hydration energy of the ions also play a role. Higher ionic charges tend to increase the solubility by making the compound more soluble, while larger ionic radii can decrease the solubility due to weaker attractions between the ions and water molecules.

Entropy change, ΔS
Entropy: arrangement and energy possibilities

Entropy, denoted by the symbol S, is a fundamental concept in chemistry that describes the degree of disorder or randomness in a system. It is a measure of the number of possible arrangements that particles and their energy can take within a given system. In simpler terms, entropy can be understood as the level of chaos or unpredictability within a system.

To grasp the concept of entropy, it is helpful to visualize a collection of particles, such as molecules. These particles can exist in various arrangements and possess different amounts of energy. The more ways these particles can be arranged and distributed with different energy levels, the higher the entropy of the system. Consider a single particle initially confined to a specific location. In this scenario, there is only one arrangement possible, namely the particle being in that particular spot. As a result, the entropy is low. However, if the constraints on the particle's position are removed, it is now free to move around in an expanded space. This expanded freedom allows for a greater number of possible arrangements, resulting in an increase in entropy. Expanding this concept to a system with multiple particles, the total number of possible arrangements and energy distributions increases exponentially. Every possible arrangement and energy allocation contributes to the overall entropy of the system. For instance, if particles can move, rotate, and occupy different energy levels, the number of ways they can arrange themselves becomes significantly larger, resulting in higher entropy. Entropy is closely related to the second law of thermodynamics, which states that the entropy of an isolated system tends to increase over time. This means that, left to spontaneous changes, systems tend to become more disordered and have a higher number of possible arrangements.

It's important to note that entropy is not limited to the physical layout of particles but also takes energy into account. The energy distribution among particles adds further complexity to the number of arrangements available. Therefore, when considering entropy, both the spatial arrangement of particles and their energy levels influence the total number of possibilities and contribute to the overall entropy of the system.

Entropy has significant implications in many chemical and physical processes. For example, when a solid substance dissolves in a solvent, the system's entropy generally increases. The solid particles separate and disperse, leading to a greater number of arrangements and energy possibilities. Conversely, in processes where order or structure is established, such as when a gas condenses into a liquid, the system's entropy typically decreases due to the reduction in the number of possible arrangements.

Problem 1: Explain the concept of entropy and its relationship to the number of possible arrangements in a system.

Solution: Entropy measures the disorder or randomness in a system. It is directly related to the number of ways particles and their energy can be arranged. As the number of possible arrangements increases, so does the entropy of the system.

Problem 2: Compare and contrast the entropy values of a system with highly ordered and highly disordered arrangements.

Solution: A system with highly ordered arrangements has low entropy because there are fewer ways the particles and their energy can be distributed. In contrast, a system with highly disordered arrangements has high entropy due to the abundance of possible arrangements.

Problem 3: Calculate the entropy change when a gas expands into a larger volume.

Solution: When a gas expands into a larger volume, the number of possible arrangements of its particles and their energy increases. Consequently, the system experiences an increase in entropy.

Problem 4: Explain why melting a solid increases its entropy.

Solution: Melting a solid increases its entropy because the particles gain more freedom to move around and occupy different positions. The number of possible arrangements of particles and their energy subsequently increases, leading to a higher entropy value.

Problem 5: Describe the relationship between entropy and the second law of thermodynamics.

Solution: The second law of thermodynamics states that the entropy of an isolated system tends to increase over time. This law suggests that natural processes tend to move towards states of higher disorder, reflecting an increase in the number of possible arrangements and energy distributions.

Problem 6: Compare the entropy values of a solid and a gas at the same temperature.

Solution: In general, gases have higher entropies compared to solids at the same temperature. This is because gases have a greater number of possible arrangements due to the increased freedom of movement and energy distribution among particles.

Problem 7: Explain why mixing two substances can lead to an increase in entropy.

Solution: Mixing two substances can increase entropy because it allows for a greater number of possible arrangements and energy distributions. The particles from each substance become intermingled, resulting in increased disorder and randomness.

Problem 8: Calculate the entropy change when a gas is compressed into a smaller volume.

Solution: When a gas is compressed into a smaller volume, the number of possible arrangements decreases as the particles become more confined. Consequently, the entropy of the system decreases.

Problem 9: Discuss the relationship between entropy and the spontaneity of a chemical reaction.

Solution: The spontaneity of a chemical reaction is related to the change in entropy. Generally, reactions tend to be spontaneous if the overall change in entropy is positive, indicating an increase in disorder. However, this relationship can be influenced by other factors such as enthalpy changes.

Problem 10: Explain why a reversible process results in no net change in the total entropy of a system.

Solution: In a reversible process, the system undergoes a series of equilibrium states. Since the system returns to its initial state, there is no net change in the number of possible arrangements or the entropy of the system.

Entropy Change Prediction & Explanation

During a change in state, such as melting, boiling, and dissolving, the sign of entropy changes can be predicted and explained. In simple terms, entropy refers to the measure of disorder or randomness within a system. When a substance changes its state, the arrangement of its particles also changes, affecting the level of disorder in the system.

1. Melting: When a solid substance melts, it transitions from a highly ordered, rigid structure to a less ordered, more fluid state. This change leads to an increase in entropy. The particles gain more freedom of movement, resulting in a higher degree of randomness.

2. Freezing: The reverse process of melting is freezing. When a substance freezes, it transitions from a liquid to a solid state. This change reduces the disorder in the system, resulting in a decrease in entropy. The particles become more organized and fixed in position.

3. Boiling: Boiling occurs when a liquid substance reaches its boiling point and transitions into a gas phase. Throughout this procedure, the system undergoes an augmentation in entropy. The liquid molecules gain enough energy to overcome intermolecular forces, leading to a more disordered arrangement of particles in the gas phase.

4. Condensation: The opposite of boiling is condensation. When a gas cools down and loses energy, it transforms into a liquid state. The condensation process reduces the disorder in the system, resulting in a decrease in entropy. The gas molecules become more closely packed and less randomly distributed.

5. Dissolving: When a solute dissolves in a solvent, the resulting mixture exhibits an increase in entropy. The solute molecules disperse throughout the solvent, increasing the disorder of the system. The greater the number of solute particles in the solvent, the higher the entropy change.

6. Precipitation: The reverse process of dissolving is precipitation. It occurs when a solute comes out of a solution and forms a solid. Precipitation leads to a decrease in entropy as the particles become more closely packed, and the disorder in the system decreases.

7. Sublimation: Sublimation is the transition of a solid directly into a gas phase without passing through the liquid state. During sublimation, the entropy of the system increases, as the solid particles gain greater freedom of movement in the gas phase.

8. Deposition: Deposition is the reverse process of sublimation. It involves the direct transition of a gas into a solid state. The deposition process reduces the disorder in the system, resulting in a decrease in entropy.

9. Mixing: When two or more substances mix together, the total entropy increases. This is because the individual particles become more randomly distributed, leading to a higher level of disorder in the system.

10. Separation: Conversely, when a mixture is separated into its individual components, the entropy of the system decreases. The particles become more organized and less randomly distributed, reducing the overall disorder.

11. Increased Complexity: Reactions or processes that lead to an increase in the complexity or number of molecules tend to have a positive entropy change. The introduction of new species or the formation of more intricate structures increases the disorder within the system.

12. Decreased Complexity: Conversely, reactions or processes that result in a decrease in complexity or the breakdown of complex molecules tend to have a negative entropy change. The reduction in the number of species or the disintegration of intricate structures decreases the disorder in the system.

13. Temperature Effects: Higher temperatures generally lead to greater entropy changes. As the temperature increases, the kinetic energy of particles also increases, allowing for more extensive motion and disorder.

14. Phase Transitions: Phase transitions, such as melting, boiling, freezing, and condensation, involve significant entropy changes due to the rearrangement of particles and changes in intermolecular forces.

15. Energy Transfer: Processes that involve energy transfer, such as absorption or release of heat, can impact entropy changes. Energy transfer can affect the motion and randomness of particles, thereby influencing entropy.

16. Solubility and Solutions: Soluble substances dissolving in solvents usually result in positive entropy changes due to the mixing of particles and increased disorder. The solubility of a substance in a particular solvent depends on the balance of entropy and energy considerations.

17. Reaction Rates: In some cases, the rate of a reaction can influence the sign and magnitude of the entropy change. The dynamics of the reaction, including the order of reactants and products, can impact entropy.

18. Equilibrium: When a system reaches equilibrium, there is no further change in entropy. The forward and reverse processes occur at equal rates, resulting in a balance between disorder and order within the system.

19. Entropy and Entropy Change: Entropy change is a measure of how much disorder or randomness changes during a process or reaction. Positive entropy change signifies an increase in disorder, while negative entropy change indicates a decrease in disorder.

Problem 1: What is the sign of entropy change during the melting of a solid substance?

Solution: The melting of a solid substance involves a change from a highly ordered state to a less ordered state. Therefore, the sign of the entropy change during melting is positive (+).

Problem 2: What is the sign of entropy change during the freezing of a liquid?

Solution: The freezing of a liquid involves a change from a less ordered state to a more ordered state. Consequently, the sign of the entropy change during freezing is negative (-).

Problem 3: What is the sign of entropy change during boiling?

Solution: Boiling is a process where a liquid transitions to a gas. This transition results in an increase in disorder, making the entropy change during boiling positive (+).

Problem 4: What is the sign of entropy change during condensation?

Solution: Condensation is the process where a gas transforms into a liquid, reducing disorder. Thus, the entropy change during condensation is negative (-).

Problem 5: What is the sign of entropy change during dissolving of a solute in a solvent?

Solution: Dissolving involves the mixing of solute particles within a solvent, leading to an increase in disorder. Therefore, the entropy change during dissolving is positive (+).

Problem 6: What is the sign of entropy change during precipitation?

Solution: Precipitation occurs when a dissolved solute comes out of the solution and forms a solid. This process reduces disorder and results in a negative entropy change (-).

Problem 7: What is the sign of entropy change during sublimation?

Answer: Sublimation refers to the transformation in which a solid undergoes a direct change into a gaseous state. Since this transition increases disorder, the entropy change during sublimation is positive (+).

Problem 8: What is the sign of entropy change during deposition?

Solution: Deposition is the process where a gas transforms directly into a solid. This transition reduces disorder, leading to a negative entropy change (-).

Problem 9: How does temperature affect the sign of entropy change during phase transitions?

Solution: Higher temperatures generally lead to greater entropy changes during phase transitions. An increase in temperature enhances the motion and randomness of particles, resulting in positive entropy changes (+).

Problem 10: How does the complexity of a reaction affect the sign of entropy change?

Solution: Reactions that increase complexity or involve the formation of more molecules tend to have positive entropy changes (+). Conversely, reactions that decrease complexity or involve the breakdown of more complex molecules tend to have negative entropy changes (-).

Problem 11: How does mixing affect the sign of entropy change?

Solution: When substances mix, the total entropy of the system increases due to the increased disorder resulting from the random distribution of particles. Therefore, mixing generally leads to positive entropy changes (+).

Problem 12: How does separation of a mixture affect the sign of entropy change?

Solution: Separation of a mixture decreases the disorder in the system as the individual components become more organized. Consequently, separation typically leads to negative entropy changes (-).

Problem 13: Can the rate of a reaction affect the sign of entropy change?

Solution: Yes, the rate of a reaction can influence the sign and magnitude of entropy change. The dynamics of the reaction, including the order of reactants and products, can affect entropy. However, the overall impact depends on specific reaction conditions and cannot be generalized.

Problem 14: How does energy transfer impact the sign of entropy change?

Solution: Energy transfer, such as heat absorption or release, can affect the entropy change. The impact depends on the extent to which energy transfer affects the motion and randomness of particles. In general, energy transfer can influence entropy changes.

Problem 15: What is the relationship between the sign of entropy change and equilibrium?

Solution: At equilibrium, the forward and reverse processes occur at equal rates, resulting in no further entropy change. Therefore, the sign of entropy change at equilibrium is zero (0).

Entropy Change during Temperature Change

In the world of chemistry, temperature plays a crucial role in understanding the behavior of substances. When a temperature change occurs, it can have a significant impact on the entropy, which measures the disorder or randomness within a system. Let's explore the sign of entropy changes during a temperature change, such as heating or cooling, in simple yet precise terms.

1. Heating a Substance: When a substance is heated, its temperature increases, and so does its entropy. The additional energy provided by heating causes the particles to move more vigorously and randomly, increasing the disorder within the system. As a result, the entropy change during heating is positive (+).

2. Cooling a Substance: Conversely, when a substance is cooled, its temperature decreases, leading to a decrease in entropy. The reduction in temperature decreases the kinetic energy of the particles, causing them to move with less random motion and reducing the disorder within the system. Therefore, the entropy change during cooling is negative (-).

3. Temperature and Disorder: The relationship between temperature and entropy is based on the principle that disorder tends to increase with higher temperatures. At higher temperatures, particles have more energy, resulting in greater molecular motion and increased randomness. Thus, decreasing the temperature reduces the disorder and, subsequently, the entropy.

4. Absolute Zero: At the temperature of absolute zero (-273.15°C or 0 Kelvin), the entropy of any pure crystalline substance is zero. This is because, at absolute zero, all molecular motion ceases, and the system reaches its most ordered state.

5. Phase Transitions: Temperature changes also influence entropy during phase transitions, such as melting, boiling, freezing, and condensation. As temperature increases, the entropy generally increases due to the greater energy available for molecular motion and increased disorder associated with the transition to a higher energy state.

6. Energy Transfer: Temperature changes often involve energy transfer, such as the absorption or release of heat. Energy transfer, which alters the motion of particles, affects the entropy change accordingly. Absorbing heat generally increases entropy, while releasing heat decreases entropy.

7. Disorderly Gases: In the gaseous state, higher temperatures lead to greater molecular motion and increased disorder. As the temperature rises, gas particles move more rapidly and occupy a larger volume, resulting in increased entropy.

8. Solids and Liquids: Temperature changes also impact entropy in solids and liquids. When a solid is heated, it transitions to a liquid, and the entropy generally increases due to the increased disorder associated with the particles' increased freedom of movement.

9. Temperature and Reaction Rate: It's vital to note that while temperature changes affect entropy, they are also closely related to reaction rate. In general, increasing the temperature enhances reaction rates by providing more energy for reactant particles to collide, leading to a greater number of successful collisions and higher reaction rates.

10. Absolute Entropy: Entropy itself is temperature-dependent, and substances have different absolute entropy values at a given temperature. The absolute entropy of a substance generally increases as the temperature rises because higher temperatures allow for more molecular motion and disorder.

11. Entropy Change in Chemical Reactions: Temperature changes can influence entropy changes in chemical reactions. When a reaction produces more gas molecules or involves the formation of products with higher entropy than the reactants, an increase in temperature generally has a positive impact on the entropy change.

12. Entropy Change in Endothermic Reactions: Endothermic reactions, which absorb heat from the surroundings, generally exhibit positive entropy changes with increasing temperature. The increase in energy input leads to greater molecular motion and disorder, resulting in higher entropy.

13. Entropy Change in Exothermic Reactions: Exothermic reactions release heat energy to the surroundings. In such reactions, a decrease in temperature generally has a negative impact on the entropy change since lowering the temperature decreases the disorder and molecular motion, resulting in reduced entropy.

14. Resolving Entropy Confusion: It is crucial to distinguish between entropy change and absolute entropy. Entropy change refers to the difference in entropy between initial and final states, while absolute entropy refers to the total entropy of a substance at a particular temperature.

15. Entropy and Equilibrium: Entropy considerations are essential in understanding systems at equilibrium. At equilibrium, entropy change is typically zero since the rate of the forward and reverse processes becomes equal, resulting in no net change in entropy.

16. Reversible Processes: In reversible processes, temperature changes have a significant influence on entropy change. For reversible changes, the entropy change can be calculated using the equation $\Delta S = q/T$, where q represents the heat transferred, and T is the temperature.

17. Entropy in Real Processes: In real processes, especially those involving irreversible changes, calculating entropy change becomes more complex. The dynamic nature of real-world systems often involves factors that go beyond simple temperature changes, making the entropy calculation more challenging.

18. Entropy and Nature: The increase in entropy with temperature generally aligns with the trend of nature seeking higher disorder. The second law of thermodynamics states that the entropy of an isolated system tends to increase over time, leading to a more disordered state.

19. Important Considerations: While temperature change often correlates with entropy change, it is crucial to analyze the specific conditions and molecular-level interactions of a system to accurately determine entropy changes and predict their effect on the behavior of substances.

20. Experimental Determination: Determining entropy changes experimentally involves measuring temperature changes, tracking heat flow, and analyzing other factors affecting disorder or randomness. Laboratory techniques and data analysis aid in understanding entropy changes during temperature variations.

During a temperature change, the sign of the entropy change can be predicted using the following principles:

1. For an endothermic process (heat absorbed from the surroundings), the entropy change is usually positive. This is because an increase in temperature leads to an increase in the randomness of particle motions and energy distribution, resulting in a higher degree of disorder.

2. For an exothermic process (heat released to the surroundings), the entropy change is usually negative. This is because a decrease in temperature causes the particles to move with less randomness and energy distribution, resulting in a lower degree of disorder.

3. For phase transitions (such as melting, vaporization, or condensation), the sign of the entropy change depends on the nature of the transition and the temperature conditions. Generally, solid to liquid or liquid to gas transitions have positive entropy changes, indicating an increase in disorder. On the other hand, gas to liquid or liquid to solid transitions have negative entropy changes, indicating a decrease in disorder.

4. For reactions involving gases, changes in volume or pressure can impact the entropy change. An increase in volume or a decrease in pressure leads to a positive entropy change, as the gas molecules have more freedom of movement and higher disorder. Conversely, a decrease in volume or an increase in pressure leads to a negative entropy change, as the gas molecules have less space to move and lower disorder.

Here are some sample problems based on questions that students might face in a chemistry exam:

1. Question: Does the entropy of water increase or decrease when it freezes at 0°C?

Solution: The entropy decreases when water freezes because the transition from a liquid to a solid state results in a decrease in disorder.

2. Question: What happens to the entropy of a gas when its pressure is increased?

Solution: The entropy of a gas decreases when the pressure is increased because the available volume for the gas molecules to move decreases, resulting in lower disorder.

3. Question: Does the entropy of a system increase or decrease when it absorbs heat from the surroundings at constant temperature?

Solution: The entropy of the system increases when it absorbs heat because the transfer of energy leads to an increase in the randomness of particle motions and energy distribution, resulting in higher disorder.

4. Question: Is the entropy change positive or negative when a liquid boils and converts to a gas?

Solution: The entropy change is positive when a liquid boils because the transition from a liquid to a gas state results in an increase in disorder.

Entropy Changes with Gaseous Molecules

Relationship between entropy and gas molecules: Gaseous molecules have higher entropy compared to liquids and solids because they can move freely and occupy a larger volume.

Significance of changes in the number of gaseous molecules: Reactions in which the number of gaseous molecules changes can have a significant impact on the overall entropy change.

Increase in the number of gaseous molecules: When the reactants have fewer gaseous molecules than the products, an increase in entropy occurs. This is because the system transitions from a more ordered state (fewer gas molecules) to a more disordered state (more gas molecules).

Example: Consider the reaction $N_2(g) + 3H_2(g) \rightarrow 2NH_3(g)$. In this reaction, the number of gas molecules increases from 4 (2 + 3) to 6 (2 * 3), resulting in an increase in entropy.

Simplification: In simple terms, if the reaction produces more gas molecules, the entropy change is positive, indicating an increase in disorder.

Decrease in the number of gaseous molecules: Conversely, if the reactants have more gaseous molecules than the products, a decrease in entropy occurs. This indicates a transition from a more disordered state (more gas molecules) to a more ordered state (fewer gas molecules).

Example: Let's consider the reaction $2NO_2(g) \rightarrow N_2O_4(g)$. Here, the number of gas molecules decreases from 2 to 1, resulting in a decrease in entropy.

Simplification: In simple terms, if the reaction produces fewer gas molecules, the entropy change is negative, indicating a decrease in disorder.

Equilibrium and entropy change: It's important to note that an entropy change alone does not dictate the direction of a reaction. The reaction will proceed in the direction that minimizes the free energy of the system.

Factors influencing entropy change: Other factors such as temperature and pressure can also affect the entropy change in a reaction involving gases.

Temperature increase: Increasing the temperature generally increases the disorder of a system, leading to an increase in entropy.

Pressure increase: Increasing the pressure tends to decrease the volume occupied by the gas molecules, resulting in a decrease in entropy.

Significance of entropy change: The sign of the entropy change provides insights into the spontaneity and feasibility of a reaction. Positive entropy changes favor spontaneous reactions, while negative entropy changes can hinder spontaneity.

Connection to the second law of thermodynamics: The second law of thermodynamics states that the entropy of the universe always increases for a spontaneous process. The sign of the entropy change helps determine if a reaction aligns with this law.

Application in chemical industry: Understanding the sign of entropy change is crucial in industrial processes, such as the Haber-Bosch process for ammonia synthesis, where the entropy change affects reaction efficiency.

Connection to entropy and energy dispersal: Entropy is related to the dispersal of energy in a system. Reactions involving changes in the number of gas molecules impact the dispersal of energy and, hence, the overall entropy change.

Summarizing the concept: In summary, reactions with an increase in the number of gaseous molecules generally have a positive entropy change, indicating an increase in disorder. Reactions with a decrease in the number of gaseous molecules result in a negative entropy change, indicating a decrease in disorder.

Question: Predict the sign of the entropy change in the reaction $A(g) + 2B(g) \rightarrow 3C(g)$.

Solution: In this reaction, the number of gaseous molecules increases from 3 (1 + 2) to 3, resulting in an increase in entropy. Therefore, the entropy change is positive.

Question: Determine the sign of the entropy change when one mole of $NH_3(g)$ decomposes to form $N_2(g)$ and $3H_2(g)$.

Solution: The number of gaseous molecules decreases from 4 (1 + 3) to 3, resulting in a decrease in entropy. Hence, the entropy change is negative.

Question: What is the sign of the entropy change when 2 moles of $H_2(g)$ react with 1 mole of $O_2(g)$ to produce 2 moles of $H_2O(g)$?

Solution: The number of gaseous molecules decreases from 3 (2 + 1) to 2, resulting in a decrease in entropy. Therefore, the entropy change is negative.

Question: Consider the reaction $CO_2(g) + H_2(g) \rightarrow CO(g) + H_2O(g)$. Predict the sign of the entropy change.

Solution: The number of gaseous molecules remains the same before and after the reaction (2 + 1 = 1 + 1). Therefore, there is no change in the entropy, and the entropy change is zero.

Question: Determine the sign of the entropy change in the reaction $2SO_2(g) + O_2(g) \rightarrow 2SO_3(g)$.

Solution: The number of gaseous molecules decreases from 4 (2 + 1) to 2, resulting in a decrease in entropy. Hence, the entropy change is negative.

Question: What is the sign of the entropy change when N2H4(g) decomposes into N2(g) and 2H2(g)?

Solution: The number of gaseous molecules increases from 1 to 3 (1 = 1 + 2), resulting in an increase in entropy. Therefore, the entropy change is positive.

Question: Predict the sign of the entropy change when 2NH4Cl(s) decomposes to produce 2NH3(g) + H2(g) + Cl2(g).

Solution: The reactant is a solid while the products are gases, resulting in an increase in the number of gaseous molecules. Therefore, the entropy change is positive.

Question: Determine the sign of the entropy change when 2 moles of HF(g) react with 1 mole of H2O(l) to produce 1 mole of H3O+(aq) and 1 mole of F-(aq).

Solution: The number of gaseous molecules remains the same (2) while the number of particles in solution increases. Overall, there is an increase in entropy, so the entropy change is positive.

Question: In the reaction CO(g) + H2O(g) -> CO2(g) + H2(g), predict the sign of the entropy change.

Solution: The number of gaseous molecules remains the same before and after the reaction (1 + 1 = 1 + 1). Therefore, there is no change in the entropy, and the entropy change is zero.

Question: What is the sign of the entropy change when CaCO3(s) decomposes to form CaO(s) and CO2(g)?

Solution: The reactant is a solid, while one product is a solid and the other is a gas. The change from solid to gas leads to an increase in entropy, so the entropy change is positive.

Entropy Change Calculation Formula

Here are a few practice problems related to calculating the entropy change (ΔS) for chemical reactions, along with their solutions:

Calculate the entropy change for the reaction: $2H_2(g) + O_2(g) \rightarrow 2H_2O(g)$. Given standard entropies: $S^{\diamond}(H_2)$ = 130.7 J/mol•K, $S^{\diamond}(O_2)$ = 205.0 J/mol•K, $S^{\diamond}(H_2O)$ = 188.8 J/mol•K.

Solution: $\Delta S^{\diamond} = \Sigma S^{\diamond}$(products) - $\Sigma S^{\diamond}$(reactants)

= 2×$S^{\diamond}(H_2O)$ - [2×$S^{\diamond}(H_2)$ + $S^{\diamond}(O_2)$]

= 2×188.8 J/mol•K - [2×130.7 J/mol•K + 205.0 J/mol•K] = 377.6 J/mol•K - 466.4 J/mol•K

= -88.8 J/mol•K

Therefore, the entropy change for the given reaction is -88.8 J/mol•K.

Calculate the entropy change for the reaction: C(graphite) + 2H_2(g) → CH_4(g).

Given standard entropies: $S^{\diamond}(C)$ = 5.7 J/mol•K, $S^{\diamond}(H_2)$ = 130.7 J/mol•K, $S^{\diamond}(CH_4)$ = 186.2 J/mol•K.

Solution: $\Delta S^{\diamond} = \Sigma S^{\diamond}$(products) - $\Sigma S^{\diamond}$(reactants)

= $S^{\diamond}(CH_4)$ - [$S^{\diamond}(C)$ + 2×$S^{\diamond}(H_2)$]

= 186.2 J/mol•K - [5.7 J/mol•K + 2×130.7 J/mol•K]

= 186.2 J/mol•K - 267.1 J/mol•K = -80.9 J/mol•K

Therefore, the entropy change for the given reaction is -80.9 J/mol•K.

Calculate the entropy change for the reaction: $2NH_3(g) + 3O_2(g) \rightarrow N_2(g) + 6H_2O(g)$.

Given standard entropies: $S^{\diamond}(NH_3)$ = 192.8 J/mol•K, $S^{\diamond}(O_2)$ = 205.0 J/mol•K, $S^{\diamond}(N_2)$ = 191.5 J/mol•K, $S^{\diamond}(H_2O)$ = 188.8 J/mol•K.

Solution: $\Delta S^{\diamond} = \Sigma S^{\diamond}$(products) - $\Sigma S^{\diamond}$(reactants)

= $S^{\diamond}(N_2)$ + 6×$S^{\diamond}(H_2O)$ - [2×$S^{\diamond}(NH_3)$ + 3×$S^{\diamond}(O_2)$]

= 191.5 J/mol•K + 6×188.8 J/mol•K - [2×192.8 J/mol•K + 3×205.0 J/mol•K]

= 191.5 J/mol•K + 1132.8 J/mol•K - 385.6 J/mol•K - 615.0 J/mol•K

= 1339.2 J/mol•K - 1000.6 J/mol•K

= 338.6 J/mol•K

Gibbs free energy change, ΔG

Gibbs Equation: Energy Changes

The Gibbs equation, $\Delta G^\diamond = \Delta H^\diamond - T\Delta S^\diamond$, is a fundamental equation in thermodynamics that relates the change in Gibbs free energy to the change in enthalpy, temperature, and entropy at standard conditions.

To fully comprehend the significance of the Gibbs equation, it is imperative to understand the components it comprises. $\Delta G^\diamond$ represents the change in Gibbs free energy under standard conditions, $\Delta H^\diamond$ denotes the change in enthalpy also at standard conditions, T symbolizes temperature, and $\Delta S^\diamond$ stands for the change in entropy, again at standard conditions.

Gibbs free energy, a key concept in thermodynamics, provides information regarding the spontaneity and equilibrium of chemical reactions. By measuring the difference between the enthalpy and entropy, the Gibbs equation quantifies the available energy in a system.

In a chemical reaction, $\Delta G^\diamond$ can assume three distinct values: positive, negative, or zero. A positive $\Delta G^\diamond$ indicates that the reaction is non-spontaneous and requires an input of energy to occur. Conversely, a negative $\Delta G^\diamond$ denotes that the reaction is spontaneous and releases energy. Finally, a $\Delta G^\diamond$ of zero signifies that the system is at equilibrium, with no net change in Gibbs free energy.

The enthalpy, $\Delta H^\diamond$, represents the total heat exchanged with the surroundings during a reaction at constant pressure. It accounts for the changes in chemical bonds and intermolecular forces. A positive $\Delta H^\diamond$ indicates an endothermic reaction where heat is absorbed, while a negative $\Delta H^\diamond$ signifies an exothermic reaction where heat is released.

Temperature, denoted by T in the equation, is expressed in Kelvin and represents the average kinetic energy of the molecules in a system. It influences the rate of reactions and determines their spontaneity. As temperature increases, the likelihood of molecules possessing sufficient energy to overcome the activation energy barrier also increases.

Entropy, $\Delta S^\diamond$, quantifies the degree of disorder or randomness in a system. It characterizes the distribution and arrangements of particles. A positive $\Delta S^\diamond$ denotes an increase in disorder, while a negative $\Delta S^\diamond$ suggests a decrease. Spontaneous reactions tend to have a positive $\Delta S^\diamond$, as they result in a more disordered system.

By combining these thermodynamic parameters in the Gibbs equation, we can assess the feasibility of a chemical reaction under standard conditions. If $\Delta G^\diamond$ is negative, the reaction is spontaneous and will proceed without external intervention. If $\Delta G^\diamond$ is positive, the reaction will not occur spontaneously and requires an input of energy. If $\Delta G^\diamond$ is zero, the reaction is at equilibrium, with no net change in the system.

It is important to note that the Gibbs equation is only applicable to reactions taking place under standard conditions. It provides valuable insight into the thermodynamics of chemical reactions but does not account for factors such as concentration, pressure, or non-standard temperatures.

Problem 1: Given the values of $\Delta H^\diamond$ and $\Delta S^\diamond$, calculate $\Delta G^\diamond$ at a given temperature, T.

Solution 1: Use the equation $\Delta G^\diamond = \Delta H^\diamond - T\Delta S^\diamond$ and substitute the given values of $\Delta H^\diamond$, $\Delta S^\diamond$, and T to calculate $\Delta G^\diamond$.

Problem 2: Determine the spontaneity of a reaction based on the values of $\Delta H^\diamond$ and $\Delta S^\diamond$.

Solution 2: Evaluate the sign of $\Delta G^\diamond$. If $\Delta G^\diamond$ is negative, the reaction is spontaneous. If $\Delta G^\diamond$ is positive, the reaction is non-spontaneous. If the standard Gibbs free energy change ($\Delta G^\diamond$) is equal to zero, it indicates that the reaction has reached a state of equilibrium.

Problem 3: Calculate the standard enthalpy change, $\Delta H^\diamond$, given the values of $\Delta G^\diamond$ and $\Delta S^\diamond$ at a given temperature, T.

Solution 3: Rearrange the Gibbs equation to solve for ΔH° as $\Delta H^{\circ} = \Delta G^{\circ} + T\Delta S^{\circ}$. Substitute the values of ΔG°, ΔS°, and T to find the standard enthalpy change.

Problem 4: Estimate the temperature at which a reaction becomes spontaneous based on the values of ΔH° and ΔS°.

Solution 4: Rearrange the Gibbs equation to solve for the temperature, T, as $T = \Delta H^{\circ}/\Delta S^{\circ}$. Substitute the values of ΔH° and ΔS° to find the temperature at which the reaction becomes spontaneous.

Problem 5: Given the value of ΔG°, determine the range of temperatures at which the reaction is spontaneous.

Solution 5: Rearrange the Gibbs equation to solve for the temperature, T, as $T = (\Delta H^{\circ} - \Delta G^{\circ})/\Delta S^{\circ}$. Substitute the values of ΔH°, ΔG°, and ΔS° to find the temperature range for the reaction to be spontaneous.

Problem 6: Compare the values of ΔH° and ΔS° to predict the spontaneity of a reaction.

Solution 6: If ΔH° is negative (exothermic) and ΔS° is positive (increase in disorder), the reaction is likely to be spontaneous. However, other factors such as the temperature need to be considered to make a conclusive assessment.

Problem 7: Given the values of ΔH° and ΔG°, determine the value of ΔS° at a given temperature, T.

Solution 7: Rearrange the Gibbs equation to solve for ΔS° as $\Delta S^{\circ} = (\Delta H^{\circ} - \Delta G^{\circ})/T$. Substitute the values of ΔH°, ΔG°, and T to calculate the standard entropy change.

Problem 8: Estimate the change in Gibbs free energy, ΔG°, based on the values of ΔH°, ΔS°, and temperature, T.

Solution 8: Use the Gibbs equation $\Delta G^{\circ} = \Delta H^{\circ} - T\Delta S^{\circ}$ and substitute the given values of ΔH°, ΔS°, and T to calculate the change in Gibbs free energy, ΔG°.

Problem 9: Given the values of ΔG° and ΔH°, find the temperature, T, at which the reaction is at equilibrium.

Solution 9: Rearrange the Gibbs equation to solve for the temperature, T, as $T = \Delta H^{\circ}/\Delta S^{\circ}$. Substitute the values of ΔH° and ΔS° to find the temperature at which the reaction is at equilibrium.

Problem 10: Determine the spontaneity of a reaction based on the values of ΔG° and temperature, T.

Solution 10: Evaluate the sign of ΔG° at the given temperature, T. If ΔG° is negative, the reaction is spontaneous. If ΔG° is positive, the reaction is non-spontaneous. If ΔG° is zero, the reaction is at equilibrium.

Calculating ΔG° with Gibbs Equation

Here are 20 chemistry exam-style problems with solutions based on the equation $\Delta G^{\circ} = \Delta H^{\circ} - T\Delta S^{\circ}$:

Question 1: Calculate the standard Gibbs free energy change (ΔG°) for a reaction with a standard enthalpy change (ΔH°) of -150 kJ/mol and a standard entropy change (ΔS°) of 75 J/(mol·K) at a temperature (T) of 298 K.

Solution: $\Delta G^{\circ} = \Delta H^{\circ} - T\Delta S^{\circ}$

$\Delta G^{\circ} = -150 \text{ kJ/mol} - (298 \text{ K})(75 \text{ J/(mol·K)})$

$\Delta G^{\circ} = -150 \text{ kJ/mol} - 22.35 \text{ kJ/mol}$

$\Delta G^{\circ} = -172.35 \text{ kJ/mol}$

Question 2: A reaction has a standard entropy change (ΔS°) of 60 J/(mol·K) and a standard Gibbs free energy change (ΔG°) of 40 kJ/mol at 298 K. Determine the standard enthalpy change (ΔH°) of the reaction.

Solution: $\Delta G^{\circ} = \Delta H^{\circ} - T\Delta S^{\circ}$

$40 \text{ kJ/mol} = \Delta H^{\circ} - (298 \text{ K})(60 \text{ J/(mol·K)})$

$40 \text{ kJ/mol} = \Delta H^{\circ} - 17.88 \text{ kJ/mol}$

$\Delta H^{\circ} = 57.88 \text{ kJ/mol}$

Question 3: Given that a reaction has a standard Gibbs free energy change (ΔG°) of -100 kJ/mol and a standard enthalpy change (ΔH°) of -50 kJ/mol, calculate the standard entropy change (ΔS°) for the reaction at 298 K.

Solution: $\Delta G^\circ = \Delta H^\circ - T\Delta S^\circ$

-100 kJ/mol = -50 kJ/mol - (298 K)(ΔS°)

-100 kJ/mol = -50 kJ/mol - (298 K)(ΔS°)

-50 kJ/mol = (298 K)(ΔS°)

ΔS° = -0.167 J/(mol·K)

Question 4: Calculate the temperature at which the standard Gibbs free energy change (ΔG°) for a reaction becomes negative, given a standard enthalpy change (ΔH°) of -200 kJ/mol and a standard entropy change (ΔS°) of 100 J/(mol·K).

Solution: $\Delta G^\circ = \Delta H^\circ - T\Delta S^\circ$

0 = -200 kJ/mol - T(100 J/(mol·K))

T = -200 kJ/mol / (100 J/(mol·K))

T = 2000 K

Question 5: A reaction has a standard entropy change (ΔS°) of -50 J/(mol·K) and a standard Gibbs free energy change (ΔG°) of -30 kJ/mol at 298 K. Compute the standard enthalpy change (ΔH°) associated with the reaction.

Solution: $\Delta G^\circ = \Delta H^\circ - T\Delta S^\circ$

-30 kJ/mol = ΔH° - (298 K)(-50 J/(mol·K))

-30 kJ/mol = ΔH° + 14.9 kJ/mol

ΔH° = -44.9 kJ/mol

Question 6: Calculation of ΔG°, ΔH°, and ΔS° is based on which conditions?

Solution: ΔG°, ΔH°, and ΔS° are calculated under standard conditions, which include a temperature of 298 K, a pressure of 1 atm, and a concentration of 1 M for all species involved.

Question 7: For a reaction with a standard Gibbs free energy change (ΔG°) of -20 kJ/mol and a standard enthalpy change (ΔH°) of 40 kJ/mol, calculate the standard entropy change (ΔS°) at 298 K.

Solution: $\Delta G^\circ = \Delta H^\circ - T\Delta S^\circ$

-20 kJ/mol = 40 kJ/mol - (298 K)(ΔS°)

-60 kJ/mol = (298 K)(ΔS°)

ΔS° = -0.201 J/(mol·K)

Question 8: At what temperature will a reaction with a standard enthalpy change (ΔH°) of 80 kJ/mol and a standard entropy change (ΔS°) of 40 J/(mol·K) have a standard Gibbs free energy change (ΔG°) of zero?

Solution: $\Delta G^\circ = \Delta H^\circ - T\Delta S^\circ$

0 = 80 kJ/mol - T(40 J/(mol·K))

T = 80 kJ/mol / (40 J/(mol·K))

T = 2000 K

Question 9: Given that a reaction has a standard entropy change (ΔS°) of -80 J/(mol·K) and a standard Gibbs free energy change (ΔG°) of 60 kJ/mol, calculate the standard enthalpy change (ΔH°) for the reaction at 298 K.

Solution: $\Delta G^\circ = \Delta H^\circ - T\Delta S^\circ$

60 kJ/mol = ΔH° - (298 K)(-80 J/(mol·K))

60 kJ/mol = ΔH° + 23.84 kJ/mol

ΔH° = 36.16 kJ/mol

Question 10: Calculate the standard Gibbs free energy change ($\Delta G^{\Diamond}$) for a reaction with a standard enthalpy change ($\Delta H^{\Diamond}$) of 120 kJ/mol and a standard entropy change ($\Delta S^{\Diamond}$) of 20 J/(mol·K) at a temperature (T) of 298 K.

Solution: $\Delta G^{\Diamond} = \Delta H^{\Diamond} - T\Delta S^{\Diamond}$

$\Delta G^{\Diamond} = 120$ kJ/mol - (298 K)(20 J/(mol·K))

$\Delta G^{\Diamond} = 120$ kJ/mol - 5.96 kJ/mol

$\Delta G^{\Diamond} = 114.04$ kJ/mol

Question 11: A reaction has a standard entropy change ($\Delta S^{\Diamond}$) of 40 J/(mol·K) and a standard Gibbs free energy change ($\Delta G^{\Diamond}$) of -10 kJ/mol at 298 K. Calculate the standard enthalpy change ($\Delta H^{\Diamond}$) for the reaction.

Solution: $\Delta G^{\Diamond} = \Delta H^{\Diamond} - T\Delta S^{\Diamond}$

-10 kJ/mol = $\Delta H^{\Diamond}$ - (298 K)(40 J/(mol·K))

-10 kJ/mol = $\Delta H^{\Diamond}$ + 11.92 kJ/mol

$\Delta H^{\Diamond} = -21.92$ kJ/mol

Question 12: Given that a reaction has a standard Gibbs free energy change ($\Delta G^{\Diamond}$) of -60 kJ/mol and a standard enthalpy change ($\Delta H^{\Diamond}$) of -20 kJ/mol, calculate the standard entropy change ($\Delta S^{\Diamond}$) for the reaction at 298 K.

Solution: $\Delta G^{\Diamond} = \Delta H^{\Diamond} - T\Delta S^{\Diamond}$

-60 kJ/mol = -20 kJ/mol - (298 K)($\Delta S^{\Diamond}$)

-40 kJ/mol = (298 K)($\Delta S^{\Diamond}$)

$\Delta S^{\Diamond} = -0.134$ J/(mol·K)

Question 13: Calculate the temperature at which the standard Gibbs free energy change ($\Delta G^{\Diamond}$) for a reaction becomes zero, given a standard enthalpy change ($\Delta H^{\Diamond}$) of -30 kJ/mol and a standard entropy change ($\Delta S^{\Diamond}$) of 20 J/(mol·K).

Solution: $\Delta G^{\Diamond} = \Delta H^{\Diamond} - T\Delta S^{\Diamond}$

0 = -30 kJ/mol - T(20 J/(mol·K))

T = -30 kJ/mol / (20 J/(mol·K))

T = 1500 K

Question 14: A reaction has a standard entropy change ($\Delta S^{\Diamond}$) of -20 J/(mol·K) and a standard Gibbs free energy change ($\Delta G^{\Diamond}$) of -40 kJ/mol at 298 K. Calculate the standard enthalpy change ($\Delta H^{\Diamond}$) for the reaction.

Solution: $\Delta G^{\Diamond} = \Delta H^{\Diamond} - T\Delta S^{\Diamond}$

-40 kJ/mol = $\Delta H^{\Diamond}$ - (298 K)(-20 J/(mol·K))

-40 kJ/mol = $\Delta H^{\Diamond}$ + 5.96 kJ/mol

$\Delta H^{\Diamond} = -45.96$ kJ/mol

Question 15: At what temperature will a reaction with a standard enthalpy change ($\Delta H^{\Diamond}$) of 60 kJ/mol and a standard entropy change ($\Delta S^{\Diamond}$) of 30 J/(mol·K) have a standard Gibbs free energy change ($\Delta G^{\Diamond}$) of zero?

Solution: $\Delta G^{\Diamond} = \Delta H^{\Diamond} - T\Delta S^{\Diamond}$

0 = 60 kJ/mol - T(30 J/(mol·K))

T = 60 kJ/mol / (30 J/(mol·K))

T = 2000 K

Question 16: Given that a reaction has a standard entropy change ($\Delta S^{\Diamond}$) of -30 J/(mol·K) and a standard Gibbs free energy change ($\Delta G^{\Diamond}$) of 30 kJ/mol, calculate the standard enthalpy change ($\Delta H^{\Diamond}$) for the reaction at 298 K.

Solution: $\Delta G^{\Diamond} = \Delta H^{\Diamond} - T\Delta S^{\Diamond}$

30 kJ/mol = $\Delta H^{\Diamond}$ - (298 K)(-30 J/(mol·K))

$$30 \text{ kJ/mol} = \Delta H^\ominus + 8.94 \text{ kJ/mol}$$
$$\Delta H^\ominus = 21.06 \text{ kJ/mol}$$

Question 17: Calculate the standard Gibbs free energy change ($\Delta G^\ominus$) for a reaction with a standard enthalpy change ($\Delta H^\ominus$) of -80 kJ/mol and a standard entropy change ($\Delta S^\ominus$) of 50 J/(mol·K) at a temperature (T) of 298 K.

Solution: $\Delta G^\ominus = \Delta H^\ominus - T\Delta S^\ominus$
$$\Delta G^\ominus = -80 \text{ kJ/mol} - (298 \text{ K})(50 \text{ J/(mol·K)})$$
$$\Delta G^\ominus = -80 \text{ kJ/mol} - 14.9 \text{ kJ/mol}$$
$$\Delta G^\ominus = -94.9 \text{ kJ/mol}$$

Question 18: A reaction has a standard entropy change ($\Delta S^\ominus$) of 50 J/(mol·K) and a standard Gibbs free energy change ($\Delta G^\ominus$) of -60 kJ/mol at 298 K. Calculate the standard enthalpy change ($\Delta H^\ominus$) for the reaction.

Solution: $\Delta G^\ominus = \Delta H^\ominus - T\Delta S^\ominus$
$$-60 \text{ kJ/mol} = \Delta H^\ominus - (298 \text{ K})(50 \text{ J/(mol·K)})$$
$$-60 \text{ kJ/mol} = \Delta H^\ominus + 14.9 \text{ kJ/mol}$$
$$\Delta H^\ominus = -74.9 \text{ kJ/mol}$$

Question 19: For a reaction with a standard Gibbs free energy change ($\Delta G^\ominus$) of -30 kJ/mol and a standard enthalpy change ($\Delta H^\ominus$) of 60 kJ/mol, calculate the standard entropy change ($\Delta S^\ominus$) at 298 K.

Solution: $\Delta G^\ominus = \Delta H^\ominus - T\Delta S^\ominus$
$$-30 \text{ kJ/mol} = 60 \text{ kJ/mol} - (298 \text{ K})(\Delta S^\ominus)$$
$$-90 \text{ kJ/mol} = (298 \text{ K})(\Delta S^\ominus)$$
$$\Delta S^\ominus = -0.302 \text{ J/(mol·K)}$$

Question 20: Calculate the temperature at which the standard Gibbs free energy change ($\Delta G^\ominus$) for a reaction becomes negative, given a standard enthalpy change ($\Delta H^\ominus$) of 40 kJ/mol and a standard entropy change ($\Delta S^\ominus$) of 20 J/(mol·K).

Solution: $\Delta G^\ominus = \Delta H^\ominus - T\Delta S^\ominus$
$$0 = 40 \text{ kJ/mol} - T(20 \text{ J/(mol·K)})$$
$$T = 40 \text{ kJ/mol} / (20 \text{ J/(mol·K)})$$
$$T = 2000 \text{ K}$$

Feasibility Based on ΔG

Determining the feasibility of a reaction or process is a fundamental concept in chemistry. One approach to assess the feasibility is by analyzing the sign of ΔG, which represents the change in Gibbs free energy. In a system with constant temperature and pressure, the Gibbs free energy represents the energy that can be utilized to perform useful work.

An understanding of ΔG can provide insights into whether a reaction or process will occur spontaneously or if external energy input is required.

In chemical thermodynamics, a reaction or process is deemed feasible if ΔG is negative. A negative value indicates that the system tends to move towards a lower energy state and that the reaction or process can occur spontaneously. This implies that the products of the reaction or the final state of the process have lower Gibbs free energy than the reactants or initial state. Such reactions or processes are referred to as exergonic.

On the other hand, if ΔG is positive, it indicates that the system is in a higher energy state and that the reaction or process is not favored to occur spontaneously. In this case, external energy input is required to overcome the energy barrier and drive the reaction or process forward. Reactions or processes with positive ΔG are called endergonic.

In some cases, ΔG may be zero, indicating a balanced system with no net change in Gibbs free energy. This occurs when the reactants and products are in equilibrium, and the forward and reverse reactions proceed at the same rate. The system is at a thermodynamic steady state, where the free energy is at a minimum.

It is important to note that the sign of ΔG alone does not provide information about the rate at which a reaction or process occurs. While a negative ΔG suggests spontaneity, the kinetics of the reaction or process must also be considered to determine the actual rate of the transformation. Factors such as activation energy and reaction mechanisms play a role in kinetics.

Additionally, ΔG is influenced by temperature. The equation relating ΔG to temperature, $\Delta G = \Delta H - T\Delta S$, where ΔH represents the change in enthalpy and ΔS represents the change in entropy, allows for a more precise analysis of thermodynamic feasibility. At low temperatures, a reaction or process may have a positive ΔG despite having a negative ΔH due to the entropic contribution being unfavorable. Understanding the temperature dependence of ΔG is crucial for accurately assessing the feasibility of a reaction or process.

Problem 1: Consider the reaction $A + B \rightarrow C$. Calculate the ΔG for the reaction, given that the standard Gibbs free energy change ($\Delta G°$) is -50 kJ/mol.

Solution 1: $\Delta G = \Delta G° + RT \ln(Q)$, where Q is the reaction quotient and R is the gas constant. Assuming the reaction is taking place under standard conditions, Q = 1. Therefore, $\Delta G = \Delta G°$, which is -50 kJ/mol.

Since ΔG is negative, the reaction is feasible and will occur spontaneously under standard conditions.

Problem 2: For the reaction $2A + B \rightarrow C + D$, the value of $\Delta G°$ is +100 kJ/mol. If the reaction mixture has a reaction quotient (Q) of 1000, calculate the value of ΔG.

Solution 2: Again, we use the formula $\Delta G = \Delta G° + RT \ln(Q)$. Given that Q = 1000, we can substitute the values:

$$\Delta G = 100 \text{ kJ/mol} + (8.314 \text{ J/mol·K}) * (298 \text{ K}) * \ln(1000)$$

Calculating the value of the natural logarithm:

$$\Delta G = 100 \text{ kJ/mol} + (8.314 \text{ J/mol·K}) * (298 \text{ K}) * 6.907$$

Simplifying the expression:

$$\Delta G = 100 \text{ kJ/mol} + 16263 \text{ J/mol}$$

Converting the units:

$$\Delta G = 100 \text{ kJ/mol} + 16.263 \text{ kJ/mol}$$

$$\Delta G = 116.263 \text{ kJ/mol}$$

Since ΔG is positive, the reaction is not feasible under the given conditions. External energy input is required for the reaction to occur.

Effect of Temperature on Feasibility

The effect of temperature change on the feasibility of a reaction can be predicted using the Gibbs free energy equation:

$$\Delta G = \Delta H - T\Delta S$$

Where ΔG is the change in Gibbs free energy, ΔH is the change in enthalpy, ΔS is the change in entropy, and T is the temperature in Kelvin. In this equation, ΔH represents the heat energy transferred during the reaction, while ΔS represents the change in randomness or disorder.

When considering the impact of temperature on a reaction, it's important to examine the sign of ΔH and ΔS. If ΔH is negative (exothermic reaction), the formation of products releases heat, and if ΔS is positive, the reaction results in an increase in randomness or disorder.

When temperature increases, the value of $-T\Delta S$ term in the equation also increases. If this term becomes greater than ΔH, the value of ΔG becomes negative, indicating a spontaneous reaction. On the contrary, if the $-T\Delta S$ term becomes smaller than ΔH, the value of ΔG becomes positive, indicating a non-spontaneous reaction.

158

In summary, when the temperature increases:
- If the reaction is exothermic (ΔH is negative) and ΔS is positive, the reaction becomes more feasible as the value of ΔG becomes more negative.
- If the reaction is exothermic (ΔH is negative) and ΔS is negative, the reaction becomes less feasible as the value of ΔG becomes less negative or positive.
- If the reaction is endothermic (ΔH is positive) and ΔS is positive or negative, the reaction becomes less feasible as the value of ΔG becomes positive.

Problem 1: Given a reaction with $\Delta H = -50$ kJ/mol and $\Delta S = 100$ J/(mol·K), determine whether increasing the temperature would make this reaction more or less feasible.

Solution 1: To determine the feasibility, $\Delta G = \Delta H - T\Delta S$, where ΔG represents the change in Gibbs free energy, ΔH is the change in enthalpy, T represents temperature, and ΔS is the change in entropy. When increasing temperature, the $-T\Delta S$ term becomes more negative, resulting in a more negative ΔG. Thus, the reaction becomes more feasible.

Problem 2: For an exothermic reaction with $\Delta H = -60$ kJ/mol and $\Delta S = -150$ J/(mol·K), explain the effect of temperature change on the feasibility of this reaction.

Solution 2: Since ΔH is negative for an exothermic reaction, increasing the temperature will make the $-T\Delta S$ term less negative. If the magnitude of ΔH is greater than the magnitude of $-T\Delta S$, the reaction will become less feasible as ΔG becomes positive.

Problem 3: A reaction has $\Delta H = 40$ kJ/mol and $\Delta S = 80$ J/(mol·K). Determine whether the reaction is feasible at 300 K and at 400 K.

Solution 3: $\Delta G = \Delta H - T\Delta S$, where ΔG represents the change in Gibbs free energy, ΔH is the change in enthalpy, T represents temperature, and ΔS is the change in entropy. At 300 K, if ΔG is negative, the reaction is feasible; if positive, it is not. Repeat the calculation for 400 K.

Problem 4: For an endothermic reaction with $\Delta H = 80$ kJ/mol and $\Delta S = -120$ J/(mol·K), explain the effect of temperature change on the feasibility of this reaction.

Solution 4: With an endothermic reaction, increasing the temperature will make the $-T\Delta S$ term more negative. A larger $-T\Delta S$ term can offset the positive ΔH value, making the reaction more feasible as ΔG becomes negative.

Problem 5: A reaction has $\Delta H = -30$ kJ/mol and $\Delta S = -60$ J/(mol·K). Determine the temperature at which the reaction becomes energetically favorable.

Solution 5: Rearrange the equation $\Delta G = \Delta H - T\Delta S$ to solve for temperature. Substitute ΔH and ΔS values and solve for T. The obtained temperature will determine the feasibility of the reaction.

Problem 6: Given a reaction with $\Delta H = 20$ kJ/mol and $\Delta S = 50$ J/(mol·K). If the temperature is increased, would the reaction become less feasible or more feasible?

Solution 6: Increasing the temperature makes the $-T\Delta S$ term more negative. If the magnitude of $-T\Delta S$ is greater than the magnitude of ΔH, the reaction becomes more feasible as ΔG is negative.

Problem 7: Explain the effect of temperature change on the feasibility of a reaction when both ΔH and ΔS are positive.

Solution 7: When both ΔH and ΔS are positive, increasing the temperature results in a more negative $-T\Delta S$ value. If the magnitude of $-T\Delta S$ is greater than the magnitude of ΔH, the reaction becomes more feasible.

Problem 8: A reaction has $\Delta H = -25$ kJ/mol and $\Delta S = 80$ J/(mol·K). Calculate the temperature at which the reaction becomes non-feasible.

Solution 8: Calculate ΔG using the equation $\Delta G = \Delta H - T\Delta S$. Solve for T when ΔG is equal to zero. The obtained temperature will indicate when the reaction becomes non-feasible.

Problem 9: Given a reaction with ΔH = 50 kJ/mol and ΔS = 100 J/(mol·K), determine the effect of temperature change on the feasibility of this reaction.

Solution 9: Increasing the temperature makes the -TΔS term more negative. Depending on the magnitude of -TΔS compared to ΔH, the reaction may become more or less feasible as ΔG changes.

Problem 10: Explain the effect of temperature change on the feasibility of a reaction when ΔH is negative and ΔS is positive.

Solution 10: With a negative ΔH and positive ΔS, increasing the temperature makes the -TΔS term more negative. If the magnitude of -TΔS is greater than the magnitude of ΔH, the reaction becomes more feasible.

Problem 11: A reaction has ΔH = 60 kJ/mol and ΔS = -100 J/(mol·K). Calculate the temperature at which the reaction becomes feasible.

Solution 11: Calculate ΔG using the equation ΔG = ΔH - TΔS. Rearrange the equation to solve for T and substitute ΔH and ΔS values. Solve for T to find the temperature at which the reaction becomes feasible.

Problem 12: A reaction has ΔH = -20 kJ/mol and ΔS = 50 J/(mol·K). At which temperature would the reaction have the greatest feasibility?

Solution 12: Calculate ΔG at different temperatures using the equation ΔG = ΔH - TΔS. Note the signs of ΔG values as ΔT changes. The temperature corresponding to the most negative ΔG value indicates the greatest feasibility.

Problem 13: Explain the effect of temperature change on the feasibility of a reaction when ΔH and ΔS are both negative.

Solution 13: When both ΔH and ΔS are negative, increasing the temperature makes the -TΔS term more negative. If the magnitude of -TΔS is greater than the magnitude of ΔH, the reaction becomes more feasible.

Problem 14: A reaction has ΔH = 40 kJ/mol and ΔS = -80 J/(mol·K). Calculate the temperature at which the reaction becomes non-feasible.

Solution 14: Calculate ΔG using the equation ΔG = ΔH - TΔS. Determine the temperature at which ΔG becomes positive, indicating the reaction becomes non-feasible.

Problem 15: Given a reaction with ΔH = -50 kJ/mol and ΔS = 150 J/(mol·K), determine the effect of temperature change on the feasibility of this reaction.

Solution 15: Increasing the temperature makes the -TΔS term more negative. Depending on the magnitude of -TΔS compared to ΔH, the reaction may become more or less feasible as ΔG changes.

Problem 16: A reaction has ΔH = 30 kJ/mol and ΔS = 60 J/(mol·K). Calculate the temperature at which the reaction becomes non-feasible.

Solution 16: Calculate ΔG using the equation ΔG = ΔH - TΔS. Solve for T when ΔG becomes positive, indicating the temperature at which the reaction becomes non-feasible.

Problem 17: Explain the effect of temperature change on the feasibility of a reaction when ΔH is positive and ΔS is negative.

Solution 17: With a positive ΔH and negative ΔS, increasing the temperature makes the -TΔS term more negative. The reaction becomes more feasible if the magnitude of -TΔS is greater than the magnitude of ΔH.

Problem 18: A reaction has ΔH = -40 kJ/mol and ΔS = -120 J/(mol·K). Calculate the temperature at which the reaction becomes non-feasible.

Solution 18: Calculate ΔG using the equation ΔG = ΔH - TΔS. Find the temperature at which ΔG becomes positive, indicating the temperature at which the reaction becomes non-feasible.

Problem 19: Given a reaction with ΔH = 60 kJ/mol and ΔS = 100 J/(mol·K), determine whether increasing the temperature would make this reaction more or less feasible.

Solution 19: Increasing the temperature makes the -TΔS term more negative. Depending on the magnitude of -TΔS compared to ΔH, the reaction may become more or less feasible as ΔG changes.

Problem 20: For an exothermic reaction with ΔH = -80 kJ/mol and ΔS = -150 J/(mol·K), explain the effect of temperature change on the feasibility of this reaction.

Solution 20: With an exothermic reaction, increasing the temperature makes the -TΔS term less negative. If the magnitude of ΔH is greater than the magnitude of -TΔS, the reaction becomes less feasible as ΔG becomes positive.

Electrochemistry

Electrolysis
Predicting Substances in Electrolysis: State, Electrode Potential, Concentration

During electrolysis, substances are liberated or produced at the electrodes based on various factors such as the state of the electrolyte (molten or aqueous), position in the redox series (electrode potential), and concentration. The process involves the flow of electric current through an electrolyte, which is a substance that conducts electricity when dissolved in water or when molten. In molten electrolytes, the substances liberated during electrolysis tend to be the elements that make up the compound. For example, in the electrolysis of molten sodium chloride (NaCl), sodium metal (Na) is produced at the cathode, while chlorine gas (Cl_2) is liberated at the anode. This is because sodium ions (Na^+) migrate towards the cathode and gain electrons to form sodium atoms, while chloride ions (Cl^-) migrate towards the anode and lose electrons to form chlorine molecules. In the case of aqueous electrolytes, the substances liberated during electrolysis can vary based on the concentration of ions in the solution. This is due to the presence of water molecules and the potential for other reactions to occur. The position in the redox series also plays a role in determining which substances are produced at the electrodes. At the cathode, reduction reactions occur, meaning that cations (positively charged ions) gain electrons to form neutral elements or compounds. The most easily reduced species tend to be the ones with the highest electrode potential, or those that are lower in the redox series. For example, if an aqueous electrolyte solution contains both copper ions (Cu^{2+}) and hydrogen ions (H^+), copper ions with a higher reduction potential will be reduced to form metallic copper (Cu) at the cathode.

On the other hand, at the anode, oxidation reactions occur, where anions (negatively charged ions) lose electrons to form neutral elements or compounds. The most easily oxidized species tend to be those with the lowest electrode potential or those higher in the redox series. For instance, if an aqueous electrolyte solution contains chloride ions (Cl^-) and hydroxide ions (OH^-), the chloride ions will be oxidized to form chlorine gas (Cl_2) at the anode.

The concentration of ions in the electrolyte can also affect the liberation of substances. In general, higher concentrations of ions promote greater liberation of the corresponding substances at the electrodes. However, it is important to note that the concentration of other species and presence of impurities can also influence the electrolysis process.

In summary, the substances liberated during electrolysis depend on several factors, including the state of the electrolyte, position in the redox series, and concentration of ions. The specific substances produced at the electrodes can be determined by considering the reduction and oxidation potentials of the species involved.

Problem 1: In the electrolysis of molten magnesium chloride ($MgCl_2$), what substances are liberated at the electrodes?

Solution 1: At the cathode: Magnesium metal (Mg) is liberated.

At the anode: Chlorine gas (Cl_2) is liberated.

Problem 2: In the electrolysis of aqueous sodium chloride (NaCl) solution, what substances are liberated at the electrodes?

Solution 2: At the cathode: Sodium metal (Na) is liberated.

At the anode: Chlorine gas (Cl2) is liberated.

Problem 3: During the electrolysis of molten lead(II) bromide (PbBr2), which substances are liberated at the electrodes?

Solution 3: At the cathode: Lead metal (Pb) is liberated.

At the anode: Bromine gas (Br2) is liberated.

Problem 4: What is liberated at the cathode and anode during the electrolysis of water?

Solution 4: At the cathode: Hydrogen gas (H2) is liberated.

At the anode: Oxygen gas (O2) is liberated.

Problem 5: In the electrolysis of aqueous copper(II) sulfate (CuSO4) solution, what substances are liberated at the electrodes?

Solution 5: At the cathode: Copper metal (Cu) is liberated.

At the anode: Oxygen gas (O2) is liberated.

Problem 6: What is liberated at the cathode and anode when molten calcium chloride (CaCl2) undergoes electrolysis?

Solution 6:

At the cathode: Calcium metal (Ca) is liberated.

At the anode: Chlorine gas (Cl2) is liberated.

Problem 7: During the electrolysis of molten aluminum oxide (Al2O3), which substances are liberated at the electrodes?

Solution 7: At the cathode: Aluminum metal (Al) is liberated.

At the anode: Oxygen gas (O2) is liberated.

Problem 8: In the electrolysis of aqueous potassium iodide (KI) solution, what substances are liberated at the electrodes?

Solution 8: At the cathode: Potassium metal (K) is liberated.

At the anode: Iodine gas (I2) is liberated.

Problem 9: What is liberated at the cathode and anode during the electrolysis of molten zinc chloride (ZnCl2)?

Solution 9: At the cathode: Zinc metal (Zn) is liberated.

At the anode: Chlorine gas (Cl2) is liberated.

Problem 10: In the electrolysis of aqueous sulfuric acid (H2SO4) solution, what substances are liberated at the electrodes?

Solution 10: At the cathode: Hydrogen gas (H2) is liberated.

At the anode: Oxygen gas (O2) is liberated.

Problem 11: During the electrolysis of molten sodium fluoride (NaF), what substances are liberated at the electrodes?

Solution 11: At the cathode: Sodium metal (Na) is liberated.

At the anode: Fluorine gas (F2) is liberated.

Problem 12: In the electrolysis of aqueous silver nitrate (AgNO3) solution, what substances are liberated at the electrodes?

Solution 12: At the cathode: Silver metal (Ag) is liberated.

At the anode: Oxygen gas (O2) is liberated.

Problem 13: What is liberated at the cathode and anode during the electrolysis of molten potassium bromide (KBr)?

Solution 13: At the cathode: Potassium metal (K) is liberated.

At the anode: Bromine gas (Br2) is liberated.

Problem 14: During the electrolysis of aqueous nickel(II) chloride (NiCl2) solution, what substances are liberated at the electrodes?

Solution 14: At the cathode: Nickel metal (Ni) is liberated.

At the anode: Chlorine gas (Cl2) is liberated.

Problem 15: In the electrolysis of molten iron(III) oxide (Fe2O3), what substances are liberated at the electrodes?

Solution 15: At the cathode: Iron metal (Fe) is liberated.

At the anode: Oxygen gas (O2) is liberated.

Problem 16: What is liberated at the cathode and anode during the electrolysis of aqueous lithium chloride (LiCl) solution?

Solution 16: At the cathode: Lithium metal (Li) is liberated.

At the anode: Chlorine gas (Cl2) is liberated.

Problem 17: During the electrolysis of molten potassium iodide (KI), what substances are liberated at the electrodes?

Solution 17: At the cathode: Potassium metal (K) is liberated.

At the anode: Iodine gas (I2) is liberated.

Problem 18: In the electrolysis of aqueous zinc sulfate (ZnSO4) solution, what substances are liberated at the electrodes?

Solution 18: At the cathode: Zinc metal (Zn) is liberated.

At the anode: Oxygen gas (O2) is liberated.

Problem 19: What is liberated at the cathode and anode when molten lead(II) chloride (PbCl2) undergoes electrolysis?

Solution 19: At the cathode: Lead metal (Pb) is liberated.

At the anode: Chlorine gas (Cl2) is liberated.

Problem 20: During the electrolysis of aqueous potassium permanganate (KMnO4) solution, what substances are liberated at the electrodes?

Solution 20: At the cathode: Manganese dioxide (MnO2) is formed.

At the anode: Oxygen gas (O2) is liberated.

Relationship Between Faraday Constant and Charge on Electron: F = Le

In the field of chemistry, the relationship between the Faraday constant (F), the Avogadro constant (L), and the charge on the electron (e) is of utmost importance. Express and utilize this connection as F = Le, where F denotes the Faraday constant, L represents the Avogadro constant, and e signifies the charge carried by an electron.

The Faraday constant (F) represents the charge of one mole of electrons, which is equivalent to 96,485.3383 coulombs. It is named after the renowned scientist Michael Faraday, who made significant contributions to the field of electrochemistry.

On the other hand, the Avogadro constant (L) is a fundamental constant that represents the number of atoms or molecules in one mole of a substance. Its value is approximately 6.022×10^{23} particles per mole.

The charge on the electron (e) is a fundamental property of subatomic particles and is equal to approximately -1.602×10^{-19} coulombs. It represents the magnitude of the elementary charge carried by an electron.

The relationship F = Le provides a link between these three fundamental constants. It states that the Faraday constant (F) is equal to the product of the Avogadro constant (L) and the charge on the electron (e).

This relationship is derived from the fact that one mole of electrons carries a charge equal to the charge on a single electron. Since there are L particles (atoms or molecules) in one mole of a substance, the total charge carried by these particles would be Le.

By expressing the relationship F = Le, we can understand the interplay between these fundamental constants and their implications in various electrochemical reactions and calculations.

For instance, in electrolysis, where the passage of an electric current causes a chemical reaction, the Faraday constant can be utilized to determine the quantity of substance produced or consumed during the reaction. It allows us to relate the charge passed, given in coulombs, to the amount of substance involved, measured in moles.

Moreover, the relationship F = Le finds application in quantifying the charge transfer in redox reactions, where the transfer of electrons between species occurs. By knowing the value of the Faraday constant, one can determine the number of electrons involved in the reaction by dividing the charge passed by the magnitude of the elementary charge (e).

Additionally, the Faraday constant can be employed to calculate the electrical work done in an electrochemical cell. The product of the Faraday constant and the electromotive force (EMF) of the cell provides the energy change associated with the transfer of one mole of electrons.

Here are 20 problems based on the relationship between the Faraday constant, the Avogadro constant, and the charge on the electron:

Problem 1: Calculate the charge on a mole of electrons given that the Faraday constant is 96,485 C/mol.

Solution: Since the Faraday constant (F) represents the charge on one mole of electrons, we can say that F = 1 mol × e. Rearranging the equation, we get e = F/L. Substituting the values for F = 96,485 C/mol and L = 6.022 × 10^23 mol^-1, we find e = 1.602 × 10^-19 C, which is the charge on a single electron.

Problem 2: The Avogadro constant (L) is a value that defines the quantity of entities present in one mole of a substance. Calculate the value of L in terms of the Faraday constant (F) and the charge on the electron (e).

Solution: Rearranging the equation for e = F/L, we find L = F/e. Substituting the values for F = 96,485 C/mol and e = 1.602 × 10^-19 C, we can calculate L as L = 6.022 × 10^23 mol^-1.

Problem 3: If the charge on an electron is 1.602 × 10^-19 C, what is the value of the Faraday constant?

Solution: Using the relationship F = Le, we can substitute e = 1.602 × 10^-19 C. Thus, F = L × e = (6.022 × 10^23 mol^-1) × (1.602 × 10^-19 C) = 9.649 × 10^4 C/mol.

Problem 4: A certain reaction requires the reduction of 3 moles of electrons. Calculate the total charge involved in this reaction.

Solution: Since one mole of electrons has a charge of 1.602 × 10^-19 C, the total charge involved in the reduction of 3 moles of electrons can be calculated as 3 moles × (1.602 × 10^-19 C/mole) = 4.806 × 10^-19 C.

Problem 5: The electric charge carried by a single electron is equal to 1.602 × 10^-19 Coulombs (C). How many electrons are necessary to form 1 mole of electrons?

Solution: Since one mole of electrons has a charge of F = 96,485 C, we can find the number of electrons needed by dividing the charge for one mole by the charge on a single electron. Thus, 96,485 C / (1.602 × 10^-19 C/electron) = 6.022 × 10^23 electrons.

Problem 6: If a reaction involves the transfer of 5.5 × 10^20 electrons, how many moles of electrons are involved?

Solution: Using the relationship e = F/L, we can rearrange the equation to find L = F/e. Substituting the values for F = 96,485 C/mol and e = 1.602 × 10^-19 C, we can find L as L = 6.022 × 10^23 mol^-1. To find the number of moles of electrons, we divide the total charge involved (5.5 × 10^20 electrons) by L. Thus, 5.5 × 10^20 electrons / (6.022 × 10^23 mol^-1) = 9.127 × 10^-4 mol.

Problem 7: If the Avogadro constant is 6.022 × 10^23 mol^-1, what is the value of the charge on a single electron?

Solution: Substituting the values for F = 96,485 C/mol and L = 6.022 × 10^23 mol^-1 into the equation e = F/L, we can calculate e as e = 1.602 × 10^-19 C.

Problem 8: What is the relationship between the Faraday constant, the Avogadro constant, and the charge on the electron?

Solution: Expressed by the equation F = Le, the relationship involves the Faraday constant (F), the Avogadro constant (L), and the charge carried by a single electron (e).

Problem 9: Given that the charge on a single electron is 1.602 × 10^-19 C, what is the value of the Avogadro constant?

Solution: Substituting the values for F = 96,485 C/mol and e = 1.602 × 10^-19 C into the equation L = F/e, we can calculate L as L = 6.022 × 10^23 mol^-1.

Problem 10: Calculate the charge on 0.5 mole of electrons using the relationship F = Le.

Solution: Since F represents the charge on one mole of electrons, we can say that F = 1 mol × e. Using the equation e = F/L and substituting the values for F = 96,485 C/mol, L = 6.022 × 10^23 mol^-1, and n = 0.5 mol, we find the charge on 0.5 mol of electrons as 96,485 C/mol × 0.5 mol = 48,242.5 C.

Quantity of Charge Passed

1. Problem: Calculate the quantity of charge passed when a current of 2.5 A flows for 3.5 minutes.
Solution: Q = It = (2.5 A)(3.5 min)(60 s/min)
Q = 525 C

2. Problem: If a current of 0.8 A passes through a circuit for 2 hours, what is the quantity of charge passed?
Solution: Q = It = (0.8 A)(2 h)(3600 s/h)
Q = 5,760 C

3. Problem: A current of 5 mA passes for 30 seconds. Calculate the quantity of charge passed.
Solution: Q = It = (5 mA)(30 s)(0.001 C/mA)
Q = 0.15 C

4. Problem: How much charge is passed through a circuit in 10 minutes if the current is 0.6 A?
Solution: Q = It = (0.6 A)(10 min)(60 s/min)
Q = 360 C

5. Problem: A current of 1.2 A is applied for 45 minutes. Determine the quantity of charge passed.
Solution: Q = It = (1.2 A)(45 min)(60 s/min)
Q = 3,240 C

6. Problem: Calculate the quantity of charge passed if a current of 3 A is applied for 1 hour and 20 minutes.
Solution: Q = It = (3 A)(1 h 20 min)(60 s/min)
Q = 7,200 C

7. Problem: If a current of 500 mA flows for 5 seconds, what is the quantity of charge passed?
Solution: Q = It = (500 mA)(5 s)(0.001 C/mA)
Q = 2.5 C

8. Problem: A current of 0.4 A is applied for 15 minutes. Calculate the quantity of charge passed.
Solution: Q = It = (0.4 A)(15 min)(60 s/min)
Q = 360 C

9. Problem: How much charge is passed through a circuit in 20 minutes if the current is 1.5 A?
Solution: Q = It = (1.5 A)(20 min)(60 s/min)
Q = 1,800 C

10. Problem: A current of 800 mA is applied for 2 hours and 30 minutes. Determine the quantity of charge passed.

Solution: $Q = It = (800 \text{ mA})(2 \text{ h } 30 \text{ min})(60 \text{ s/min})$

$Q = 7,200 \text{ C}$

11. Problem: Calculate the quantity of charge passed when a current of 1.2 A flows for 2 minutes and 30 seconds.

Solution: $Q = It = (1.2 \text{ A})(2 \text{ min } 30 \text{ s})(60 \text{ s/min})$

$Q = 180 \text{ C}$

12. Problem: If a current of 0.6 A passes through a circuit for 1 hour and 45 minutes, what is the quantity of charge passed?

Solution: $Q = It = (0.6 \text{ A})(1 \text{ h } 45 \text{ min})(60 \text{ s/min})$

$Q = 3,420 \text{ C}$

13. Problem: A current of 4 mA passes for 20 seconds. Calculate the quantity of charge passed.

Solution: $Q = It = (4 \text{ mA})(20 \text{ s})(0.001 \text{ C/mA})$

$Q = 0.08 \text{ C}$

14. Problem: How much charge is passed through a circuit in 5 minutes if the current is 1.8 A?

Solution: $Q = It = (1.8 \text{ A})(5 \text{ min})(60 \text{ s/min})$

$Q = 540 \text{ C}$

15. Problem: A current of 1.5 A is applied for 30 minutes. Determine the quantity of charge passed.

Solution: $Q = It = (1.5 \text{ A})(30 \text{ min})(60 \text{ s/min})$

$Q = 2,700 \text{ C}$

16. Problem: Calculate the quantity of charge passed if a current of 2.5 A is applied for 45 minutes.

Solution: $Q = It = (2.5 \text{ A})(45 \text{ min})(60 \text{ s/min})$

$Q = 6,750 \text{ C}$

17. Problem: If a current of 350 mA flows for 10 seconds, what is the quantity of charge passed?

Solution: $Q = It = (350 \text{ mA})(10 \text{ s})(0.001 \text{ C/mA})$

$Q = 0.035 \text{ C}$

18. Problem: A current of 0.8 A is applied for 25 minutes. Calculate the quantity of charge passed.

Solution: $Q = It = (0.8 \text{ A})(25 \text{ min})(60 \text{ s/min})$

$Q = 1,200 \text{ C}$

19. Problem: How much charge is passed through a circuit in 15 minutes if the current is 1.2 A?

Solution: $Q = It = (1.2 \text{ A})(15 \text{ min})(60 \text{ s/min})$

$Q = 1080 \text{ C}$

20. Problem: A current of 650 mA is applied for 1 hour and 15 minutes. Determine the quantity of charge passed.

Solution: $Q = It = (650 \text{ mA})(1 \text{ h } 15 \text{ min})(60 \text{ s/min})$

$Q = 17,550 \text{ C}$

Mass and/or Volume of Substance Liberated

Here are 20 problems related to the calculation of the mass and/or volume of a substance liberated during electrolysis, along with their solutions:

1. Problem: Calculate the mass of copper liberated when a current of 2.5 A is passed through a solution of copper sulfate for 30 minutes.

Solution: First, calculate the charge (Q) using $Q = I \times t$. Mass can be calculated using the formula $M = (Q \times \text{Molar mass}) / (n \times F)$, where n is the number of electrons involved in the reaction.

Answer: Mass of copper liberated = (Q × Molar mass) / (n × F).

2. Problem: Determine the volume of hydrogen gas evolved at STP when a current of 0.5 A is passed through water for 10 minutes.

Solution: Calculate the moles of hydrogen gas evolved using the formula n = Q / (2 × F). Next, employ the ideal gas equation to determine the volume at Standard Temperature and Pressure (STP).

Answer: Volume of hydrogen gas evolved = n × 22.4 L.

3. Problem: Given that 2.4 g of aluminum is liberated by the passage of 1.2 A current for a certain period, calculate the time taken.

Solution: Rearrange the formula M = (Q × Molar mass) / (n × F) to find the time (t).

Answer: Time taken = (M × n × F) / (Q × Molar mass).

4. Problem: Calculate the mass of chlorine gas liberated when a current of 3 A is passed through a solution of hydrochloric acid for 20 minutes.

Solution: Use the same formula as in problem 1 to calculate the mass of chlorine gas liberated.

Answer: Mass of chlorine gas liberated = (Q × Molar mass) / (n × F).

5. Problem: Determine the volume of oxygen gas evolved at STP when a current of 2 A is passed through potassium chlorate for 15 minutes.

Solution: Calculate the moles of oxygen gas evolved using the formula n = Q / (4 × F). Then, use the ideal gas equation to calculate the volume at STP.

Answer: Volume of oxygen gas evolved = n × 22.4 L.

6. Problem: If 0.5 g of nickel is released by a 0.3 A current for a specific duration, determine the time required.

Solution: Rearrange the formula M = (Q × Molar mass) / (n × F) to find the time (t).

Answer: Time taken = (M × n × F) / (Q × Molar mass).

7. Problem: Calculate the mass of silver liberated when a current of 2.5 A is passed through a solution of silver nitrate for 30 minutes.

Solution: Use the same formula as in problem 1 to calculate the mass of silver liberated.

Answer: Mass of silver liberated = (Q × Molar mass) / (n × F).

8. Problem: Determine the volume of chlorine gas evolved at STP when a current of 1 A is passed through a solution of sodium chloride for 25 minutes.

Solution: Calculate the moles of chlorine gas evolved using the formula n = Q / (2 × F). Next, employ the ideal gas equation to compute the volume at Standard Temperature and Pressure.

Answer: Volume of chlorine gas evolved = n × 22.4 L.

9. Problem: Given that 1.2 g of zinc is liberated by the passage of 0.6 A current for a certain period, calculate the time taken.

Solution: To find the time (t), you need to rearrange the formula M = (Q × Molar mass) / (n × F).

Answer: Time taken = (M × n × F) / (Q × Molar mass).

10. Problem: Calculate the mass of oxygen gas liberated when a current of 4 A is passed through water for 40 minutes.

Solution: Use the same formula as in problem 1 to calculate the mass of oxygen gas liberated.

Answer: Mass of oxygen gas liberated = (Q × Molar mass) / (n × F).

11. Problem: Determine the volume of hydrogen gas evolved at STP when a current of 0.8 A is passed through hydrochloric acid for 12 minutes.

Solution: Calculate the moles of hydrogen gas evolved using the formula n = Q / (2 × F). Then, use the ideal gas equation to calculate the volume at STP.

Answer: Volume of hydrogen gas evolved = n × 22.4 L.

12. Problem: Given that 1.8 g of copper is liberated by the passage of 0.9 A current for a certain period, calculate the time taken.

Solution: Rearrange the formula M = (Q × Molar mass) / (n × F) to find the time (t).

Answer: Time taken = (M × n × F) / (Q × Molar mass).

13. Problem: Calculate the mass of chlorine gas liberated when a current of 3.5 A is passed through a solution of hydrochloric acid for 25 minutes.

Solution: Use the same formula as in problem 1 to calculate the mass of chlorine gas liberated.

Answer: Mass of chlorine gas liberated = (Q × Molar mass) / (n × F).

14. Problem: Determine the volume of oxygen gas evolved at STP when a current of 1.5 A is passed through potassium chlorate for 20 minutes.

Solution: Calculate the moles of oxygen gas evolved using the formula n = Q / (4 × F). Then, use the ideal gas equation to calculate the volume at STP.

Answer: Volume of oxygen gas evolved = n × 22.4 L.

15. Problem: Calculate the time taken for a 0.5 A current to liberate 0.8 g of nickel.

Solution: Rearrange the formula M = (Q × Molar mass) / (n × F) to find the time (t).

Answer: Time taken = (M × n × F) / (Q × Molar mass).

16 Problem: Determine the mass of silver that is freed when a 4.5 A current is run through a silver nitrate solution for a duration of 35 minutes.

Solution: Use the same formula as in problem 1 to calculate the mass of silver liberated.

Answer: Mass of silver liberated = (Q × Molar mass) / (n × F).

17. Problem: Determine the volume of chlorine gas evolved at STP when a current of 2 A is passed through a solution of sodium chloride for 30 minutes.

Solution: Calculate the moles of chlorine gas evolved using the formula n = Q / (2 × F). Then, use the ideal gas equation to calculate the volume at STP.

Answer: Volume of chlorine gas evolved = n × 22.4 L.

18. Problem: Given that 1.5 g of zinc is liberated by the passage of 0.8 A current for a certain period, calculate the time taken.

Solution: Rearrange the formula M = (Q × Molar mass) / (n × F) to find the time (t).

Answer: Time taken = (M × n × F) / (Q × Molar mass).

19. Problem: Calculate the mass of oxygen gas liberated when a current of 5 A is passed through water for 50 minutes.

Solution: Use the same formula as in problem 1 to calculate the mass of oxygen gas liberated.

Answer: Mass of oxygen gas liberated = (Q × Molar mass) / (n × F).

20. Problem: Determine the volume of hydrogen gas evolved at STP when a current of 1.2 A is passed through hydrochloric acid for 15 minutes.

Solution: Calculate the moles of hydrogen gas evolved using the formula n = Q / (2 × F). Then, use the ideal gas equation to calculate the volume at STP.

Answer: Volume of hydrogen gas evolved = n × 22.4 L.

***

Electrolytic Method: Determining Avogadro Constant

Electrolysis is the process of using an electric current to induce a chemical reaction, typically in an electrolyte solution. It involves the movement of ions towards the respective electrodes, where reduction and oxidation reactions take place. In the context of determining the Avogadro constant, electrolysis is employed in combination with Faraday's laws.

The first law of Faraday indicates that in the process of electrolysis, the amount of a substance that is either released or accumulated at an electrode is directly related to the amount of electricity that flows through the cell. This forms the basis for determining the quantity of a substance participating in the electrolysis process.

To establish a relationship between the Avogadro constant and electrolysis, a compound containing a known number of atoms or ions per unit formula weight is selected as the electrolyte. Consider the example of copper(II) sulfate ($CuSO_4$) as the electrolyte. It is chosen due to its known formula weight and the fixed ratio of copper and sulfate ions in the compound.

The electrolysis of copper(II) sulfate involves the passage of an electric current through an electrolytic cell containing a copper(II) sulfate solution. Two electrodes, namely the anode and cathode, are immersed in the electrolyte and connected to a direct current (DC) power source.

As the electric current flows through the cell, copper ions (Cu^{2+}) travel towards the cathode, where they undergo reduction. At the cathode, each copper ion gains two electrons and is discharged as metallic copper (Cu). The amount of copper deposited can be determined by weighing the cathode before and after the electrolysis process.

According to Faraday's first law, the mass of copper deposited is directly proportional to the quantity of electricity passed through the cell. This relationship can be expressed mathematically as:

$$m = (zFQ) / M$$

Where:

m = mass of copper deposited

z = number of electrons involved in the reduction of each copper ion (2 in this case)

F = Faraday's constant (96,485 C/mol)

Q = quantity of electricity passed through the cell (Coulombs)

M = molar mass of copper (63.55 g/mol)

By numerically determining the mass of copper deposited and measuring the quantity of electricity passed through the cell (in Coulombs), this equation can be rearranged to calculate the value of the Avogadro constant (N_A).

$$N_A = (zFQ) / (mM)$$

To further refine the estimation of the Avogadro constant, multiple determinations are conducted under varying conditions. By averaging the results and considering the uncertainty associated with each measurement, a more precise value can be obtained.

It is worth noting that the determination of the Avogadro constant through the electrolytic method requires careful experimental design and accurate measurements. Factors such as the purity of the electrolyte, precise measurement of the quantity of electricity, accurate weighing of the deposited substance, and control of experimental conditions play critical roles in obtaining reliable results.

Additionally, the electrolytic method for determining the Avogadro constant is just one of several approaches employed in metrology. Other methods include X-ray crystallography, the measurement of the speed of sound in a gas, and the use of the Mössbauer effect, among others. These diverse methods provide complementary and cross-validated results, ensuring the accuracy of the determined value of the Avogadro constant.

Here are 20 sample problems related to the determination of the Avogadro constant using the electrolytic method, along with their solutions that address common questions students might face in a chemistry exam:

Problem 1:

A student is asked to describe the process of electrolysis and its relevance to the determination of the Avogadro constant. How would you respond?

Solution:

Electrolysis is the process of using an electric current to induce a chemical reaction. In the context of the Avogadro constant determination, electrolysis is employed to quantify the mass of a substance deposited during the reaction. This allows us to establish a relationship between the quantity of electricity passed and the Avogadro constant.

Problem 2:

What is Faraday's first law? How does it relate to the determination of the Avogadro constant?

Solution:

Faraday's first law states that the mass of a substance liberated or deposited at an electrode during electrolysis is directly proportional to the quantity of electricity passed through the cell. This law is utilized in the electrolytic method to determine the Avogadro constant by relating the mass of the substance deposited to the quantity of electricity and molar mass of the substance.

Problem 3:

Why is copper(II) sulfate chosen as the electrolyte in determining the Avogadro constant?

Solution:

Copper(II) sulfate is chosen due to its known formula weight and a fixed ratio of copper and sulfate ions in the compound. This ensures that a consistent number of copper ions will be reduced at the cathode during electrolysis, providing a basis for estimating the Avogadro constant.

Problem 4:

How does the Avogadro constant relate to the electrolytic method? Explain.

Solution:

The Avogadro constant is related to the electrolytic method by the quantity of electricity passed through the cell during electrolysis. By quantifying the mass of the substance deposited and measuring the quantity of electricity, the Avogadro constant can be estimated using the equation $N_A = (zFQ) / (mM)$, where z represents the number of electrons involved in the reduction, F is Faraday's constant, Q is the quantity of electricity, m is the mass of the substance deposited, and M is the molar mass of the substance.

Problem 5:

What precautions should be taken during the experimental setup to ensure accurate determination of the Avogadro constant through the electrolytic method?

Solution:

Precautions include using a pure electrolyte, ensuring accurate measurement of the quantity of electricity passed, precise weighing of the deposited substance, and controlling experimental conditions such as temperature and pressure. These precautions help minimize errors and ensure reliable results.

Problem 6:

Which factors influence the accuracy of the determination of the Avogadro constant through the electrolytic method?

Solution:

Factors such as the purity of the electrolyte, accuracy in measuring the quantity of electricity, precision in weighing the deposited substance, and control of experimental conditions like temperature and pressure affect the accuracy of the determination of the Avogadro constant.

Problem 7:

What are some alternative methods, apart from electrolysis, that can be used to determine the Avogadro constant?

Solution:

Alternative methods include X-ray crystallography, the measurement of the speed of sound in a gas, and the use of the Mössbauer effect, among others. These methods provide different approaches to validate and complement the results obtained through the electrolytic method.

Problem 8:

What is the significance of conducting multiple determinations to estimate the value of the Avogadro constant accurately?

Solution:

Conducting multiple determinations helps obtain more reliable results by considering the average value and taking into account the uncertainty associated with each measurement. This improves the overall precision of the estimated value of the Avogadro constant.

Problem 9:

Explain the role of the cathode in the electrolysis of copper(II) sulfate solution.

Solution:

The cathode is where reduction takes place during electrolysis. In the case of copper(II) sulfate, copper ions ($Cu2+$) gain two electrons at the cathode and are discharged as metallic copper (Cu).

Problem 10:

How can the mass of copper deposited during electrolysis be measured accurately?

Solution:

The mass of copper deposited can be measured accurately by weighing the cathode before and after electrolysis. The difference in mass gives the amount of copper deposited.

Problem 11:

Why is it important to consider the uncertainty associated with each measurement in the determination of the Avogadro constant?

Solution:

Considering the uncertainty associated with each measurement is important because it helps to account for the variations and imperfections in experimental procedures. It gives a more realistic estimate of the Avogadro constant and reflects the range within which the value lies.

Problem 12:

What is the general equation used to calculate the value of the Avogadro constant using the electrolytic method?

Solution:

The general equation is $NA = (zFQ) / (mM)$, where NA is the Avogadro constant, z is the number of electrons involved in the reduction, F is Faraday's constant, Q is the quantity of electricity passed through the cell, m is the mass of the substance deposited, and M is the molar mass of the substance.

Problem 13:

What are some potential sources of error in the determination of the Avogadro constant using the electrolytic method?

Solution:

Some potential sources of error include impurities in the electrolyte, incomplete deposition of the substance at the electrode, errors in measuring the quantity of electricity passed, and inaccuracies in weighing the deposited substance. Environmental factors like temperature and pressure can also contribute to errors.

Problem 14:

How does the value of the Avogadro constant obtained through the electrolytic method compare to values obtained using other methods?

Solution:

The value of the Avogadro constant obtained through the electrolytic method is expected to be consistent with values obtained using other reliable and validated methods. Multiple determination methods help ensure agreement among different approaches, reducing uncertainties in the estimated value.

Problem 15:

Can the electrolytic method be used to determine the Avogadro constant for any substance, or are there specific requirements?

Solution:

The electrolytic method can be employed to determine the Avogadro constant for substances that can undergo appropriate reduction or oxidation reactions during electrolysis. It is essential to choose an electrolyte with known properties and a fixed ratio of ions to establish a reliable relationship for the determination.

Problem 16:

What are some real-world applications of the Avogadro constant determination?

Solution:

Determining the Avogadro constant has real-world applications in fields like metrology, material science, and pharmaceuticals. It allows for precise measurement, understanding atomic and molecular structures, and calculating quantities in chemical reactions and formulations.

Problem 17:

How can the Avogadro constant be used in practical calculations within different chemistry subfields?

Solution:

The Avogadro constant, when coupled with the mole concept, allows for the conversion of mass to moles and vice versa, facilitating calculations in various areas of chemistry, such as stoichiometry, gas laws, solution chemistry, and more.

Problem 18:

What are the limitations of the electrolytic method for determining the Avogadro constant?

Solution:

The limitations of the electrolytic method include the necessity of a suitable electrolyte, errors associated with the experimental setup and measurements, and the need for careful control of experimental conditions. These factors can introduce uncertainty and affect the accuracy of the results.

Problem 19:

What would happen if impurities were present in the electrolyte during the determination of the Avogadro constant?

Solution:

Impurities in the electrolyte can interfere with the deposition of the substance at the electrode, leading to inaccurate measurements of mass. This can result in an erroneous estimation of the Avogadro constant.

Problem 20:

Describe a step-by-step procedure for determining the Avogadro constant using the electrolytic method.

Solution:

1. Choose a suitable electrolyte with known properties and a fixed ratio of ions.
2. Set up an electrolytic cell with two electrodes (anode and cathode) immersed in the electrolyte.
3. Connect the electrodes to a direct current (DC) power source.
4. Measure and record the mass of the cathode before electrolysis.

5. Pass a known quantity of electricity (measured in Coulombs) through the cell.

6. Terminate the electrolysis and weigh the cathode to determine the mass of the deposited substance.

7. Calculate the Avogadro constant using the equation NA = (zFQ) / (mM), where z is the number of electrons involved in the reduction, F is Faraday's constant, Q is the quantity of electricity passed, m is the mass of the substance deposited, and M is the molar mass of the substance.

8. Conduct multiple determinations and calculate the average value to improve accuracy and reliability.

❦

Standard electrode potentials $E^{\diamond}$; standard cell potentials $E^{\diamond}$ cell and the Nernst equation
Standard electrode (reduction) potential

A standard electrode (reduction) potential refers to the measure of the ability of an electrode to gain or accept electrons during a chemical reaction, specifically a reduction reaction, under standard conditions. It is determined by comparing the potential of the electrode to a standard hydrogen electrode (SHE) at 25 degrees Celsius, with a hydrogen ion activity of 1 mol/L and a hydrogen gas pressure of 1 atmosphere.

In simpler terms, when two electrodes are connected through a solution and a current flows, the standard electrode (reduction) potential tells us how likely an electrode is to gain or accept electrons compared to the standard hydrogen electrode. It is a way to measure the "strength" of an electrode in terms of its ability to undergo reduction reactions.

Standard electrode (reduction) potential is an important concept in chemistry because it helps to predict the direction and feasibility of chemical reactions. If an electrode has a more positive standard electrode potential, it means that it is more likely to gain electrons and undergo reduction. Conversely, if an electrode has a more negative standard electrode potential, it is less likely to gain electrons and undergo reduction.

The standard electrode potentials of different electrodes can be arranged in a series called the electrochemical series. This series allows us to compare the relative strengths of different electrodes and predict which reactions are more favorable or spontaneous.

The standard electrode (reduction) potentials can also be used to calculate the cell potential, which is a measure of the electromotive force (EMF) or voltage of an electrochemical cell. By subtracting the potential of the anode (oxidation) from the potential of the cathode (reduction), we can determine the overall potential of the cell.

It is important to note that standard conditions refer to a set of conditions that are commonly used for comparison in chemistry experiments. These conditions include a temperature of 25 degrees Celsius, a pressure of 1 atmosphere, and a concentration of 1 mol/L for any dissolved species.

The measurement of standard electrode (reduction) potentials is typically carried out using a reference electrode, such as the standard hydrogen electrode or the saturated calomel electrode. These reference electrodes provide a known and stable potential against which other electrodes can be compared.

Standard cell potential

The standard cell potential, also known as the standard electromotive force (EMF), is a measure of the voltage or electrical potential difference between the two half-cells of an electrochemical cell under standard conditions. It is a key concept in electrochemistry and provides valuable information about the feasibility and direction of a redox (reduction-oxidation) reaction.

In simple terms, the standard cell potential tells us how strongly a redox reaction will occur in an electrochemical cell. It is determined by the difference in the standard electrode potentials of the two half-cells involved in the cell. The higher the positive standard cell potential, the stronger the driving force for the reaction to occur.

The standard cell potential is typically symbolized as E°cell. It is measured in volts (V) or millivolts (mV). The standard conditions for measuring E°cell include a temperature of 25 degrees Celsius, a pressure of 1 atmosphere, and concentrations of 1 mol/L for any dissolved species involved in the reaction.

To understand the concept of standard cell potential, it is important to grasp the notion of half-cells. A half-cell consists of an electrode immersed in a solution of its corresponding ion. In an electrochemical cell, two different half-cells are connected by a conducting material, allowing the transfer of electrons from one half-cell to the other.

The standard cell potential can be calculated by subtracting the standard electrode potential of the anode (oxidation half-reaction) from that of the cathode (reduction half-reaction). The anode is where oxidation occurs, resulting in the loss of electrons, while the cathode is where reduction occurs and involves the gain of electrons.

The sign of the standard cell potential indicates the direction of electron flow, with a positive value indicating that the reaction proceeds spontaneously in the forward direction. A negative standard cell potential implies a non-spontaneous reaction, meaning that it would require an external voltage or electrical energy input to occur.

The standard cell potential allows chemists and scientists to predict the feasibility and spontaneity of redox reactions. If the calculated E°cell is positive, it suggests that the redox reaction is thermodynamically favored and will occur without any external assistance. Conversely, a negative E°cell value indicates that the reaction will not occur spontaneously under standard conditions.

The standard cell potential is influenced by factors such as temperature, pressure, and concentration. However, under standard conditions, these factors are fixed, allowing for consistent and comparable measurements and predictions.

It is important to note that the standard cell potential does not provide information about the rate at which a reaction occurs. It solely indicates the thermodynamic favorability of the reaction. Reaction rates are determined by factors such as the nature of the reactants, surface area, catalysts, and temperature.

The standard cell potential plays a significant role in various applications, including batteries, fuel cells, corrosion analysis, and electrolysis processes. By understanding the standard cell potential, scientists can design and optimize electrochemical systems for practical and efficient use.

The standard hydrogen electrode

The Standard Hydrogen Electrode (SHE) is a crucial component in electrochemistry used as a reference point for measuring electrode potentials. It allows for the determination of absolute electrode potentials by comparing the potential of the electrode under investigation to that of the SHE. This reference electrode is defined as having a potential of zero volts and is typically an essential tool in the study of electrochemical reactions and the development of electrochemical cells.

The development and adoption of the Standard Hydrogen Electrode have significantly contributed to the advancement of the field of electrochemistry. It provides a standardized and reproducible reference point that facilitates the comparison of different electrode potentials. By establishing a universally recognized reference electrode, scientists can more accurately measure and characterize electrochemical systems. This enables the prediction and understanding of chemical reactions, as well as the development of practical applications such as batteries, fuel cells, and corrosion prevention strategies.

The SHE consists of two major components: a platinum electrode and an aqueous solution of hydrochloric acid (HCl). The platinum electrode is coated with a layer of finely divided platinum black, which provides a large surface area for the electrode to interact with the electrolyte solution. The platinum black enhances the electrode's efficiency in facilitating the redox reactions involved in the electrochemical measurements. It also minimizes any potential effects of impurities or contaminants present in the HCl solution.

The aqueous hydrochloric acid solution acts as the electrolyte in the SHE system. It ensures the flow of charge by carrying ions necessary for the electrochemical reaction. The concentration of HCl is typically around 1 M, providing sufficient conductivity for the measurement while maintaining a stable and reproducible electrode potential.

A half-cell setup is used to assemble the SHE, with the platinum electrode immersed in the HCl solution. The reaction occurring at the SHE is:

$$2H^+ (aq) + 2e^- \rightarrow H_2 (g)$$

This reaction involves the reduction of two protons (H+) to form hydrogen gas (H2) at the platinum electrode. The platinum electrode acts as a catalyst for this reaction, facilitating the transfer of electrons from the electrode to the protons in the solution.

To maintain a constant and reproducible potential for the SHE, the H2 gas is maintained at a pressure of 1 bar. This ensures that the partial pressure of H2 at the electrode surface remains constant, allowing for consistent electrode potential measurements. The barometric pressure and temperature also need to be taken into account during measurements to ensure accurate results.

The electrode potential of the SHE is defined as zero volts by convention. This convention allows for the determination of the standard electrode potentials of other half-cells using the Nernst equation. The Nernst equation relates the electrode potential (E) of a half-cell to the standard electrode potential (E°), the molar concentration of reactants and products, and the temperature (T):

$$E = E° - [(RT)/(nF)] * \ln(Q)$$

Where R is the ideal gas constant (8.314 J/(mol·K)), T is the temperature in Kelvin, n is the number of electrons involved in the reaction, F is Faraday's constant (96,485 C/mol), and Q is the reaction quotient.

The standard electrode potential (E°) of a half-cell can be determined by measuring the potential of the half-cell against the SHE under standard conditions. These conditions include a concentration of 1 M for any solutes involved in the half-cell reaction and 1 bar pressure of any gases. This allows for consistent and reproducible measurements of the electrode potential, which can then be used to calculate the standard electrode potential.

The SHE has found widespread applications in various fields of chemistry. One of its primary uses is the determination of the standard electrode potentials of other half-cells. By comparing the potential of a half-cell of interest to that of the SHE, the standard electrode potential of the half-cell can be determined. This information is crucial for understanding the thermodynamics of electrochemical reactions and designing efficient electrochemical systems.

Additionally, the SHE plays a vital role in the development and characterization of electrochemical cells such as batteries and fuel cells. These devices rely on redox reactions to generate and store electrical energy. The electrode potentials of the different cell components are essential in optimizing cell performance and understanding the underlying electrochemical processes.

Furthermore, the SHE is used in corrosion studies to evaluate the tendency of different materials to corrode. By measuring the corrosion potential of a material relative to the SHE, researchers can assess its susceptibility to oxidation and degradation. This information is crucial for the design and selection of materials that will be exposed to corrosive environments, ensuring the longevity and reliability of various structures and devices.

Methods for Measuring Standard Electrode Potentials
Metals and non-metals in contact with their ions in aqueous solutions

Measurement of standard electrode potentials is a pivotal process in electrochemistry, providing a quantitative description of the redox potential of an electrode. The process involves a metal or non-metal in contact with its ions in an aqueous solution. The standard electrode potential, denoted as E°, serves as a measure of the tendency of an electrode to lose or gain electrons, hence undergoing oxidation or reduction. It is measured under standard conditions, that is, 298 K temperature, one molar solute concentration, and one bar pressure. The standard electrode potential is measured using a method known as the half-cell method. In this process, the electrode under consideration is connected to a standard hydrogen electrode (SHE), which serves as a reference electrode. The SHE is assigned a potential of 0 volts by convention. The system is connected through a salt bridge, which maintains electrical neutrality within the solution by allowing ions to migrate between the two solutions. The electrode of interest is immersed in a 1M solution of its ions. For instance, if the electrode under investigation is copper, it is immersed in a 1M solution of Cu^{2+} ions. The copper electrode acts as the working electrode while the SHE acts as the reference electrode in this electrochemical cell. The potential difference between the working and reference electrode is then measured using a voltmeter. This potential difference is the standard electrode potential of the working electrode. It is important to note that the sign of the potential difference determines whether the electrode is more prone to oxidation (losing electrons) or reduction (gaining electrons).

For non-metals, a similar procedure is followed. However, the non-metal is usually incorporated into the electrode in a different manner. For instance, a platinum electrode could be used, with the non-metal, such as chlorine, dissolved in an aqueous solution along with its ions. The standard electrode potential is a useful property because it allows the prediction of the direction of redox reactions. If the standard electrode potential of the oxidizing agent is higher than that of the reducing agent, the reaction is spontaneous. Conversely, if the standard electrode potential of the oxidizing agent is lower than that of the reducing agent, the reaction is non-spontaneous. One limitation of this method is that it is only applicable for reactions occurring at standard conditions.

In practical situations, the temperature, pressure, and concentration may vary. To account for this, the Nernst equation can be used to calculate the electrode potential under non-standard conditions. The Nernst equation provides a relationship between the electrode potential, the standard electrode potential, the reaction quotient, and the temperature. It allows the calculation of the electrode potential for any given concentration of ions, not just for 1M solutions. In measuring standard electrode potentials, it is important to ensure that the voltmeter used is of high impedance. This is to avoid drawing current from the cell, which could alter the potential difference being measured. The measurement of standard electrode potentials has a wide range of applications. In batteries and fuel cells, it allows the determination of the cell potential, which is the driving force for the electrochemical reaction occurring within the cell. It also aids in the understanding of corrosion processes and the selection of suitable materials to prevent or minimize corrosion. In the field of environmental chemistry, the standard electrode potentials can be used to understand the behavior of metals and non-metals in natural waters. This is particularly important in the study of heavy metal contamination, where the standard electrode potentials can help predict the mobility and bioavailability of these metals.

Problem 1: What is the method used to measure the standard electrode potential for metals in contact with their ions in aqueous solution?

Solution: The standard electrode potential for metals is usually measured using an electrochemical cell. This involves setting up two half-cells: one with the metal and its ions in aqueous solution, and the other with a standard hydrogen electrode (SHE) which has a potential of 0 volts. The metal electrode is connected to the hydrogen electrode via a salt bridge, which allows ions to pass between the two half-cells. The voltage difference between the two electrodes is then measured using a voltmeter. This difference in potential is the standard electrode potential of the metal.

Problem 2: Sometimes, the value obtained for the standard electrode potential of a metal may be negative. What does this signify?

Solution: A negative standard electrode potential means that the metal is more likely to be oxidized (lose electrons) than hydrogen. This provides information about the reactivity of the metal. The reactivity of the metal increases as the standard electrode potential becomes more negative.

Problem 3: How can we measure the standard electrode potential of a non-metal?

Solution: For non-metals, one common way to measure the standard electrode potential is to use a redox reaction with a known reactant. For example, the standard electrode potential of chlorine can be measured by setting up a half-cell with chlorine gas and chloride ions, and another half-cell with the standard hydrogen electrode. The potential difference between the two electrodes, measured under standard conditions, is the standard electrode potential of chlorine.

Problem 4: Why is it important to keep the conditions standard when measuring the standard electrode potential?

Solution: Standard conditions (temperature of 298K, pressure of 1 atm, and concentration of 1M for all species in solution) ensure that the measured potential is consistent and can be compared to other potentials measured under the same conditions. If the conditions are not standard, the measured potential will be affected and may not accurately represent the electrode potential.

Problem 5: Why do we use the standard hydrogen electrode (SHE) as a reference when measuring standard electrode potentials?

Solution: The standard hydrogen electrode is used as a reference because its potential is defined as 0 volts under standard conditions. This allows the potentials of other electrodes to be measured relative to this zero point, providing a consistent scale for comparison.

Ions of the same element in different oxidation states

Standard electrode potentials, also known as redox potentials, are crucial in electrochemistry as they provide vital information about the equilibrium position of redox reactions. They are typically measured using an electrochemical cell, involving an oxidation half-reaction and a reduction half-reaction. The potentials of ions in different oxidation states are especially interesting because they provide insight into the relative stability of these states. The first step in measuring the standard electrode potential is to set up a standard hydrogen electrode (SHE), which acts as the reference electrode. The SHE is a hydrogen electrode set to a pressure of 1 bar and a temperature of 298 K. It is composed of hydrogen gas bubbling over a platinum electrode immersed in a solution of 1M HCl. The potential of the SHE is arbitrarily set to 0 V, and all other potentials are measured relative to it. To measure the electrode potential of an ion of an element in a certain oxidation state, a half-cell is set up. This half-cell consists of a metal electrode immersed in a 1M solution of a salt of the metal. If the ion in question is insoluble or a gas, a platinum electrode is used, and the ion is co-deposited on the electrode surface with a suitable conducting material, usually graphite. The two half-cells, the SHE and the ion half-cell, are then connected with a salt bridge, which completes the circuit and allows migration of ions, thus maintaining electrical neutrality in the solutions. The potential difference between the two half-cells is measured using a high impedance voltmeter. The electrode potential of the ion is then determined using the Nernst equation, which relates the potential of an electrode to the concentrations of the oxidized and reduced forms of the species. The standard electrode potential, $E°$, is the potential when all species are at unit activity, usually approximated as 1M concentration for solutes. The process is repeated for each oxidation state of the element. It's important to note that the sign of the potential can provide valuable information about the spontaneity of the reaction. A positive potential indicates a spontaneous reaction, while a negative potential suggests a non-spontaneous reaction. The measurement of standard electrode potentials must be conducted under standard conditions, i.e., at a temperature of 298 K, a pressure of 1 bar, and concentrations of 1M for all ions involved in the redox reaction. Deviations from these conditions can lead to errors in the obtained results. It should be noted, however, that the

178

standard electrode potentials are not absolute values; they are relative values. This is because the electrode potential is a measure of the tendency of the species to lose or gain electrons, relative to the hydrogen electrode. The technique described above is the most common method for measuring the standard electrode potential of ions in different oxidation states. However, other methods can also be used, depending on the specific requirements of the experiment. For example, the cyclic voltammetry technique can be used to measure the standard electrode potentials of electroactive species in solution, while the potentiostatic method can be used for species that are difficult to oxidize or reduce.

Problem 1: How can we measure the standard electrode potentials of ions of the same element in different oxidation states?

Solution 1: The method used to measure the standard electrode potentials of ions of the same element in different oxidation states is based on using an electrochemical cell. This cell is composed of two half-cells. Each half-cell consists of a metal electrode immersed in a solution of its own ions. The metal electrode of one half-cell represents the element in one oxidation state, and the electrode of the other half-cell represents the element in another oxidation state. The two half-cells are connected via a salt bridge that allows ions to flow freely between the half-cells, maintaining electrical neutrality. The potential difference between the two electrodes can be measured using a voltmeter, which gives the standard electrode potential.

Problem 2: Can we measure the standard electrode potentials of ions of the same element in different oxidation states if the element doesn't exist in the metallic state?

Solution 2: Yes, it is possible to measure the standard electrode potentials of ions of the same element in different oxidation states even if the element doesn't exist in the metallic state. In such cases, an inert electrode, like platinum, is used. The inert electrode is immersed in a solution of the ions of the element in the different oxidation states. The potential difference between the platinum electrode and the standard hydrogen electrode gives the standard electrode potential.

Problem 3: What if the ion of the element in one oxidation state is not soluble in water?

Solution 3: In this case, the soluble ion can be measured against the standard hydrogen electrode (SHE). The standard hydrogen electrode is set at 0 V by convention. The potential difference between the soluble ion and the SHE gives the standard electrode potential.

Problem 4: How can we measure the standard electrode potential if the element is in a gaseous state?

Solution 4: If the element is in a gaseous state, it can be bubbled through a solution containing a salt of the element. An inert electrode, such as platinum, can be used in this case. The electrode potential can then be measured relative to the standard hydrogen electrode.

Problem 5: What is the significance of the salt bridge in an electrochemical cell?

Solution 5: The salt bridge serves two main purposes. Firstly, it completes the electrical circuit allowing the flow of current. Secondly, it maintains the electrical neutrality of the solutions in the half-cells by allowing the movement of ions. Without the salt bridge, the solutions in the half-cells would quickly become electrically charged, stopping the reaction.

Calculate standard cell potential by adding two standard electrode potentials

Problem 1: Given the following standard electrode potentials, calculate the standard cell potential for the reaction:

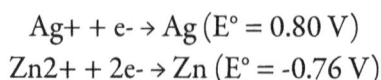

Ag+ + e- → Ag (E° = 0.80 V)

Zn2+ + 2e- → Zn (E° = -0.76 V)

Solution: The standard cell potential (E°cell) is the difference between the standard electrode potentials of the cathode and anode. The reaction with the higher E° is the reduction (cathode) and the one with the lower E° is the oxidation (anode).

E°cell = E°cathode - E°anode

E°cell = 0.80 V - (-0.76 V) = 1.56 V

Problem 2: Calculate the standard cell potential for the reaction: Cu^{2+} + 2e- → Cu (E° = 0.34 V)

Fe^{2+} + 2e- → Fe (E° = -0.44 V)

Solution: Here, the Cu^{2+} → Cu reaction has a higher E°, so it is the reduction (cathode), and the Fe^{2+} → Fe is the oxidation (anode).

E°cell = E°cathode - E°anode

E°cell = 0.34 V - (-0.44 V) = 0.78 V

Problem 3: Given the following standard electrode potentials, calculate the standard cell potential for the reaction:

Al^{3+} + 3e- → Al (E° = -1.66 V)

Mg^{2+} + 2e- → Mg (E° = -2.37 V)

Solution: The Al^{3+} → Al reaction has a higher E°, so it is the reduction (cathode), and the Mg^{2+} → Mg is the oxidation (anode).

E°cell = E°cathode - E°anode

E°cell = -1.66 V - (-2.37 V) = 0.71 V

Problem 4: Calculate the standard cell potential for the reaction: H+ + e- → H2 (E° = 0.00 V)

Sn^{4+} + 2e- → Sn^{2+} (E° = 0.15 V)

Solution: In this case, the Sn^{4+} → Sn^{2+} reaction has a higher E° and so is the reduction (cathode), while the H+ → H2 is the oxidation (anode).

E°cell = E°cathode - E°anode

E°cell = 0.15 V - 0.00 V = 0.15 V

Determine each electrode's polarity and from that understand the direction of electron flow in a basic cell's external circuit

The use of standard cell potentials in determining the polarity of each electrode and the direction of electron flow in a simple cell is a fundamental principle in electrochemistry. This principle lays the foundation for understanding the operation of batteries, electrolysis, and many other applications. Cell potentials, also known as electromotive force (EMF), refer to the maximum potential difference, or voltage, that a galvanic cell can generate. The standard cell potential is calculated under standard conditions, that is, at a temperature of 25°C, with a pressure of 1 atmosphere and a concentration of 1 M for all species in the cell. In a galvanic cell, chemical reactions occur at the electrodes that lead to the flow of electrons from the anode to the cathode. The anode is the electrode where oxidation occurs, and it is here that electrons are lost. The cathode, on the other hand, is the electrode where reduction takes place, and it is here that electrons are gained. The standard cell potentials can be used to deduce the polarity of each electrode. The electrode with a higher standard reduction potential is the cathode and is positively charged. Conversely, the electrode with a lower standard reduction potential is the anode and is negatively charged. These polarities are crucial in determining the direction of electron flow in the external circuit of a simple cell. Electrons are naturally inclined to move from areas of high electron density (negative charge) to areas of low electron density (positive charge). Therefore, in a simple cell, the electrons in the external circuit will flow from the anode (negative electrode) to the cathode (positive electrode). This flow of electrons is essentially what constitutes electric current, with the direction of current being conventionally defined as the direction in which positive charges move. Thus, in the external circuit, electric current flows from the cathode to the anode, opposite to the direction of electron flow. The difference in the standard reduction potentials of the two electrodes determines the magnitude of the cell potential.

A larger difference implies a larger cell potential and, consequently, a larger driving force for the flow of electrons. It's also worth noting that the cell potential is a measure of the cell's ability to do work. A cell with a higher potential can do more work, driving a larger current through a load, than a cell with a lower potential. Furthermore, the standard cell potential is a state function, meaning it is path-independent and depends only on the initial and final states of the system. This property allows for the convenient calculation of cell potentials for reactions that can be divided into multiple steps. In summary, the use of standard cell potentials allows for a detailed understanding of the operation of a simple cell. By identifying the anode and cathode and their respective polarities, one can deduce the direction of electron flow in the external circuit. This knowledge is fundamental to the design and operation of batteries and other electrochemical devices. Ultimately, the cell potential provides a quantitative measure of the driving force for electron flow, offering insights into the cell's ability to do work and the amount of current it can produce. As such, the concepts of cell potential and electron flow are intrinsically linked and form a cornerstone of electrochemistry.

Problem 1: If a simple cell is constructed using zinc and copper with their respective sulfate solutions, determine the polarity of each electrode and the direction of electron flow in the external circuit.

Solution: The standard reduction potentials(E°) for Zn2+/Zn and Cu2+/Cu are -0.76V and +0.34V, respectively. Since copper has a higher reduction potential, it will act as the cathode (where reduction occurs) and will be the positive electrode. Zinc, with the lower reduction potential, will be the anode (where oxidation occurs) and will be the negative electrode. Electrons flow from the anode to the cathode in the external circuit, so in this case, they will move from the zinc electrode to the copper electrode.

Problem 2: In a simple cell using silver and lead, identify the polarity of each electrode and the direction of electron flow in the external circuit.

Solution: The standard reduction potentials for Ag+/Ag and Pb2+/Pb are +0.80V and -0.13V respectively. Silver, with the higher reduction potential, will be the cathode (positive electrode) and lead, with the lower reduction potential, will be the anode (negative electrode). Electrons in the external circuit will flow from the anode to the cathode, so they will move from the lead electrode to the silver electrode.

Problem 3: Determine the polarity of each electrode and the direction of electron flow in the external circuit for a simple cell constructed with magnesium and nickel.

Solution: The standard reduction potentials for Mg2+/Mg and Ni2+/Ni are -2.37V and -0.23V respectively. Nickel, with the higher reduction potential, will act as the cathode (positive electrode) and magnesium, with the lower reduction potential, will act as the anode (negative electrode). The electron flow in the external circuit will be from the anode to the cathode, meaning they will flow from the magnesium electrode to the nickel electrode.

Problem 4: If a simple cell is made using aluminum and iron, what would be the polarity of each electrode and the direction of electron flow in the external circuit?

Solution: The standard reduction potentials for Al3+/Al and Fe2+/Fe are -1.66V and -0.44V respectively. Iron, having the higher reduction potential, will serve as the cathode (positive electrode) and aluminum, with the lower reduction potential, will be the anode (negative electrode). In the external circuit, electrons flow from the anode to the cathode, implying they will move from the aluminum electrode to the iron electrode.

Predicting the viability of the reaction

The concept of standard cell potentials is an essential tool in predicting the feasibility of a chemical reaction. This provides a quantitative method for determining whether a reaction will occur spontaneously or not. The standard cell potential, also known as the standard electromotive force (emf), is the potential difference between the two electrodes of an electrochemical cell when all substances are in their standard states at 25 degrees Celsius. The standard cell potential is calculated by subtracting the standard reduction potential of the anode (the electrode where oxidation occurs) from the standard reduction potential of the cathode (the electrode where reduction takes place). The sign of the standard cell potential provides valuable insight into the spontaneity of the reaction. A spontaneous reaction is indicated by a positive standard cell potential. According to the laws of thermodynamics, a process will occur spontaneously if the Gibbs free energy change (ΔG) for the process is negative. The relationship between Gibbs free energy change and standard cell potential is given by the following equation: $\Delta G = -nFE$, where n is the number of moles of electrons transferred in the reaction, F is Faraday's constant, and E is the standard cell potential. Therefore, a positive standard cell potential results in a negative free energy change, suggesting a spontaneous reaction. Conversely, if the standard cell potential is negative, the reaction does not occur spontaneously under standard conditions. A negative standard cell potential means that the Gibbs free energy change is positive, and the reaction does not occur spontaneously. Non-spontaneous reactions require an external energy source to proceed. However, it's crucial to remember that standard cell potentials are calculated under standard conditions. In real-world applications, the conditions may not be standard, and the cell potential may differ from the standard cell potential. Therefore, the actual feasibility of the reaction under non-standard conditions must be calculated using the Nernst equation. It's also important to note that while cell potentials predict the spontaneity of a reaction, they provide no information about the reaction rate. A reaction with a positive cell potential may still proceed very slowly without a catalyst. Finally, the standard cell potentials also allow us to build an electrochemical series, ranking various redox couples according to their standard reduction potentials. This series is instrumental in predicting the outcome of displacement reactions and the feasibility of a redox reaction.

Problem 1:

Consider the reaction:

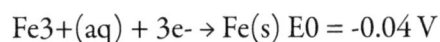

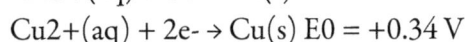

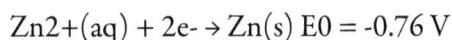

$$Fe^{3+}(aq) + 3e^- \rightarrow Fe(s) \quad E^0 = -0.04 \text{ V}$$
$$Cu^{2+}(aq) + 2e^- \rightarrow Cu(s) \quad E^0 = +0.34 \text{ V}$$

Would the reaction between Fe^{3+} and Cu be feasible in a standard state?

Solution: The reaction between Fe^{3+} and Cu would be:

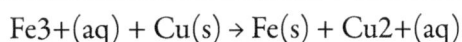

$$Fe^{3+}(aq) + Cu(s) \rightarrow Fe(s) + Cu^{2+}(aq)$$

The standard cell potential for this reaction, E^0, is the difference between the reduction potentials of the half-reactions:

$$E^0 = E^0(Cu^{2+}/Cu) - E^0(Fe^{3+}/Fe) = 0.34 \text{ V} - (-0.04 \text{ V}) = 0.38 \text{ V}$$

Since E^0 is positive, the reaction is feasible in a standard state.

Problem 2:

Consider the reaction:

$$Zn^{2+}(aq) + 2e^- \rightarrow Zn(s) \quad E^0 = -0.76 \text{ V}$$
$$Ag^+(aq) + e^- \rightarrow Ag(s) \quad E^0 = +0.80 \text{ V}$$

Would the reaction between Zn^{2+} and Ag be feasible in a standard state?

Solution: The reaction between Zn^{2+} and Ag would be: $Zn^{2+}(aq) + 2Ag(s) \rightarrow Zn(s) + 2Ag^+(aq)$

The standard cell potential for this reaction, E^0, is the difference between the reduction potentials of the half-reactions:

$$E^0 = E^0(Ag^+/Ag) - E^0(Zn^{2+}/Zn) = 0.80 \text{ V} - (-0.76 \text{ V})$$
$$= 1.56 \text{ V}$$

Since E^0 is positive, the reaction is feasible in a standard state.

Problem 3: Consider the reaction:

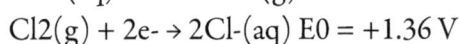

$$H^+(aq) + e^- \rightarrow \tfrac{1}{2}H_2(g) \quad E^0 = 0.00 \text{ V}$$
$$Cl_2(g) + 2e^- \rightarrow 2Cl^-(aq) \quad E^0 = +1.36 \text{ V}$$

Would the reaction between H^+ and Cl_2 be feasible in a standard state?

Solution: The reaction between H+ and Cl2 would be:

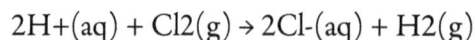

$$2H+(aq) + Cl2(g) \rightarrow 2Cl-(aq) + H2(g)$$

The standard cell potential for this reaction, E0, is the difference between the reduction potentials of the half-reactions:

$$E0 = E0(Cl2/2Cl-) - E0(H+/1/2H2) = 1.36\,V - 0.00\,V = 1.36\,V$$

Since E0 is positive, the reaction is feasible in a standard state.

Determine the reactivity of elements, compounds and ions as oxidising or reducing agents using E$\diamondsuit$ values

E$\diamondsuit$ values, or standard electrode potentials, provide a quantitative measure of the relative reactivity of various elements, compounds, and ions as oxidising or reducing agents. The E$\diamondsuit$ value represents the ability of a species to accept or donate electrons - the fundamental process in redox reactions. By using these values, we can predict the spontaneity of redox reactions and thus, the reactivity of the involved species. Oxidising agents are species that accept electrons, and their strength as an oxidant is directly proportional to their E$\diamondsuit$ values. A higher E$\diamondsuit$ value indicates a stronger tendency to gain electrons, thus making the species a more potent oxidising agent. For instance, Fluorine (F2) with an E$\diamondsuit$ value of +2.87V, is one of the strongest oxidising agents. On the other hand, reducing agents are species that donate electrons, and their strength as a reductant is inversely proportional to their E$\diamondsuit$ values. A lower E$\diamondsuit$ value (or more negative) indicates a stronger tendency to lose electrons, thus making the species a more potent reducing agent. For instance, Lithium (Li) with an E$\diamondsuit$ value of -3.04V, is a strong reducing agent. It is essential to note that the E$\diamondsuit$ value is dependent on the standard state conditions i.e., 298 K temperature, 1 atm pressure, and 1M concentration. Deviations from these conditions can affect the E$\diamondsuit$ values and consequently, the reactivity of the species. E$\diamondsuit$ values can be used to predict the direction of a redox reaction. If the E$\diamondsuit$ for the overall reaction is positive, the reaction is spontaneous, indicating that the oxidising agent and reducing agent are suitably paired. However, if the E$\diamondsuit$ for the reaction is negative, the reaction is non-spontaneous, indicating a mismatch in the reactivity of the participating species. Compounds and ions also exhibit oxidising and reducing behaviour based on their E$\diamondsuit$ values. For instance, permanganate ion (MnO4-) is a strong oxidising agent due to its high E$\diamondsuit$ value. Similarly, ferrous ion (Fe2+) acts as a reducing agent due to its low E$\diamondsuit$ value. E$\diamondsuit$ values also provide insights into the relative stability of different oxidation states of an element. A more positive E$\diamondsuit$ value suggests a more stable oxidation state. It is also important to acknowledge the role of overpotential in the reactivity of species. Overpotential is the extra voltage required to drive a non-spontaneous reaction. Species with high overpotential may exhibit lower reactivity than predicted by their E$\diamondsuit$ values alone. While E$\diamondsuit$ values are a useful tool in predicting reactivity, they do not provide information about the reaction kinetics i.e., the speed of the reaction. Other factors such as activation energy, temperature, and catalysts also influence the reaction rate.

Problem 1: Given the E$\diamondsuit$ values for the following half-reactions: Fe2+ $\rightarrow$ Fe3+ + e-; E$\diamondsuit$ = 0.77V

Fe3+ + 3e- $\rightarrow$ Fe; E$\diamondsuit$ = -0.036V

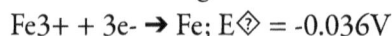

Which species is the stronger oxidizing agent and which is the stronger reducing agent?

Solution: The species with the more positive E$\diamondsuit$ value is the stronger oxidizing agent. Therefore, Fe3+ is a stronger oxidizing agent. The species with the more negative E$\diamondsuit$ value is the stronger reducing agent. Therefore, Fe is a stronger reducing agent.

Problem 2: Given the standard electrode potentials for the following half-reactions: Zn2+ + 2e- $\rightarrow$ Zn; E$\diamondsuit$ = -0.76V

Cu2+ + 2e- $\rightarrow$ Cu; E$\diamondsuit$ = +0.34V

Which element is more likely to be oxidized and which is more likely to be reduced?

Solution: The element with the more negative E⬦ value is more likely to be oxidized. Therefore, Zn is more likely to be oxidized. The element with the more positive E⬦ value is more likely to be reduced. Therefore, Cu2+ is more likely to be reduced.

Problem 3: Considering the half-cell reactions:

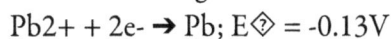

Pb2+ + 2e- ➔ Pb; E⬦ = -0.13V

Sn2+ + 2e- ➔ Sn; E⬦ = -0.14V

Which metal ion is a better oxidizing agent?

Solution: The ion with the more positive E⬦ value is the better oxidizing agent. Therefore, Pb2+ is a better oxidizing agent than Sn2+.

Problem 4: Given the standard electrode potentials:

Ag+ + e- ➔ Ag; E⬦ = +0.80V

Ni2+ + 2e- ➔ Ni; E⬦ = -0.25V

Which ion is a more powerful oxidizing agent and which metal is a better reducing agent?

Solution: The ion with the more positive E⬦ value is the more powerful oxidizing agent. Therefore, Ag+ is a more powerful oxidizing agent. The metal with the more negative E⬦ value is a better reducing agent. Therefore, Ni is a better reducing agent.

Redox equations: half-equations

Problem 1: Consider the reaction between zinc and hydrochloric acid. Zinc (Zn) reacts with hydrochloric acid (HCl) to produce zinc chloride (ZnCl2) and hydrogen gas (H2). Write the redox equation using the relevant half-equations.

Solution: First, write down the half-reactions.

Zn → Zn2+ + 2e- (Oxidation half-reaction)

2H+ + 2e- → H2 (Reduction half-reaction)

Then, combine the two half-reactions to get the overall reaction. Zn + 2H+ → Zn2+ + H2

Finally, balance the charges by adding the chloride ions.

Zn + 2HCl → ZnCl2 + H2

Problem 2: Consider the reaction between copper(II) sulfate and iron. Iron (Fe) reacts with copper(II) sulfate (CuSO4) to produce iron(II) sulfate (FeSO4) and copper (Cu). Write the redox equation using the relevant half-equations.

Solution: First, write down the half-reactions.

Fe → Fe2+ + 2e- (Oxidation half-reaction)

Cu2+ + 2e- → Cu (Reduction half-reaction)

Then, combine the two half-reactions to get the overall reaction. Fe + Cu2+ → Fe2+ + Cu

Finally, balance the charges by adding the sulfate ions.

Fe + CuSO4 → FeSO4 + Cu

Problem 3: Consider the reaction between sodium and chlorine. Sodium (Na) reacts with chlorine (Cl2) to produce sodium chloride (NaCl). Write the redox equation using the relevant half-equations.

Solution: First, write down the half-reactions.

2Na → 2Na+ + 2e- (Oxidation half-reaction)

Cl2 + 2e- → 2Cl- (Reduction half-reaction)

Then, combine the two half-reactions to get the overall reaction. 2Na + Cl2 → 2Na+ + 2Cl-

Finally, balance the charges by combining the ions.

2Na + Cl2 → 2NaCl

Electrode potential E varies with ion concentration

The electrode potential, denoted as E, is a critical parameter in electrochemistry. It refers to the ability of an electrode in an electrochemical cell to lose or gain electrons, thus indicating its redox potential. The electrode potential is influenced by a number of factors, one of them being the concentrations of aqueous ions. This relationship is governed by the Nernst equation, a fundamental principle in electrochemistry that links the reduction potential of an electrochemical reaction (specifically, the half-cell potential) to the standard electrode potential, temperature, and reactant/product concentrations. The Nernst equation can be expressed as $E = E° - (RT/nF) \ln Q$. In this equation, $E°$ represents the standard electrode potential, R stands for the gas constant, T denotes the absolute temperature, n is the number of electrons transferred during the redox reaction, F corresponds to Faraday's constant, and Q is the reaction quotient. When the concentrations of the aqueous ions change, the reaction quotient, Q, changes, which effectively impacts the electrode potential, E. If the concentration of the reactant species decreases, Q decreases, causing the natural logarithm term in the Nernst equation to become negative. This leads to an increase in the overall electrode potential, E. Conversely, if the concentration of the product species decreases, Q decreases, resulting in a positive natural logarithm term and a subsequent decrease in E. On the other hand, if the concentration of the reactant species increases, Q increases, causing the natural logarithm term in the Nernst equation to become positive. This results in a decrease in the overall electrode potential, E. Similarly, if the concentration of the product species increases, Q increases, and the natural logarithm term becomes negative, leading to an increase in E. It's important to note that these changes in the electrode potential due to changes in aqueous ion concentrations are valid only when the temperature and pressure are kept constant. Variations in these parameters can also significantly impact the electrode potential. In electrochemical cells, the manipulation of ion concentrations is a common strategy to control and optimize the cell's performance. For instance, in batteries, the concentrations of the ions in the electrolyte solution can be adjusted to optimize the cell's energy density, power density, and cycle life. In fuel cells, the concentrations of the proton and oxygen ions can be controlled to manage the cell's performance, including its power output and efficiency. Similarly, in electrochemical sensors, the concentrations of the target ions can be adjusted to tune the sensor's sensitivity, selectivity, and response time. Understanding the complexities of this relationship is not just theoretically significant, but also practically important. For example, in industrial applications, managing ion concentrations can be used to control electrochemical processes, such as electroplating and electrowinning. In environmental applications, changes in ion concentrations can be used to monitor the health of aquatic ecosystems. In medical applications, changes in ion concentrations can be used to diagnose and treat various diseases.

Problem 1: Given that the reaction at an electrode is $Zn^{2+} (aq) + 2e^- \rightarrow Zn(s)$, predict how the electrode potential (E) will change if the concentration of Zn^{2+} ions increases. Solution: In this case, the Nernst equation describes the relationship between the electrode potential and the ion concentration. The Nernst equation is: $E = E_0 - (RT/nF)\ln Q$ where E_0 is the standard electrode potential, R is the gas constant, T is the temperature, n is the number of electrons transferred in the reaction, F is the Faraday constant, and Q is the reaction quotient, which in this case is the concentration of Zn^{2+} ions. As the concentration of Zn^{2+} ions increases, the value of $\ln Q$ (natural log of Q) increases. Since this term is subtracted in the Nernst equation, an increase in ion concentration will cause the electrode potential to decrease. Therefore, the electrode potential E decreases as the Zn^{2+} ion concentration increases.

Problem 2: How does the electrode potential E change if the concentration of Ag^+ ions decreases in the following reaction: $Ag^+ (aq) + e^- \rightarrow Ag(s)$?

Solution: Using the Nernst equation, we can understand that if the concentration of Ag^+ ions decreases, the value of $\ln Q$ also decreases. Because $\ln Q$ is subtracted in the Nernst equation, a decrease in Ag^+ ion concentration would result in an increase in the electrode potential. Therefore, the electrode potential E increases as the Ag+ ion concentration decreases. Problem 3: What is the effect on the electrode potential E when the concentration of Cu2+ ions remains constant in the reaction: Cu2+ (aq) + 2e- → Cu(s)?

Solution: In this case, if the concentration of Cu2+ ions remains constant, the value of Q in the Nernst equation also stays constant. If there is no change in Q, then there is no change in lnQ. Since the value of lnQ is not changing, it also means that there is no change in the electrode potential E. Therefore, the electrode potential E remains constant when the Cu2+ ion concentration is constant.

Use the Nernst equation, e.g. $E = E^⦵ + (0.059/z) \log [\text{oxidised species}]/[\text{reduced species}]$

The Nernst equation is an essential tool in predicting how electrode potentials vary with changes in the concentrations of aqueous ions. It not only provides a quantitative measure of the potential but also offers insights into the redox reactions taking place at the electrode surface. In this analysis, we will consider two widely studied reactions in electrochemistry: the reduction of Cu2+(aq) to Cu(s) and the reduction of Fe3+(aq) to Fe2+(aq).

Electrochemical cells, such as those found in batteries, derive their energy from redox reactions. The ability of these cells to produce electricity is largely dependent on the electrode potentials, which are influenced by the concentrations of the ions involved in the redox reactions. The Nernst equation provides a mathematical relationship to quantify this dependency.

The equation, $E = E^⦵ + (0.059/z) \log [\text{oxidised species}]/[\text{reduced species}]$, relates the electrode potential (E) to the standard electrode potential ($E^⦵$), the number of electrons transferred in the reaction (z). Here, 0.059 is the Nernst factor (at room temperature) which adjusts for the effect of temperature on the reaction.

Let's consider the first reaction, Cu2+(aq) + 2e− ⇌ Cu(s). Here, the copper ion (Cu2+) is the oxidised species, while copper metal (Cu) is the reduced species. The standard reduction potential, $E^⦵$, for this reaction is +0.34 volts. By varying the concentration of Cu2+ ions and keeping the concentration of Cu constant (as it's a pure solid), we can predict the effect on the electrode potential using the Nernst equation.

As the concentration of Cu2+ ions increases, the ratio [Cu2+]/[Cu] in the Nernst equation also increases. This results in a more significant logarithmic term, which overall increases the value of the electrode potential (E). Conversely, decreasing the concentration of Cu2+ ions decreases the electrode potential. This reveals a direct relationship between the concentration of the oxidised species and the electrode potential in a reduction reaction.

The same approach applies to the second reaction, Fe3+(aq) + e− ⇌ Fe2+(aq). Here, the iron (III) ion (Fe3+) is the oxidised species, while iron (II) ion (Fe2+) is the reduced species. The standard reduction potential, $E^⦵$, for this reaction is +0.77 volts. By manipulating the concentrations of Fe3+ and Fe2+ ions, we can similarly predict the effect on the electrode potential.

In this case, if we increase the concentration of Fe3+ ions while keeping the concentration of Fe2+ ions constant, the ratio [Fe3+]/[Fe2+] increases. This results in an increase in the electrode potential (E). Conversely, decreasing the concentration of Fe3+ ions while keeping the concentration of Fe2+ constant will decrease the electrode potential. This again shows a direct relationship between the concentration of the oxidised species and the electrode potential.

Problem 1: Calculate the electrode potential for the copper half-cell reaction Cu2+(aq) + 2e− ⇌ Cu(s) if the concentration of Cu2+ ions is 0.5 mol L-1. (Given: The standard electrode potential, $E^⦵$ for the copper half-cell reaction is +0.34 V)

Solution: The Nernst equation is $E = E^⦵ + (0.059/z) \log [\text{oxidised species}]/[\text{reduced species}]$

For the copper half-cell reaction, z = 2 (the number of electrons transferred in the reaction)

The equation becomes $E = E^⦵ + (0.059/2) \log [Cu2+]$

Inserting the given values, we get $E = 0.34 V + (0.059/2) \log (0.5)$

Solving this, we get E = 0.34 V - 0.015 V = 0.325 V

Problem 2: Calculate the electrode potential for the iron half-cell reaction $Fe^{3+}(aq) + e^- \rightleftharpoons Fe^{2+}(aq)$ if the concentration of Fe^{3+} ions is 0.1 mol L-1 and the concentration of Fe^{2+} ions is 0.3 mol L-1. (Given: The standard electrode potential, $E^{\diamondsuit}$ for the iron half-cell reaction is +0.77 V)

Solution: The Nernst equation is E = $E^{\diamondsuit}$ + (0.059/z) log [oxidised species]/[reduced species]

For the iron half-cell reaction, z = 1 (the number of electrons transferred in the reaction)

The equation becomes E = $E^{\diamondsuit}$ + (0.059/1) log [Fe^{3+}]/[Fe^{2+}]

Inserting the given values, we get E = 0.77 V + 0.059 log (0.1/0.3)

Solving this, we get E = 0.77 V - 0.03 V = 0.74 V

Problem 3: Calculate the change in the electrode potential for the copper half-cell reaction $Cu^{2+}(aq) + 2e^- \rightleftharpoons Cu(s)$ when the concentration of Cu^{2+} ions is increased from 0.1 mol L-1 to 1 mol L-1. (Given: The standard electrode potential, $E^{\diamondsuit}$ for the copper half-cell reaction is +0.34 V)

Solution: The Nernst equation is E = $E^{\diamondsuit}$ + (0.059/z) log [oxidised species]

For the copper half-cell reaction, z = 2 (the number of electrons transferred in the reaction)

Calculating the electrode potential at the initial concentration: E1 = 0.34 V + (0.059/2) log (0.1) = 0.34 V - 0.0295 V = 0.3105 V

Calculating the electrode potential at the final concentration: E2 = 0.34 V + (0.059/2) log (1) = 0.34 V

Therefore, the change in the electrode potential is E2 - E1

= 0.34 V - 0.3105 V = 0.0295 V.

$$\Delta G^{\diamondsuit} = -nE^{\diamondsuit}_{cell} F$$

The Gibbs Free Energy equation, $\Delta G^{\diamondsuit} = -nE^{\diamondsuit}cellF$, is a fundamental equation in the field of electrochemistry. This equation is used to determine the spontaneity of a redox reaction under standard conditions. It is a fundamental concept in thermodynamics and its accurate application is crucial to understanding electrochemical reactions. The symbol ΔG in the equation represents the change in Gibbs Free Energy, a thermodynamic potential that measures the maximum reversible work that a system can perform at constant temperature and pressure. If the value of ΔG is negative, it indicates that the reaction is spontaneous, while a positive value suggests that the reaction is non-spontaneous. The symbol n in the equation denotes the number of moles of electrons transferred in the redox reaction. This value is obtained from the balanced chemical equation for the reaction. It is critical to correct calculation of the Gibbs free energy change. The symbol $E^{\diamondsuit}cell$ in the equation represents the standard cell potential, measured in volts. This value is calculated from the standard reduction potentials of the two half-reactions involved in the overall redox reaction, with the cell potential for the reaction being the difference between the reduction potential of the anode and the cathode. The symbol F in the equation signifies the Faraday constant, named after the famous scientist Michael Faraday. The Faraday constant represents the electric charge carried by one mole of electrons, approximately 96,485 coulombs. The negative sign in the equation is of paramount importance as it indicates the direction of the flow of electrons. Electrons flow from the anode (where oxidation occurs) to the cathode (where reduction occurs). This flow of electrons generates an electric potential which is harnessed in batteries and fuel cells. The Gibbs Free Energy equation is not only used to determine the spontaneity of a reaction, but it also provides valuable information about the equilibrium of the reaction. If ΔG is zero, this signifies that the reaction has achieved equilibrium under standard conditions. Accurate calculation of the Gibbs free energy change requires careful attention to the stoichiometry of the redox reaction. The balanced chemical equation must correctly represent the number of electrons transferred in the reaction for the calculation of n. The cell potential $E^{\diamondsuit}cell$, is typically determined experimentally and tabulated for convenience. The use of these tabulated standard reduction potentials allows chemists to calculate the cell potential for a wide range of redox reactions. The Faraday constant F is a fundamental constant in electrochemistry. Its value is derived from the charge of an electron and Avogadro's number, the number of particles in a mole.

Problem 1: A cell is composed of a standard copper electrode and a standard hydrogen electrode. If the standard reduction potential for the hydrogen electrode is 0.00 V and for the copper electrode is +0.34 V, calculate the standard free energy change ($\Delta G°$) for the cell. 2 electrons are transferred in the reaction. (Take Faraday's constant, F = 96485 C mol−1)

Solution 1: Firstly, the standard cell potential (E⬦ cell) is calculated by subtracting the standard reduction potential of the anode (E⬦ anode) from that of the cathode (E⬦ cathode).

E⬦ cell = E⬦ cathode - E⬦ anode = +0.34 V - 0.00 V

= +0.34 V

Then, the standard free energy change ($\Delta G°$) can be calculated using the equation ΔG⬦ = -nFE⬦ cell.

ΔG⬦ = -2 × 96485 C mol−1 × 0.34 V = -65529.4 J mol−1, or -65.53 kJ mol−1.

Problem 2: The standard cell potential for a galvanic cell is -1.10 V. The reaction in the cell involves the transfer of 3 moles of electrons. What is the standard free energy change (ΔG⬦) for this cell reaction? (Take Faraday's constant, F = 96485 C mol−1)

Solution 2: We can use the equation ΔG⬦ = -nFE⬦ cell to calculate the standard free energy change.

ΔG⬦ = -3 × 96485 C mol−1 * -1.10 V = +318089.5 J mol−1, or +318.09 kJ mol−1.

Problem 3: A galvanic cell has a standard cell potential of 0.76 V. If the cell reaction involves the transfer of 5 moles of electrons, calculate the standard free energy change (ΔG⬦) for this cell. (Take Faraday's constant, F = 96485 C mol−1)

Solution 3: Again, we use the equation ΔG⬦ = -nFE⬦ cell to calculate the standard free energy change.

ΔG⬦ = -5 × 96485 C mol−1 × 0.76 V = -366445.4 J mol−1, or -366.45 kJ mol−1.

Equilibria

Acids and bases
Conjugate acid and Conjugate base

In the realm of chemistry, understanding the concept of acids and bases is fundamental. However, to further delve into the complexities of acid-base chemistry, one needs to comprehend the terms "conjugate acid" and "conjugate base". These terms play a significant role in the Brønsted-Lowry theory of acids and bases. The Brønsted-Lowry theory defines an acid as a substance that can donate a proton $(H+)$ and a base as a substance that can accept a proton. This concept contrasts with the older Arrhenius definition that limited acids to substances that produce hydrogen ions in water and bases to substances that produce hydroxide ions in water. The Brønsted-Lowry theory is more comprehensive and explores proton transfer reactions in a broader context. A conjugate acid, in the Brønsted-Lowry theory, is a species formed by the reception of a proton $(H+)$ by a base—in other words, it is the base with a hydrogen ion added to it. On the contrary, a conjugate base is what is left over after an acid has donated a proton during a chemical reaction. Hence, a conjugate base is a species formed by the removal of a proton $(H+)$ from an acid. To illustrate, consider the reaction between ammonia $(NH3)$, a base, and water. In this reaction, ammonia accepts a proton from a water molecule, transforming into its conjugate acid, the ammonium ion $(NH4+)$. Concurrently, the water molecule, after donating a proton, becomes its conjugate base, the hydroxide ion $(OH-)$. The concept of conjugate acid-base pairs is crucial for understanding many chemical reactions, especially those in biological systems. It allows us to predict the direction of proton transfer reactions and their equilibrium positions. Moreover, it aids in understanding the strength of acids and bases. The strength of a conjugate acid is directly related to the strength of its conjugate base. A strong acid will have a weak conjugate base, while a weak acid will have a strong conjugate base. This relationship is due to the fact that a strong acid will readily donate its proton, leaving behind a base that has a low propensity to re-accept the proton. On the other hand, a weak acid does not readily donate its proton, implying that its conjugate base has a higher propensity to accept a proton. In essence, the stronger an acid, the weaker is its conjugate base, and vice versa. This principle is known as the leveling effect. To determine the pH of solutions, understanding conjugate acid-base pairs is critical. The pH of a solution is determined by the concentration of hydronium ions $(H3O+)$. This concentration can be influenced by the presence of a conjugate acid or base, which can act as a proton donor or acceptor, respectively. Conjugate acid-base pairs also play a significant role in buffer solutions. A buffer solution has the ability to withstand pH alterations when minor quantities of an acid or a base are introduced. This property is due to the presence of a weak acid and its conjugate base, or a weak base and its conjugate acid, in significant amounts.

Problem 1: For the reaction: $NH3 + H2O \rightarrow NH4+ + OH-$

Identify the conjugate acid and the conjugate base.

Solution: In this reaction, NH3 is the base because it accepts a proton $(H+)$ from water to become NH4+. Therefore, NH4+ is the conjugate acid of NH3. Similarly, H2O is the acid because it donates a proton to NH3. After donating a proton, it forms OH-, so OH- is the conjugate base of H2O.

Problem 2: Given the reaction: $HCl + H2O \rightarrow H3O+ + Cl-$

Find the base, acid, conjugate base, and conjugate acid

Solution: In this reaction, HCl is the acid because it donates a proton (H+) to H2O. After donating a proton, it forms Cl-, so Cl- is the conjugate base of HCl. On the other hand, H2O is the base because it accepts a proton from HCl to become H3O+. So, H3O+ is the conjugate acid of H2O.

Problem 3: Consider the reaction:

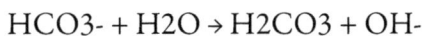

HCO3- + H2O → H2CO3 + OH-

Which one is the conjugate base and which one is the conjugate acid?

Solution: In this reaction, HCO3- is the base because it donates a proton to H2O and becomes H2CO3. So, H2CO3 is the conjugate acid of HCO3-. Similarly, H2O is the acid because it accepts a proton from HCO3- and forms OH-. Therefore, OH- is the conjugate base of H2O.

Problem 4: Given the reaction: H2O + H2O → H3O+ + OH-

What is the acid, base, conjugate acid, and conjugate base?

Solution: In this reaction, one of the water molecules acts as an acid by donating a proton (H+) to the other water molecule. After donating a proton, it forms OH-, so OH- is the conjugate base of water. The other water molecule acts as a base by accepting a proton to become H3O+. Therefore, H3O+ is the conjugate acid of water.

Conjugate acid–base pairs

In the realm of acid-base chemistry, a significant concept is the notion of conjugate acid–base pairs. To provide a clear definition, a conjugate acid refers to a substance formed when a base gains a proton (H+). Conversely, a conjugate base is a substance that forms when an acid loses a proton. This fundamental principle of acid-base chemistry hinges on the interchangeability of acids and bases through the gain or loss of a proton. To elaborate further, a conjugate acid-base pair consists of two substances related to each other by the transfer of a proton. The only difference between each member of a pair is one H+ ion. The acid in the pair is the molecule or ion that has one more proton than its conjugate base. It's essential to note that any reaction that involves an acid and a base, known as an acid-base reaction, will also involve both of their conjugates. When an acid donates a proton, it becomes its conjugate base. Simultaneously, when a base accepts a proton, it transforms into its conjugate acid. This phenomenon is a direct representation of the Bronsted-Lowry theory of acids and bases, which postulates that acids are proton donors, and bases are proton acceptors. To identify such pairs in reactions, one must recognize the substances that are transformed into one another by the gain or loss of a proton. For instance, in the reaction between ammonia (NH3) and water (H2O), where NH3 + H2O → NH4+ + OH-, ammonia acts as the base and water as the acid. Upon accepting a proton from water, ammonia becomes its conjugate acid, ammonium (NH4+), and water becomes its conjugate base, hydroxide (OH-). This example demonstrates the ease with which conjugate acid-base pairs can be identified in reactions. The key is to track the proton (H+). Observing its transfer between reactants allows us to define the conjugate pairs. In this case, NH3/NH4+ and H2O/OH- are the conjugate base/acid pairs. However, it's important to mention that the strength of an acid or base is inversely related to the strength of its conjugate. Weak conjugate bases are possessed by strong acids, while weak conjugate acids are associated with strong bases. This inverse relationship is a direct consequence of the equilibrium dynamics of acid-base reactions. Moreover, the concept of conjugate acid-base pairs is crucial in understanding the behavior of buffer solutions. Buffers are solutions that prevent alterations in pH levels when small quantities of acid or base are introduced. They consist of substantial amounts of a weak acid and its associated base (or a weak base and its related acid). Another critical application of the concept of conjugate acid-base pairs is in the biological system. For instance, the bicarbonate buffering system in the blood, where carbonic acid (H2CO3) and bicarbonate ion (HCO3-) act as a conjugate acid-base pair, helps maintain the pH of the blood. On a closing note, the concept of conjugate acid-base pairs forms the backbone of acid-base chemistry. By providing a framework for understanding proton transfer, it allows chemists to predict the direction of reactions, understand buffer systems, and comprehend the behavior of biological systems. As such, it is an indispensable tool in the chemist's arsenal.

Problem 1: Identify the conjugate acid-base pairs in the following reaction: $H_2O + NH_3 \rightarrow OH^- + NH_4^+$

Solution: In this reaction, the H_2O acts as an acid as it donates a proton (H^+) to NH_3. The NH_3 acts as a base as it accepts the proton from H_2O. After the reaction, OH^- and NH_4^+ are formed. So, we have two acid-base pairs here: (H_2O, OH^-) and (NH_3, NH_4^+). The pair (H_2O, OH^-) is a conjugate acid-base pair in which H_2O is the acid (proton donor) and OH^- is the conjugate base (proton acceptor). Conversely, the pair (NH_3, NH_4^+) is another conjugate acid-base pair in which NH_3 is the base (proton acceptor) and NH_4^+ is the conjugate acid (proton donor).

Problem 2: Identify the acid, base, conjugate acid, and conjugate base in the following reaction:
$$HNO_3 + H_2O \rightarrow H_3O^+ + NO_3^-$$

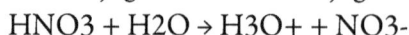

Solution: In this reaction, HNO_3 donates a proton (H^+) to H_2O and thus acts as an acid. H_2O accepts the proton and thus acts as a base. After the reaction, H_3O^+ and NO_3^- are formed. H_3O^+ is the conjugate acid (proton donor) of the base H_2O, and NO_3^- is the conjugate base (proton acceptor) of the acid HNO_3.

Problem 3: Determine the conjugate acid-base pairs in the subsequent reaction: $HCl + H_2O \rightarrow H_3O^+ + Cl^-$

Solution: In this reaction, HCl donates a proton (H^+) to H_2O and thus acts as an acid. H_2O accepts the proton and thus acts as a base. After the reaction, H_3O^+ and Cl^- are formed. So, we have two acid-base pairs here: (HCl, Cl^-) and (H_2O, H_3O^+). The pair (HCl, Cl^-) is a conjugate acid-base pair in which HCl is the acid (proton donor) and Cl^- is the conjugate base (proton acceptor). Conversely, the pair (H_2O, H_3O^+) is another conjugate acid-base pair in which H_2O is the base (proton acceptor) and H_3O^+ is the conjugate acid (proton donor).

pH, K_a, pK_a and K_w

pH is a key concept in chemistry that quantifies the acidity or alkalinity of a solution. Mathematically, it is defined as the negative base-10 logarithm of the activity of the hydrogen ion in a solution. It is represented as $pH = -\log[H^+]$, where $[H^+]$ is the molar concentration of hydrogen ions in the solution. pH is a dimensionless quantity and ranges from 0 to 14 in water at 25 degrees Celsius, with values less than 7 indicating acidity, values greater than 7 indicating alkalinity, and 7 indicating neutrality. K_a, also known as the acid dissociation constant, is a measure of the strength of an acid in solution. It is defined as the equilibrium constant for the reaction of an acid (HA) dissociating into its ions (H^+ and A^-). Mathematically, it is expressed as $K_a = [H^+][A^-]/[HA]$, where $[H^+]$, $[A^-]$, and $[HA]$ are the molar concentrations of the ions and the acid at equilibrium, respectively.

The pK_a is a logarithmic scale used to express the acid dissociation constant. It is defined as the negative logarithm of the acid dissociation constant, represented as $pK_a = -\log K_a$. The pK_a value provides a convenient way to express the acidity of a solution, with lower pK_a values indicating stronger acids. The autoionization of water, or the self-ionization of water, is represented by the equation $2H_2O \rightleftharpoons H_3O^+ + OH^-$, which implies that water molecules can donate and accept protons to and from each other, yielding hydronium (H_3O^+) and hydroxide (OH^-) ions. The equilibrium constant for this reaction is represented as K_w, also known as the ion product of water. K_w is a crucial concept in acid-base chemistry. It is defined as the product of molar concentrations of hydronium and hydroxide ions in water at a specific temperature. Mathematically, it is expressed as $K_w = [H_3O^+][OH^-]$. At 25 degrees Celsius, the value of K_w is 1.0×10^{-14}. The relationship between pH, K_a, and pK_a is critical in understanding the acidity or basicity of a solution. For instance, the pK_a can be used to calculate the pH of a solution. By rearranging the expression for pK_a, we get $K_a = 10^{-pK_a}$. Once K_a is known, it can be substituted into the expression for the acid dissociation ($K_a = [H^+][A^-]/[HA]$) to solve for the unknown $[H^+]$, which can then be used to calculate pH using the equation $pH = -\log[H^+]$. Furthermore, the pH and pOH of a solution can be linked through K_w. Since $K_w = [H_3O^+][OH^-]$, taking the negative logarithm of both sides gives -log Kw = -log [H3O+] + -log [OH-]. Because -log Kw equals 14 at 25 degrees Celsius, -log [H3O+] is equivalent to pH, and -log [OH-] equates to pOH, we can also establish that pH + pOH = 14.

1) pH: The pH of a solution is a gauge of the amount of hydrogen ions present in a solution.
It is defined mathematically as: pH = -log[H+]
Where [H+] is the concentration of hydrogen ions in moles per liter.

2) Ka: Ka is the acid dissociation constant. This represents the potency of an acid when dissolved in a solution.
It is defined as: Ka = [H+][A-]/[HA]
Where [H+] is the concentration of hydrogen ions, [A-] is the concentration of the acid's conjugate base, and [HA] is the concentration of the undissociated acid.

3) pKa: pKa is the negative logarithm of the Ka value. It is used to express the acidity of a solution in a more understandable way. It is defined as: pKa = -log Ka

4) Kw: Kw is the ion product of water This refers to the result of multiplying the concentrations of hydrogen ions and hydroxide ions in water. It is defined as: Kw = [H+][OH-]

Now let's look at some problems:

Problem 1: If the concentration of hydrogen ions in a solution is 1×10^{-7} M, what is the pH of the solution?
Solution: Using the formula pH = -log[H+], we substitute the given concentration into the formula:
pH = $-\log(1 \times 10^{-7})$ = 7

Problem 2: A solution of hydrochloric acid has a [H+] of 0.01 M. What is the pH of the solution?
Solution: Using the formula pH = -log[H+]
pH = $-\log(0.01)$ = 2

Problem 3: The Ka of acetic acid is 1.8×10^{-5}. What is the pKa of acetic acid?
Solution: Using the formula pKa = -log Ka
pKa = $-\log(1.8 \times 10^{-5})$ = 4.74

Problem 4: Calculate the [H+] in a solution if the pH is 3.
Solution: Since pH = -log[H+], we can rearrange the formula to find [H+]: [H+] = 10^{-pH}
[H+] = 10^{-3} = 0.001 M

Problem 5: If the pH of a solution is 6, what is the [OH-]?
Solution: First, we find [H+] by using the formula [H+]
= 10^{pH}
[H+] = 10^{-6} = 1×10^{-6} M
Then, we use the Kw equation to find [OH-]: Kw = [H+][OH]
Rearranging the formula, we get [OH-] = Kw/[H+]
Assuming that Kw is 1×10^{-14} at room temperature, we substitute the values into the formula:
[OH-] = $(1 \times 10^{-14})/(1 \times 10^{-6})$ = 1×10^{-8} M.

[H$^+$(aq)] and pH values

Strong Acids: In the realm of acid-base chemistry, strong acids hold a unique position due to their complete ionization in aqueous solutions. The determination of the hydronium ion concentration, [H+(aq)], and the calculation of the pH value are fundamental aspects of understanding the behavior of these strong acids. Strong acids, such as hydrochloric acid (HCl), nitric acid (HNO3), and sulfuric acid (H2SO4), are known for their ability to donate protons (H+) readily. When these acids are dissolved in water, they dissociate entirely into their constituent ions. For example, hydrochloric acid dissociates into H+ and Cl- ions. The concentration of H+(aq) is directly proportional to the concentration of the strong acid itself. The calculation of [H+(aq)] for a strong acid is straightforward. If we know the molarity (M) of the strong acid, that is also the [H+(aq)] since strong acids completely dissociate into their ions. For example, if we have a 0.1 M HCl solution, the $[H^+(aq)]$ will also be 0.1 M. The pH value, derived from the German term "potenz Hydrogen" which means the power of hydrogen, is a measure of the acidity or basicity of a solution. It is defined as the negative logarithm (base 10) of the $[H^+(aq)]$. In mathematical terms, $pH = -\log[H^+(aq)]$. To calculate the pH value of a strong acid, we need to take the negative logarithm of the $[H^+(aq)]$. Using our previous example, the pH of a 0.1 M HCl solution would be $pH = -\log(0.1) = 1$. Thus, the lower the pH, the higher the acidity of the solution. We should remember that the pH scale operates on a logarithmic basis. This implies that every unit on the scale signifies a difference in acidity by a factor of ten. Therefore, a solution having a pH of 2 is ten times more acidic than another solution with a pH of 3. This logarithmic nature allows the pH scale to compress the wide range of $[H^+(aq)]$ found in aqueous solutions into a more manageable scale ranging from 0 to 14. In the case of sulfuric acid, a diprotic acid, the calculation is slightly more complex due to the presence of two acidic protons. However, the first ionization is considered strong, and the second is weak. Typically, for most practical purposes, only the first dissociation is considered in the calculation of $[H^+(aq)]$ and pH. In the professional realm of chemistry, precise calculations of $[H^+(aq)]$ and pH values play a pivotal role in various fields, from environmental science to industrial processes, biochemistry, and pharmaceuticals. The behavior of many chemical reactions, biological processes, and even the quality of natural waters can be profoundly influenced by the pH.

Problem 1: Calculate the $[H^+(aq)]$ and pH of a 0.005 M solution of HCl. (HCl is a strong acid)

Solution 1: For strong acids like HCl, it completely ionises in water. Therefore, $[H^+(aq)]$ is equal to the initial concentration of the acid. So, $[H^+(aq)] = 0.005$ M

To calculate the pH, we use the formula:

$$pH = -\log[H^+(aq)]$$
$$pH = -\log(0.005)$$
$$pH = 2.3$$

Problem 2: Calculate the $[H^+(aq)]$ and pH of a 0.00001 M solution of HNO3. (HNO3 is a strong acid)

Solution 2: For strong acids like HNO3, it completely ionises in water. Therefore, $[H^+(aq)]$ is equal to the initial concentration of the acid. So, $[H^+(aq)] = 0.00001$ M

To calculate the pH, we use the formula:

$$pH = -\log[H^+(aq)]$$
$$pH = -\log(0.00001)$$
$$pH = 5$$

Problem 3: Calculate the $[H^+(aq)]$ and pH of a 0.1 M solution of H2SO4. (H2SO4 is a strong acid)

Solution 3: For strong acids like H2SO4, it completely ionises in water. Therefore, $[H^+(aq)]$ is equal to the initial concentration of the acid. So, $[H^+(aq)] = 0.1$ M

To calculate the pH, we use the formula:

$$pH = -\log[H^+(aq)]$$
$$pH = -\log(0.1)$$
$$pH = 1$$

Keep in mind that the pH scale extends from 0 to 14, where 7 is considered neutral. Substances with a pH below 7 are acids, whereas those with a pH above 7 are bases. The lower the pH, the stronger the acid.

Strong Alkalis: Strong alkalis, also known as strong bases, are substances that have the ability to completely dissociate into their ions in solution, resulting in a high concentration of hydroxide ions (OH-) and a very low concentration of hydrogen ions (H+). This characteristic of strong alkalis classifies them as strong electrolytes, and they generally have a pH value greater than 7, indicating their basic nature. Examples of strong alkalis include sodium hydroxide (NaOH), potassium hydroxide (KOH), and barium hydroxide (Ba(OH)2). Let's explore the calculation of the [H+(aq)] concentration and the pH of strong alkalis, using sodium hydroxide as an example. Sodium hydroxide fully dissociates in water to form sodium ions (Na+) and hydroxide ions (OH-). The concentration of OH- ions is directly proportional to the initial concentration of the sodium hydroxide solution. The ion product of water, Kw, is a constant at a given temperature and is equal to the product of the concentrations of H+ and OH- ions. At 25 degrees Celsius, Kw is 1.0×10^{-14}. Given that strong alkalis have a high OH- concentration and a low H+ concentration, we can use the relationship Kw = [H+][OH-] to calculate the [H+(aq)] concentration. If we assume a 1.0 M solution of NaOH, the [OH-] would also be 1.0 M, due to complete dissociation. Substituting this into the equation, we can solve for [H+], giving us [H+] = Kw / [OH-] = 1.0×10^{-14} / 1.0 = 1.0×10^{-14} M. The pH of a solution is defined as the negative logarithm (base 10) of the hydrogen ion concentration, or pH = -log[H+]. Substituting the calculated [H+] value gives us pH = -log(1.0×10^{-14}), which results in a pH value of 14. This means that a 1.0 M solution of sodium hydroxide, a strong alkali, has a pH of 14, which is the maximum pH value possible at 25 degrees Celsius. This is an extreme example, and it is important to note that the pH of real-world strong alkalis may vary depending on their concentration and the specific temperature. The same principles can be applied to other strong alkalis, such as potassium hydroxide and barium hydroxide. These strong bases also completely dissociate into their respective ions, resulting in a high concentration of OH- ions and a correspondingly low concentration of H+ ions. The [H+(aq)] and pH can be calculated using the same equations, with the specific concentrations of the alkalis used.

Problem 1: Calculate [H+(aq)] and pH values for a 0.01 M solution of NaOH (a strong alkali).

Solution: NaOH is a strong alkali, so it will dissociate completely in water to form Na+ and OH-. The concentration of OH- ions will also be 0.01 M.

We know that [OH-] [H+] = 1.00×10^{-14} M² (this is the ion product of water). We can rearrange this to find [H+]:

$$[H+] = 1.00 \times 10^{-14} \text{ M}^2 / [OH-]$$
$$[H+] = 1.00 \times 10^{-14} \text{ M}^2 / 0.01 \text{ M}$$
$$[H+] = 1.00 \times 10^{-12} \text{ M}$$

To find the pH, we use the formula pH = -log[H+].

$$pH = -\log(1.00 \times 10^{-12})$$
$$pH = 12$$

Problem 2: Calculate [H+(aq)] and pH values for a 0.1 M solution of KOH (a strong alkali).

Solution: KOH is a strong alkali, so it will dissociate completely in water to form K+ and OH-. The OH- ions concentration will also be 0.1 M.

We use the ion product of water to find [H+]:

$$[H+] = 1.00 \times 10^{-14} \text{ M}^2 / [OH-]$$
$$[H+] = 1.00 \times 10^{-14} \text{ M}^2 / 0.1 \text{ M}$$
$$[H+] = 1.00 \times 10^{-13} \text{ M}$$

To find the pH, we use the formula pH = -log[H+].

$$pH = -\log(1.00 \times 10^{-13})$$
$$pH = 13$$

Note: In these calculations, we assume that the temperature is 25°C. The pH and the ion product of water can vary depending on the temperature.

Weak Acids: Understanding the concentration of hydrogen ions [H+(aq)] and the pH values of weak acids is an essential component of chemistry. It requires knowledge of concepts such as acid dissociation constants, the auto-ionization of water, and the mathematical calculations involving logarithms. Weak acids only partially dissociate in water, providing it with a specific characteristic of having both the undissociated form and the dissociated form present in the solution. The degree of dissociation of a weak acid is described by its acid dissociation constant (Ka). The larger the Ka, the more the acid dissociates and, consequently, the greater the [H+(aq)] in the solution. The [H+(aq)] in a weak acid solution can be determined using the equation for the acid dissociation constant,

$Ka = [H+(aq)][A-(aq)]/[HA(aq)]$. The weak acid (HA) dissociates into H+(aq) and A-(aq) ions. However, because the acid is weak, the concentration of HA is almost equal to the initial concentration of the acid. By setting up and solving this equation, one can find the [H+(aq)]. Once the [H+(aq)] is known, the pH of the solution can be calculated using the definition of pH. The pH is the negative logarithm (base 10) of the hydrogen ion concentration, $pH = -\log[H+(aq)]$. So, by substituting the calculated [H+(aq)] into this equation, the pH can be found. The pH scale ranges from 0 to 14, with 7 indicating a neutral solution. Values less than 7 indicate an acidic solution, while values greater than 7 indicate a basic solution. Thus, the pH value provides a measure of the acidity or alkalinity of a solution. It's important to note that the calculation of [H+(aq)] and pH for weak acids typically involves making an assumption that the [H+(aq)] is much smaller than the initial concentration of the weak acid. This assumption simplifies the mathematics considerably but also introduces a small error. In situations where precision is required, one may need to use more complex methods, such as the quadratic formula. In addition, temperature plays a key role in these calculations. Most pH calculations are based on the properties of water at 25 degrees Celsius. At different temperatures, the ion product of water (Kw) changes, affecting the [H+(aq)] and thus the pH. Weak acids are found in various daily life situations, from the food we eat to the functioning of our bodies. For example, vinegar is a solution of acetic acid, a weak acid, and has a pH of about 2.4. Human blood, which contains carbonic acid, another weak acid, maintains a pH of approximately 7.4 for proper physiological function. Understanding the [H+(aq)] and pH of weak acids is not only critical for chemists but also for biologists, environmental scientists, pharmacists, and many other professionals. For instance, maintaining the correct pH is vital in many industrial processes, and slight deviations can lead to inferior product quality or even dangerous situations.

Problem 1: A 0.20 M solution of acetic acid (CH_3COOH, $Ka = 1.8 \times 10^{-5}$) is prepared. Calculate the [H+(aq)] and pH of this solution.

Solution 1: To solve this problem, we use the formula for the ionization of a weak acid: $CH_3COOH \rightleftharpoons H+ + CH_3COO-$

The equilibrium expression for this reaction is:

$Ka = [H+(aq)][CH_3COO-]/[CH_3COOH]$

Since the acid is initially un-ionized, [H+] and [CH3COO-] are both 0, and [CH3COOH] is 0.20. Let x be the amount of CH3COOH that ionizes, thus increasing [H+] and [CH3COO-] by x and decreasing [CH3COOH] by x. The equilibrium expression becomes:

$$1.8 \times 10^{-5} = x^2/(0.20 - x)$$

Assuming x is small compared to 0.20, we can approximate 0.20 - x as 0.20: $1.8 \times 10^{-5} = x^2 / 0.20$

Solving for x, we find that $x = [H+] = 1.9 \times 10^{-3}$ M. The pH is then calculated as $pH = -\log[H+] = -\log(1.9 \times 10^{-3}) = 2.72$.

Problem 2: Calculate the [H+(aq)] and pH of a 0.15 M solution of formic acid ($HCOOH$, $Ka = 1.8 \times 10^{-4}$).

Solution 2: We use the same method as above:

$$HCOOH \rightleftharpoons H+ + HCOO-$$

Let $x = [H+] = [HCOO-]$. The equilibrium expression is:

$$1.8 \times 10^{-4} = x^2 / (0.15 - x)$$

Buffer Solution

A buffer solution, in the field of chemistry, can be defined as an aqueous solution made up of a mixture of a weak acid and its conjugate base, or a weak base and its conjugate acid.

The main role of a buffer solution is to keep the pH level stable when minor quantities of an acid or a base are introduced. This function is of paramount importance in various chemical and biological applications, hence the prominence of buffer solutions in these fields. The unique property of buffer solutions can be attributed to Le Chatelier's Principle. When an acid (or base) is added to the buffer solution, the equilibrium shifts to counteract the pH change, thereby maintaining the pH level. This reaction is facilitated by the presence of the weak acid and its conjugate base, or vice versa, in the buffer solution. The composition of a buffer solution is crucial in determining its pH. The pH of a buffer solution is controlled by the pKa of the weak acid or weak base, and the ratio of the concentrations of the acid/base and its conjugate. Increasing the concentration of the acid in relation to its conjugate base will increase the acidity of the solution, and vice versa. This relationship is defined by the Henderson-Hasselbalch equation. The capacity of a buffer solution, or its ability to resist changes in pH, is also determined by the concentrations of the acid/base and its conjugate. Buffer capacity is highest when the concentrations of the weak acid/base and its conjugate are equal. This is because the solution can absorb the greatest amount of acid or base before the pH starts to change significantly. Buffer solutions are widely used in biological systems. The human body, for instance, uses buffer systems to maintain a stable pH in the blood and other bodily fluids. This is critical for the proper functioning of enzymes and other biochemical processes. Buffer solutions are also used in the preservation of food and in fermentation processes. In the field of analytical chemistry, buffer solutions are vital in a variety of techniques. They are used in titrations to maintain a constant pH, in chromatography to control the pH of the mobile phase, and in spectrophotometry for the preparation of samples. In pharmaceutical science, buffer solutions are used in the formulation of drugs to maintain the stability of the active ingredients. They are also used in the development of cosmetic products, where they prevent changes in pH that could affect the product's quality or safety. Buffer solutions are also used in environmental science, where they are used in the treatment of wastewater and the monitoring of water quality. The stability of the pH in these systems is crucial for the survival of aquatic life and the prevention of corrosion in water treatment equipment. In the field of biochemistry, buffer solutions are used in the isolation and purification of proteins. They are also used in DNA extraction and polymerase chain reaction (PCR) processes. In agriculture, buffer solutions are used to maintain the pH of soil, which is critical for the growth of plants. They are also used in the formulation of pesticides and fertilizers. The preparation of a buffer solution requires careful consideration of the desired pH and buffer capacity. The choice of a suitable weak acid or base, and its conjugate, is critical. The concentrations of these components must then be adjusted to achieve the desired pH and buffer capacity. Despite their numerous applications, buffer solutions do have limitations. They can only maintain a stable pH within a certain range, and their buffering capacity is limited. This means that they can be overwhelmed by the addition of large amounts of acid or base. Buffer solutions can also interact with other chemical species in a solution, which can affect their buffering capacity. For example, if a buffer solution is used in a reaction that produces or consumes hydrogen ions, this can lead to changes in the pH of the solution. The stability of buffer solutions can also be affected by temperature. Changes in temperature can shift the equilibrium of the acid-base reaction, leading to changes in the pH of the solution.

A buffer solution is a type of solution that has the ability to maintain its pH level when minor amounts of either an acid or a base are introduced. It is typically made up of a weak acid and its conjugate base, or a weak base and its conjugate acid. Buffer solutions are extremely important in many biological systems and industrial processes.

Q: If you add a small amount of hydrochloric acid to a buffer solution made from ammonia and ammonium chloride, what will happen to the pH of the solution?

A: In this case, the ammonia in the buffer solution would react with the hydrochloric acid to form more ammonium ions. This would limit the change in pH, demonstrating the buffer solution's ability to resist changes in pH.

Q: How would you prepare a buffer solution with a pH of 4.5 using acetic acid and sodium acetate?

A: To prepare a buffer solution of a specific pH, you would need to use the Henderson-Hasselbalch equation, which relates the pH, pKa, and the ratio of the concentrations of the acid and its conjugate base. In this case, the pKa of acetic acid is approximately 4.76. You would then solve the equation for the ratio of the concentrations of sodium acetate to acetic acid, which should be approximately 0.6. This means you would need to mix these two substances in this ratio to obtain a buffer solution with a pH of 4.5.

Q: Why is it important to use a buffer solution in a biochemical experiment involving enzymes?

A: Enzymes are very sensitive to changes in pH. A slight change in pH can alter the shape of the enzyme and affect its ability to function. Therefore, in biochemical experiments, it is important to use a buffer solution to maintain a constant pH and ensure that the enzymes can function optimally.

A buffer solution, in essence, is a solution that effectively maintains or resists changes in pH when certain quantities of acids or bases are added. The creation of a buffer solution is an essential process in the field of chemistry. It is used widely in biochemical processes, industrial applications, and in calibration of pH meters. There are multiple methods to prepare a buffer solution, all of which require a deep understanding of the principles of acid-base chemistry. The first significant step in creating a buffer solution is the selection of appropriate buffer components. Usually, the buffer consists of a weak acid and its corresponding base, or a weak base and its corresponding acid. The choice of buffer components is determined by the desired pH of the buffer solution. The pKa value of the weak acid or base used should be close to the desired pH for the buffer to be effective. One common method to prepare a buffer solution is by direct addition of an acid and its salt. For example, a buffer at pH 4.74 could be prepared by dissolving acetic acid and sodium acetate in water. The desired ratio of acid to salt depends on the intended pH of the buffer solution, as dictated by the Henderson-Hasselbalch equation. This equation, $pH = pKa + \log([A-]/[HA])$, is a key tool in buffer preparation as it allows calculation of the required ratio of acid to conjugate base. Another method of preparing a buffer solution is by partial neutralization of an acid or base. This involves adding a strong acid to a solution of a weak base (or vice versa) until the desired pH is reached. For instance, if one starts with a weak base such as ammonia and adds hydrochloric acid gradually until the pH is 9.25, the resulting solution will be a buffer composed of ammonia and ammonium chloride. The third common method for preparing a buffer solution is by the combination of a weak base and a strong acid, or a weak acid and a strong base. This method involves the addition of a strong acid to a solution of a weak base to create a buffer of the weak base and its conjugate acid. Alternatively, a strong base can be added to a solution of a weak acid to create a buffer of the weak acid and its conjugate base. The preparation of a buffer solution is not merely a matter of combining the correct components. There are various factors that need to be carefully considered, such as the concentration of the buffer components, the volume of the solution, temperature, and the presence of other ions or compounds that might interfere with the buffer system. The concentration of the buffer components is a crucial factor in the effectiveness of the buffer solution. High concentrations of buffer components will result in a buffer with a high buffering capacity, meaning it can neutralize larger amounts of acids or bases without a significant change in pH. The volume of the solution is also important, as it determines the total amount of acid or base the buffer can neutralize. A larger volume means a higher buffering capacity, but it also requires more buffer components. Temperature plays a role as well because it can affect the ionization of the weak acid or base, thereby influencing the pH of the buffer solution. Generally, the pKa value of the weak acid or base decreases as the temperature increases, resulting in a lower pH for the buffer solution. The presence of other ions or compounds can affect the buffer system, either by reacting with the buffer components or by altering their ionization. It is therefore important to ensure that the buffer components are compatible with any other substances present in the solution.

A buffer solution can be made by mixing a weak acid and its salt or a weak base and its salt. For example, a common buffer solution is made by mixing acetic acid (a weak acid) and sodium acetate (its salt). As another example, a buffer can be made by mixing ammonia (a weak base) and ammonium chloride (its salt). The acid and base in the buffer solution react with added acids or bases to maintain a nearly constant pH.

Problem 1: Why does adding a small amount of hydrochloric acid (HCl) to a buffer solution made from acetic acid and sodium acetate not significantly change the pH of the solution?

Solution: The acetic acid in the buffer solution can react with the added HCl to produce water and acetate ions. Because acetic acid is a weak acid, it does not fully dissociate, so the reaction with HCl does not significantly change the pH of the solution.

Problem 2: How can you prepare a buffer solution with a pH of 9.25 using ammonia and ammonium chloride?

Solution: First, calculate the ratio of $[NH4+]/[NH3]$ using the Henderson-Hasselbalch equation: $pH = pKa + log([A-]/[HA])$. The pKa of ammonia is 9.25, so the ratio of $[NH4+]/[NH3]$ should be 1. Then, mix equal amounts of ammonia and ammonium chloride to make the buffer.

Problem 3: Why is it important to use a weak acid or base to make a buffer solution instead of a strong acid or base?

Solution: A buffer solution is designed to resist changes in pH. A strong acid or base fully dissociates in water, allowing it to react with any added acid or base and change the pH. A weak acid or base only partially dissociates, so it can absorb the added acid or base without significantly changing the pH.

Buffer solutions are pivotal in controlling pH due to their ability to resist changes in pH when small amounts of an acid or base are added or when the solution is diluted. A buffer solution's capacity to maintain a constant pH is based on the equilibrium established between a weak acid and its conjugate base, or a weak base and its conjugate acid. The equilibrium of a weak acid, HA, can be represented as follows: $HA \leftrightarrow H+ + A-$. When an acid, H+, is added to this system, the equilibrium shifts to the left according to Le Chatelier's principle, absorbing the extra H+ ions and forming more HA. This reduces the increase in acidity and keeps the pH relatively constant. Similarly, when a base, OH-, is added, it reacts with the H+ ions to form water. This removal of H+ ions shifts the equilibrium to the right, resulting in the dissociation of more HA into H+ and A-. This replenishes the H+ ions, neutralizes the base, and keeps the pH relatively constant. The equilibrium of a weak base, B, and its conjugate acid, HB+, can be represented as follows: $B + H2O \leftrightarrow HB+ + OH-$. When a base, OH-, is added to this system, the equilibrium shifts to the left, absorbing the extra OH- ions, and forming more B. This reduces the increase in alkalinity and keeps the pH relatively constant. Similarly, when an acid, H+, is added, it reacts with the OH- ions to form water. This removal of OH- ions shifts the equilibrium to the right, resulting in the dissociation of more B into HB+ and OH-. This replenishes the OH- ions, neutralizes the acid, and keeps the pH relatively constant. The pH of a buffer solution can be calculated using the Henderson-Hasselbalch equation: $pH = pKa + log([A-]/[HA])$ for a weak acid and its conjugate base, or $pOH = pKb + log([HB+]/[B])$ for a weak base and its conjugate acid. The pH or pOH of a buffer solution is thus determined by the ratio of the concentrations of the conjugate base to the weak acid, or the conjugate acid to the weak base, respectively. The buffer capacity, referring to the quantity of acid or base that a buffer solution can soak up without a substantial shift in pH, is contingent on the total concentrations of either the weak acid and its corresponding base, or the weak base and its corresponding acid. The buffer capacity is maximum when $[A-] = [HA]$ or $[HB+] = [B]$.

A buffer solution can be created by combining a weak acid with its corresponding conjugate base, or a weak base with its matching conjugate acid. Alternatively, it can also be made by slightly neutralizing a weak acid with a strong base, or a weak base with a strong acid. The choice of the weak acid or base and its conjugate depends on the desired pH of the buffer solution. Buffer solutions play a crucial role in various chemical and biological processes where the pH needs to be maintained at a constant value. For example, human blood is a buffer solution consisting of carbonic acid and bicarbonate ion. This buffer system maintains the pH of blood around 7.4, which is vital for the proper functioning of biological processes.

Buffer solutions are designed to maintain a specific pH level, even when small amounts of acid or base are added. They are able to do this because they contain both a weak acid and its conjugate base (or a weak base and its conjugate acid).

When an acid (H+) is added to the buffer solution, it is neutralized by the base component of the buffer solution, thus preventing a significant pH decrease. For example, let's consider a buffer solution of acetic acid (CH_3COOH) and its conjugate base, acetate (CH_3COO^-).

The reaction would be: $CH_3COOH \rightleftharpoons CH_3COO^- + H^+$

When an acid is added, the equilibrium shifts to the left, consuming the added H+ ions and keeping the pH relatively constant. Similarly, when a base (OH-) is added to the buffer solution, it is neutralized by the acidic component of the buffer, preventing a significant pH increase.

In the same buffer solution, the reaction would be:

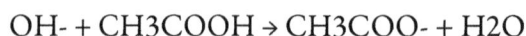

$OH^- + CH_3COOH \rightarrow CH_3COO^- + H_2O$

This reaction consumes the added OH- ions, again keeping the pH relatively constant.

Now, for the problems and solutions:

Problem: What happens if you add NaOH to a buffer solution of CH_3COOH and CH_3COO^-?

Solution: NaOH is a strong base that will dissociate completely in water to Na+ and OH-. The OH- will react with the acetic acid (CH_3COOH) in the buffer solution to form water and acetate (CH_3COO^-), neutralizing the added base and keeping the pH relatively constant.

Problem: What would happen if you added HCl to the same buffer solution?

Solution: HCl is a strong acid that will dissociate completely in water to H+ and Cl-. The H+ will react with the acetate (CH_3COO^-) in the buffer solution to form acetic acid (CH_3COOH), neutralizing the added acid and keeping the pH relatively constant.

Problem: How would the pH of the buffer solution change if you added a large amount of acid or base?

Solution: If a large amount of acid or base is added, the buffer capacity can be exceeded. This means that there are not enough weak base or weak acid molecules in the buffer to neutralize the added acid or base. Therefore, the pH of the buffer solution would significantly change.

Buffers are used extensively in laboratories, particularly in biochemical experiments involving enzymes, as most enzymes are highly sensitive to pH changes. They are also used to calibrate pH meters, which would otherwise provide inaccurate readings due to the volatile nature of pH in unbuffered solutions. Additionally, buffers are used in many industrial processes. For instance, in the fermentation industry, optimal pH conditions are crucial for the growth of microorganisms, which is maintained using buffer solutions. Understanding the concept of buffer solutions necessitates a comprehension of the acid-base equilibrium. In a typical buffer solution, a weak acid and its conjugate base (or a weak base and its conjugate acid) exist in equilibrium. When an external acid or base is added, the equilibrium shifts to counteract the pH change.

For example, if an acid is added, the base component of the buffer will neutralize it, shifting the equilibrium to produce more weak acid and thereby minimizing the pH change.

The bicarbonate buffer system (HCO3-) is a prime example of a naturally occurring buffer system, which plays a vital role in controlling the pH of human blood. Blood pH is tightly regulated by the body, as even slight deviations can have serious consequences, including coma or death. Bicarbonate ions (HCO3-) and carbonic acid (H2CO3) constitute the bicarbonate buffer system. This system operates based on Le Chatelier's principle, where the equilibrium adjusts itself to nullify the effects of added substances. When the blood becomes too acidic, the equilibrium shifts to the right, producing more HCO3- ions to neutralize the excess H+ ions. Conversely, when the blood is too basic, the equilibrium shifts to the left, producing more H+ ions to neutralize the excess OH- ions. Another aspect of the bicarbonate buffer system is its interaction with the respiratory and renal systems. When blood pH drops (acidosis), the respiratory system compensates by increasing the breathing rate to expel more CO2, a component of the buffer system. This shifts the equilibrium to the left, decreasing the H+ concentration and thus, increasing the pH. The kidneys also contribute by excreting more H+ ions and reabsorbing HCO3- ions. Alternatively, in a state of alkalosis (high blood pH), the respiratory system decreases the breathing rate, retaining more CO2 and shifting the equilibrium to the right, thus increasing the H+ concentration and lowering the pH. The kidneys also respond by excreting more HCO3- ions and reabsorbing H+ ions. Despite these compensatory mechanisms, the bicarbonate buffer system alone is not sufficient to maintain blood pH within the narrow range required for normal physiological functions. Other buffer systems, such as the phosphate buffer system and protein buffer system, also contribute to pH regulation in the body. Buffer solutions play a crucial role in chemistry, biology, and medicine due to their ability to maintain a stable pH level.

They are used in various experimental and practical applications where the pH must be accurately controlled.

1) Use in Biological Systems: Buffer solutions are crucial in the human body. For instance, the bicarbonate buffer system (HCO3-) in our blood helps to maintain a steady blood pH level of about 7.4. If the blood pH shifts below or above this level, it can lead to life-threatening conditions like acidosis or alkalosis. This buffer system is based on the equilibrium between carbonic acid (H2CO3) and bicarbonate ion (HCO3-) in the blood. If there is an increase in H+ ions (i.e., the blood becomes more acidic), the bicarbonate ion can neutralize them by forming more carbonic acid. Conversely, if the blood becomes too alkaline (i.e., a decrease in H+ ions), carbonic acid can donate H+ ions to restore the balance.

2) Use in Lab Experiments: Buffer solutions are used in various chemical and biochemical laboratory procedures, including DNA extraction, protein purification, and enzyme activity studies. They are also used in the preparation of certain types of biological samples for microscopy.

3) Use in Industrial Processes: Many industrial processes, like fermentation and dye production, require a specific pH to work optimally. Buffer solutions are used to maintain this pH.

Problems and Solutions:

Problem 1: Explain why the pH of pure water is 7.

Solution: Pure water is neutral because it has equal concentrations of H+ (acidic) and OH- (basic) ions. When water self-ionizes, it forms equal amounts of these ions, so the pH is 7, which is the middle of the pH scale.

Problem 2: How does the bicarbonate buffer system work to control the pH of blood?

Solution: The bicarbonate buffer system maintains the pH of blood by either absorbing excess H+ ions or releasing H+ ions when needed. When the blood becomes too acidic, HCO3- ions combine with H+ ions to form H2CO3, reducing the concentration of H+ ions and increasing the pH. When the blood becomes too basic, H2CO3 dissociates to form HCO3- and H+, adding more H+ ions to the blood and lowering the pH.

Problem 3: What would happen if the buffer system in the human body failed?

Solution: If the buffer system in the human body failed, the pH of the blood could fluctuate drastically, leading to a condition called acidosis (if the blood becomes too acidic) or alkalosis (if the blood becomes too basic). Both conditions can be life-threatening if not corrected quickly.

Buffer solutions are crucial in a wide array of chemical and biological applications. These solutions resist changes in pH when small amounts of strong acids or bases are added. The pH of a buffer solution can be calculated using the Henderson-Hasselbalch equation, which is pH = pKa + log ([A-]/[HA]), where [A-] is the concentration of the base in moles per liter and [HA] is the concentration of the acid in moles per liter. Consider a solution containing a weak acid, HA, and its conjugate base, A-. If an acid is added to the system, the added H+ ions will be consumed by the base, A-, forming more HA and thereby resisting a change in pH. Conversely, if a base is added to the system, the added OH- ions will react with the acid, HA, to form water and A-, again resisting a change in pH. The pKa value, a measure of the strength of the acid in solution, comes from the Ka value, the acid dissociation constant. It's the equilibrium constant for the reaction of the acid with water to form H+ and A-. The pKa is simply the negative logarithm base 10 of Ka, meaning a lower pKa value indicates a stronger acid. To calculate the pH of a buffer solution, we first need to know the pKa of the acid and the ratio of the concentrations of A- to HA. The pH is then calculated by substitifying these values into the Henderson-Hasselbalch equation. For example, if we have a buffer solution with a pKa of 4.74 (acetic acid) and the concentrations of A- and HA are equal, the pH of the solution would be 4.74 + log(1), which simplifies to 4.74. The reason is that the logarithm of 1 equals 0. If the concentration of A- is ten times greater than the concentration of HA, the pH would be 4.74 + log(10), which simplifies to 4.74 + 1 = 5.74. Conversely, if the concentration of HA is ten times greater than the concentration of A-, the pH would be 4.74 + log(0.1), which simplifies to 4.74 - 1 = 3.74. It's important to note that while the Henderson-Hasselbalch equation is a useful tool for estimating the pH of buffer solutions, it makes several assumptions that may not hold in all situations. For example, it assumes that the concentrations of A- and HA do not change significantly when the acid or base is added, which may not be true for large additions of acid or base. Moreover, the equation assumes that all A- and HA are in the aqueous phase, which may not be the case if the acid or base is partially insoluble. It also assumes that the addition of acid or base does not significantly change the volume of the solution, which may not be true if large volumes of acid or base are added. In these cases, the pH of the buffer solution may need to be determined experimentally, or using a more complex mathematical model that takes these factors into account. However, for most practical applications, the Henderson-Hasselbalch equation provides a good estimate of the pH of a buffer solution, given appropriate data.

Problem 1: What would be the pH of a buffer solution created by combining 50.0 mL of 0.300 M NH3 and 50.0 mL of 0.300 M NH4Cl, given that the Kb for NH3 is 1.8×10^{-5}?

Solution: First, calculate the moles of NH3 and NH4Cl in the solution:

Moles of NH3 = volume in L concentration = 0.050 L 0.300 M = 0.015 moles

Moles of NH4Cl = 0.050 L * 0.300 M = 0.015 moles

As they are in equal amounts, we can see that [NH3] = [NH4+] in the buffer.

Using the Henderson-Hasselbalch equation (pH = pKa + log([A-]/[HA]), we can find the pH of the buffer solution. However, we are given a Kb (base dissociation constant) not a Ka (acid dissociation constant). We can convert Kb to Ka using the ion product of water (Kw = Ka * Kb) where Kw is 1.0×10^{-14} at 25°C.

Ka = Kw / Kb = 1.0×10^{-14} / 1.8×10^{-5}

= 5.56×10^{-10}.

Then, calculate the pKa = -log(Ka) = 9.25.

Now, we can plug into the Henderson-Hasselbalch equation:

pH = pKa + log([NH3]/[NH4+]) = 9.25 + log(1) = 9.25

Problem 2: Determine the pH of a buffer solution that is created by incorporating 0.10 mol of acetic acid (CH3COOH) and 0.10 mol of sodium acetate (CH3COONa) into sufficient water to yield 1.0 L of solution. The acidity constant (Ka) for acetic acid is 1.8×10^{-5}.

Solution: First, recognize that CH3COOH is a weak acid and CH3COONa is its conjugate base. The concentration of both is 0.10 M (0.10 mol/ 1 L).

The Ka expression for CH3COOH is Ka = [CH3COO-][H+]/[CH3COOH]. Since CH3COOH and CH3COO- are in equal concentrations, they will cancel out in the Henderson-Hasselbalch equation (pH = pKa + log([A-]/[HA]).

Calculate pKa = -log(Ka) = -log(1.8×10^{-5}) = 4.74.

Then, plug into the Henderson-Hasselbalch equation:

pH = pKa + log([CH3COO-]/[CH3COOH])

= 4.74 + log(1) = 4.74.

Solubility product, K_{sp}

The solubility product, often denoted as Ksp, is a fundamental concept in the field of physical chemistry. It is a constant of equilibrium that provides a quantitative measure of the solubility of a sparingly soluble substance. It is the product of the concentrations of the ions in a saturated solution of an ionic compound. In simple terms, solubility product is a measure of how much of a particular substance can dissolve in a certain volume of solvent at a specific temperature. The concept of solubility product is especially useful in predicting the extent to which an ionic compound is soluble in a particular solvent. The solubility product is thus a crucial parameter in understanding the behavior of sparingly soluble substances and their interactions with various solvents. To understand the concept of Ksp more clearly, consider the dissolution of an ionic compound, for instance, AB, in water. It dissociates into its ions, A+ and B-, according to the equation AB ⇌ A+ + B-. The Ksp of this reaction is given by the equation Ksp = [A+][B-], where [A+] and [B-] are the molar concentrations of the ions at equilibrium. The solubility product is temperature-dependent. As the temperature increases, so does the solubility of most solids in water, thus increasing the Ksp value. However, some substances, like cerium sulfate, exhibit a decrease in solubility with increasing temperature, leading to a decreased Ksp value. The concept of Ksp is also closely related to the common-ion effect. When a common ion is added to a solution of an ionic compound, the solubility of the compound decreases. This decrease in solubility can be explained by the Le Chatelier's principle, which states that the system will shift in a direction that counteracts the change, in this case, towards the left, favoring the formation of the solid compound and decreasing its solubility. The solubility product, Ksp, is also instrumental in predicting the outcome of precipitation reactions. By comparing the ionic product (IP), the product of the molar concentrations of the ions in the solution, with the Ksp, one can predict if a precipitate will form. A precipitate will form and the solution will become supersaturated if the IP exceeds the Ksp. If the IP equals the Ksp, the solution is saturated, and if the IP is less than the Ksp, the solution is unsaturated, and no precipitate will form. A thorough understanding of solubility product principles is essential in various fields of chemistry, such as analytical chemistry, environmental chemistry, and geochemistry. For instance, in water treatment processes, the solubility product is used to determine the optimum conditions for the precipitation of harmful heavy metal ions. In pharmaceutical chemistry, the solubility product is used to optimize the solubility of drugs, thereby enhancing their bioavailability. In geochemistry, the solubility product is used to predict the mobility of metal ions in ores and the formation of mineral deposits.

Problem 1: Solubility Product; The solubility product constant (Ksp) for the salt AgCl is 1.6×10^{-10} at 25°C. At this temperature, how soluble is AgCl in water?

Solution: The dissociation of AgCl in water is represented by the chemical equation: $AgCl(s) \rightleftharpoons Ag^+(aq) + Cl^-(aq)$

The expression for the solubility product constant is:

$$K_{sp} = [Ag^+][Cl^-]$$

Since the stoichiometry of the reaction is 1:1, the concentration of Ag^+ is equal to the concentration of Cl^-, let's say S. Therefore, we can express the Ksp as: $K_{sp} = S^2$

We can solve for S (the solubility of AgCl) by taking the square root of Ksp:

$$S = \sqrt{K_{sp}}$$
$$S = \sqrt{1.6 \times 10^{-10}}$$
$$S = 4 \times 10^{-6} \text{ M}$$

So, the solubility of AgCl in water at 25°C is 4×10^{-6} M.

Problem 2: Solubility Product

A saturated solution of PbI2 has $[Pb^{2+}] = 1.2 \times 10^{-3}$ M. If the Ksp of PbI2 is 7.1×10^{-9}, what is the concentration of I- in the solution?

Solution: The dissociation of PbI2 in water is represented by the chemical equation: $PbI_2(s) \rightleftharpoons Pb^{2+}(aq) + 2I^-(aq)$

The expression for the solubility product constant is:

$$K_{sp} = [Pb^{2+}][I^-]^2$$

We can solve for [I-] by rearranging the equation:

$$[I^-] = \sqrt{K_{sp} / [Pb^{2+}]}$$
$$[I^-] = \sqrt{(7.1 \times 10^{-9}) / (1.2 \times 10^{-3})}$$
$$[I^-] = \sqrt{5.92 \times 10^{-6}}$$
$$[I^-] = 2.43 \times 10^{-3} \text{ M}$$

So, the concentration of I- in the solution is 2.43×10^{-3} M.

Problem 3: Solubility Product; The Ksp of BaSO4 is 1.1×10^{-10}. If a solution contains $[Ba^{2+}] = 1.0 \times 10^{-5}$ M, at what concentration of SO_4^{2-} will BaSO4 start to precipitate?

Solution: The dissociation of BaSO4 in water is represented by the chemical equation: $BaSO4(s) \rightleftharpoons Ba^{2+}(aq) + SO_4^{2-}(aq)$

The expression for the solubility product constant is:

$$K_{sp} = [Ba^{2+}][SO_4^{2-}]$$

We can solve for $[SO_4^{2-}]$ by rearranging the equation:

$$[SO_4^{2-}] = K_{sp} / [Ba^{2+}]$$
$$[SO_4^{2-}] = (1.1 \times 10^{-10}) / (1.0 \times 10^{-5})$$
$$[SO_4^{2-}] = 1.1 \times 10^{-5} \text{ M}$$

So, BaSO4 will start to precipitate when the concentration of SO_4^{2-} exceeds 1.1×10^{-5} M.

Again, assuming x is small compared to 0.15:

$$1.8 \times 10^{-4} = x^2 / 0.15$$

Solving for x, we find that $x = [H^+] = 1.6 \times 10^{-2}$ M. The pH is then calculated as $pH = -\log[H^+] = -\log(1.6 \times 10^{-2}) = 1.80$.

Problem 3: What are the $[H^+(aq)]$ and pH of a 0.10 M solution of hydrocyanic acid (HCN, Ka = 4.9×10^{-10})?

Solution 3: The equilibrium reaction is: $HCN \rightleftharpoons H^+ + CN^-$

Let $x = [H^+] = [CN^-]$, and the equilibrium expression is:

$$4.9 \times 10^{-10} = x^2 / (0.10 - x)$$

Assuming x is small compared to 0.10: $4.9 \times 10^{-10} = x^2 / 0.10$

Solving for x, we find that x = [H+] = 7.0×10^{-6} M. The pH is then pH = -log[H+] = -log(7.0×10^{-6}) = 5.15.

An expression for Ksp: The expression for Ksp provides a quantitative measure of the solubility of a given compound. The solubility product constant is specific to the chemical reaction at hand, which generally involves a solid ionic compound dissolving into its respective ions. For a general reaction of an ionic compound, such as AB(s) ↔ A+(aq) + B-(aq), the Ksp expression would be written as Ksp = [A+][B-]. The brackets denote the molar concentration of the ions at equilibrium. In this expression, each concentration is raised to the power of its stoichiometric coefficient from the balanced chemical equation. For a more complex compound, the expression for Ksp would need to account for the different stoichiometric coefficients. For instance, if we consider the dissolution of a hypothetical compound AB2, the reaction would be AB2(s) ↔ A2+(aq) + 2B-(aq). The Ksp expression for this reaction would be Ksp = $[A2+][B-]^2$. The concentration of B- is squared due to the stoichiometric coefficient of 2 in the balanced chemical equation. It is important to note that only the concentrations of the products (ions) are included in the Ksp expression. The concentration of the solid reactant is omitted because it remains constant. Furthermore, the concentrations of ions in a Ksp expression are equilibrium concentrations. These are the concentrations of the ions when the forward and reverse reactions occur at the same rate. The solubility product is an extremely useful tool in predicting the extent of a salt dissolution in a solution, as well as in predicting whether a precipitate will form when two ionic solutions are mixed. It can also be used to calculate the molar solubility or mass solubility of a compound. In equilibrium calculations, Ksp can be used in conjunction with other equilibrium constants to establish a complete picture of the system's behavior. For instance, in the presence of complex ion formation or acid-base reactions, Ksp can be combined with the respective equilibrium constants to reflect the overall process. The value of Ksp varies with temperature, hence solubility measurements must always specify the temperature at which they were made. Typically, as temperature increases, so does the Ksp, meaning that more solid will dissolve in solution at higher temperatures. It is important to note that Ksp is not the same as solubility. While they are related, solubility is a measure of the maximum amount of a substance that can dissolve in a given amount of solvent, while Ksp is a measure of how much a compound tends to dissolve based on the concentrations of the ions in equilibrium with the solid.

Problem 1: A saturated solution of barium fluoride, BaF2, has a concentration of 1.1×10^{-3} M. Calculate the Ksp of BaF2.

Solution: BaF2 ⇌ Ba^2+ + $2F^-$

The concentration of Ba^2+ is 1.1×10^{-3} M.

The concentration of F^- will be $2 \times 1.1 \times 10^{-3}$ M

= 2.2×10^{-3} M.

The Ksp = $[Ba^2+][F^-]^2$ = (1.1×10^{-3}) * (2.2×10^{-3})^2

= 5.29×10^{-9}.

Problem 2: The Ksp of lead(II) chloride, PbCl2, is 1.6×10^{-5}.

What is the solubility of PbCl2 in moles per liter in pure water?

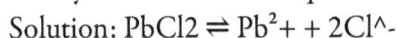

Solution: PbCl2 ⇌ Pb^2+ + $2Cl^-$

Let the solubility of PbCl2 be 's' M. Then [Pb^2+]

= s, and [Cl^-] = 2s.

The Ksp = $[Pb^2+][Cl^-]^2$ = s * $(2s)^2$ = $4s^3$.

Solving the equation $4s^3 = 1.6 \times 10^{-5}$ gives s = 0.018 M. Hence, the molar solubility of PbCl2 is 0.018 M.

Problem 3: The Ksp of silver bromide, AgBr, is 5.0×10^{-13}.

Determine the solubility of AgBr in grams for every liter.

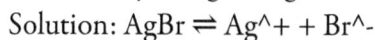

Solution: $AgBr \rightleftharpoons Ag^+ + Br^-$

Let the solubility of AgBr be 's' M. Then $[Ag^+]$
= s, and $[Br^-]$ = s.

The $Ksp = [Ag^+][Br^-] = s * s = s^2$.

Solving the equation $s^2 = 5.0 \times 10^{-13}$ gives

$s = 7.07 \times 10^{-7}$ M.

The molar mass of AgBr is 187.77 g/mol. Hence, the solubility of AgBr is 7.07×10^{-7} M * 187.77 g/mol = 1.33×10^{-4} g/L.

Calculate K_{sp} from concentrations and vice versa:

In essence, Ksp is the equilibrium constant for the dissolution of a sparingly soluble ionic compound in water. It is defined for the following general dissolution reaction: $AB(s) \rightleftharpoons A+(aq) + B-(aq)$. The Ksp expression is: Ksp = [A+][B−], where [A+] and [B−] represent the equilibrium concentrations of the dissolved ions. The concentrations of solid AB do not appear in the expression as the concentration of a pure solid is a constant.

To calculate Ksp from concentrations, we need to know the equilibrium concentrations of the ions in the solution. These can typically be determined from experimental data. Once we know these concentrations, they are simply plugged into the Ksp expression and the calculation is straightforward.

Let's consider an example: Suppose we have a solution of AgCl in equilibrium with solid AgCl. The dissolution reaction is $AgCl(s) \rightleftharpoons Ag+(aq) + Cl-(aq)$. If we know the equilibrium concentrations of Ag+ and Cl− are both 1.0×10^{-5} M, then the Ksp for AgCl is $(1.0 \times 10^{-5})(1.0 \times 10^{-5}) = 1.0 \times 10^{-10}$.

Now, let's consider the inverse problem: calculating concentrations from Ksp. If we know the Ksp of a compound, we can calculate the maximum concentration of ions that can be in solution at equilibrium. This process often involves setting up and solving a quadratic equation.

For example, consider the sparingly soluble compound PbI2. The dissolution reaction is $PbI2(s) \rightleftharpoons Pb2+(aq) + 2I-(aq)$. The Ksp expression is Ksp = $[Pb2+][I-]^2$. If we know the Ksp for PbI2 is 7.1×10^{-9}, we can set up the following equilibrium: let x = [Pb2+] = the concentration of Pb2+ that dissolves. Because there are two I− ions for every Pb2+ ion that dissolves, 2x = [I−]. Substituting these into the Ksp expression gives: $(7.1 \times 10^{-9}) = x(2x)^2$. Solving this equation yields $x = 1.2 \times 10^{-3}$ M, which is the equilibrium concentration of Pb2+ and 2x = 2.4×10^{-3} M is the equilibrium concentration of I−.

Problem 1: The solubility of silver chloride (AgCl) in water at 25°C is 1.3×10^{-5} mol/L. Calculate the Ksp of AgCl at this temperature.

Solution: Silver chloride dissociates according to the following equation: $AgCl (s) \leftrightarrow Ag+ (aq) + Cl- (aq)$

Because the solubility of AgCl is 1.3×10^{-5} mol/L, the concentrations of Ag+ and Cl- are both 1.3×10^{-5} M. The Ksp expression for this reaction is: Ksp = [Ag+][Cl-]

Substituting the values into the Ksp expression gives:

Ksp = $(1.3 \times 10^{-5})(1.3 \times 10^{-5}) = 1.69 \times 10^{-10}$

Problem 2: The solubility product constant (Ksp) for lead(II) iodide (PbI2) at 25°C is 7.1×10^{-9}. What is the solubility of PbI2 in mol/L at this temperature?

Solution: Lead(II) iodide dissociates according to the following equation: $PbI2 (s) \leftrightarrow Pb2+ (aq) + 2I- (aq)$

Because one PbI2 molecule provides one Pb2+ ion and two I- ions, if we let the solubility of PbI2 be represented by 's', then the concentration of Pb2+ is 's', and the concentration of I- is '2s'. The Ksp expression for this reaction is:

Ksp = $[Pb2+][I-]^2$

Substituting the values into the Ksp expression gives:

$$7.1 \times 10^{-9} = s(2s)^2$$

Solving for 's' gives s = 1.2×10^{-3} M. So, the solubility of PbI2 is 1.2×10^{-3} mol/L.

Problem 3: The Ksp of barium sulfate (BaSO4) is 1.5×10^{-10}. What is the concentration of Ba2+ ions in a saturated solution?

Solution: BaSO4 dissociates as follows:

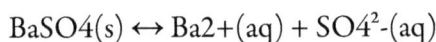

$$BaSO4(s) \leftrightarrow Ba2+(aq) + SO4^{2-}(aq)$$

The Ksp expression for this reaction is: Ksp = $[Ba2+][SO4^{2-}]$

In a saturated solution, $[Ba2+] = [SO4^{2-}]$. Therefore, we can write the Ksp expression as: $1.5 \times 10^{-10} = [Ba2+]^2$

Solving for $[Ba2+]$ gives $[Ba2+] = 1.2 \times 10^{-5}$ M. So, the concentration of Ba2+ ions in a saturated solution is 1.2×10^{-5} M.

Explain solubility variance using common ion effect

The common ion effect is a phenomenon widely used to explain the changes in the solubility of a compound when in a solution containing a common ion. This principle, which is a direct result of Le Chatelier's principle, plays a pivotal role in various scientific fields, particularly in environmental science, pharmacology, and industrial applications. At its core, the common ion effect refers to the decrease in the solubility of a salt when added to a solution that already contains one of its ions. The common ion effect is an application of equilibrium dynamics in chemical reactions. In a solution at equilibrium, the rates of forward and reverse reactions are the same. When a common ion is added, the equilibrium shifts to decrease the concentration of that ion. To illustrate, consider the dissolution of silver chloride (AgCl). In the absence of any common ions, AgCl dissociates into Ag+ and Cl- ions. However, if the solution already contains Cl- ions (the common ion), this will shift the equilibrium to the left, making AgCl less soluble. The increased concentration of Cl- ions leads to an increase in the rate of the reverse reaction (precipitation), thus decreasing the overall solubility of AgCl. This is an application of Le Chatelier's principle, which states that if a change in conditions is imposed on a system at equilibrium, the equilibrium will shift to counteract that change. In this case, the increase in the concentration of Cl- ions (the change) causes the equilibrium to shift to the left (the counteraction), thus decreasing the solubility of AgCl. The common ion effect also affects pH values, which are a measure of the acidity or alkalinity of a solution. It is a key factor to consider in buffers, which are solutions that resist changes in pH when small amounts of an acid or a base are added. Buffers usually consist of a weak acid and its conjugate base, and the common ion effect is a key factor in their ability to resist changes in pH. It is worth noting that the common ion effect is temperature-dependent. Generally, as temperature increases, solubility increases. However, the common ion effect can still decrease the solubility of a compound even at higher temperatures. Furthermore, the common ion effect has significant implications in the field of environmental chemistry. For example, it can influence the solubility of harmful metal ions in water sources, affecting their toxicity and environmental impact. In pharmacology, the common ion effect is considered when formulating drugs to ensure their bioavailability. Some drugs are better absorbed in the body in their ionized form, while others are better absorbed in their non-ionized form. By manipulating the common ion effect, pharmacologists can increase the efficacy of drug delivery. In industry, the common ion effect is used in various processes such as water softening. Hard water contains calcium and magnesium ions, which can be precipitated out by adding a common ion.

Problem 1: Calculate the solubility of silver chloride (AgCl) in a 0.10 M solution of silver nitrate (AgNO3). (Ksp for AgCl

$$= 1.8 \times 10^{-10})$$

Solution 1: In this problem, silver nitrate is the common ion. The solubility product constant expression for AgCl is Ksp = [Ag+][Cl-]. However, because of the presence of AgNO3, the concentration of Ag+ is already 0.10 M. Therefore, the AgCl will dissolve until the product of the ion concentrations equals the Ksp. Solving for the chloride ion concentration,

[Cl-] = Ksp / [Ag+] = (1.8 × 10^-10) / 0.10 M = 1.8 × 10^-9 M

The solubility of AgCl in a 0.10 M solution of AgNO3 is 1.8 × 10^-9 M.

Problem 2: A solution has a Ca2+ concentration of 0.010 M. What is the maximum fluoride ion concentration this solution can hold without resulting in the precipitation of CaF2?

(Ksp for CaF2 = 3.9 × 10^-11)

Solution 2: The Ksp expression for CaF2 is Ksp = [Ca2+][F-]². Since the concentration of the calcium ion is already 0.010 M, we can substitute this into the Ksp expression and solve for the fluoride ion concentration.

[F-] = √(Ksp / [Ca2+]) = √((3.9 × 10^-11) / 0.010 M)

= 6.2 × 10^-5 M

Therefore, the maximum fluoride ion concentration that this solution can have without causing precipitation of CaF2 is 6.2 × 10^-5 M.

Problem 3: The solubility of BaSO4 in pure water is 1.0 × 10^-5 M. What is its solubility in a 0.10 M solution of Ba(NO3)2? (Ksp for BaSO4 = 1.1 × 10^-10)

Solution 3: The Ksp expression for BaSO4 is

Ksp = [Ba2+][SO4 2-]. The concentration of Ba2+ ions is already 0.10 M due to the presence of Ba(NO3)2. Thus, we can substitute this into the Ksp expression and solve for the sulfate ion concentration.

[SO4 2-] = Ksp / [Ba2+] = (1.1 × 10^-10) / 0.10 M = 1.1 × 10^-9 M

Therefore, the solubility of BaSO4 in a 0.10 M solution of Ba(NO3)2 is 1.1 × 10^-9 M.

Calculate using Ksp values, ion concentration

The solubility product constant (Ksp) is an essential parameter in determining the solubility of a salt in a solution. It denotes the maximum concentration of a substance that can dissolve in a solution without forming a precipitate. The Ksp is calculated using the concentrations of the ions that make up the substance.

The common ion effect comes into play when a salt is dissolved in a solution already containing one of the ions of the salt. The solubility of a salt decreases with the existence of a common ion, a concept that originates from Le Chatelier's principle. To illustrate this, consider a salt AB which dissolves in water according to the equation AB(s) ⇌ A+(aq) + B-(aq). The Ksp expression for this reaction would be Ksp = [A+][B-]. If an ion A+ or B- is already present in the solution, it shifts the equilibrium to the left, reducing the solubility of the salt.

Now, let's take an example to understand the practical application of this concept.

Suppose we want to calculate the concentration of Ag+ ions in a saturated solution of AgCl, having a Ksp of 1.8 × 10^-10, in a 0.1 M solution of NaCl. The reaction for the dissolution of AgCl is AgCl(s) ⇌ Ag+(aq) + Cl-(aq). The Cl- ions in NaCl represent the common ion.

If we assume 's' to be the solubility of AgCl in the NaCl solution, the concentration of Ag+ ions will also be 's'. However, the concentration of Cl- ions would be 0.1 + s (0.1 from NaCl and s from AgCl). Substituting these in the Ksp expression gives: Ksp = s * (0.1 + s).

As the Ksp is very small, the 's' in '0.1 + s' can be neglected. This simplifies the equation to Ksp = 0.1s. Solving for 's' gives the concentration of Ag+ ions in the solution.

Therefore, the common ion effect can significantly influence the solubility of a substance. This is a critical concept in various chemical processes, including chemical synthesis, industrial production, environmental science, and medical applications.

Problem 1: The solubility product constant (Ksp) of silver chloride (AgCl) is 1.8×10^{-10}. If a solution of AgCl is prepared by dissolving AgCl in water containing 0.10 M Cl-, what is the solubility of AgCl in this solution?

Solution 1: The solubility of AgCl can be represented by the equation: AgCl(s) ⇔ Ag+(aq) + Cl-(aq)

The Ksp expression for the dissolution of AgCl is:

$$Ksp = [Ag+][Cl-]$$

If the solubility of AgCl is represented by "s", the concentration of Ag+ ions will also be "s" (because for every molecule of AgCl that dissolves, one Ag+ ion is produced). However, the concentration of Cl- ions in the solution is 0.10 + s (the Cl- ions from the AgCl plus the Cl- ions that were already in the solution).

Substituting the known values into the Ksp expression, we get:

$$1.8 \times 10^{-10} = s(0.10 + s)$$

Since Ksp is a very small number, we can assume that s is small compared to 0.10 and therefore 0.10 + s is approximately 0.10. So the equation simplifies to: $1.8 \times 10^{-10} = s(0.10)$

or $s = 1.8 \times 10^{-10} / 0.10 = 1.8 \times 10^{-9}$ M

Problem 2: The Ksp of calcium fluoride (CaF2) is 4×10^{-11}. What is the fluoride ion concentration in a saturated solution of CaF2?

Solution 2: The dissolution of CaF2 can be represented by the equation: CaF2(s) ⇔ Ca2+(aq) + 2F-(aq)

The Ksp expression for the dissolution of CaF2 is:

$$Ksp = [Ca2+][F-]^2$$

Since for every molecule of CaF2 that dissolves, one Ca2+ ion and two F- ions are produced, the concentrations of Ca2+ and F- in a saturated solution are "s" and "2s" respectively.

Substituting these into the Ksp expression, we get:

$$4 \times 10^{-11} = s(2s)^2 = 4s^3$$

or $s^3 = 4 \times 10^{-11} / 4 = 1 \times 10^{-11}$

or $s = (1 \times 10^{-11})^{(1/3)} = 2.15 \times 10^{-4}$ M

Since the concentration of F- ions is 2s, the fluoride ion concentration in the solution is $2(2.15 \times 10^{-4}$ M$) = 4.3 \times 10^{-4}$ M.

Partition coefficients

The partition coefficient, often denoted by the symbol Kpc, is a fundamental concept in the field of chemistry, specifically in areas like biochemistry, pharmacology, environmental science, and chemical engineering. Essentially, it is a ratio that describes the distribution of a compound between two immiscible phases at equilibrium. It is a measure of the differential solubility of a substance in these two phases, typically a hydrophobic (non-polar) phase and a hydrophilic (polar) phase.

The partition coefficient is a very important parameter as it provides key information about the solubility, bioavailability, bioaccumulation, and environmental fate of molecules. It gives an insight into the behavior of a substance when it comes into contact with a system consisting of two immiscible phases, such as oil and water. A partition coefficient can be seen as a quantitative expression of the affinity that a substance has for each of the two phases.

The larger the partition coefficient, the more the substance favors the hydrophobic phase. Conversely, a smaller partition coefficient indicates that the substance is more soluble in the hydrophilic phase. This characteristic can be used to predict the distribution of a substance in a biological system, for instance, where the two phases could be blood and fat tissue.

The partition coefficient is a crucial parameter in pharmacokinetics. It can determine the distribution of a drug within the body, affecting its absorption, distribution, metabolism, and elimination (ADME). A drug with a high partition coefficient may be extensively stored in fatty tissues, potentially leading to longer-lasting effects, but also the risk of toxicity.

In environmental chemistry, the partition coefficient is used to predict the behavior of pollutants in the environment. For example, pesticides with high partition coefficients tend to accumulate in the soil and sediments, posing a risk to the environment and potentially entering the food chain.

The calculation of the partition coefficient can be done experimentally, by measuring the concentration of the substance in both phases once equilibrium is reached. Alternatively, it can also be estimated using various predictive methods, often based on the molecular structure of the substance.

The partition coefficient is typically expressed in its logarithmic form, log P, to reduce the wide range of values into a more manageable scale. The log P value is particularly valuable in the drug discovery process, where it is used as a key predictor of a compound's potential as a drug candidate.

The term "partition coefficient" is sometimes used interchangeably with "distribution coefficient". However, the latter term is more general and can refer to the distribution of a substance between any two phases, not necessarily immiscible.

It's important to note that the partition coefficient is temperature-dependent. This means that the ratio of the concentrations of a substance in the two phases can change with temperature. Consequently, temperature must be controlled and reported when determining the partition coefficient.

The partition coefficient is also pH-dependent for ionizable compounds. The ionization state of a compound can significantly affect its solubility in the two phases, and therefore its partition coefficient.

Question: If a compound has a partition coefficient of 4 in a water/octanol system, where will the majority of the compound be found?

Solution: The partition coefficient (Kpc) indicates the ratio of the compound in each of the two solvents. A Kpc of 4 means that the compound is 4 times more likely to be in the octanol than in the water. Therefore, the majority of the compound will be found in the octanol.

Question: What does it mean if a compound has a partition coefficient less than 1 in a dichloromethane/water system?

Solution: If a compound has a partition coefficient (Kpc) less than 1, it means that the compound is more soluble in water than in dichloromethane. Therefore, at equilibrium, more of the compound will be found in the water phase.

Question: A drug has a partition coefficient of 0.5 in an octanol/water system. Would this drug be more effective if administered orally or via IV?

Solution: A drug with a low partition coefficient (Kpc) like 0.5 in an octanol/water system is more hydrophilic (water-loving), meaning it is more soluble in water than in fats/oils. Such a drug would likely be more effective if administered via IV, as it can be more readily absorbed and distributed in the body's largely water-based systems.

Question: How does the partition coefficient affect the distribution of a drug within the body?

Solution: The partition coefficient determines how a drug will distribute between water and lipid environments in the body. A drug with a high partition coefficient is more lipid-soluble and may travel more readily through cell membranes, distributing more widely throughout the body. Conversely, a drug with a low partition coefficient is more water-soluble and may remain more localized where water is abundant.

Question: What might be the implication if a toxic compound has a high partition coefficient in a fat/water system?

Solution: If a toxic compound has a high partition coefficient in a fat/water system, it means that the compound is more soluble in fat than in water. This can be problematic as the compound may accumulate in fatty tissues, leading to potential tissue damage or other toxic effects.

Let's take a hypothetical system in which the solute exists in the same physical state in both solvents. For instance, a solute could be dissolved in both an aqueous phase (water) and an organic phase (such as dichloromethane). The partition coefficient for this system can be calculated using the ratio of concentrations of the solute in the two solvents at equilibrium.

The partition coefficient can be experimentally determined by dissolving the solute in a mixture of the two solvents, allowing the system to reach equilibrium, and then separately measuring the concentrations of the solute in each of the solvents. The partition coefficient is then calculated as the ratio of these two concentrations. However, it's important to note that the partition coefficient is not simply a measure of solubility, but a measure of preference. A solute with a high partition coefficient is not necessarily more soluble in the organic phase than in the aqueous phase; it simply prefers the organic phase.

In the context of drug discovery and design, the partition coefficient is an essential parameter. It influences a compound's pharmacokinetics, including absorption, distribution, metabolism, and excretion. A compound with a high partition coefficient might be well absorbed, but it might also be extensively metabolized and rapidly excreted, limiting its therapeutic effectiveness.

Moreover, the partition coefficient can play a significant role in environmental chemistry. It can determine the fate and transport of chemicals in the environment, including their potential to contaminate groundwater or accumulate in organisms.

In terms of analytical chemistry, the partition coefficient can affect the efficiency of extraction and separation techniques. For instance, in liquid-liquid extraction, a substance with a high partition coefficient for the organic phase would be efficiently extracted into that phase.

While the partition coefficient is a useful parameter, it's also important to consider other factors, such as the pH of the system and the specific interactions between the solute and the solvents. For instance, a solute might have a high partition coefficient in a particular system because it forms strong hydrogen bonds with the solvent in the organic phase.

Understanding the partition coefficient and how to calculate it is fundamental for chemists working in a variety of fields, from drug design to environmental science. It allows us to predict and control the behavior of chemicals in diverse contexts, contributing to the development of new medicines, the remediation of environmental contamination, and the advancement of our understanding of natural processes.

Problem 1: A compound X has a partition coefficient of 3 between water and hexane. If you have 200 mg of X, and you mix it in a system containing equal volumes of water and hexane, how much X will be in the water and how much will be in the hexane after equilibrium is reached?

Solution: The partition coefficient (K_d) is the ratio of concentrations of X in the two solvents at equilibrium. It is given by $K_d = [X]hexane / [X]water$.

If we let $[X]water = x$, then $[X]hexane = 3x$ (because the partition coefficient is 3). The total amount of X is given by the volume of the solvents multiplied by the concentration of X in each solvent. In this case, the volumes of water and hexane are equal, so the total amount of X is $x + 3x = 200$ mg. Solving for x gives $x = 50$ mg. Therefore, there are 50 mg of X in the water and 150 mg of X in the hexane.

Problem 2: The partition coefficient of a solute between benzene and water is 0.2. If 100 g of the solute is dissolved in a liter of benzene, how much of the solute will be transferred to a liter water when the two are contacted and allowed to reach equilibrium?

Solution: The partition coefficient (K_d) represents the equilibrium concentration ratio of a solute between two solvents. It is given by $K_d = [solute]benzene / [solute]water$. Thus $[solute]water = [solute]benzene / K_d$. The initial concentration of solute in benzene is 100 g/L. Therefore, at equilibrium, $[solute]water = (100 \text{ g/L}) / 0.2$

= 500 g/L. However, the total amount of solute is conserved, so 100 g - 500 g = -400 g. This is not possible, so all 100 g of the solute transfers to the water.

Problem 3: A drug has a partition coefficient of 4 between octanol and water. If you dissolve 120 mg of the drug in a system containing equal volumes of octanol and water, how much of the drug will be in each solvent after equilibrium is reached?

Solution: The partition coefficient (Kd) is the ratio of concentrations of the drug in the two solvents at equilibrium. It is given by Kd = [drug]octanol / [drug]water. If we let [drug]water = x, then [drug]octanol = 4x (because the partition coefficient is 4). The total amount of the drug is given by the volume of the solvents multiplied by the concentration of the drug in each solvent. In this case, the volumes of octanol and water are equal, so the total amount of drug is x + 4x = 120 mg. Solving for x gives x = 24 mg. Therefore, there are 24 mg of the drug in the water and 96 mg of the drug in the octanol.

The numerical value of the partition coefficient can be influenced by several factors, one of which is the polarity of the solute and the solvents used. A solute's polarity refers to the distribution of electrical charges across its molecules, which determines its ability to form bonds with other substances. It is widely known that "like dissolves like" - polar solutes tend to dissolve in polar solvents, and nonpolar solutes in nonpolar solvents. This is because the intermolecular forces that hold the solute and solvent molecules together are similar in nature, making the dissolution process energetically favorable.

Consequently, the partition coefficient can be significantly affected by the polarity match between the solute and the solvents. For instance, if a solute is polar and one of the solvents is polar while the other is nonpolar, the solute will preferentially dissolve in the polar solvent, leading to a higher partition coefficient. Conversely, if the solute is nonpolar, it will preferentially dissolve in the nonpolar solvent, resulting in a lower partition coefficient.

Beyond the polarity of the solute and solvents, the partition coefficient is also affected by the specific interactions that can occur between them. These can include hydrogen bonding, dipole-dipole interactions, and London dispersion forces. For example, a solute capable of forming hydrogen bonds with a solvent will be more soluble in that solvent, increasing the partition coefficient.

Temperature is another critical factor influencing the partition coefficient. As the temperature increases, the solubility of the solute can either increase or decrease depending on the nature of the solute and solvent. For instance, a solute that forms strong intermolecular forces with a solvent may become less soluble as the temperature increases.

The pH of the solvents can also impact the partition coefficient, especially for solutes that are weak acids or bases. Changes in pH can alter the degree of ionization of these solutes, affecting their solubility in the solvents and hence the partition coefficient. Typically, an increase in the ionization of a solute leads to an increase in its solubility in polar solvents and a decrease in its solubility in nonpolar solvents.

Size and shape of the molecules also play a role. Larger or more complex molecules might have different regions of polarity, allowing them to interact with both polar and non-polar solvents, which would affect the partition coefficient.

The presence of other solutes in the solvents can also modify the partition coefficient. This is because the added solutes can compete with the original solute for interactions with the solvent molecules, potentially decreasing the solubility of the original solute and thereby lowering the partition coefficient.

Problem 1: A drug molecule is soluble in both water and octanol. The partition coefficient (P) of the drug is 3. What does this value indicate about the solubility of the drug in water and octanol?

Solution: The partition coefficient (P) is the ratio of concentrations of a compound in the two solvents (octanol and water in this case) at equilibrium. It is used to measure the distribution of the compound between water and lipids. The value of P=3 indicates that the drug is 3 times more soluble in octanol than in water. This suggests that the drug molecule is relatively non-polar since octanol is a non-polar solvent.

Problem 2: Consider two solvents: water (polar) and hexane (non-polar). A solute has a partition coefficient of 0.2. Can you predict the solute's polarity based on this information?

Solution: The partition coefficient (P) is a measure of the solute's preference between the two solvents. A P value of 0.2 means that the solute is more soluble in water than in hexane. This implies that the solute is likely to be polar because "like dissolves like" - polar solutes dissolve better in polar solvents (water in this case), and nonpolar solutes dissolve better in nonpolar solvents (hexane in this case).

Problem 3: A solute has a partition coefficient of 5 in a water-oil system. If the solute's polarity is changed so that it becomes less polar, how would you expect the partition coefficient to change?

Solution: The partition coefficient (P) is a measure of how a solute distributes itself between a polar solvent (water in this case) and a nonpolar solvent (oil in this case). A higher P value indicates that the solute prefers the nonpolar solvent. If the solute becomes less polar, it will prefer the nonpolar solvent (oil) even more, and thus the partition coefficient will increase.

Reaction Kinetics

The rate equation is a fundamental concept in the study of chemical kinetics. It is an algebraic equation that describes the rate of a chemical reaction in terms of the concentration of the reactants. The equation includes a rate constant and the concentrations of the reactants, each raised to a power, known as the order of the reaction with respect to that reactant. The rate equation provides valuable insights into the mechanism of a reaction, revealing the relationship between the rate of reaction and the concentrations of the reactants at any given point in time.

The order of a reaction is a term used to describe the effect of the concentration of a particular reactant on the rate of the reaction. It is determined experimentally and is usually a whole number, although it can be zero or a fraction. The order of a reaction with respect to a specific reactant is determined by the exponent of its concentration in the rate equation. For example, a reaction that is first order with respect to a reactant has a rate that is directly proportional to the concentration of that reactant.

It provides a measure of how the rate of a reaction changes with changes in the concentration of all reactants. For instance, a reaction could be second order overall, meaning that the rate of the reaction depends on the square of the concentration of one reactant, or the product of the concentrations of two reactants, each raised to the first power.

The rate constant, often denoted by 'k', is a proportionality constant in the rate equation that is specific to a particular reaction at a given temperature. It does not depend on the concentrations of the reactants, but it does change with temperature. The magnitude and units of the rate constant depend on the overall order of the reaction. The rate constant provides a measure of the inherent speed of a reaction, with larger rate constants indicating faster reactions.

Half-life is a term used to describe the time it takes for half of a reactant in a reaction to be consumed. It is particularly useful in reactions that are first order, where the half-life is constant and does not depend on the initial concentration of reactants. The concept of half-life is widely used beyond chemistry, in fields such as nuclear physics and pharmacology.

The rate-determining step is the slowest step in a multi-step reaction mechanism. It is the step that determines the overall rate of the reaction as it has the highest activation energy. In the rate equation, the rate-determining step is represented by the rate constant. Understanding the rate-determining step is crucial in controlling the rate of chemical reactions in various industrial processes.

An intermediate in a chemical reaction is a species that is formed in an early step of the reaction mechanism and consumed in a later step. Intermediates do not appear in the overall chemical equation for the reaction, as they are consumed during the reaction. However, they can play a crucial role in the mechanism of the reaction.

Rate Equation: This is an equation that describes the rate of a chemical reaction in terms of the concentration of its reactants. For a reaction $A + B \rightarrow C$, the rate equation could be represented as: Rate $= k [A]^m [B]^n$, where $[A]$ and $[B]$ are the concentrations of the reactants, the rate constant is represented by 'k', while 'm' and 'n' denote the reaction orders in relation to $[A]$ and $[B]$ respectively.

Order of Reaction: This refers to the power to which the concentration of a reactant is raised in the rate equation. It indicates how the rate of reaction is affected by the concentration of that reactant.

Overall Order of Reaction: This is the sum of the orders of all reactants in the rate equation. For the rate equation Rate = k [A]^m [B]^n, the overall order of reaction would be m+n.

Rate Constant (k): This is a proportionality constant in the rate equation. It is a measure of the reaction speed and it depends on factors like temperature and catalysts.

Half-Life: This refers to the duration needed for a reactant's concentration to be reduced by half in a reaction. It is commonly used to describe radioactive decay but applies to all reactions.

Rate Determining Step: This refers to the slowest phase in a reaction mechanism. It determines the overall rate of the reaction because a reaction cannot proceed faster than its slowest step.

Intermediate: This is a species formed during a reaction that is consumed in a later step. It does not appear in the overall reaction equation because it is produced and then consumed during the reaction process.

Problem 1: For a reaction A + B → C, the rate equation is found to be: Rate = k [A]^2 [B]. What is the order of the reaction with respect to A, B, and the overall order?

Solution: With respect to A, the order is 2. With respect to B, the order is 1. The overall order is the sum of individual orders, i.e., 2 + 1 = 3.

Problem 2: If the rate constant (k) for a first order reaction is 0.693 min^-1, what is the half-life of this reaction?

Solution: For a first order reaction, the half-life (t1/2) is given by the formula t1/2 = 0.693/k. Substituting the given value of k, we get t1/2 = 0.693/0.693 min = 1 min.

Problem 3: In a reaction mechanism, the slowest step is the second step: A + B → C + D. What is the rate-determining step of this mechanism?

Solution: The rate-determining step is the slowest step in a reaction mechanism. Therefore, in this case, the rate-determining step is A + B → C + D.

Rate equations are fundamental concepts in chemical kinetics, a branch of physical chemistry that deals with the speed or rate at which chemical reactions occur. A rate equation, or a rate law, uses mathematical terms to explain how the speed of a reaction is influenced by the concentration of its reactants.

Rate equations are typically of the form: rate = k [A]^m [B]^n.

In this equation, 'rate' is the observed speed of the reaction, 'k' is the rate constant, '[A]' and '[B]' represent the molar concentrations of the reactants, and 'm' and 'n' are the reaction orders with respect to reactants A and B, respectively. The reaction orders (m and n) are usually 0, 1, or 2, and they are determined experimentally; they are not based on the stoichiometric coefficients of the balanced chemical equation.

The rate constant 'k' is a proportionality constant that reflects the intrinsic reactivity of the reactants under the given conditions (temperature, pressure, catalyst presence, etc.). It is important to note that the rate constant is not affected by the concentration of the reactants.

When 'm' or 'n' is 0, it implies that the rate of reaction is independent of the concentration of that specific reactant. This is known as a zero-order reaction. For instance, if m=0, then rate = k [A]^0 [B]^n simplifies to rate = k [B]^n.

When 'm' or 'n' is 1, the rate of reaction is directly proportional to the concentration of that reactant. This is referred to as a first-order reaction. For instance, if n=1, then rate = k [A]^m [B]^1 simplifies to rate = k [A]^m [B].

When 'm' or 'n' is 2, the rate of reaction is proportional to the square of the concentration of that reactant. This is called a second-order reaction. For example, if m=2, then rate = k [A]^2 [B]^n simplifies to rate = k [A]^2 [B]^n. It's worth noting that the overall order of the reaction is the sum of the individual orders. Meaning, if a reaction is first order with respect to reactant A and second order with respect to reactant B, the overall order of the reaction is 1+2=3.

The rate equation is extremely useful in predicting the behavior of a chemical system over time. By knowing the rate law, rate constant, and concentrations of the reactants, one can predict the rate of the reaction under different conditions.

In practice, the form of the rate equation and the values of 'm' and 'n' are determined through a series of experiments, where concentrations of the reactants are carefully measured over time. The rate at which the concentration of a product increases or a reactant decreases provides the reaction rate.

In addition, the rate constant 'k' can be influenced by temperature. As the temperature increases, the rate constant also increases, causing the reaction to proceed at a faster rate. This relationship is described by the Arrhenius equation.

Problem 1: The rate of a certain chemical reaction is given by the rate equation: rate = k $[A]^2[B]$. If the concentration of A is doubled while the concentration of B remains the same, by what factor does the rate of the reaction change?

Solution: In the given situation, m equals 2 and n equals 1. If the concentration of A is doubled, the rate of the reaction will increase by a factor of $2^2 = 4$. Therefore, the rate of the reaction changes by a factor of 4.

Problem 2: The rate of a reaction is given by the rate equation: rate = k $[A][B]^2$. If the concentration of A remains the same and the concentration of B is halved, by what factor does the rate of the reaction change?

Solution: In this case, m = 1 and n = 2. If the concentration of B is halved, the rate of the reaction will decrease by a factor of $(1/2)^2 = ¼$. Therefore, the rate of the reaction changes by a factor of ¼.

Problem 3: A reaction follows the rate equation: rate = k $[A]^2[B]^0$. If the concentration of A is tripled and the concentration of B is doubled, by what factor does the rate of the reaction change?

Solution: In this case, m = 2 and n = 0. Even if the concentration of B is doubled, the rate of the reaction will not change because n = 0. However, if the concentration of A is tripled, the rate of the reaction will increase by a factor of $3^2 = 9$. Therefore, the rate of the reaction changes by a factor of 9.

Problem 4: The equation for the rate of a reaction is represented as:

rate = k $[A][B]$. If both the concentrations of A and B are doubled, by what factor does the rate of the reaction change?

Solution: In this case, m = 1 and n = 1. If the concentrations of both A and B are doubled, the rate of the reaction will increase by a factor of $2 \times 2 = 4$. Therefore, the rate of the reaction changes by a factor of 4.

Understanding the order of a reaction is a critical aspect of the field of chemical kinetics. It is the power to which the concentration of reactant is raised in the rate equation, and it can be deduced from concentration-time graphs or from experimental data relating to the initial rates or half-life methods.

Let's start with concentration-time graphs. A zero-order reaction is characterized by a straight line when plotting concentration against time, with the slope of the line being negative due to the reduction in concentration over time. The rate of reaction is independent of the concentration of the reactant, meaning that the reaction progresses at a constant rate.

A first-order reaction, on the other hand, exhibits a curve when plotting concentration against time. However, when plotting the natural logarithm of concentration against time, a straight line is obtained. This indicates that the rate of reaction is directly proportional to the concentration of the reactant. As the reactant depletes, the rate of the reaction decreases.

A second-order reaction, when plotted on a concentration-time graph, also produces a curve, similar to the first-order reaction. But when one over the concentration of the reactant is plotted against time, a straight line results, indicating that the rate of reaction is proportional to the square of the concentration of the reactant.

Turning our attention to the initial rates method, this strategy involves conducting a series of experiments where the concentration of one reactant is varied while others are kept constant. The initial rates of the reaction for each experiment are measured and used to calculate the order of reaction.

The reaction is considered first order with respect to a certain reactant if the initial rate doubles when the reactant's concentration doubles. However, if the rate increases fourfold, the reaction is regarded as second order. The half-life method is another valuable tool in determining the order of a reaction. The half-life of a reaction is the time required for half of the reactant to be consumed. For a zero-order reaction, the half-life is dependent on the initial concentration of the reactant. This means as the reaction progresses, the half-life increases.

In contrast, for a first-order reaction, the half-life is constant and independent of the initial concentration of the reactant. This implies that no matter the concentration of the reactant, it always takes the same amount of time for half of it to be used up.

For a second-order reaction, the half-life increases as the reaction progresses, meaning it is dependent on the initial concentration of the reactant. However, unlike the zero-order reaction, the half-life for a second-order reaction gets longer as the reaction continues.

Problem 1: Given the following concentration-time data for the reaction

$A \rightarrow B$:

Time (s) [A] (mol/L)

0 0.500

10 0.250

20 0.125

30 0.063

Deduce the order of the reaction.

Solution: From the data, it can be seen that the concentration of A is halved every 10 seconds. This suggests that the reaction is first order. The reaction rate in a first order reaction is directly related to the concentration of the substances involved in the reaction. Hence, when the concentration of A is halved, the time taken for the reaction to occur remains constant.

Problem 2: Given the following initial rates data for the reaction $2A \rightarrow B$:

[A] (mol/L) Rate (mol/L.s)

0.1 0.2

0.2 0.4

0.3 0.6

Deduce the order of the reaction.

Solution: From the data, it can be seen that the rate of reaction doubles when the concentration of A is doubled, and triples when the concentration of A is tripled. This suggests that the reaction is first order. The reaction rate in a first order reaction is directly related to the concentration of the substances involved in the reaction. Therefore, if the concentration of A is increased two or three times, the reaction rate will also increase by two or three times, respectively.

Problem 3: A reaction has a half-life of 10 seconds at an initial concentration of 0.1 mol/L and a half-life of 20 seconds at an initial concentration of 0.05 mol/L. Deduce the order of the reaction.

Solution: In this case, the half-life of the reaction is increasing as the concentration is decreasing. This is characteristic of a second order reaction. Hence, when the initial concentration of the reactant is halved, the half-life of the reaction doubles.

Problem 4: Given the following reaction:

2NO + O2 → 2NO2. The initial rate of reaction was measured at different initial concentrations of NO and O2.

[NO] (mol/L) [O2] (mol/L) Rate (mol/L.s)

0.1 0.1 0.2

0.2 0.1 0.4

0.1 0.2 0.4

Deduce the order of the reaction with respect to NO and O2.

Solution: From the data, it can be seen that when the concentration of NO is doubled, the rate of reaction doubles, suggesting that the reaction is first order with respect to NO. When the concentration of O2 is doubled, the rate of reaction also doubles, suggesting that the reaction is first order with respect to O2. Hence, the overall order of the reaction is

1 + 1 = 2. Interpreting experimental data in graphical form is integral to the field of chemistry. Graphs provide a visual representation of data, enabling more straightforward comprehension and analysis of complex information. Two types of graphs frequently used in chemistry are concentration-time and rate-concentration graphs.

Concentration-time graphs, also known as reaction progress curves, plot the concentration of a reactant or product against time. These graphs demonstrate the rate at which a reaction occurs, which can be crucial for understanding reaction kinetics. By examining the slope at any given point on a concentration-time graph, one can determine the instantaneous rate of the reaction, which can provide valuable insights into the reaction's mechanism.

A steeper slope indicates a faster reaction rate, whereas a flatter slope corresponds to a slower rate. A declining slope on a concentration-time graph signifies the reactant's concentration decreasing with time, typical in a forward reaction. Conversely, an increasing slope represents an increase in product concentration over time. The shape of a concentration-time graph can also provide information about the reaction order. For a zero-order reaction, the graph produces a straight line, indicating that the reaction rate is independent of the concentration of the reactants. For a first-order reaction, the graph is a curve that gets less steep over time, implying that the reaction rate is directly proportional to the reactant concentration. For a second-order reaction, the graph is also a curve, but one that gets steeper over time, signifying that the reaction rate is proportional to the square of the reactant concentration.

Rate-concentration graphs, on the other hand, plot the rate of reaction against the concentration of one reactant while keeping the concentration of other reactants constant. This type of graph is particularly useful for determining a reaction's order with respect to a specific reactant.

For zero-order reactions, a rate-concentration graph will yield a horizontal line, indicating that the reaction rate does not change with varying reactant concentration. For first-order reactions, the graph will produce a straight line with a positive slope, signifying that the reaction rate increases linearly with an increase in reactant concentration. For second-order reactions, the graph will yield a curve that gets steeper as the concentration increases, indicating that the reaction rate increases exponentially with an increase in reactant concentration. By conducting multiple experiments and generating rate-concentration graphs for each reactant, chemists can determine the overall reaction order and the reaction order with respect to each reactant. This information is crucial for understanding the reaction mechanism and predicting how changes in reactant concentrations will impact the reaction rate.

Problem 1: A graph shows the concentration of a reactant in a chemical reaction over time. At the start of the reaction, the concentration of the reactant is 0.1 M. After 2 minutes, the concentration is 0.08 M. After 6 minutes, the concentration is 0.04 M. What is the average rate of reaction between 2 and 6 minutes?

Solution: The rate of reaction can be calculated by the change in concentration over the change in time. Between 2 and 6 minutes, the change in concentration is 0.04 M - 0.08 M = -0.04 M. The change in time is 6 minutes - 2 minutes = 4 minutes. Therefore, the rate of reaction is -0.04 M/4 min = -0.01 M/min.

Problem 2: A graph shows the rate of a chemical reaction versus the concentration of the reactant. At a concentration of 0.05 M, the rate of the reaction is 0.01 M/s. At a concentration of 0.10 M, the rate of the reaction is 0.02 M/s. What is the rate constant (k) of this reaction assuming it is a first-order reaction?

Solution: In a first-order reaction, the reaction rate is directly linked to the reactant's concentration. This implies that the rate constant (k) can be determined by dividing the reaction rate by the concentration of the reactant. Using the data at the concentration of 0.05 M, k = 0.01 M/s ÷ 0.05 M = 0.2 s^-1.

Problem 3: A graph depicts a reaction where the concentration of the reactant decreases from 0.15 M to 0.05 M over 10 minutes. What is the rate of reaction?

Solution: The reaction rate is determined by the variation in concentration over the time change. The change in concentration is 0.05 M - 0.15 M = -0.1 M. The change in time is 10 minutes. Therefore, the rate of reaction is -0.1 M/10 min = -0.01 M/min.

Problem 4: In a rate-concentration graph, the rate of a reaction doubles when the concentration of the reactant doubles. What is the order of this reaction?

Solution: If the rate of the reaction doubles when the concentration of the reactant doubles, it means that the reaction is of first order.

Calculating an initial rate using concentration data is a fundamental component of chemical kinetics, which is the study of reaction rates. The initial rate of a chemical reaction is defined as the instantaneous rate of reaction just as it begins, i.e., at t=0. The rate of a reaction is the speed at which the concentration of reactant decreases or the concentration of product increases with time.

To calculate the initial rate, you would need concentration data of the reactants at the start of the reaction. This data can be obtained from experimental observations. The initial rate is usually calculated from the slope of the tangent to the concentration-time curve at t=0.

The rate of reaction is often expressed in terms of 'rate law'. The rate law of a chemical reaction is an equation that connects the speed of a reaction to the levels of the reactants. The rate law cannot be predicted from the stoichiometric equation and must be determined experimentally. The general form of a rate law is given by Rate = k [A]^m [B]^n, where [A] and [B] are the concentrations of the reactants, m and n are the orders of the reaction with respect to the reactants A and B respectively, and k is the rate constant.

To calculate the initial rate, you would substitute the initial concentrations of the reactants into the rate law. If the reaction is first order with respect to a reactant A and zero order with respect to a reactant B, the rate of reaction at any time would be proportional to the concentration of A only. Therefore, the initial rate would be equal to the rate constant k times the initial concentration of A.

The reaction rate at any given time is proportional to the square of the concentration of reactant A, if the reaction is second order in relation to A, and proportional to the concentration of reactant B, if it is first order in relation to B. Therefore, the initial rate would be equal to the rate constant k times the square of the initial concentration of A times the initial concentration of B.

If the reaction is zero order, the rate of reaction is constant and does not depend on the concentration of the reactants. Therefore, the initial rate is equal to the rate constant k.

It is noteworthy to mention that the initial rate is an approximation. It assumes that the concentration of the reactants does not change significantly during the initial stages of the reaction. However, this may not always be the case, especially for fast reactions. Therefore, the initial rate should be used with caution and should not be used to extrapolate the rate of reaction at later times.

Problem 1: In an experiment, the concentration of reactant A decreases from 0.50 M to 0.30 M in 10 minutes. Calculate the initial rate of reaction.

Solution: The initial rate of reaction is calculated using the formula: The rate is equal to the negative of the change in concentration divided by the change in time. The change in concentration of A = final concentration - initial concentration = 0.30 M - 0.50 M = -0.20 M

The change in time = 10 minutes = 10/60 hours = 0.167 hours (we express time in hours for the rate to be in M/h)

Rate = - (-0.20 M) / 0.167 hours = 1.20 M/h

Problem 2: The concentration of reactant B decreases from 0.80 M to 0.40 M over a 15-minute period. What is the initial rate of the reaction?

Solution: The change in concentration of B = final concentration - initial concentration = 0.40 M - 0.80 M = -0.40 M. The change in time = 15 minutes = 15/60 hours = 0.25 hours. Rate = - (-0.40 M) / 0.25 hours = 1.60 M/h

Problem 3: During a reaction, the concentration of reactant C decreases from 1.00 M to 0.60 M in 20 minutes. Calculate the initial rate of reaction.

Solution: The change in concentration of C = final concentration - initial concentration = 0.60 M - 1.00 M = -0.40 M. The change in time = 20 minutes = 20/60 hours = 0.33 hours. Rate = - (-0.40 M) / 0.33 hours = 1.21 M/h

Remember, the negative sign indicates the decrease in concentration over time. In the rate equation, we use the absolute value, which gives us the rate of disappearance of the reactant.

A rate equation is a mathematical representation that describes the rate of reaction in terms of the concentration of the reactants. It is a crucial component in the study of chemical kinetics as it provides a quantitative relation between the rate of a reaction and the concentrations of the reactants. This document seeks to construct a rate equation, elucidating the process in a systematic, comprehensive manner, akin to the style of a professional chemistry writer.

The first step in the construction of a rate equation is the identification of the reaction under examination. For the sake of this discourse, let's consider a simple hypothetical reaction where reactants A and B combine to form a product C. Thus, the reaction can be written as $A + B \rightarrow C$.

The rate of this reaction can be represented as the change in the concentration of the product with respect to time ($d[C]/dt$). However, in practical scenarios, it is often more convenient to express the rate as the reduction in the concentrations of the reactants, since they are being consumed in the process. Hence, the rate can be defined as

$$-d[A]/dt = -d[B]/dt = d[C]/dt.$$

The rate equation for this reaction can be generally expressed as Rate = $k[A]^m[B]^n$, where 'k' is the rate constant, and 'm' and 'n' are the orders of the reaction regarding A and B.

These values (m, n) reflect how the rate is affected by the concentration of each reactant. The determination of the values of 'm' and 'n' is empirical, meaning they are determined experimentally. Various methods such as the method of initial rates or the isolation method can be used. These methods involve measuring the rate of reaction at different initial concentrations of the reactants and analyzing the data to deduce the order of the reaction with respect to each reactant.

219

Once 'm' and 'n' are determined, they are substituted into the rate equation. It's important to note that the sum of 'm' and 'n' gives the overall order of the reaction. For instance, if for our hypothetical reaction, 'm' is found to be 1 and 'n' is 2, the rate equation becomes Rate = k [A] [B]^2, indicating a third order reaction.

The value of the rate constant 'k' can also be determined experimentally. It is a proportionality constant that reflects the inherent speed of the reaction. It is influenced by factors such as temperature, presence of a catalyst, and the nature of the reactants. It is also important to note that the rate equation does not provide any information about the mechanism of the reaction. It is purely an empirical equation, derived from experimental data, and does not necessarily mirror the stoichiometry of the balanced chemical equation.

Problem 1: The rate of a chemical reaction is known to be directly proportional to the concentration of reactant A and the square of the concentration of reactant B. If the rate of the reaction is represented by R, the concentration of A by [A] and the concentration of B by [B], write the rate equation for this reaction.

Solution 1: The rate equation for this reaction can be written as: R = k [A] [B]^2

Problem 2: The rate constant of a first order reaction is 0.23 s^{-1}. 1. If the initial concentration of the reactant is 0.5 M, what will be the concentration of the reactant after 10 seconds?

Solution 2: For a first order reaction, the rate equation is given by: $[A] = [A]_0 e^{(-kt)}$, where [A] is the concentration of the reactant at time t, $[A]_0$ is the initial concentration, k is the rate constant and t is the time. Substituting the given values, we get: $[A] = 0.5 e^{(-0.23 \times 10)} = 0.5 e^{-2.3} = 0.1$ M.

Problem 3: The rate of a chemical reaction quadruples when the concentration of the reactant is doubled. What is the order of the reaction?

Solution 3: If the rate of a reaction increases by a power of 2 (in this case, $4 = 2^2$) when the concentration of the reactant is doubled, the reaction is of second order.

Problem 4: The rate of a reaction was found to be 4.0 × 10^{-3} M/s when the concentration of the reactant was 0.2 M. What is the rate constant if the reaction is of first order?

Solution 4: For a first order reaction, the rate equation is given by: R = k [A], where R is the rate of the reaction, k is the rate constant and [A] is the concentration of the reactant. Rearranging for k, we get: k = R / [A]. Substituting the given values, we get: k = (4.0 × 10^{-3} M/s) / (0.2 M) = 2.0 × 10^{-2} s^{-1}.

Half-life is a key concept in the understanding of chemical reactions, specifically first-order reactions. In the realms of chemistry and physics, half-life is the time required for a quantity to reduce to half its initial value. The term is commonly used in nuclear physics to describe how quickly unstable atoms undergo radioactive decay, but it is also used in chemistry.

In the context of a first-order reaction, it is significant to understand that the half-life of such a reaction is independent of the initial concentration of the reactant. This concept may seem counter-intuitive at first, as one might expect that a higher concentration of reactants would lead to a faster reaction rate and thus a shorter half-life. However, in a first-order reaction, this is not the case.

First-order reactions are characterized by a rate law that is directly proportional to the concentration of a single reactant. The rate equation for such a reaction is typically expressed as Rate = k[A], where k is the rate constant, and [A] is the concentration of the reactant. The half-life of a first-order reaction is given by the equation $t_{1/2}$ = 0.693/k. Noticeably, this equation does not involve the concentration of the reactant, indicating that the half-life is independent of it.

To understand why this is the case, it is essential to delve into the integral rate law for a first-order reaction, which is $\ln[A] = -kt + \ln[A_0]$. The integral rate law provides a direct relationship between the concentration of the reactant and time. By solving this equation for time when [A] is equal to $[A_0]/2$ (which is the definition of half-life), we obtain the expression for the half-life of a first-order reaction, which only depends on the rate constant k and not on the initial concentration [A0].

This characteristic of first-order reactions has significant implications in various fields, from pharmacology to environmental science. For example, the elimination of a drug from the body often follows a first-order kinetics. Regardless of the initial dose, the time it takes for the drug concentration in the body to reduce by half is constant. This property allows medical professionals to predict the behavior of the drug in the body and adjust the dosage accordingly.

In environmental science, many natural degradation processes are also first-order reactions. The independence of half-life from initial concentration allows scientists to predict how long a pollutant will remain in the environment. The understanding that the half-life of a first-order reaction is independent of concentration also has implications in the field of chemical engineering. In the design of chemical reactors, knowing that the half-life of a first-order reaction is a constant allows for more accurate modeling and prediction of reaction progress over time.

Problem 1: The half-life of a first-order reaction is 20 minutes. If the initial concentration of the reactant is 0.5 M, what will be the concentration after 1 hour (60 minutes)?

Solution 1: For a first-order reaction, the half-life is independent of the concentration. This means that the concentration of the reactant will be halved every 20 minutes. So, after 20 minutes, the concentration will be 0.5/2 = 0.25 M, after 40 minutes it will be 0.25/2 = 0.125 M, and after 60 minutes it will be 0.125/2 = 0.0625 M.

Problem 2: The half-life of a first-order reaction is known to be 30 minutes. If the initial concentration of the reactant is 2 M, how much time will it take for the concentration to reduce to 0.25 M?

Solution 2: Again, because the half-life is independent of concentration for a first-order reaction, the concentration of the reactant will be halved every 30 minutes. So, from 2 M to 1 M will take 30 minutes, from 1 M to 0.5 M will take another 30 minutes, and from 0.5 M to 0.25 M will take another 30 minutes. Therefore, it will take a total of 30 + 30 + 30 = 90 minutes for the concentration to reduce to 0.25 M.

Problem 3: The first-order reaction has a half-life of 15 minutes. If the reaction starts with a concentration of 4 M, what will be the concentration after 45 minutes?

Solution 3: Given that the half-life is independent of concentration for a first-order reaction, the concentration of the reactant will be halved every 15 minutes. Therefore, after 15 minutes, the concentration will be 4/2 = 2 M. After 30 minutes, it will be 2/2 = 1 M. And after 45 minutes, it will be ½ = 0.5 M. Thus, the concentration of the reactant after 45 minutes will be 0.5 M.

The half-life of a first-order reaction is a fundamental concept in chemical kinetics that is often used in calculations. The half-life, often denoted as t½, is the time required for half of a given amount of a substance to undergo a reaction. It is particularly crucial in first-order reactions where the rate of the reaction is directly proportional to the concentration of only one reactant.

In first-order reactions, the half-life is independent of the initial concentration of the reactant, making it a unique characteristic of each reaction. This property makes it an invaluable tool in predicting the progress of a reaction over time. The formula for the half-life of a first-order reaction is t½ = 0.693/k, where k is the rate constant of the reaction.

To use this formula in calculations, the first step is to determine the rate constant, k. This can be done by performing experiments to measure the rate of the reaction under different conditions and fitting the data to the rate equation for a first-order reaction. Once the rate constant is known, the half-life can be easily calculated.

In pharmaceutical chemistry, the half-life of a drug is an important parameter in determining its dosing schedule. Drugs that undergo first-order kinetics will have a constant half-life, allowing pharmacists to predict how long a drug will stay in the body and when the next dose should be administered to maintain therapeutic levels.

In environmental chemistry, the half-life of a pollutant or a radioactive isotope provides vital information about how long it will remain in the environment. By using the half-life, scientists can estimate the time it takes for the pollutant to degrade to a safe level.

In chemical engineering, the half-life is used in the design of reactors. By knowing the half-life of a reaction, engineers can estimate how long it will take for a certain percentage of reactants to be converted into products, enabling them to design reactors of appropriate size and specify operating conditions.

In addition, half-life calculations also find use in areas such as enzyme kinetics, nuclear chemistry, and materials science. For instance, in enzyme kinetics, the half-life of an enzyme-substrate complex can help determine the rate at which a product is formed.

In nuclear chemistry, the half-life of a radioactive element is used to date geological samples and archaeological artifacts. This process, known as radiometric dating, relies on the predictable decay of certain radioactive isotopes over time.

In materials science, the half-life can refer to the time it takes for a material to lose half of its mechanical or physical properties, such as strength or elasticity. This information is valuable in predicting the lifespan and performance of materials under different conditions.

It is important to note that the use of half-life in calculations assumes that the reaction follows first-order kinetics. If the reaction is not first order, the half-life will not be constant and other methods must be used to predict the progress of the reaction.

Problem 1: The half-life of a certain first-order reaction is 5 hours. If we start with 10 grams of the reactant, how much will remain after 15 hours?

Solution: The formula for the amount of reactant remaining after a certain time in a first-order reaction is given by:

$$N = N0 * (1/2)^{(t/t1/2)}$$

where,

N is the final amount of the substance,

N0 is the initial amount,

t is the time elapsed, and

t1/2 is the half-life of the substance.

Substituting the given values:

$$N = 10 \, (1/2)^{(15/5)} = 10 \, (1/2)^3 = 10 \times 1/8 = 1.25g$$

So, after 15 hours, 1.25 grams of the reactant will remain.

Problem 2: The radioactive element X has a half-life of 4 days. If we start with 100 grams of element X, how much of it will decay after 8 days?

Solution: The amount of the substance remaining after time t can be calculated using the same formula as above:

$$N = N0 * (1/2)^{(t/t1/2)}$$

Substituting the given values:

$$N = 100 \, (1/2)^{(8/4)} = 100 \, (1/2)^2 = 100 \times ¼ = 25g$$

So, 25 grams of element X will remain after 8 days. The decayed amount is the difference between the initial and remaining amount, which is 100 - 25 = 75 grams. Therefore, 75 grams of element X will decay after 8 days.

Problem 3: If the half-life of a first-order reaction is 2 hours, how long will it take for 80% of the reactant to be consumed?

Solution: We can use the formula for the amount of reactant remaining after a certain time in a first-order reaction (N = N0 * (1/2)^(t/t1/2)) and rearrange it to solve for t:

t = t1/2 * log2(N0/N)

To find when 80% of the reactant is consumed, we want N to be 20% of N0 (since 100% - 80% = 20%).
Therefore:

t = 2 log2(1/0.2) = 2 log2(5) ≈ 4.64 hours

So, it will take approximately 4.64 hours for 80% of the reactant to be consumed.

Consider a reaction where A and B react to form product P with the overall reaction equation given by:

$$A + B \rightarrow P$$

The rate equation for this reaction would be given by the rate law: Rate = k[A]^m[B]^n where 'Rate' represents the rate of the reaction, 'k' is the rate constant, '[A]' and '[B]' are the concentrations of reactants A and B, and 'm' and 'n' are the reaction orders with respect to A and B. This rate equation implies that the speed of the reaction depends on the concentrations of the reactants and their orders.

To propose a reaction mechanism that is consistent with this rate equation, we have to consider both elementary and complex reactions. Elementary reactions are single-step processes while complex reactions involve multiple steps. The rate equation derived from the overall reaction equation may not necessarily depict the actual process at the molecular level. Therefore, we must dissect the overall reaction into a series of elementary steps that can lead to the formation of the product.

Let's consider a two-step reaction mechanism as an example:

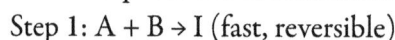

Step 1: A + B → I (fast, reversible)

Step 2: I → P (slow)

In this mechanism, A and B react quickly to form an intermediate species I, which then slowly reacts to form the product P. Because the second step is slower, it determines the overall rate of the reaction (rate-determining step).

The rate equation for the first step could be written as:

Rate = k1[A][B] - k-1[I] the rate constants for the forward and reverse reactions are represented by 'k1' and 'k-1', respectively. However, the concentration of the intermediate I is not observable and needs to be eliminated from the rate equation.

Because the first step is fast and reversible, we can assume that it quickly reaches a state of equilibrium before the second step begins. This allows us to write an equilibrium expression for the first step: K_eq = [I] / ([A][B]) = k1/k-1

From this, we can express [I] in terms of [A] and [B]:

[I] = K_eq [A][B]

Substituting this expression into the rate equation for step 2 (Rate = k2[I], where 'k2' is the rate constant for the second step), we get the rate equation for the overall reaction:

Rate = k2 K_eq [A][B]. This equation is consistent with the given rate equation if the orders of the reaction with respect to A and B are both 1 (i.e., m=n=1). This mechanism of reaction and the associated rate equation is a simple illustration and may need to be adjusted based on the actual experimental observations. For instance, if the reaction orders with respect to A and B are not 1, the mechanism may involve more than two steps or the reactants may not be involved in the rate-determining step directly.

Problem 1: Consider the reaction:

$$2NO_2(g) \rightarrow 2NO(g) + O_2(g)$$

The rate law for this reaction is Rate = k[NO2]². Suggest a reaction mechanism that is consistent with this rate law.

Solution 1: The rate law indicates that the reaction is second order with respect to NO2. This suggests that two molecules of NO2 must collide in a single step to form the products. Thus, a plausible mechanism for the reaction is:

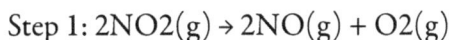

Step 1: 2NO2(g) → 2NO(g) + O2(g)

This is a bimolecular reaction that involves the collision of two NO2 molecules.

Problem 2: The reaction between hydrogen and iodine to produce hydrogen iodide can be represented by the equation: H2(g) + I2(g) → 2HI(g)

The rate law for this reaction is Rate = k[H2][I2]. Suggest a reaction mechanism that aligns with this rate law.

Solution 2: The rate law implies that the reaction is first order with respect to both H2 and I2. This suggests that one molecule of H2 must collide with one molecule of I2 in a single step to form the products. So, a possible mechanism for this reaction could be:

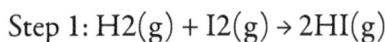

Step 1: H2(g) + I2(g) → 2HI(g)

This is a bimolecular reaction involving the collision of one H2 molecule and one I2 molecule.

Problem 3: The reaction 2NO(g) + Cl2(g) → 2NOCl(g) has the rate law Rate = k[NO]²[Cl2]. Suggest a reaction mechanism that is consistent with this rate law.

Solution 3: The rate law suggests that the rate-determining step involves the collision of two NO molecules and one Cl2 molecule. A possible mechanism that matches this rate law would be: Step 1: 2NO(g) + Cl2(g) → 2NOCl(g)

This refers to a termolecular reaction, a relatively uncommon occurrence as it necessitates the concurrent collision of three molecules. However, in this case, the rate law suggests that this is the mechanism.

The prediction of the order of a reaction based on the reaction mechanism and the rate-determining step is an integral part of chemical kinetics. Understanding this relationship aids in the development of more efficient and effective chemical processes. A reaction mechanism is a detailed description of how a chemical reaction occurs, including the steps that lead from reactants to products. Essentially, it elucidates the 'path' that molecules follow during a chemical transformation. It includes all the elementary steps, which are the simplest indivisible processes in a reaction. Each step is characterized by its molecularity, which refers to the number of molecules involved in the reaction. The rate-determining step (RDS), on the other hand, is the slowest step in a reaction mechanism. It determines the overall rate of the reaction because a reaction cannot proceed faster than its slowest step. The RDS is related to the activation energy of the reaction, which is the energy required to initiate the reaction. This step often involves the formation or consumption of intermediates, which are species that are produced in one step and consumed in another.

The reaction order is determined by adding up the powers of the reactant concentrations in the rate law equation. It defines how the rate of reaction changes with respect to the concentration of the reactants. The order of a reaction is determined experimentally and cannot be predicted solely from the balanced chemical equation. However, the order of the reaction can often be inferred from the reaction mechanism and the rate-determining step. For an elementary reaction, the order of the reaction is equal to the molecularity of the rate-determining step. For example, if the rate-determining step involves a single molecule (i.e., it's a unimolecular reaction), the reaction is first order. If it involves two molecules (a bimolecular reaction), the reaction is second order.

Consider a reaction where the slowest (rate-determining) step is bimolecular, involving a collision of two molecules to form an activated complex. In this case, the rate of the reaction would be second order overall, as the rate of formation of the activated complex (and hence the rate of the reaction) is proportional to the product of the concentrations of the two reactant molecules. In contrast, if the slowest step is unimolecular, involving the spontaneous decomposition of a single molecule, the rate of the reaction would be first order overall. This is because the rate of decomposition (and hence the rate of the reaction) is proportional to the concentration of the single reactant molecule. However, it's important to note that this rule applies strictly to elementary reactions. For complex reactions involving multiple steps, the overall reaction order may not be the same as the molecularity of the rate-determining step. The overall order of the reaction depends on the concentrations of the reactants and the rate constants of all the steps, not just the slowest one. The concentration of any intermediates is also influential, as these can impact the rates of subsequent steps.

Problem 1: Consider a reaction that proceeds according to the following mechanism:

Step 1: A + B —> C (slow)

Step 2: C + D —> E + F (fast)

What is the overall order of the reaction?

Solution: The overall rate order of the reaction is determined by the slow or rate-determining step. In this case, the slow step is A + B —> C. This step involves the reactants A and B, so the rate law is Rate = $k[A][B]$. Thus, the overall order of the reaction is 2 (first order in A and first order in B).

Problem 2: The reaction mechanism for a chemical reaction is given as follows:

Step 1: A + B —> C (fast, equilibrium)

Step 2: C + A —> D (slow)

Step 3: D + B —> E + F (fast)

What is the overall order of the reaction?

Solution: The rate-determining step is the slowest step, which is Step 2: C + A —> D. However, since C is a product of the fast equilibrium step, we can express it in terms of A and B using the equilibrium constant. Thus, our rate law becomes Rate = $k[A][B][A] = k[A]^2[B]$.

Thus, the total reaction order is 3.

Problem 3: Consider a reaction that follows the given mechanism: Step 1: 2A —> B (slow)

Step 2: B + C —> D + E (fast)

What is the overall order of the reaction?

Solution: The slow step or rate-determining step is 2A —> B. This step only involves the reactant A, so the rate law is Rate

= $k[A]^2$. Therefore, the total order of the reaction is 2.

Problem 4: The reaction mechanism for a chemical reaction is given as follows:

Step 1: A + B —> C (fast, equilibrium)

Step 2: C + B —> D (slow)

What is the overall order of the reaction?

Solution: The rate-determining step is the slowest step, which is Step 2: C + B —> D. But C is a product of the fast equilibrium step 1, so we can express it in terms of A and B. This gives us a rate law of Rate = $k[A][B]^2$.

Thus, the total reaction order is 3.

The process of deducing a rate equation from a given reaction mechanism and rate-determining step is a fundamental aspect of physical chemistry. This is an analytical approach that enables scientists to predict the speed at which a chemical reaction proceeds, and it is based on experimentally determined rate laws and theoretical reaction mechanisms.

A reaction mechanism is a proposed series of elementary steps that describe the path by which a reaction proceeds from reactants to products. Each elementary step has its own rate law that describes how the concentration of the reactants affects the rate of that particular step. The rate-determining step is the slowest step in the sequence of events, and it is this step that ultimately determines the overall rate of the reaction.

To deduce the rate equation, we first need to understand the stoichiometric coefficients of the reactants involved in the rate-determining step. The rate law for an elementary step is directly derived from its molecularity, which corresponds to the number of molecules participating in that step. For instance, for a bimolecular reaction involving two molecules A and B, the rate law would be Rate = k[A][B], where k is the rate constant. The rate-determining step is key to deducing the rate equation for the overall reaction. Since this step is the slowest, it effectively sets the pace for the entire reaction, much like how the slowest runner in a relay race determines the team's overall time. Therefore, the rate law for the rate-determining step is often the same as the rate law for the overall reaction. However, in complex reactions involving intermediates, deriving the rate equation requires additional steps. Intermediates are species that are formed in one step and consumed in a subsequent step, and they do not appear in the overall balanced equation for the reaction. Since the concentration of intermediates cannot be experimentally determined, they must be eliminated from the rate law.

This is where the steady-state approximation comes in. This concept assumes that the rate of formation and consumption of an intermediate is equal, leading to a constant concentration of the intermediate during the reaction. By setting up and solving a series of equations based on this approximation, the concentration of the intermediate can be expressed in terms of the concentrations of the stable reactants and products, thereby allowing for its elimination from the rate law.

In some cases, the pre-equilibrium approximation may be used instead. This approach assumes that the steps before the rate-determining step reach equilibrium rapidly compared to the rate-determining step itself. The equilibrium constant for these fast steps can then be used to express the concentration of the intermediates in terms of the reactants and products.

The resulting rate law, free of intermediates, describes the rate of the reaction in terms of the concentrations of the stable reactants, the order of the reaction, and the rate constant.

This provides valuable information about the reaction's kinetics, enabling scientists to predict how changes in conditions will affect the reaction rate.

Problem 1: Consider the reaction:

$2NO_2(g) \rightarrow 2NO(g) + O_2(g)$

The proposed mechanism is:

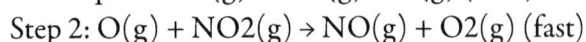

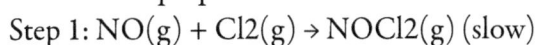

Step 1: $NO_2(g) \rightarrow NO(g) + O(g)$ (slow)

Step 2: $O(g) + NO_2(g) \rightarrow NO(g) + O_2(g)$ (fast)

Identify the rate-determining step and write the rate equation.

Solution 1: The rate-determining step is the slowest step in the reaction mechanism, which is Step 1 in this case. So, the rate equation for this reaction will be Rate = k[NO2], where k is the rate constant.

Problem 2: The reaction 2A + B → 2C + D has the following mechanism:

Step 1: A + B → C + X (slow)

Step 2: A + X → C + D (fast)

What is the rate equation?

Solution 2: The rate-determining step is Step 1. Therefore, the rate equation is Rate = k[A][B].

Problem 3: Consider the reaction:

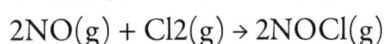

$2NO(g) + Cl_2(g) \rightarrow 2NOCl(g)$

The proposed mechanism is:

Step 1: $NO(g) + Cl_2(g) \rightarrow NOCl_2(g)$ (slow)

Step 2: $NOCl_2(g) + NO(g) \rightarrow 2NOCl(g)$ (fast)

Write the rate equation.

Solution 3: The rate-determining step is the slowest step in the reaction mechanism, which is Step 1. So, the rate equation for this reaction will be Rate = k[NO][Cl2].

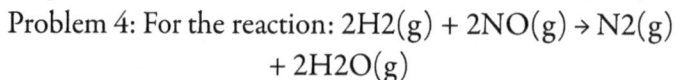

Problem 4: For the reaction: $2H_2(g) + 2NO(g) \rightarrow N_2(g) + 2H_2O(g)$

The proposed mechanism is:

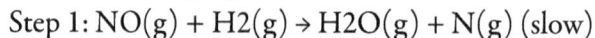

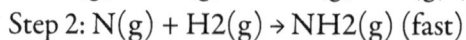

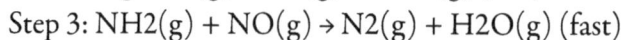

Step 1: $NO(g) + H_2(g) \rightarrow H_2O(g) + N(g)$ (slow)

Step 2: $N(g) + H_2(g) \rightarrow NH_2(g)$ (fast)

Step 3: $NH_2(g) + NO(g) \rightarrow N_2(g) + H_2O(g)$ (fast)

Deduce the rate equation.

Solution 4: Step 1, which is the slowest step, is the rate-determining step. So, the rate equation for this reaction will be

Rate = k[NO][H2].

In the realm of chemical reactions and their intricate mechanisms, identifying intermediates and catalysts plays a significant role. This is because they hold the key to understanding the course of a reaction and its speed. In order to identify an intermediate or a catalyst from a given reaction mechanism, a comprehensive understanding of the fundamental principles of chemistry is necessary.

An intermediate in a chemical reaction is a transient species which is formed during the reaction process and is consumed before the reaction reaches its final products. It is a temporary molecule that exists during the course of a reaction and is not present in either the reactants or the products. Identifying an intermediate often requires a careful analysis of the reaction steps. It involves the understanding of how reactants are transformed into products and the path they follow in this process. Catalysts, on the other hand, are substances that speed up the rate of a chemical reaction without being consumed in the process. They function by lowering the activation energy of the reaction, thereby allowing the reaction to proceed faster. Notably, catalysts are present in the start and at the end of the reaction in the same form. To identify an intermediate in a reaction mechanism, look for any species that is produced in one of the elementary steps and then consumed in a later step. These species do not appear in the overall balanced equation for the reaction, as they are formed and then consumed within the reaction mechanism. In contrast, to identify a catalyst in a reaction mechanism, look for a species that appears in the initial steps of the reaction and is regenerated in the final steps. It is noteworthy that a catalyst does not go through any permanent chemical change and hence can participate in the reaction repeatedly. For example, in the reaction between nitrogen dioxide (NO2) and carbon monoxide (CO) to produce nitrogen monoxide (NO) and carbon dioxide (CO2), the mechanism involves two steps. In the first step, two NO2 molecules react to form NO3 and NO. In the second step, NO3 reacts with CO to form NO2 and CO2. Here, NO2 can be identified as an intermediate as it is formed in the first step and consumed in the second step. Similarly, in the catalytic conversion of ozone to oxygen, chlorine acts as a catalyst. It reacts with ozone to form chlorine monoxide and oxygen. In a subsequent step, chlorine monoxide reacts with another ozone molecule to regenerate the original chlorine atom and form more oxygen. Here, chlorine remains unchanged throughout the reaction, functioning as a catalyst. Hence, to identify an intermediate or catalyst in a reaction mechanism, one must scrutinize the step-by-step progress of the reaction. It is a task that demands not just theoretical knowledge but also a deep understanding of the practical aspects of chemistry. By identifying these intermediates and catalysts, chemists can gain valuable insights into the intricacies of chemical reactions, aiding them in their continuous quest for knowledge and discovery.

Problem 1: Consider the reaction mechanism:

Step 1: $A + B \rightarrow C + D$ (slow)

Step 2: $C + B \rightarrow A + E$ (fast)

Question: Identify the intermediate in the reaction.

Solution: The intermediate in the reaction is the species that is formed in one step and consumed in another. In this case, 'C' is formed in the first step and consumed in the second step. Hence, 'C' is the intermediate.

Problem 2: Consider the reaction mechanism:

Step 1: A + B → C + D (slow)

Step 2: D + E → F + G (fast)

Question: Identify any catalyst in the reaction.

Solution: A catalyst is a substance that participates in the reaction but is regenerated by the end. In this case, 'D' is generated in the first step and consumed in the second step, hence it is not a catalyst. There are no catalysts in this reaction mechanism.

Problem 3: Consider the reaction mechanism:

Step 1: A + B → C + D

Step 2: C + E → A + F (slow)

Step 3: F + B → G + H (fast)

Question: Identify the catalyst and intermediate in the reaction.

Solution: The catalyst is a substance that participates in the reaction but is regenerated by the end. In this case, 'A' is consumed in the first step and regenerated in the second step. Hence, 'A' is the catalyst. The species that is created in one stage and used up in another is referred to as the intermediate. 'F' is formed in the second step and consumed in the third step. Hence, 'F' is the intermediate.

The step in a chemical reaction that proceeds at the slowest pace and thus sets the overall speed of the reaction is known as the rate-determining step. It is a critical concept in chemical kinetics, as it provides insights into the factors that control the speed of chemical transformations.

In a multi-step reaction, the sequence of elementary reactions (individual steps) leads to the final product. However, not all steps occur at the same speed. Some are fast, almost instantaneous, while others are slow, delaying the overall reaction. The slowest of these steps, the bottleneck in the reaction pathway, is the rate-determining step.

Determining the rate-determining step involves an understanding of the rate equation for the reaction. The rate equation is a mathematical formula connecting the speed of a reaction to the amounts of the reactants and the reaction's rate constant. The order of the reaction with respect to each reactant in the rate equation provides valuable information about the rate-determining step.

If the rate equation shows that the rate of reaction depends on the concentration of a reactant to the first power, it implies that one molecule of that reactant is involved in the rate-determining step. If the exponent is two, it indicates that either two molecules of the reactant are involved, or that the reactant is used twice in the rate-determining step.

The rate-determining step is not necessarily the first step in a reaction mechanism. For instance, in a two-step reaction, if the first step is fast and reversible, and the second step is slow, the second step is the rate-determining step. The fast first step reaches a quasi-equilibrium, and the slow second step determines the overall rate of the reaction.

It's important to note that the rate equation cannot be derived from the balanced chemical equation for the overall reaction; it must be determined experimentally. This is due to the fact that the balanced chemical equation only provides the initial and final states of the reaction, and does not provide any information about the intermediate steps.

In some cases, the rate-determining step may change as the reaction proceeds. For instance, in a reaction with a fast initial step followed by a slow step, the slow step may become faster as intermediate products build up, and a subsequent step may become the slowest, and hence the rate-determining step.

Identifying the rate-determining step is crucial for chemists, especially when designing synthetic routes in chemical manufacturing. By understanding which step is the slowest, chemists can focus their efforts on speeding up that particular step, either by adjusting the conditions of the reaction, adding a catalyst, or modifying the reactants.

In understanding complex biological systems, such as enzyme-catalyzed reactions, identifying the rate-determining steps can also offer insights into how these systems function at a molecular level. Moreover, it can help in designing drugs that can alter the rate of these reactions, providing a basis for therapeutic interventions.

Problem 1: Consider the following reaction mechanism:

Step 1: $2A \rightarrow B$ (slow)

Step 2: $B + A \rightarrow C$ (fast)

Rate equation: Rate $= k[A]^2$

Solution: In this case, the rate determining step is Step 1, because it is the slowest step and determines the overall rate of the reaction. The order of the reaction matches the stoichiometry of reactants in this slow step.

Problem 2: Given the following reaction mechanism:

Step 1: $A + B \rightarrow C$ (fast)

Step 2: $C + A \rightarrow D$ (slow)

Rate equation: Rate $= k[A][B]$

Solution: Here, the rate determining step is Step 2 because it is the slowest step. But, the rate equation doesn't match with the stoichiometry of the second step. This is because the concentration of 'C' is dependent on the first step of the reaction. Thus, the rate equation reflects both steps, but the slowest step (Step 2) is rate determining.

Problem 3: Given the following reaction mechanism:

Step 1: $A + B \rightarrow C$ (slow)

Step 2: $C + B \rightarrow D$ (fast)

Step 3: $D + A \rightarrow E$ (slow)

Rate equation: Rate $= k[A]^2[B]^2$

Solution: In this scenario, both Step 1 and Step 3 are slow steps. However, the rate equation matches the stoichiometry of reactants in Step 3. Thus, the rate determining step is Step 3.

Problem 4: Given the reaction mechanism:

Step 1: $A + B \rightarrow C + D$ (fast equilibrium)

Step 2: $C + E \rightarrow F$ (slow)

Rate equation: Rate $= k[A][B][E]$

Solution: The rate determining step is Step 2, as it is the slowest step. Although 'C' is produced in the first step, its concentration is determined by the slow step (Step 2) leading to its inclusion in the rate equation.

The relationship between temperature change and the rate constant of a chemical reaction is a fundamental principle in the field of chemical kinetics. It is well-established that an increase in temperature has a direct effect on the rate constant, and consequently, the rate of a chemical reaction. This effect is primarily due to the increased kinetic energy that molecules possess at higher temperatures, which enhances their capacity to overcome the energy barrier known as the activation energy.

The rate constant, denoted by 'k', is a proportionality factor in the rate equation that links the rate of reaction to the concentrations of the reactants. It is an intrinsic property of a reaction, which means it is independent of the concentrations of the reactants but heavily dependent on the temperature.

At a molecular level, reactions occur when particles collide with sufficient energy and proper orientation. This requisite energy is known as the activation energy (Ea), the minimum energy needed for a reaction to occur. Increasing the temperature provides more particles with this necessary energy, thus leading to an increased rate constant and subsequently, an escalated rate of reaction. This is the essence of the Collision Theory, which explains how chemical reactions occur and why reaction rates differ for different reactions.

The Arrhenius equation quantitatively describes the temperature dependence of the rate constant. According to this equation, $k = A\, e^{(-Ea/RT)}$, where A is the pre-exponential factor (also known as the frequency factor), R is the gas constant, T is the temperature in Kelvin, and e is the base of natural logarithms. The equation indicates that the rate constant increases exponentially with temperature.

When the temperature is elevated, the exponential factor in the Arrhenius equation, $e^{(-Ea/RT)}$, becomes larger because the denominator of the exponent, RT, increases. Thus, even a small rise in temperature can cause a significant increase in the rate constant. This is why reactions, especially those with high activation energies, can be drastically sped up by a slight increase in temperature. The effect of temperature on the rate constant also has significant implications for the selectivity of reactions. At higher temperatures, not only does the rate of reaction increase, but the distribution of energy among the molecules also widens. This can favor the formation of products that require higher activation energies, thus altering the selectivity of the reaction. This relationship between temperature and rate constant is ubiquitous in chemical reactions, from simple reactions in high school chemistry labs to complex reactions in industrial chemical plants. It is also exploited in biological systems. For instance, enzymes, nature's catalysts, are highly sensitive to temperature changes. Beyond an optimum temperature, the rate of enzymatic reactions declines due to the denaturation of the enzymes.

However, it's essential to note that while an increase in temperature generally accelerates reactions, it is not always desirable. For instance, in food storage and preservation, lower temperatures are preferred to slow down the rate of spoilage reactions. Similarly, in some industrial processes, excessive temperatures might lead to unwanted side reactions or degradation of the product. Conversely, in processes such as combustion in engines or industrial chemical reactions, higher temperatures are often required to ensure the reaction proceeds at a desirable rate. This temperature dependence of the rate constant is also crucial in environmental chemistry, where temperature changes can significantly impact rates of reactions that influence air quality or the greenhouse effect.

Problem 1: Q: What happens to the rate of a reaction if the temperature is increased?

A: The rate of a reaction typically increases when the temperature is increased. This is because increasing the temperature increases the kinetic energy of the particles involved in the reaction, making them move faster. As a result, the particles collide more frequently and with greater energy, which increases the likelihood of successful collisions (ones that lead to a reaction). This in turn increases the rate constant of the reaction, leading to a faster reaction rate.

Problem 2: Q: What is the effect of decreasing the temperature on a reaction's rate constant?

A: Decreasing the temperature of a reaction will cause the rate constant to decrease. This is because at lower temperatures, the particles involved in the reaction have less kinetic energy and therefore move slower. This results in less frequent and less energetic collisions between particles, reducing the likelihood of successful reactions. As a result, the rate constant decreases, slowing down the rate of reaction.

Problem 3: Q: How does temperature affect the activation energy of a reaction? A: The activation energy of a reaction is the energy barrier that must be overcome for a reaction to occur. It is not directly affected by temperature. Nonetheless, raising the temperature boosts the percentage of particles with energy exceeding the activation energy. This means that more particles can overcome the energy barrier and react, which effectively lowers the impact of the activation energy on the rate of reaction.

Problem 4: Q: If the rate constant of a reaction doubles when the temperature is increased from 25 °C to 35 °C, what can be inferred about the reaction? A: If the rate constant of a reaction doubles when the temperature is increased by 10 °C, it indicates that the reaction is sensitive to temperature changes. The particles involved in the reaction gain more kinetic energy and move faster with the rise in temperature, resulting in more frequent and more energetic collisions. This increases the likelihood of successful reactions, thus increasing the rate constant and the rate of the reaction.

Problem 5: Q: How does the concept of Arrhenius equation relate to the effect of temperature on the rate of a reaction?

A: The Arrhenius equation mathematically describes the effect of temperature on the rate constant and hence the rate of a reaction. It states that the rate constant of a reaction increases exponentially with the increase in temperature. This is because higher temperatures increase the number of particles that have energy greater than the activation energy of the reaction, leading to a higher rate of reaction.

Homogeneous and heterogeneous catalysts

A catalyst, by definition, is a material that accelerates the rate of a chemical reaction by reducing the energy barrier, while itself remaining chemically unchanged. Catalysts can be generally divided into two categories: homogeneous catalysts and heterogeneous catalysts. This distinction is based on the phase of the catalyst in relation to the reactants. Understanding the difference between these two types of catalysts is crucial to the study of chemical kinetics and catalysis. Homogeneous catalysts are in the same phase as the reactants, often as solutions, and the reactions they facilitate occur in the same phase, usually liquid. The catalyst and reactant blend at a molecular level, enabling the catalyst to readily interact with the reactant, thereby significantly increasing reaction rates. The ease of blending between the catalyst and reactants in homogeneous catalysis, resulting in high contact efficiency, is an advantage. This allows the catalyst to be used more effectively, potentially leading to increased reaction rates. Moreover, homogeneous catalysts often allow for greater selectivity and better control over the reaction conditions than heterogeneous catalysts. Conversely, heterogeneous catalysts are in a different phase from the reactants. Typically, these catalysts are solids and the reactions they facilitate take place on their surface. The reactants are typically in the gas or liquid phase. The reaction happens at the boundary between the catalyst and the reactant, implying that the surface properties of the catalyst are vital in the catalytic process. In heterogeneous catalysis, the reactants must diffuse to the catalyst's surface, react, and then the products must diffuse away. This process is intricate and can be influenced by various factors, including the size and shape of the catalyst particles, the surface area of the catalyst, and the interaction between the catalyst and the reactants. Heterogeneous catalysts have some benefits over homogeneous ones. They can be easily separated from the reaction mixture and reused, which is especially important in industrial applications. They are also generally more stable and can function under more severe conditions, including high temperatures and pressures. However, heterogeneous catalysts also have some drawbacks. One of the biggest challenges is ensuring efficient contact between the reactants and the catalyst. Unlike in homogeneous catalysis, where the catalyst and reactants are mixed at the molecular level, in heterogeneous catalysis, only the surface of the catalyst is in contact with the reactants, which can limit the catalyst's effectiveness. The differences between homogeneous and heterogeneous catalysts are not only academically important, but also have practical implications. The choice between a homogeneous or heterogeneous catalyst can significantly influence the reaction rate, efficiency, selectivity, and overall success of a chemical process. Homogeneous catalysts are often chosen in laboratory settings due to their high efficiency and selectivity. They allow for precise control over reaction conditions, making them ideal for research and development. However, their use in large-scale industrial processes can be restricted due to difficulties in separating the catalyst from the reaction mixture and the potential for catalyst loss. On the other hand, heterogeneous catalysts are commonly used in industry. They are typically more robust and can withstand harsh reaction conditions, making them suitable for continuous processes. The ease of separation and reuse also makes them a more cost-effective choice for large-scale applications.

Catalysts can either be homogeneous or heterogeneous, this is determined by if they are in the same stage as the reactants or not.

1) Homogeneous Catalysts: These are catalysts which are in the same stage as the reactants. For example, if the reactants and the catalyst are both in a gaseous or liquid state, then it's referred to as a homogeneous catalyst. An example of this is the use of sulfuric acid (liquid) in the esterification of carboxylic acids (liquid).

2) Heterogeneous Catalysts: These are catalysts that are in a different phase than the reactants. They are often solid while the reactants are either in a liquid or gaseous state. An example of this is the use of a solid catalyst in the Haber process (the conversion of nitrogen and hydrogen gases into ammonia).

Problems:

1) Problem: Why is sulfuric acid referred to as a homogeneous catalyst in the esterification of carboxylic acids?

Solution: Sulfuric acid is referred to as a homogeneous catalyst in the esterification of carboxylic acids because it's in the same phase (liquid) as the reactants. This means it can mix uniformly with the reactants, thus accelerating the reaction without altering itself or the final product.

2) Problem: Why is the catalyst used in the Haber process considered heterogeneous?

Solution: In the Haber process, the catalyst used is iron, which is a solid, while the reactants (nitrogen and hydrogen) are in the gaseous state. Since the catalyst and the reactants are in different stages, the catalyst is considered heterogeneous.

3) Problem: What would be the consequences if a catalyst used in a reaction was not in the same phase as the reactants?

Solution: If a catalyst used in a reaction was not in the same phase as the reactants, it may not interact effectively with the reactants. This could decrease the reaction rate or even stop the reaction from taking place. Therefore, it's important for the catalyst to be in the same phase as the reactants for maximum efficiency (in the case of homogeneous catalysis) or have a large surface area for contact with the reactants (in the case of heterogeneous catalysis).

Heterogeneous catalysis is a process in which catalysts and reactants exist in different phases, typically solid-gas or solid-liquid. It plays a vital role in many chemical processes, such as the Haber process. This article will focus on iron as a heterogeneous catalyst in the Haber process, paying particular attention to adsorption, bond weakening, and desorption. Adsorption, the first step in heterogeneous catalysis, involves the sticking of atoms, ions, or molecules from a gas, liquid, or dissolved solid to the catalyst's surface. In the Haber process, the reactants - nitrogen (N_2) and hydrogen (H_2) - adhere to the iron catalyst's surface. This critical step allows for increased concentration of reactants at the catalyst surface, accelerating the reaction. The nature of the adsorption can vary based on the catalyst and the reactants. In the Haber process, chemisorption takes place, wherein the reactants attach to the catalyst's surface through chemical bonds. These bonds are generally stronger than physical interactions, resulting in more efficient catalysis. Following adsorption, the reactants often experience bond weakening or bond-breaking, paving the way for the catalytic reaction. In the Haber process, the nitrogen molecule's triple bond and the hydrogen molecule's single bond weaken upon adsorption to the iron surface. Bond weakening lowers the energy needed for the reaction and enables the formation of new bonds, leading to the production of new molecules. This key process considerably reduces the activation energy barrier, increasing the reaction rate. As bond weakening continues, the reactants approach a state where they can more easily react—this state is known as a transition state. At this stage, the reactants are most likely to interact, leading to new product formation. In the Haber process, nitrogen and hydrogen react to form ammonia (NH_3), a key component in many industrial processes, including fertilizer production. Once the product forms, desorption occurs. This process involves the product molecules detaching from the catalyst's surface and being released into the surrounding environment. In the Haber process, the newly formed ammonia molecules desorb from the iron surface and are collected. Desorption is essential because it creates space on the catalyst's surface for more reactant molecules to be adsorbed and react. Without it, the catalyst would quickly become saturated with product molecules, reducing the reaction rate. The desorption process varies, depending on the product molecules' interaction with the catalyst's surface and the energy required to break these bonds. The effectiveness of a heterogeneous catalyst largely depends on its ability to facilitate adsorption, bond weakening, and desorption. In the Haber process, iron proves particularly effective due to its specific surface properties and interactions with nitrogen and hydrogen. Iron's ability to effectively adsorb nitrogen and hydrogen is a key advantage. Its surface provides a suitable environment for their adsorption and subsequent reaction. Moreover, iron's ability to weaken the bonds in the nitrogen and hydrogen molecules effectively enhances their reactivity, aiding in their conversion into ammonia. Lastly, the desorption of ammonia from the iron surface is another crucial aspect of the catalytic process. It ensures the catalyst's surface always has active sites available for additional reactions.

Problem 1: Question: What role does iron play in the Haber process for ammonia production and how does it influence the reaction as a heterogeneous catalyst?

Answer: Iron acts as a catalyst in the Haber process where nitrogen and hydrogen gases combine to form ammonia.

This catalyst enhances the reaction rate.

The process includes these steps:

Adsorption: Nitrogen and hydrogen molecules first adhere to the iron catalyst's surface.

Bond Weakening: The adsorption action weakens the bonds in the nitrogen and hydrogen molecules, making them more prone to react.

Reaction: The weakened bonds let nitrogen and hydrogen molecules react more easily together, creating ammonia molecules.

Desorption: The newly created ammonia molecules are then detached from the catalyst's surface, making room for more nitrogen and hydrogen molecules to attach and react.

Problem 2: Question: Why is adsorption crucial in heterogeneous catalysis?

Answer: Adsorption is a critical step in heterogeneous catalysis. It allows the molecules of the reactants to be in close proximity to the catalyst, easing the interaction between them. This adhesion to the catalyst's surface weakens the bonds within the reactant molecules, making them more reactive and accelerating the speed of the reaction.

Problem 3: Question: In the Haber process, why is desorption an indispensable part of the catalytic process?

Answer: Desorption is crucial as it facilitates the reaction products to depart from the catalyst's surface. This stage is vital as it creates space on the catalyst's surface for more reactant molecules to adhere and react. Without desorption, the catalyst surface would quickly become saturated with product molecules, causing the reaction to decelerate or cease.

Problem 4: Question: How does bond weakening assist the catalysis process in the Haber process?

Answer: Bond weakening in the reactant molecules (in this case, nitrogen and hydrogen in the Haber process) aids in reducing the activation energy needed for the reaction. This makes the reaction occur more readily and faster, thus increasing the efficiency of the process.

Heterogeneous catalysis is an intricate, fascinating process where a catalyst which is in a different phase from the reactants is used. In relation to exhaust gas treatment, valuable metals like palladium, platinum, and rhodium are generally employed as heterogeneous catalysts to eradicate harmful nitrogen oxides (NOx) from vehicle engines. This process is frequently known as selective catalytic reduction (SCR). The process begins with the absorption of the reactants onto the catalyst's surface, a phenomenon referred to as adsorption. Here, the reactants' molecules are drawn to and accumulate on the catalyst's surface because of its large surface area, which provides numerous active sites for attachment. Next, bond weakening happens. Here, the catalyst weakens the chemical bonds within the reactant molecules (in this case, nitrogen oxides), reducing the energy needed for the reaction to occur. This can lead to several outcomes, such as the breaking of bonds within the reactant molecules or the creation of new bonds between them and the catalyst atoms. After the weakening of the reactants' bonds, the actual reaction, typically a redox reaction, occurs. For nitrogen oxides removal, the catalyst aids in reducing the nitrogen oxides to nitrogen and oxygen. Desorption, the final step, happens after the reaction. Here, the reaction products are released from the catalyst's surface, freeing it up for further adsorption and reaction. The efficiency of the desorption process greatly influences the overall efficiency of the catalytic process. In the SCR process, ammonia or urea is often used to convert nitrogen oxides into nitrogen and water, facilitated by a catalyst, usually a mix of palladium, platinum, and rhodium. Each of these metals facilitates specific reactions, and their combination offers a comprehensive solution for removing harmful pollutants from exhaust gases. The adsorption, bond weakening, and desorption process is cyclical. Once the products are released from the catalyst's surface, it is free to absorb more reactant molecules, and the process repeats. The effectiveness of this process depends on several factors, including the catalyst's specific properties, the reactants' concentration, and the temperature and pressure conditions. Understanding heterogeneous catalysis in relation to exhaust gas treatment is crucial for building more efficient catalytic converters. Through continuous research, scientists and engineers are constantly seeking to enhance the efficiency of these essential devices.

Question 1: Can you clarify the function of platinum in the catalytic elimination of nitrogen oxides from vehicle exhaust gases? Answer: Platinum serves as a heterogeneous catalyst in the elimination of nitrogen oxides from vehicle exhaust gases. The process encompasses the following stages:

Adsorption: The nitrogen oxide molecules initially bind to the platinum catalyst's surface in a process known as adsorption.

Bond Weakening: Once adhered, the bonds inside the nitrogen oxide molecules are weakened due to the interaction with the platinum surface. Reaction: As the bonds are weakened, it permits the nitrogen oxide molecules to react with other molecules (such as oxygen or carbon monoxide) that are also adhered to the catalyst surface, resulting in the creation of less harmful substances like nitrogen and carbon dioxide.

Desorption: The final products (nitrogen and carbon dioxide) then detach from the catalyst surface, a process known as desorption, and are released into the exhaust gases.

Question 2: How does palladium function as a heterogeneous catalyst in the elimination of nitrogen oxides from exhaust gases? Answer: The process is akin to that of platinum.

Adsorption: The nitrogen oxide molecules adhere to the palladium catalyst's surface.

Bond Weakening: The bonds inside the nitrogen oxide molecules are weakened as a result of their interaction with the palladium catalyst surface.

Reaction: The weakened nitrogen oxide molecules interact with other molecules (such as oxygen or carbon monoxide) on the catalyst surface, leading to the creation of nitrogen and carbon dioxide.

Desorption: The nitrogen and carbon dioxide then detach from the catalyst surface and are emitted as less harmful exhaust gases.

Question 3: What function does rhodium serve in the catalytic elimination of nitrogen oxides from exhaust gases?

Answer: Rhodium also functions as a heterogeneous catalyst in this process.

Adsorption: Initially, the nitrogen oxide molecules adhere to the rhodium surface.

Bond Weakening: The adhered nitrogen oxide molecules experience weakened bonds due to their interaction with the rhodium catalyst surface.

Reaction: The weakened nitrogen oxide molecules then react with other adhered molecules (like oxygen or carbon monoxide) to produce nitrogen and carbon dioxide.

Desorption: Lastly, the nitrogen and carbon dioxide products detach from the rhodium surface and are emitted as exhaust gases.

The role of a homogeneous catalyst is essential to atmospheric chemistry, particularly in the reaction of sulfur dioxide (SO2) oxidation by atmospheric nitrogen oxides (NOx). Comprehending the complex stages of the catalytic cycle sheds light on the pivotal role these catalysts play in the chemical dynamics of our atmosphere. The catalyst begins its function in the initial phase by merging with one of the reactants - sulfur dioxide in this case. This interaction results in the creation of an intermediate complex. Notably, the catalyst remains chemically unchanged throughout the reaction.

Homogeneous catalysis is typically identified by its single-phase nature, where both the catalyst and reactants coexist in the same phase. This leads to a consistent interaction between the catalyst and the reactant molecules, allowing for accelerated reaction rates due to the increased successful collision frequency.

Regarding the atmospheric oxidation of sulfur dioxide, nitrogen oxides serve as catalysts, enabling sulfur dioxide's transformation into sulfur trioxide (SO3). The reaction can be depicted as: $2SO_2 (g) + O_2 (g) \rightarrow 2SO_3 (g)$.

In this reaction, the nitrogen oxides expedite the fusion of sulfur dioxide and oxygen, accelerating the sulfur trioxide conversion rate. The catalyst, nitrogen oxides in this case, participates in the reaction's initial step, forming an intermediate complex with sulfur dioxide molecules.

This intermediate complex then reacts with the available oxygen in the atmosphere, resulting in sulfur trioxide formation. The nitrogen oxides are subsequently released from the complex and can be recycled in the subsequent reaction cycle.

After fulfilling their role, the nitrogen oxides are regenerated and revert back to their original form, prepared for the next reaction cycles. This regeneration process is known as catalyst regeneration and is a vital part of the catalytic reaction.

Homogeneous catalysis's benefit, as demonstrated in this reaction, lies in the high selectivity and efficiency it brings to the reaction process. The catalyst and the reactants being in the same phase significantly enhances the reaction's efficiency.

The atmospheric oxidation of sulfur dioxide is a crucial process in acid rain formation, a significant environmental issue. Understanding nitrogen oxides' role as catalysts in this process can aid in crafting strategies to control air pollution.

The nitrogen oxide catalysts are not depleted in the reaction, a defining feature of catalysts. Instead, they are recycled and reused in the reaction process, enabling the continuous conversion of sulfur dioxide into sulfur trioxide.

The homogeneous nature of the catalysts ensures effective interaction with the reactants, thereby enhancing the reaction rate. The reactants and catalysts being in the same phase allows for optimal interaction and reaction efficiency.

The catalyst's reformation in the later step is a notable aspect of the catalytic process. After facilitating the reaction, the catalyst is regenerated and can participate in the following reaction cycles. This characteristic contributes to the catalytic process's efficiency and sustainability.

The unique feature of homogeneous catalysis is its single-phase nature, allowing for optimal interaction between the reactants and the catalyst. This characteristic significantly increases the reaction's efficiency and speed, underscoring homogeneous catalysis's critical role in many chemical reactions.

Problem 1: Can you explain the function of nitrogen oxides in the atmosphere in the oxidation of sulfur dioxide?

Solution: Nitrogen oxides in the atmosphere serve as a homogeneous catalyst in the oxidation of sulfur dioxide. They initially interact with sulfur dioxide to form an intermediate compound. This compound later reacts with another reactant, typically atmospheric oxygen, resulting in sulfur trioxide and the reformation of nitrogen oxide. This process is a key factor in the creation of acid rain. Since nitrogen oxides are not depleted in the reaction, they can persist as a catalyst.

Problem 2: Why are nitrogen oxides in the atmosphere deemed as homogeneous catalysts in the oxidation of sulfur dioxide?

Solution: A homogeneous catalyst is a catalyst that is in the same physical state (solid, liquid, or gas) as the reactants. In this scenario, both nitrogen oxides and sulfur dioxide exist in the gaseous state, hence they are referred to as homogeneous catalysts. Furthermore, they are not used up during the reaction, as they are involved in the initial stage and are reformed in the final stage.

Problem 3: What is the intermediate compound that is produced when nitrogen oxides in the atmosphere interact with sulfur dioxide?

Solution: The intermediate compound that forms when nitrogen dioxide (one type of nitrogen oxide in the atmosphere) interacts with sulfur dioxide is nitrosylsulfuric acid. This is an unstable compound that further interacts with oxygen to create sulfur trioxide and reconstruct nitrogen dioxide, the catalyst.

Problem 4: What is the importance of the reformation of the catalyst in the oxidation of sulfur dioxide?

Solution: The reformation of the catalyst (nitrogen oxides in the atmosphere) in the final stage of the reaction process is important because it means that the catalyst is not used up in the reaction. This allows the catalyst to keep promoting the reaction, enabling the transformation of more sulfur dioxide into sulfur trioxide.

Problem 5: How does the oxidation of sulfur dioxide lead to the creation of acid rain?

Solution: The oxidation of sulfur dioxide, facilitated by nitrogen oxides in the atmosphere, results in sulfur trioxide. When sulfur trioxide encounters water in the atmosphere, it forms sulfuric acid, which then descends to the earth as acid rain.

Homogeneous catalysis is a method of catalysis in which the catalyst and the reactants exist in the same phase. This process is crucial in many chemical reactions, including the I-/S2O8 2- reaction where iron ions (Fe2+ or Fe3+) serve as the catalyst. In this form of catalytic reaction, the catalyst, reactants, and products all exist together in the same phase, typically in a solution. The uniformity of this system allows for the effective transfer of energy and matter, resulting in an accelerated reaction rate. The reaction involving iodide ions (I-) and persulfate ions (S2O8 2-) is an excellent example of this process. The catalyst (Fe2+ or Fe3+) initiates the reaction through a series of steps, first interacting with the reactants to form an activated complex. In the I-/S2O8 2- reaction, the Fe2+ or Fe3+ ions react with the persulfate ions to create an intermediate complex. This complex is highly reactive and possesses a lower activation energy than the uncatalyzed reaction. This lower activation energy increases the reaction rate, allowing the reaction to move forward more quickly. The intermediate complex then interacts with the iodide ions, resulting in the creation of iodine and sulfate ions. Meanwhile, the catalyst (Fe2+ or Fe3+) is reduced. This is the product formation stage, where the reactants are converted into the desired products. An important feature of this stage is that the catalyst is not consumed but is instead transformed. After the product formation stage, the catalyst goes through a regeneration stage. Here, the reduced iron ions are re-oxidized, reverting back to their original state (Fe2+ or Fe3+). The catalyst's ability to be reused without being consumed is a key aspect of homogeneous catalysis. The re-oxidation of the catalyst is usually assisted by another reactant in the system. In the I-/S2O8 2- reaction, the iodine formed in the product formation stage can re-oxidize the Fe2+ ions back to Fe3+ ions, thus completing the catalytic cycle. This repeating process, involving the use, modification, and reformation of the catalyst, is referred to as the catalytic cycle. This cycle can happen numerous times within a reaction, significantly improving the speed and efficiency of the reaction. A major benefit of homogeneous catalysis, as shown in this reaction, is the ability to accurately control the reaction conditions. By adjusting the concentration of the catalyst, the reaction rate can be effectively controlled. Furthermore, homogeneous catalysts typically exhibit high selectivity, resulting in the production of specific products with minimal by-products. This is particularly useful in industrial processes, where yield and product quality are critical. However, it should be noted that while the catalyst is not consumed during the reaction, it can be deactivated over time due to various factors such as temperature changes, pressure changes, or the presence of impurities. Therefore, maintaining optimal reaction conditions is essential for the catalyst's consistent performance.

Problem 1: What function do Fe^{2+} or Fe^{3+} ions play as a homogeneous catalyst in the $I^-/S_2O_8^{2-}$ reaction?

Solution: The interaction between iodide ions (I^-) and persulfate ions ($S_2O_8^{2-}$) is a redox reaction, which can be sluggish at ambient temperature. A catalyst is employed to expedite this reaction, where the iron(II) or iron(III) ions act as the homogeneous catalyst.

In the initial phase, Fe^{2+} or Fe^{3+} ions interact with $S_2O_8^{2-}$ ions, resulting in Fe^{3+} or Fe^{2+} ions and SO_4^{2-} ions. This reaction occurs quickly. In the subsequent phase, Fe^{3+} or Fe^{2+} ions interact with I^- ions to restore Fe^{2+} or Fe^{3+} ions and generate I_2. This reaction is slower.

Overall, the catalyst (Fe^{2+} or Fe^{3+}) is utilized in the first phase and restored in the second phase, thus it is not depleted in the reaction. This is a typical trait of a catalyst.

Problem 2: Can you define a homogeneous catalyst and its significance in the $I^-/S_2O_8^{2-}$ reaction?

Solution: A homogeneous catalyst refers to a catalyst that shares the same phase as the reactants. In this instance, Fe^{2+} or Fe^{3+} ions are in the aqueous phase, similar to the I^- and $S_2O_8^{2-}$ ions.

Homogeneous catalysts are crucial as they can directly interact with the reactants, thereby enhancing the reaction rate. In the $I^-/S_2O_8^{2-}$ reaction, the Fe^{2+} or Fe^{3+} ions boost the reaction rate by offering an alternative reaction pathway with a reduced activation energy.

Problem 3: Is it feasible for the $I^-/S_2O_8^{2-}$ reaction to occur without a catalyst? If it is possible, what impact would it have on the reaction rate?

Solution: Yes, the $I^-/S_2O_8^{2-}$ reaction can occur without a catalyst. Nevertheless, the absence of a catalyst would significantly slow down the reaction rate. This is because the catalyst offers an alternative reaction pathway with a reduced activation energy, thereby speeding up the reaction. Without the catalyst, the reactants need to overcome a higher activation energy, which slows the reaction.

Group 2

Part 2: Inorganic Chemistry
Group 2 metals properties similarities and trends

Thermal stability, the ability of compounds to resist heat-induced breakdown or decomposition, is a significant focus in inorganic and physical chemistry, particularly with regards to nitrates and carbonates. The thermal stability of these compounds, when exposed to heat, varies across the periodic table and is influenced by factors like the cation size, charge density, and the anion's polarizability. The ionic radius plays a key role in the thermic stability of nitrates and carbonates. Generally, compounds with larger cation's ionic radius tend to be more thermally stable. This stability is due to the inverse relationship between the cation size and its charge density. Charge density, the ratio of a cation's charge to its volume, is greater in smaller cations than larger ones with the same charge. The high charge density of smaller cations results in a strong electrostatic attraction with the compound's anion. This strong attraction makes it challenging for the compound to break down under heat, thereby reducing its thermal stability.

The ionic radius also impacts the polarizability of the large anion, another factor affecting the thermal stability of nitrates and carbonates. Polarizability describes how easily an atom or ion's electron cloud can be distorted by an external electric field, such as that created by a cation. Large anions like nitrate and carbonate ions are more polarizable due to their larger electron clouds. When a small ionic radius and high charge density cation is present, it can more effectively polarize the large anion. This polarization weakens the anion's bonds, making the compound more prone to heat-induced decomposition. When considering nitrates, moving down the periodic table group increases the cationic radius, reducing the cation's polarizing power, and making nitrates more thermally stable. The thermal stability of carbonates follows a similar trend. With carbonates, moving down the group in the periodic table makes the carbonate ion less polarizable due to the cation's increasing size and decreasing charge density. This reduced polarizability makes the carbonate ion less susceptible to heat-induced decomposition, thus increasing the compound's thermal stability. Nevertheless, other factors like lattice energy and hydration enthalpy also influence the thermal stability of nitrates and carbonates. The interaction of these factors leads to complex trends and exceptions, which are still a topic of current research. Studying the thermal stability of nitrates and carbonates offers valuable insights into their properties, with both theoretical and practical implications. For instance, this understanding can aid in creating heat-resistant materials or predicting these compounds' behavior under varying conditions. Furthermore, understanding the thermal stability of nitrates and carbonates has environmental chemistry implications. Many nitrates and carbonates are crucial in fertilizers, and their thermal stability can influence soil chemistry and crop growth.

1: Can you illustrate the thermal stability trend of Group 2 metal nitrates and clarify the reason behind this?
Answer: As you move down the Group 2 metal nitrates, thermal stability increases. This is attributed to the growth in the ionic size of the metal cation. As this size expands, the cation's polarizing power reduces, leading to less distortion of the nitrate ion's electron cloud and making it harder to decompose, thereby enhancing thermal stability.

2: How does the ionic radius influence the polarization of the large anion in a metal carbonate?

Answer: The more extensive the ionic radius of the cation in a metal carbonate, the less it polarizes. This is because the positive charge of larger ions is dispersed over an increased area, reducing their capacity to distort the anion's electron cloud. Consequently, larger cations lead to less polarization of the large carbonate anion, increasing the carbonate's thermal stability.

3: What causes the thermal stability of Group 1 metal carbonates to be higher than that of Group 2 metal carbonates?

Answer: Group 1 metal cations have a larger ionic radius compared to Group 2 metal cations. This means that Group 1 cations possess less polarizing power than Group 2 cations, leading to lesser distortion of the carbonate anion's electron cloud. This complexity in decomposition increases the thermal stability of Group 1 metal carbonates in comparison to Group 2 metal carbonates.

4: Between magnesium nitrate and calcium nitrate, which one has a higher thermal stability, and why?

Answer: Calcium nitrate exhibits a higher thermal stability than magnesium nitrate. The reason for this is that calcium has a larger ionic radius than magnesium, which decreases its polarizing power. Consequently, the nitrate ion in calcium nitrate is less polarized and less prone to decompose compared to that in magnesium nitrate.

5: Why does the thermal stability of Group 2 metal carbonates increase as you move down the group?

Answer: The thermal stability of Group 2 metal carbonates increases as you move down the group, and this is due to the expansion of the metal cation's ionic size. As the ionic size grows, the cation's polarizing power reduces, resulting in less distortion of the carbonate ion's electron cloud. This makes decomposition more difficult and thus enhances thermal stability.

Two primary factors influence the solubility of compounds and the enthalpy change of solution, or $\Delta H^\diamond$ sol: the enthalpy change associated with hydration, and the compound's lattice energy. This discussion will explore how these two elements impact the solubility and $\Delta H^\diamond$ sol of hydroxides and sulfates, with a focus on qualitative changes. The enthalpy change of solution represents the total energy change that occurs when a solute dissolves in a solvent. This change is determined by a balance between the energy needed to break the ionic lattice of the solute (known as lattice energy) and the energy released upon the ions' hydration (enthalpy change of hydration). Lattice energy is the energy needed to separate one mole of a solid ionic compound into its gaseous ions, and it provides a measure of the strength of the forces within an ionic compound. The stronger the lattice energy, the stronger the forces binding the ions, and thus, the compound is less likely to be soluble. Conversely, the enthalpy change of hydration is the energy change that occurs when one mole of gaseous ions is fully surrounded by water molecules. This process releases energy, making it exothermic. The greater the enthalpy change of hydration, the more energy is released upon ion hydration, making the compound more likely to be soluble. Hydroxides and sulfates are both ionic compounds, consisting of positive and negative ions. However, the ions' size and charge can influence their solubility and $\Delta H^\diamond$ sol. In hydroxides, as we descend a group in the periodic table, the ions increase in size, reducing the lattice energy due to the increased ion separation. This reduction means less energy is needed to separate the ions, making the hydroxides more soluble. However, the enthalpy change of hydration decreases as the ions grow larger, resulting in less energy released upon ion hydration. Despite this, hydroxide solubility increases down the group, from lithium to cesium. This suggests that the decrease in lattice energy outweighs the decrease in hydration enthalpy change, resulting in increased solubility overall. Now, turning to sulfates, their solubility decreases as we move down Group 2 in the periodic table, from beryllium to barium. This is because the decrease in lattice energy is less than the decrease in hydration enthalpy change, resulting in a net reduction in solubility. This is explained by the sulfate ion, SO_4^{2-}. Its size is much larger than Group 2 metal ions and doesn't change significantly as you move down the group. Consequently, the decrease in lattice energy is slight. However, the Group 2 metal ions increase in size down the group, leading to a substantial decrease in hydration enthalpy change. Moreover, both hydroxides' and sulfates' $\Delta H\diamond$ sol are affected by the balance between lattice energy and hydration enthalpy change. If the energy released during hydration exceeds the energy needed to break the lattice, the $\Delta H\diamond$ sol will be negative, indicating an exothermic process. On the other hand, if breaking the lattice requires more energy than what is released during hydration, the $\Delta H\diamond$ sol will be positive, indicating an endothermic process.

Problem 1: Can you outline and elaborate on the solubility progression of Group 2 hydroxides?

Solution: The solubility of Group 2 hydroxides heightens when you progress down the group. This occurs as the lattice enthalpy, which is the energy needed to create one mole of an ionic compound from its gaseous ions, diminishes quicker than the hydration enthalpy. As the ionic size grows larger down the group, the ions get further apart, leading to a weaker electrostatic force and thus a lesser lattice enthalpy. Conversely, the hydration enthalpy, which is the energy emitted when gaseous ions are encircled by water molecules, decreases at a steadier pace. Consequently, the resultant change in enthalpy of the solution becomes more exothermic, enhancing solubility.

Problem 2: Can you outline and elaborate on the solubility progression of Group 2 sulfates?

Solution: The solubility of Group 2 sulfates lessens as you progress down the group. This happens because the hydration enthalpy diminishes at a quicker pace than the lattice enthalpy. As the ionic size grows larger, the hydration enthalpy decreases significantly, as the water molecules cannot encircle the larger ions as effectively, leading to less energy being emitted. However, the lattice enthalpy diminishes at a steadier pace. Consequently, the resultant change in enthalpy of the solution becomes less exothermic, thus reducing solubility.

Problem 3: How do the relative magnitudes of the enthalpy change of hydration and the lattice energy influence the $\Delta H\diamond$sol?

Solution: The enthalpy change of the solution, $\Delta H\diamond$sol, is influenced by the difference between the enthalpy change of hydration and the lattice energy. If the hydration enthalpy exceeds the lattice energy, the $\Delta H\diamond$sol will be exothermic, making the compound soluble. Conversely, if the lattice energy exceeds, the $\Delta H\diamond$sol will be endothermic, making the compound less soluble.

Problem 4: Why is the solubility of magnesium hydroxide lesser than that of barium hydroxide?

Solution: Magnesium hydroxide has smaller ions compared to barium hydroxide, which results in a higher lattice energy for the former. Even though the hydration enthalpy of magnesium ions is also higher, the difference between the lattice energy and hydration enthalpy is less exothermic for magnesium hydroxide than for barium hydroxide. Therefore, magnesium hydroxide exhibits lesser solubility.

Chemistry of Transition Elements

Properties of first-row transition elements

Transition metals, also known as transition elements, are a crucial set of 38 elements found in the d-block of the periodic table, spanning from group 3 to group 12. They are recognized for their wide-ranging chemical behavior, high melting points, electrical conductivity, and malleability. A defining characteristic of transition metals, setting them apart from other elements, is their ability to form stable ions with incomplete d orbitals. The d-block, which houses transition metals, signifies elements where the d-orbitals are gradually filled across four periods. The filling of these orbitals results in unique features of transition metals such as their magnetic properties and coloration, often due to unpaired d electrons. However, it is the incomplete filling of d orbitals in one or more of their stable ions that truly characterizes a transition element. Transition metals, in their ground state or when ionized, have an incomplete set of d-orbitals, signifying that not all of their five d-orbitals are fully filled with electrons. This attribute is responsible for the variable oxidation states seen in transition metals. The oxidation state denotes the level of oxidation or reduction of an atom in a chemical compound, which then determines its ability to gain, lose or share electrons. Transition metals' flexible oxidation states stem from their ability to shed different numbers of d electrons, creating stable, partially filled d-orbital ions. This distinctive ability to create stable ions with incomplete d orbitals equips transition metals with impressive catalytic properties. Catalysts are substances that accelerate chemical reactions without being used up in the process. Transition metals and their compounds are extensively employed as catalysts in various industrial chemical reactions, thanks to the availability of d electrons for bond formation. Another fascinating feature of transition metals forming ions with incomplete d orbitals is the broad range of complex ions they can create. Complex ions are ions with a central metal ion connected to one or more molecules or ions. The number and arrangement of bonds in these complex ions often depend on the availability and distribution of d electrons. Moreover, the creation of stable ions with incomplete d orbitals contributes to the transition metals' capacity to form colored compounds. The color of these compounds originates from the absorption of light energy by the d electrons, which are then promoted to higher energy levels. The specific wavelengths of light absorbed are determined by the energy difference between the d orbitals, resulting in the compound's distinctive color. The characteristic of transition metals forming incomplete d orbital ions also contributes to their magnetic properties. The unpaired electrons in these ions can align in a magnetic field, causing the metal to be magnetically attracted. This property, called paramagnetism, is stronger in transition metals due to the larger number of unpaired electrons. The variability in coordination numbers and geometries in the complexes of transition metals is another characteristic attributed to the creation of ions with incomplete d orbitals. This variability paves the way for the rich diversity of transition metal chemistry, with an extensive range of potential structures and reactivities. Finally, the ability of transition metals to create stable ions with incomplete d orbitals also contributes to their critical roles in biological systems. Many transition metals are essential nutrients, serving as active sites in enzymes or as structural components of proteins.

Problem 1: Can you identify the element among Copper (Cu), Zinc (Zinc), Scandium (Sc), and Silver (Ag) that is not classified as a transition element and explain the reason?

Solution: Zinc does not fall under the category of transition elements. Despite its position in the d-block, Zinc does not create ions with incomplete d orbitals. It exclusively creates Zn2+ ions with a completely filled 3d10 configuration.

Problem 2: Can you explain why Iron is capable of forming both Fe2+ and Fe3+ ions while Scandium can only form Sc3+ ions? Solution: Iron, with its electron configuration of [Ar] 3d6 4s2, can lose 2 electrons to create Fe2+ with a stable configuration of [Ar] 3d6. It can also lose 3 electrons to form Fe3+ with a stable configuration of [Ar] 3d5. Both these configurations have incomplete d orbitals. In contrast, Scandium, with an electron configuration of [Ar] 3d1 4s2, can only lose 3 electrons to create Sc3+ with a stable configuration of [Ar]. This is because losing just 1 or 2 electrons would result in a less stable configuration.

Problem 3: Why are compounds formed by transition elements usually colored?

Solution: The compounds of transition elements are typically colored due to the presence of incompletely filled d orbitals. The d-d transitions involve the movement of an electron from one d-orbital to another, a process that absorbs energy corresponding to certain wavelengths of light, making the compounds appear colored.

Problem 4: Why are transition elements capable of showing variable oxidation states?

Solution: Transition elements can display variable oxidation states because of the presence of unpaired electrons in the d orbitals. Depending on the compound or reaction, these elements can lose varying numbers of electrons from the s and d orbitals, resulting in multiple oxidation states.

Problem 5: What makes transition metals often act as catalysts?

Solution: Transition metals frequently serve as catalysts because their incompletely filled d-orbitals enable them to easily donate and accept electrons. This ability to alter oxidation states makes them perfect for accelerating reactions by aiding in the creation and breaking of bonds.

The 3dxy orbital is one of the five 3d orbitals present in a d subshell. The term "3d" signifies the third energy level in an atom (represented by the number 3) and the type of orbital (represented by the letter d), which resembles a four-leaf clover. The "xy" subscript represents the orbital's orientation in a three-dimensional space. This particular orbital is located in the xy plane, perpendicular to the z-axis, and consists of two lobes that appear like a four-leaf clover when observed from the z-axis. Each lobe is situated between the x and y-axis with the orbital's center at the origin. Significantly, the lobes do not touch the x or y-axis but are located in the quadrants created by these axes. A nodal plane, a zone with zero electron probability, divides the lobes in the 3dxy orbital. This plane runs through the atom's nucleus along the z-axis. Hence, the electron density in this orbital is primarily found in the lobes, with no chance of locating an electron along the z-axis. On the other hand, the 3dz2 orbital, another 3d orbital variant, has a unique shape and orientation. The "z2" in its name indicates that this orbital is aligned with the z-axis, making it the only d-orbital with this alignment. The 3dz2 orbital's shape is distinctive, including two lobes along the z-axis and a doughnut or torus-shaped region in the middle, within the xy plane. This shape is often compared to a dumbbell with a doughnut around its midpoint. In contrast to the 3dxy orbital, the lobes of the 3dz2 orbital are directly on the z-axis, both above and below the xy plane. The doughnut region in the 3dz2 orbital, much like the lobes, is an area of electron density. However, a nodal plane in the xy plane separates it from the lobes, signifying no chance of locating an electron within this plane. When comparing the two orbitals, the 3dxy and 3dz2 orbitals have noticeable differences in their shapes and orientations. The 3dxy orbital is entirely in the xy plane, with lobes between the axes, while the 3dz2 orbital extends along the z-axis, with a doughnut-shaped region in the xy plane. Despite their contrasting features, both orbitals are vital for atomic structure and bonding. Their unique shapes and orientations enable them to accommodate electrons in various energy states, adding to the complexity and diversity of chemical behavior across different elements.

Problem 1: Question: Describe the shape of a 3dxy orbital and a 3dz2 orbital.

Solution: The 3dxy orbital consists of four lobes that lie in the xy plane. The lobes are located between the x and y axes, not along them. Each lobe is shaped like a teardrop, with the pointy end of the teardrop pointing towards the origin.

The 3dz2 orbital consists of two lobes that lie along the z-axis, and a doughnut-shaped region in the xy plane. The two lobes are parallel to the z-axis, and the doughnut-shaped region is perpendicular to it. The shape is sometimes described as a "dumbbell with a doughnut."

Problem 2: Question: What is the difference between a 3dxy orbital and a 3dz2 orbital?

Solution: The main difference between a 3dxy orbital and a 3dz2 orbital is the orientation of the lobes. For a 3dxy orbital, the lobes are positioned in between the x and y axes. On the other hand, for a 3dz2 orbital, the lobes are along the z-axis, with a doughnut-shaped region in the xy plane.

Problem 3: Question: Which atomic orbitals are involved in the formation of a sigma bond?

Solution: Sigma bonds are formed by the end-to-end overlapping of atomic orbitals. This can involve s orbitals (as in an H2 molecule), p orbitals (as in a Cl2 molecule), or an s and a p orbital (as in an HCl molecule). In general, any pair of atomic orbitals can form a sigma bond as long as they overlap along the line connecting the two atomic nuclei.

Problem 4: Question: The 3d orbitals are often depicted as having four lobes, but the 3dz2 orbital has an additional "doughnut" shape in the middle. Why is this?

Solution: The shape of the 3dz2 orbital reflects the wave function for this particular orbital. In quantum mechanics, each electron in an atom is described by a wave function, which determines the probability distribution of the electron's location. The 3dz2 orbital has a unique wave function that results in a probability distribution with two lobes along the z-axis and a doughnut-shaped region in the xy plane. This shape reflects the areas in space where there is the highest probability of finding the electron.

Also known as transition metals, transition elements are a crucial group of elements in the periodic table found in the d-block. What sets them apart is their unique ability to exhibit variable oxidation states, a characteristic not commonly seen in other element groups. This makes them particularly valuable and intriguing in the field of chemistry. Understanding oxidation states, or oxidation numbers, is fundamental to comprehend the behavior of transition elements. Simply put, the oxidation state of an atom is the charge it would possess if all its bonds to different atoms were completely ionic. Owing to their unique electronic structure, transition elements often showcase varying theoretical charges. The variable oxidation states seen in transition elements are a direct consequence of their electron configuration. These metals possess partially filled d-orbitals capable of donating or accepting electrons, a feature that starkly contrasts with s-block elements, which typically have a constant oxidation state due to their valence electrons being in s-orbitals. The ability to display different oxidation states broadens the variety of chemical reactions and compounds that transition elements can participate in. Their variable oxidation states enable them to form a multitude of complex ions and compounds, making them adaptable in several chemical processes and applications. For example, iron, a transition metal, often exists in two oxidation states: +2 and +3. These states allow iron to engage in numerous chemical reactions, hence playing a vital role in biological processes such as oxygen transport in hemoglobin and myoglobin. Manganese, another transition metal, shows an even broader range of oxidation states, from +2 to +7. This variability is crucial in many industrial uses of manganese, including the production of various alloys. The color of transition metals is also influenced by their variable oxidation states. The different oxidation states can absorb varying light wavelengths, resulting in compounds that are vivid and color-rich. Many transition metals and their compounds are used as pigments due to this reason. For instance, chromium in the +3 oxidation state is a key component of the green pigment in emeralds. The variable oxidation states of transition metals also play a significant role in catalysis, where catalysts accelerate reactions without getting consumed. Transition metals are often employed as catalysts because their variable oxidation states enable them to facilitate reactions in unique ways. Furthermore, the variable oxidation states of transition metals are essential to their role in redox reactions, which involve electron transfers between chemical species. As transition metals can exist in different oxidation states, they can readily donate or accept electrons, making them central to numerous redox reactions. In biochemistry, many enzymes include transition metals with variable oxidation states at their core. These metals can alternate between different oxidation states, allowing them to catalyze a broad range of biochemical reactions. The variable oxidation states of transition metals also enable them to form stable coordination compounds, which consist of a central transition metal atom or ion surrounded by a set of ligands, which can be ions or molecules. The study of transition metals and their variable oxidation states is a vast and intriguing field in inorganic chemistry. The complex behavior these elements display due to their variable oxidation states offers a rich platform for research and exploration.

Problem 1: Question: Can you identify the oxidation state of manganese in KMnO4?

Solution: In the compound KMnO4, the oxidation state of oxygen is -2 and that of potassium is +1. Given the formula, the sum of the oxidation states should equal 0. Therefore, if we designate the oxidation state of Mn as x, we can write the equation: $(+1) + x + 4(-2) = 0$. After solving for x, we find out that x = +7. Hence, the oxidation state of manganese in KMnO4 is +7.

Problem 2: Question: Can you provide an explanation for the variable oxidation states of transition metals?

Solution: The variable oxidation states of transition metals are due to their partially filled d-orbital which allows them to donate or accept a varied number of electrons, leading to different oxidation states. For instance, iron can have either +2 or +3 oxidation states as it can lose 2 or 3 electrons from its 3d orbital.

Problem 3: Question: Can you determine the oxidation state of iron in Fe2O3?

Solution: In the compound Fe2O3, the oxidation state of oxygen is -2. If we assign x as the oxidation state of Fe, then we can write: $2x + 3(-2) = 0$. After solving the equation, we find x = +3. Hence, the oxidation state of iron in Fe2O3 is +3.

Problem 4: Question: Can you explain why the oxidation states of transition metals influence their color?

Solution: The color of transition metals is influenced by their oxidation states as different oxidation states correspond to different electron configurations. When light is absorbed, electrons are elevated to higher energy levels, then drop back down, emitting light. The specific wavelengths of light emitted are determined by the energy differences between levels, which are influenced by the electron configuration. Therefore, different oxidation states yield different colors.

The group of elements known as transition elements or transition metals, which are found in the Periodic Table, possess unique chemical and physical characteristics. They can serve as catalysts in different chemical reactions due to their distinct properties. Catalysts are substances that lower the energy required to start a chemical reaction and speed up the reaction rate without being used up in the process. This unique function of transition elements is due to their specific attributes. The electronic makeup of transition metals significantly influences their catalytic nature. These elements have d orbitals that are only partially filled, which lets them easily accept or donate electrons. This ability to shift between oxidation states allows these metals to form complex compounds with other elements and molecules, enhancing their effectiveness as catalysts. Transition elements also have high melting and boiling points due to the strong metallic bonding that results from the delocalized electrons from the d orbitals. This makes them suitable for use in high-temperature industrial procedures without falling apart or melting, thereby maintaining their effectiveness as catalysts. For instance, Iron (Fe), a transition metal, serves as a catalyst in the Haber process, which produces ammonia from nitrogen and hydrogen. It accelerates the reaction without being used up, demonstrating a typical function of a catalyst. Nickel (Ni) is another example, used in the hydrogenation of alkenes. Nickel acts as a catalyst by providing a surface for the reactants to bind to, thus increasing the reaction rate. This is a typical example of heterogeneous catalysis, where the catalyst is in a different phase (solid) from the reactants (gaseous). Other transition metals like Platinum (Pt) and Palladium (Pd) are employed as catalysts in catalytic converters in vehicles. They expedite the reaction that converts harmful gases like carbon monoxide and nitrogen oxides into less harmful substances such as nitrogen, carbon dioxide, and water. Transition metals' catalytic properties extend beyond simple reactions. They also play a vital role in more complex biological systems. For example, the transition metal Manganese (Mn) is a critical part of the enzyme arginase, which catalyzes the final step in the urea cycle, a biochemical pathway in mammals that removes harmful ammonia from the body. Another transition element, Zinc (Zn), is found in the enzyme carbonic anhydrase's active site. This enzyme catalyzes the reaction that converts carbon dioxide and water into bicarbonate and protons, a crucial reaction for maintaining our blood and tissues' pH balance. Copper (Cu), another transition element, is a vital catalyst in numerous enzymatic reactions, including those involved in the electron transport chain, a vital process for cellular energy production. In synthetic chemistry, transition metal catalysts have led to the creation of several new reactions and synthetic methods. For instance, Palladium-catalyzed cross-coupling reactions have transformed the way complex organic molecules are synthesized, leading to significant progress in the pharmaceuticals and materials science fields. In industrial settings, transition metals like Titanium (Ti) and Vanadium (V) are used as catalysts in the production of polyethylene and polypropylene. These plastics are commonly used in various products, ranging from packaging materials to automotive parts. Less common transition metals like Ruthenium (Ru) and Rhodium (Rh) are also used as catalysts in specialized chemical reactions. For example, the Monsanto process for producing acetic acid uses a rhodium-based catalyst.

Problem 1: Question: Provide an explanation for the catalytic behavior of transition metals using a specific example.

Solution: Transition metals frequently serve as catalysts due to their capacity to assume various oxidation states and to form complexes. For instance, in the Haber process used for ammonia production, iron acts as a catalyst. This process involves nitrogen and hydrogen reacting to create ammonia. Iron boosts this reaction by providing a surface for the molecules to unite, thereby reducing the energy needed for the reaction.

Problem 2: Question: What makes transition metals a common choice as catalysts in industrial processes?

Solution: Transition metals possess partially filled (n-1)d orbitals that can either donate or accept electrons, thereby aiding in the execution of chemical reactions. Their ability to alter their oxidation state allows them to assist reactions without being consumed or modified. Additionally, their extensive surface area accelerates the reaction rate.

Problem 3: Question: Identify a transition metal used as a catalyst in oil hydrogenation and explain its use.

Solution: Nickel, a transition metal, is employed as a catalyst in the hydrogenation of oils. This is because nickel can readily adsorb hydrogen and carbon atoms onto its surface, bringing them closer together and decreasing the activation energy required for the reaction.

Problem 4: Question: Why are transition metals such as platinum, palladium, and rhodium used in car catalytic converters?

Solution: Transition metals like platinum, palladium, and rhodium are utilized in car catalytic converters because they facilitate the transformation of harmful gases into less harmful substances. They expedite this reaction without being used up. For instance, they transform carbon monoxide, a poisonous gas, into carbon dioxide.

Problem 5: Question: Can you explain how the vanadium(V) oxide catalyst functions in the Contact process for sulfuric acid production?

Solution: During the Contact process, vanadium(V) oxide serves as a catalyst to hasten the conversion of sulfur dioxide into sulfur trioxide. This can then be absorbed in water to form sulfuric acid. The catalyst provides a platform for the reaction, reduces the reaction's activation energy, and boosts the reaction rate.

Transition metals, found on the Periodic Table, possess a unique set of characteristics, including the ability to form complex ions. These ions are charged entities made up of a central metal ion encompassed by a specific number of ligands, which are ions or molecules capable of donating an electron pair to the central metal ion. This unique property is a result of the transition metals' electronic structure, particularly the availability of empty d-orbitals that can accept these electron pairs. Unlike s-block and p-block elements, transition metals have partially filled d-orbitals. This provides them with a flexible oxidation state, allowing them to form a broad spectrum of complex ions. The capacity of transition metals to accept varying numbers of ligands and display different geometries in their complexes stems from the unique spatial configuration of the d-orbitals. Take iron (Fe), a transition metal, for instance. In its +2 oxidation state, as seen in the compound $FeCl_2$, it can react with ammonia (NH_3) to create a complex ion of $Fe(NH_3)_6^{2+}$. In this case, the iron atom forms six coordinate bonds with ammonia ligands, each one donating an electron pair to the metal ion. The resulting complex ion takes on an octahedral shape, a common geometry for complex ions. Another distinctive characteristic of transition metals' complex ions is their striking colors. The varying energy levels of the d-orbitals in transition metals can match the energy of visible light when these metals form complex ions. This absorption of specific light colors results in the remaining light giving the complex its unique color.

The ability of transition metals to form complex ions has significant applications in various areas. In biochemistry, for example, these metals form complex ions vital to life, such as the iron-containing complex ion in hemoglobin, which is responsible for oxygen transport in the body.

In environmental chemistry, the formation of complex ions with transition metals plays a key role. For example, these metals can form complex ions with natural ligands in the environment, affecting their solubility and reactivity, which influences their bioavailability and can have significant environmental effects.

In catalysis, transition metal complexes function as efficient catalysts due to their ability to switch between different oxidation states, enabling their participation in a wide range of chemical reactions and aiding in the production of various chemical compounds.

In analytical chemistry, transition metals' complex ions are used to identify and measure different substances. The distinct colors of these complex ions form the basis for colorimetric analysis, a technique commonly used in chemical analysis.

Question 1: Determine the transition metal and its oxidation state in the complex ion $[Fe(CN)_6]^{4-}$.

Answer: The transition metal in this complex ion is Iron, represented as Fe. The oxidation state can be figured out by taking into account that the total charge of the complex ion is -4 and the charge given by the six cyanide ions (CN-) is -6. Therefore, the oxidation state of Fe must be +2 to equalize the charge.

Question 2: Why do transition metals often form complex ions?

Answer: Transition metals frequently form complex ions because they possess empty d-orbitals that can accept pairs of electrons from ligands (molecules or ions that offer pairs of electrons). This capability enables them to form coordinate bonds and thus complex ions.

Question 3: Predict the color for a solution containing Cu2+ ions and provide an explanation for your prediction.

Answer: A solution that contains Cu2+ ions is expected to be blue. The reason for this is that Cu2+ ions absorb light in the red part of the spectrum; the color that complements red is blue, hence, the solution appears blue to our eyes.

Question 4: Identify the ligand and its type in the complex ion $[Cu(NH_3)_4]^{2+}$.

Answer: The ligand in this complex ion is Ammonia, represented as NH3. It falls under the category of monodentate ligands, meaning it has one donor atom that can form a coordinate bond with the central metal atom.

Question 5: How many unpaired electrons are present in the [Fe(H2O)6]2+ complex ion?

Answer: In the [Fe(H2O)6]2+ complex ion, the Fe2+ ion has an electron configuration of [Ar]3d6. Hence, it possesses four unpaired electrons.

The elements in the middle of the periodic table, known as transition metals, are recognized for their unique ability to form colored compounds. This characteristic is linked to their electronic structure and the presence of d-orbitals. An essential trait of transition metals is their capacity to create stable configurations with partially filled d-orbitals. These complex structures account for the vibrant colors seen in their compounds.

Transition metals are visually striking due to their color, a feature both in their elemental and compound forms. This characteristic is dictated by the d-d electronic transitions, which are essentially the shifts of electrons between two energy levels in the d orbital. The energy difference between these levels corresponds to certain light frequencies in the visible spectrum, resulting in the observed color.

Copper sulphate, a common compound of the transition metal copper, exemplifies this by forming bright blue crystals. Similarly, potassium permanganate, a compound of transition metal manganese, is dark purple. The diverse colors observed in transition metal compounds are a direct result of the specific electronic transitions occurring within their d-orbitals.

However, not all transition metals form colored compounds. For example, Scandium and Zinc, while classified as transition metals, do not form colored compounds due to their d orbitals being completely empty or full, preventing d-d transitions that absorb light and produce color.

The capability of transition metals to form colored compounds is significant across various fields. In chemistry education, transition metal compounds are frequently used for visually engaging demonstrations of chemical reactions. The color change from one compound to another clearly indicates a chemical reaction.

In industry, the formation of colored compounds by transition metals is practical. Many paints and pigments owe their vibrant colors to transition metal compounds. Chromium, for instance, is widely used in producing a range of pigments, from bright yellows to deep reds.

In biochemistry, transition metals and their colored compounds are essential. The bright red color of blood is due to iron, a transition metal, in hemoglobin, while the green color of chlorophyll, the pigment responsible for photosynthesis, is attributed to magnesium, another transition metal.

Transition metals can also form compounds that absorb and emit light in the ultraviolet and infrared spectrum, not just the visible spectrum. This property is utilized in designing solar panels and other light-energy harnessing devices.

In catalysis, transition metals and their compounds are crucial due to their ability to change oxidation states, which correlates to their ability to form colored compounds. The various colors of a transition metal compound often represent different oxidation states of the metal ion, acting as catalysts for diverse chemical reactions.

The environmental impact of colored compounds formed by transition metals is significant. When released into water bodies, many transition metal ions can color the water, signaling their presence and aiding in environmental monitoring and pollution control.

Lastly, the formation of colored compounds by transition metals contributes greatly to the beauty of gemstones and jewelry. The stunning variety of colors seen in gemstones like rubies, sapphires, and emeralds results from the presence of transition metal ions.

Problem 1: What causes transition elements to produce coloured compounds?

Answer: Transition elements generate coloured compounds due to the existence of unpaired d-electrons. These electrons, when they absorb a certain amount of energy, can be excited to move to a higher level of energy. The energy difference between the initial and final states of these electrons matches a particular light wavelength, often within the visible range, thus providing the compound with colour.

Problem 2: Why does the colour of a transition metal complex vary based on its ligands?

Answer: The colour variation of a transition metal complex based on its ligands is because the ligands affect the energy gap between the d orbitals. This is attributed to the crystal field splitting effect, where the ligands around the transition metal ion cause the d orbitals to diverge into two energy levels: a lower energy level (t2g) and a higher energy level (eg).

The color of the complex is determined by the absorbed light wavelength, which is governed by this energy gap.

Problem 3: What causes transition metal complexes of the same metal but with different ligands to have different colours?

Answer: The colour changes in transition metal complexes of the same metal with different ligands are due to the varying abilities of the ligands to divide the d orbitals. This is referred to as the spectrochemical series, which ranks ligands based on their ability to cause crystal field splitting. Ligands that cause a larger split will absorb higher energy (shorter wavelength) light, resulting in a different colour compared to those causing a smaller split.

Problem 4: Why do certain transition metal complexes appear colourless?

Answer: Some transition metal complexes appear colourless because there are no unpaired d-electrons to be excited. This implies that no visible light is absorbed, making the compound appear colourless. This is common in d10 complexes, such as $[Zn(H_2O)_6]^{2+}$, where all the d orbitals are full.

Problem 5: How does the transition metal's oxidation state influence the colour of its complex?

Answer: The oxidation state of a transition metal impacts the number of available d-electrons. Hence, different oxidation states will lead to different d orbital configurations, resulting in varying energy gaps between the orbitals. These energy differences correspond to different absorbed light wavelengths, and thus different observed colours.

Also known as transition elements, transition metals are the d-block elements in the periodic table, covering groups 3 to 12. They are unique in their ability to display multiple oxidation states, a quality not commonly seen in other periodic table groups. This unique trait is due to their electron configuration, especially the relative energy levels of the 3d and 4s orbitals. In atom's energy, the type of orbital it resides in determines an electron's energy. In transition metals, the 3d and 4s orbitals are very close in energy, allowing for easy promotion of electrons from one to the other. This results in transition metals' capacity to display variable oxidation states, as electrons can be removed from either the 3d or the 4s orbital during oxidation. In a neutral atom, the 4s orbital is filled before the 3d due to its lower energy level. However, when these atoms are ionized, the energy levels of the 3d and 4s orbitals shift, with the 3d orbital becoming lower in energy than the 4s. During ionization, electrons are first removed from the 4s orbital, then the 3d, contributing to the ability of transition metals to adopt different oxidation states. It's crucial to note that the energy difference between the 3d and 4s orbitals is minimal, allowing for an easy exchange of electrons. This flexibility contributes to why transition metals exhibit variable oxidation states. The partially filled d orbitals of transition metals provide another route for electron loss, enabling the formation of positive oxidation states. For instance, iron can be in a +2 or +3 oxidation state, depending on if one or two 3d electrons are lost in addition to the two 4s electrons. Another key factor is the shielding effect of the inner core electrons. The 3d electrons in transition metals are not effectively shielded by the inner electrons, resulting in an increased effective nuclear charge, promoting the loss of 4s electrons and contributing to variable oxidation states. The stability of different oxidation states also relies on the crystal field splitting energy, the energy difference between d orbital levels in a crystal field. Depending on the ligands surrounding the transition metal ion, some oxidation states may be more stable than others, adding another layer of complexity. Environmental conditions, such as temperature, pressure, and the presence of other elements or compounds, can also influence the most stable oxidation state for a transition metal. These conditions can shift the energy levels of the 3d and 4s orbitals, impacting electron loss or gain and thus the oxidation state. These mechanisms, although complex, are vital to many chemical reactions and processes. The ability of transition metals to change oxidation states is central to their roles as catalysts in various industrial processes, their use in pigments and dyes due to their vibrant colors, and their essential roles in biological systems, such as the iron in hemoglobin.

The variability of oxidation states in transition elements is largely attributed to the presence of partially filled d-orbitals. The small energy difference between the 3d and 4s orbitals allows for the easy removal of electrons from either of these orbitals. In the process of ion formation, transition metals can shed electrons from both the 4s and 3d sub-shells, leading to different oxidation states. For example, iron (Fe) can exist in both +2 and +3 oxidation states due to its ability to lose electrons from both the 4s and 3d sub-shells.

Question 1: What causes transition metals to often exhibit more than one oxidation state? Solution: The small energy difference between the 3d and 4s orbitals in transition metals allows for the easy removal of electrons from either orbital. This capability results in the potential for several oxidation states.

Question 2: How can iron (Fe) have both +2 and +3 oxidation states? Solution: Iron (Fe) can exist in both +2 and +3 oxidation states because it has the ability to lose electrons from both its 4s and 3d sub-shells. After shedding two electrons from the 4s sub-shell, it can lose an additional electron from the 3d sub-shell, leading to a +3 oxidation state.

Question 3: How does the energy of the 3d and 4s sub-shells impact the oxidation states of transition elements? Solution: The similar energy levels of the 3d and 4s sub-shells in transition elements facilitate the easy removal of electrons from either sub-shell during the process of ion formation. This process enables transition elements to display several oxidation states.

Question 4: Why do transition elements such as copper (Cu) exhibit variable oxidation states? Solution: Copper (Cu) has an electron configuration [Ar]3d10 4s1. It can shed one electron from the 4s orbital to form a Cu+ ion or lose one electron from 4s and one from 3d to form a Cu2+ ion. Thus, copper's ability to shed electrons from both 4s and 3d orbitals results in variable oxidation states.

Also known as d-block elements, transition metals are unique in their ability to serve as effective catalysts in a variety of chemical reactions. This distinctive capability is due to two main features: their capacity to exhibit multiple oxidation states and the presence of empty d-orbitals that can be used to establish dative bonds with ligands. This paper seeks to explore these features in more detail, highlighting their crucial role in making transition metals efficient catalysts. The capacity of transition metals to demonstrate various stable oxidation states is a crucial aspect of their catalytic behavior. This property is due to the existence of partially filled d-orbitals. For example, iron can show +2 and +3 oxidation states, manganese can range from +2 to +7, and chromium can vary from +2 to +6. This flexibility in changing oxidation states enables transition metals to serve as efficient catalysts, as they can support redox reactions by accepting or donating electrons. Redox reactions are vital to many chemical processes and biochemical pathways. In these instances, transition metals can serve as an electron bridge, facilitating the transfer of electrons from a reducing agent to an oxidizing agent. Consequently, they reduce the activation energy of the reaction, thereby accelerating the reaction rate without being consumed. The second crucial feature that makes transition metals effective catalysts is the presence of empty d-orbitals that can form dative bonds with ligands. A ligand is an ion or molecule that donates one or more pairs of electrons to the central metal atom, creating a coordinate bond. The capacity of transition metals to establish these coordinate bonds is key to their catalytic behavior. Transition metals' vacant d-orbitals are energetically accessible, meaning they can readily accept electrons from ligands. When a ligand creates a coordinate bond with the transition metal, it effectively 'activates' the ligand, making it more reactive. This ligand activation can facilitate various chemical reactions, including those involved in catalysis. Moreover, the ability to establish coordinate bonds with ligands enables transition metals to bind to a broad range of substrates, thereby increasing their catalytic versatility. This is particularly crucial in homogeneous catalysis, where the catalyst and reactants are in the same phase. In these scenarios, the ability of the transition metal to bind to and activate the substrate towards reaction is a key aspect of their catalytic behavior. Additionally, transition metals can adapt to a reaction's changing environment due to their variable oxidation states and ability to form dative bonds. They can alternate between different oxidation states and establish new coordinate bonds as the reaction progresses, thereby facilitating the reaction and enhancing their catalytic efficiency. In heterogeneous catalysis, which involves catalysts and reactants in different phases, the same principles apply. The transition metal catalyst, often in solid form, can adsorb reactant molecules onto its surface via coordinate bond formation. This adsorption process activates the reactant molecules, making them more reactive and thereby facilitating the chemical reaction. In biochemistry, transition metals play a vital role as catalysts in numerous biological processes. For instance, many enzymes, which are biological catalysts, contain transition metal ions at their active sites. These metal ions can alternate between different oxidation states, facilitating redox reactions, and form coordinate bonds with substrates, activating them towards reaction.

Problem 1: Why does iron, a transition element, serve as a catalyst in ammonia synthesis?

Solution: Iron, a transition metal, can exist in multiple oxidation states and has vacant d orbitals. In the Haber process for ammonia synthesis, iron provides a surface for nitrogen and hydrogen molecules to adsorb onto, promoting their interaction and lowering the reaction's energy barrier. This accelerates the reaction without depleting the catalyst, therefore it acts as a catalyst.

Problem 2: Why can transition metals produce colored compounds? Solution: Transition metals can produce colored compounds due to the presence of vacant d orbitals. When light is shone on these compounds, electrons in the lower energy d orbitals absorb energy and move to the higher energy d orbitals. The observed color is the complementary color of the light that has been absorbed. The specific color depends on the energy difference between the d orbitals, which can be influenced by the metal's oxidation state and the type of ligands attached to it.

Problem 3: Why are transition metals commonly used in industrial catalysts? Solution: Transition metals are widely used in industrial catalysts because they can exist in various oxidation states and form dative bonds with ligands through their empty d orbitals. This allows them to accelerate reactions by offering an alternative reaction pathway with a lower activation energy. Additionally, the catalyst can revert to its original state after the reaction, enabling it to be reused, making it cost-effective for industrial applications.

Transition metals, or transition elements, are unique in their capability to form complex ions. This attribute is largely due to their electronic configuration, specifically the presence of vacant d orbitals that are energetically accessible. To fully comprehend this phenomena, it is important to delve into the atomic structure of transition metals, the nature of d orbitals, and the mechanism of complex ion formation. Transition metals belong to the d-block of the periodic table, which includes groups 3 to 12. These elements have an incomplete d orbital in their penultimate energy level. This is a key characteristic that differentiates them from other elements. The d-orbitals are of higher energy than s and p orbitals, and are thus able to accept pairs of electrons from a donor species, a process that is pivotal in the formation of complex ions. The d orbitals in transition metals are well-suited for the formation of these ions due to their unique spatial orientation and available energy levels. They possess five energetically equivalent d-orbitals, which are shaped differently and oriented along different axes in three-dimensional space. This shape and orientation provide an environment conducive to accepting and holding pairs of electrons from other species. In terms of energy, the d-orbitals are more energetically accessible compared to other orbitals in the transition elements. This means that they can be easily occupied by electrons.

When a ligand, which is a molecule or an ion that donates an electron pair, comes into proximity with the transition metal ion, its electrons are able to jump into the vacant, energetically accessible d-orbitals. This forms a coordinate covalent bond, contributing to the formation of a complex ion. A complex ion consists of a central metal ion surrounded by several ligands. The metal ion acts as a Lewis acid (electron pair acceptor), while the ligands function as Lewis bases (electron pair donors). The ability of transition metals to form complex ions is intrinsically linked to their ability to act as Lewis acids, which is facilitated by the presence of vacant d-orbitals. The number of complex ions a transition metal can form is also influenced by the d-orbitals. The five d-orbitals in a transition metal ion can accommodate up to ten electrons, allowing the metal ion to form multiple coordinate bonds with ligands. This leads to the formation of complex ions with various coordination numbers, typically ranging from two to nine. The formation of complex ions is also influenced by the ability of d-orbitals to hybridize. Hybridization involves the mixing of atomic orbitals to form new hybrid orbitals that are identical and suitable for the formation of covalent bonds. In transition metals, the d orbitals can hybridize with s and p orbitals, resulting in d2sp3 or sp3d2 hybridization, and forming octahedral or square planar complexes respectively. The unique electronic configuration of transition metals not only allows them to form complex ions but also imparts these ions with distinctive colors. When light is absorbed by these complex ions, an electron in the d-orbital is excited to a higher energy level. The specific wavelength of light absorbed corresponds to a particular color, which is complementary to the color observed. The formation of complex ions is not only a fascinating aspect of transition metal chemistry, but it also has significant practical implications. Transition metal complexes are used in a wide array of applications, from catalysis in industrial processes to the development of anti-cancer drugs.

Problem 1: Q: What causes transition metals to commonly create colored compounds?
A: The reason transition metals frequently generate colored compounds is due to the non-identical energy of d-orbitals. The light that hits the compound may be consumed to shift an electron from a d-orbital of lower energy to one of higher energy. The perceived color is the opposite of the absorbed light.

Problem 2: Q: How does the creation of complex ions with transition metals impact biological systems?
A: The creation of complex ions is crucial in biological systems as they often have significant roles in different biological functions. For instance, the hemoglobin molecule in red blood cells includes an iron(II) ion that can establish a complex ion with oxygen molecules, allowing oxygen to be distributed around the body.

Problem 3: Q: What makes transition metals more prone to forming complex ions compared to other elements?
A: Transition metals are more inclined to form complex ions compared to other elements due to their compact sizes, high nuclear charges, and the availability of empty d orbitals for bond formation. These traits enable transition metals to have a strong attraction for lone pairs of electrons, promoting the formation of complex ions.

Problem 4: Q: Why are transition metals capable of forming diverse oxidation states?
A: Transition metals have the ability to form diverse oxidation states because they have partially filled d orbitals. These orbitals can lose electrons to form positive oxidation states. Moreover, these metals can also accept electrons into their d orbitals, allowing them to display negative oxidation states as well.

Problem 5: Q: How do the empty d orbitals in a transition metal ion contribute to the formation of complex ions?

A: The empty d orbitals in a transition metal ion offer a space where electrons from ligands can be housed. This enables the metal ion to form bonds with multiple ligands at the same time, resulting in the creation of complex ions.

Chemical properties of first transition elements

Metals found in the d-block of the periodic table are known as transition elements. These are identified by their partially filled d-orbitals in either their ground state or most stable oxidation states. The partial filling of these d-orbitals enables transition elements to create complex ions with a range of ligands - molecules or ions that can provide a pair of electrons to form a dative or coordinate bond with the central metal atom or ion.

This capacity to generate complexes is due to the presence of empty d-orbitals in the transition metals that can accept pairs of electrons from ligands. These complexes can either be monodentate, where a single ligand is attached to the metal ion, or polydentate, where multiple ligands are attached. The coordination number refers to the number of ligand atoms that can attach to a central metal ion.

Both copper(II) and cobalt(II) ions are transition metals that can create complexes with different ligands. Copper(II) ion has a 3d9 configuration, while cobalt(II) ion has a 3d7 configuration. Copper(II) ions can react with water molecules to create a complex where water acts as the ligand, providing its lone pair of electrons to the copper(II) ion to form a coordinate bond. This results in a copper(II) aqua complex with the chemical formula $[Cu(H_2O)_6]^{2+}$, surrounded by six water molecules, giving it a coordination number of six.

Cobalt(II) ions can also form a similar complex with water, resulting in a cobalt(II) aqua complex with the chemical formula $[Co(H_2O)_6]^{2+}$ and a coordination number of six. Copper(II) and cobalt(II) ions can form complexes with ammonia as well, with the copper(II) ammonia complex being $[Cu(NH_3)_4]^{2+}$ and the cobalt(II) ammonia complex being $[Co(NH_3)_6]^{2+}$.

Copper(II) and cobalt(II) ions can form complexes with the strong field ligand, the hydroxide ion, resulting in a copper(II) hydroxide complex $[Cu(OH)_2]$ and a cobalt(II) hydroxide complex $[Co(OH)_2]$. Chloride ions can also act as ligands and form complexes with these ions, resulting in copper(II) chloride complex $[CuCl_4]^{2-}$ and cobalt(II) chloride complex $[CoCl_4]^{2-}$.

These complexes are dynamic and can experience ligand exchange, where one ligand is replaced by another. The color of these complexes is often due to d-d electron transitions. The energy gap between the d-orbitals in these complexes aligns with visible light energy. When this light is absorbed, an electron moves from a lower energy d-orbital to a higher one, causing the complex to display color.

The formation of these complexes is crucial in inorganic chemistry and has significant biological implications. Many enzymes and proteins contain metal ion complexes that are essential for their function.

When complexes are formed, the transition metal ion is situated at the center, encircled by ligands. Typically, the number of ligands that a transition metal ion can bind with equals its coordination number, which is often 6.

1) Interaction of Copper(II) ions with water molecules:

The Cu2+ ion forms a complex with water molecules, with water serving as a ligand that donates its lone pair of electrons to form coordinate bonds with the Cu2+ ion. The complex ion that results is $[Cu(H_2O)_6]^{2+}$.

2) Interaction of Copper(II) ions with ammonia molecules:

When copper(II) ions interact with ammonia, the ammonia molecules displace the water molecules in the complex ion. This occurs because ammonia is a more potent ligand than water. The resulting complex ion is $[Cu(NH_3)_4(H_2O)_2]^{2+}$.

3) Interaction of Copper(II) ions with hydroxide ions:

Hydroxide ions can also serve as ligands, displacing the water molecules in the copper(II) complex to form $[Cu(OH)_4]^{2-}$.

4) Interaction of Cobalt(II) ions with chloride ions:

When cobalt(II) ions interact with chloride ions, the chloride ions behave as ligands and form a complex with the cobalt ion. The resulting complex is $[CoCl_4]^{2-}$.

Problem 1: What complex is formed when Cobalt(II) ions interact with water molecules?

Solution: When cobalt(II) ions interact with water molecules, they form the complex $[Co(H_2O)_6]^{2+}$, with water molecules serving as ligands that donate electron pairs to the cobalt ion.

Problem 2: What complex is formed when Copper(II) ions interact with ammonia?

Solution: When copper(II) ions interact with ammonia, the ammonia molecules displace the water molecules in the complex ion. The resulting complex ion is $[Cu(NH_3)_4(H_2O)_2]^{2+}$.

Problem 3: Explain the function of ligands in complex formation.

Solution: Ligands are vital in complex formation. They are the molecules or ions that contribute a pair of electrons to the transition metal ion, forming a coordinate bond. The transition metal ions possess empty d orbitals that can accept these electron pairs.

Problem 4: What color change is observed when copper(II) sulfate solution interacts with ammonia?

Solution: When a copper(II) sulfate solution interacts with ammonia, the solution transitions from a light blue color to a dark blue color. This change in color is attributed to the formation of the $[Cu(NH_3)_4(H_2O)_2]^{2+}$ complex.

In the field of inorganic chemistry, a ligand refers to a molecule or ion that attaches to a central atom, often a transition metal, to create a complex. This attachment happens through coordinate bonding or dative covalent bonding. The ligand provides a pair of electrons, also known as a lone pair of electrons, to the central metal atom or ion, forming a dative covalent bond where both electrons originate from the ligand.

The term 'ligand' stems from the Latin word 'ligandum,' meaning 'that which binds,' aptly describing the ligand's main function of binding to the central metal atom or ion. Depending on the number of donor sites they have for the central atom, ligands can be monodentate, bidentate, or polydentate. Monodentate ligands have one attachment site, bidentate ligands have two, and polydentate ligands have several.

The bond between a ligand and a metal ion can vary, being purely covalent, purely ionic, or somewhere in between, depending on the electronegativity difference between the ligand and the metal ion. The greater this difference, the more ionic the bond becomes, while a smaller difference shifts the bond towards covalency.

Ligands can be more than just simple ions or molecules, they can also be larger entities such as proteins and enzymes that bind to metal ions to form complexes. These bound ligands are called coordinated ligands. The structure and reactivity of the complex are significantly affected by the properties of the ligand, like its size, charge, and the number of donor sites.

The type of complexes formed and their properties are often determined by the nature and character of the ligand. For example, strong field ligands lead to low-spin complexes while weak field ligands lead to high-spin complexes. Additionally, ligands can affect the color of the compound due to the ligand field splitting effect. The study of ligands and their complexes is a vital aspect of inorganic chemistry. They are involved in many biological processes and have extensive industrial use. For instance, in biology, many enzymatic processes involve metal-ligand coordination complexes. In the industry, ligands are used in catalysis to accelerate reaction rates. Ligands can also be categorized based on their charge. Anionic ligands carry a negative charge, cationic ligands carry a positive charge, and neutral ligands carry no charge, each with a unique binding method with the central metal atom or ion.

It's important to note that ligands can also affect the stability of the complex. Chelating ligands, for example, can increase the stability of a complex due to the chelate effect as they form a ring structure with the central metal atom or ion, making the complex more resistant to decomposition.

Lastly, the identity of the ligand can also influence the geometry of the complex. Certain ligands can encourage the formation of tetrahedral, square planar, or octahedral complexes, highlighting the crucial role ligands play in determining the structure of a complex.

Question 1: What is a chelating ligand? Can you give an example?

Answer 1: A chelating ligand is a specific type of ligand that has the ability to bind to a metal atom at multiple points, creating a ring-like structure. The term "chelating" comes from the Greek word "chelos", meaning claw, symbolizing the way these ligands clamp onto the metal atom. Ethylenediamine can serve as an example of a chelating ligand, as it possesses two nitrogen atoms that are each able to donate a pair of electrons to a central metal atom, thereby forming two dative covalent bonds.

Question 2: Is it possible for a molecule that doesn't have any lone pairs to function as a ligand? Please explain.

Answer 2: No, it is not possible for a molecule without any lone pairs to serve as a ligand. The reason behind this is that ligands form dative covalent bonds with a central metal atom through the donation of an electron pair. Therefore, if a molecule lacks any lone pairs, it would not have any electrons to donate, preventing it from forming a dative covalent bond.

Question 3: What is the ligand in the following complex ion: $[Cu(NH_3)_4]^{2+}$?

Answer 3: The ligand in this complex ion is ammonia, or NH_3. Each of the ammonia molecules donates a lone pair of electrons to the copper ion, resulting in the formation of dative covalent bonds.

Question 4: Can you differentiate between monodentate and polydentate ligands?

Answer 4: Monodentate ligands are those that can bind to the central metal atom at one point only as they donate a single pair of electrons. Water (H_2O) is an example of a monodentate ligand. In contrast, polydentate ligands can bind to the central metal atom at multiple points, donating more than a single pair of electrons. Ethylenediamine ($NH_2CH_2CH_2NH_2$) serves as an example of a polydentate ligand.

Question 5: What function do ligands serve in biological systems?

Answer 5: Ligands play key roles in biological systems by helping or hindering certain biochemical reactions. They can function as signal molecules that bind to cell receptors, thus initiating a response. Additionally, they can serve as enzyme substrates or as cofactors that assist enzymes in performing their tasks.

Monodentate ligands, whose name comes from the Greek words "mono" for one and "dente" for tooth, belong to a group of ligands that bind to the central metal ion in a complex compound through one bond. A "ligand" is an ion or molecule that connects to a central atom creating a coordination complex. Monodentate ligands have just one donor atom for bond formation with the central metal ion.

Water (H_2O) is a well-known example of a monodentate ligand. Water molecules can form a bond with a central metal ion using one of the lone pairs of electrons on the oxygen atom, which serves as the donor atom, contributing electron density to the metal ion and creating a coordinate bond.

Another monodentate ligand is ammonia (NH_3). Here, the nitrogen atom in the ammonia molecule is the donor atom, using its lone pair of electrons to bond with the central metal ion. The electron-donating capacity of these ligands makes them Lewis bases.

Chloride ions (Cl^-) are also monodentate ligands. They can donate a pair of electrons from their outermost shell to a central metal ion forming a coordinate bond. Chloride ions are common ligands in many inorganic and bioinorganic complexes.

Cyanide (CN^-) is also a monodentate ligand. The carbon atom in CN^- uses its triple bond with the nitrogen atom to bond with the central metal ion. Even though cyanide ions can act as bidentate ligands under some conditions, they are mostly seen as monodentate ligands.

The study of monodentate ligands is critical in coordination chemistry. This understanding helps chemists to predict and manage the characteristics of complex compounds, including their reactivity, magnetism, and color.

Monodentate ligands are frequently used in the creation of coordination complexes, which have applications in fields such as catalysis, materials science, and medicine. Some chemotherapy drugs, for instance, are coordination complexes of platinum with monodentate ligands.

The notion of monodentate ligands is also crucial for understanding the biological function of metalloproteins. Many metalloproteins, proteins that contain a metal ion cofactor, are coordinated by monodentate ligands. Despite their simplicity, monodentate ligands can create a wide variety of complex structures. The type of ligand and the metal ion, as well as the conditions they react under, can all influence the resulting complex.

A crucial idea in understanding monodentate ligands is the 'coordination number', the number of coordinate bonds a metal ion forms in a complex ion. For monodentate ligands, this number is typically one, as they only have one donor atom.

The concept of monodentate ligands isn't just limited to simple molecules like H_2O or NH_3. Many larger organic molecules can also act as monodentate ligands, binding to the metal ion through one atom while the rest of the molecule stays unbound.

Problem 1: Determine if the ligands CO, NH_2^-, OH^- and CN^- are monodentate or polydentate.

Solution: These particular ligands - CO, NH_2^-, OH^- and CN^- are all monodentate. They each contain a single atom with an unshared electron pair that can create a coordinate bond with a central metal ion or atom.

Problem 2: Why is NH_3 categorised as a monodentate ligand?

Solution: NH_3, also known as ammonia, is identified as a monodentate ligand because it contains one atom (nitrogen) with an unshared electron pair that can make a coordinate covalent bond with a central metal atom or ion. The term "monodentate" translates to "one-toothed" in Latin, which signifies the ligand's capability to attach to the metal ion at one point.

Problem 3: How many attachment points does water (H_2O), a common ligand in coordination chemistry, have to a metal ion?

Solution: Water (H_2O) is a monodentate ligand, meaning it has one attachment point to a metal ion. The oxygen atom in water possesses an unshared electron pair that can create a coordinate bond with a metal ion.

Problem 4: Which ligand among Cl^-, H_2O, CN^-, NH_3 is not monodentate?

Solution: All the mentioned ligands: Cl^-, H_2O, CN^-, NH_3 are monodentate ligands. Each one has a single atom with an unshared electron pair capable of forming a coordinate bond with a central metal ion or atom.

Problem 5: In your own words, describe what a monodentate ligand is and give an example.

Solution: A monodentate ligand is an ion or molecule that attaches to a central metal atom or ion at just one point. It has a single atom with an unshared electron pair capable of bonding with the metal. An example of a monodentate ligand is ammonia (NH_3), where the nitrogen atom possesses an unshared electron pair that can bond with the metal.

Bidentate ligands are terms specific to coordination chemistry, signifying ligands with two atoms able to form coordinate covalent bonds with one metal ion. The dual-binding ability of bidentate ligands makes them more stable and gives them a higher affinity for metal ions than monodentate ligands, which can only form a single bond. The phenomenon, known as the chelate effect, shows that multidentate ligands form more stable complexes than their monodentate equivalents.

The word "bidentate" comes from Latin, with "bis" meaning twice, and "dentatus" implying toothed. Metaphorically, this suggests the ligand can 'bite' the metal ion at two points, made possible by two donor atoms in the ligand that can share a pair of electrons with the metal ion, creating a coordinate bond.

1,2-diaminoethane, often shortened to en, is a typical bidentate ligand. Its chemical formula is $H_2NCH_2CH_2NH_2$. Each nitrogen atom in 1,2-diaminoethane has a standalone pair of electrons, which can be donated to a metal ion to form two coordinate bonds.

Another prevalent bidentate ligand is the ethanedioate ion, $C_2O_4^{2-}$. In this ion, the two oxygen atoms at opposite ends can each donate a pair of electrons, forming two coordinate bonds with a metal ion. This ion is the conjugate base of oxalic acid, a common organic acid.

When a bidentate ligand like 1,2-diaminoethane or ethanedioate ion bonds with a metal ion in a chemical reaction, it creates a ring. This ring structure, called a chelate ring, is highly stable, contributing to the higher affinity bidentate ligands have for metal ions compared to monodentate ligands.

The stability of these chelate rings is influenced by their size, with five and six-membered rings being particularly stable due to their low ring strain. This makes bidentate ligands with these ring sizes prime candidates for creating stable complexes.

Bidentate ligands' ability to form these stable chelate rings has significant implications in various chemistry areas. For instance, in biochemistry, many essential metal ions coordinate to proteins through bidentate ligands. In environmental chemistry, bidentate ligands play a vital role in complexation and sequestration of toxic heavy metal ions, helping mitigate these pollutants' environmental impact.

Moreover, bidentate ligands have substantial industrial uses. In catalysis, they're often used to produce highly effective and selective catalysts, used in diverse chemical processes, from making pharmaceuticals to refining petroleum.

Question 1: Can you explain what a bidentate ligand is and give an example?

Answer: A bidentate ligand refers to a molecule or ion capable of forming two coordinate bonds with a central metal atom or ion within a complex. This is facilitated by the ligand's two donor atoms, which can share a pair of electrons with the central atom or ion. 1,2-diaminoethane ($H_2NCH_2CH_2NH_2$), also known as ethylenediamine or "en", is an example of a bidentate ligand.

Question 2: Can you describe the structure of the ethanedioate ion ($C_2O_4^{2-}$)?

Answer: The ethanedioate ion, represented by the formula $C_2O_4^{2-}$, is a bidentate ligand capable of binding to metal ions at two points. Its structure comprises two carbon atoms, each bonded to two oxygen atoms. These carbon-oxygen bonds are double bonds, and the carbon atoms are bonded to each other. As a result, the two oxygen atoms attached to each carbon atom can donate a pair of electrons, making the ion a bidentate ligand.

Question 3: What distinguishes a monodentate ligand from a bidentate ligand?

Answer: The key difference between a monodentate and a bidentate ligand is the number of donor atoms available to form coordinate bonds with a central metal ion. While monodentate ligands have just one donor atom capable of forming a coordinate bond with the metal ion, bidentate ligands have two such atoms. For instance, NH_3 (ammonia) is a monodentate ligand, while 1,2-diaminoethane ($H_2NCH_2CH_2NH_2$) is a bidentate ligand.

Question 4: Why do we classify the ethanedioate ion ($C_2O_4^{2-}$) as a bidentate ligand?

Answer: The ethanedioate ion is classified as a bidentate ligand as it has two oxygen atoms that can each donate a pair of electrons to form coordinate bonds with a central metal ion. These oxygen atoms are connected to the same carbon atom, which is in turn doubly bonded to another oxygen atom. This configuration allows the ion to bind to the metal ion at two different points, hence its classification as a bidentate ligand.

A polydentate ligand like EDTA4- is an intriguing entity in the field of coordination chemistry. The term "polydentate" denotes a ligand's capability to attach itself to a central atom at multiple points, making these ligands unique and essential in chemistry.

EDTA4- (Ethylenediaminetetraacetic acid), a typical example of hexadentate ligands, can bind to a central atom at six different locations. Its extraordinary ability to form complex structures with metal cations deems it an essential instrument in numerous chemical applications.

The EDTA4- ligand is equipped with four carboxylate groups and two amine groups, allowing it to form complexes with many metal cations. The octahedral complex is the most common formation, where the ligand completely encases the metal cation, creating a stable complex.

In coordination chemistry, the formation of such complexes carries considerable importance. A ligand like EDTA4-, that can bind at multiple points, tends to form more stable complexes than monodentate ligands, which can only attach at one point. This stability is due to the chelate effect, a thermodynamic phenomenon where the creation of chelate rings leads to more stable complexes.

EDTA4-'s versatility lies in its capability to form complexes with a vast range of metal cations, for instance, transition metals like copper, nickel, and cobalt, and alkali and alkaline earth metals like sodium and calcium.

EDTA4- finds broad application in various chemistry fields, such as analytical chemistry, where it helps determine the concentration of metal cations in a solution. It also helps stabilize dissolved metals in solution, making it valuable in environmental chemistry.

In biochemistry, EDTA4- chelates metal ions that could potentially disrupt enzymatic processes. This property is useful in molecular biology, where it prevents DNA degradation by chelating divalent cations necessary for nuclease activity.

In medicine, EDTA4- is used in chelation therapy to expel toxic metal ions from the body. It binds to metal ions such as lead, mercury, and copper, aiding their excretion from the body and reducing toxicity.

In industrial applications, EDTA4- is used in food product stabilization, water treatment processes, and textile industry for dyeing processes. Its ability to form stable complexes with metal ions assists in controlling these ions' levels in various processes.

In environmental science, EDTA4- is used to enhance the phytoremediation process. It increases the solubility and bioavailability of heavy metal ions in the soil by chelating with them, making it easier for plants to absorb and remove them.

However, EDTA4- also presents environmental challenges due to the high stability of its formed complexes, which are difficult to degrade, leading to its persistence in the environment. This persistence can disrupt the bioavailability of metal ions to organisms, potentially impacting the ecological balance.

In recent years, biodegradable alternatives to EDTA4- have been developed to address these environmental issues. These alternatives strive to retain the beneficial properties of EDTA4-, such as its chelating ability, while enhancing its biodegradability.

Problem 1: What is the highest number of coordination bonds that the EDTA4- ligand can form with a central metal ion?

Solution: The EDTA4- ligand, being a hexadentate ligand, can create up to six coordination bonds with a central metal ion. It has six donor atoms, two nitrogen and four oxygen atoms, which can each donate a pair of electrons to the metal ion.

Problem 2: What is the importance of EDTA4- as a hexadentate ligand?

Solution: The importance of EDTA4- as a hexadentate ligand lies in its ability to form a highly stable complex with a metal ion. Thanks to the 'chelate effect', multidentate ligands like EDTA4- form more stable complexes than their monodentate counterparts. This makes EDTA4- particularly useful in various applications, such as water softening, where it binds to metal ions in hard water, and in medicine, by binding to toxic metal ions.

Problem 3: Why does EDTA4- form a 1:1 complex with metal ions?

Solution: EDTA4- forms a 1:1 complex with metal ions as it has the exact quantity of donor atoms needed to satisfy the coordination number of most metal ions. The coordination number is the total number of coordination bonds a metal ion can form, which for most metal ions is six. Being a hexadentate ligand, EDTA4- can form six coordination bonds with a metal ion, resulting in a 1:1 complex.

Problem 4: EDTA4- is frequently used in titrations to find the concentration of metal ions. How would you determine the concentration of metal ions in a solution if you know the volume of EDTA4- solution used?

Solution: To determine the concentration of metal ions in a solution during a titration with EDTA4-, the stoichiometry of the reaction is used. Since EDTA4- forms a 1:1 complex with most metal ions, the moles of EDTA4- used in the titration equals the moles of metal ions in the solution. Hence, knowing the volume and concentration of EDTA4- solution used, you can calculate the moles of EDTA4-, and consequently the concentration of metal ions in the solution.

The following formula is used:

Moles of EDTA4- = volume (in litres) x concentration (in moles per litre)

Concentration of metal ions = moles of EDTA4- / volume of solution (in litres)

The term "complex" in chemistry pertains to a molecule or ion that consists of a central metal atom or ion, typically a transition metal, encircled by one or more ligands. This structure, known as a complex, is made up of multiple components, each performing a specific function, resulting in a more intricate structure than a single atom or ion. The central metal atom or ion forms the heart of the complex, which is then encompassed by ligands, usually ions or molecules capable of donating an electron pair to the metal atom, creating a coordinate bond.

Studying these intricate molecules or ions is a crucial aspect of inorganic chemistry, as it uncovers the various interactions between the central metal atom or ion and the surrounding ligands. Essentially, these complexes are the outcome of a Lewis acid-base interaction, where the metal atom or ion acts as the Lewis acid (electron pair acceptor), and the ligands serve as Lewis bases (electron pair donors).

The formation of a complex is often motivated by the requirement of the central metal atom or ion to attain a stable electron configuration. By receiving electron pairs from the ligands, the metal atom or ion can achieve a state of electron stability. This process of electron acceptance and donation is made possible by the presence of empty orbitals in the metal atom or ion, which are then occupied by the donated electron pairs from the ligands. Complexes can be of various sizes and shapes, regulated by the number and type of ligands linked to the central metal atom or ion. The complex's geometry, whether linear, square planar, tetrahedral, or octahedral, is dictated by the coordination number, which is the count of ligand atoms connected to the metal atom or ion.

The ligands' characteristics significantly influence the complex's properties. Some ligands, referred to as monodentate ligands, can only donate one electron pair to the metal atom or ion. In contrast, other ligands, known as polydentate or chelating ligands, can donate multiple electron pairs, forming a ring-like structure around the metal atom or ion.

The creation of complexes is not an isolated event; it has substantial implications in various scientific and industrial processes. For example, complexes play a role in biological systems, notably in enzyme catalysis and electron transport. In industry, complexes are utilized in catalysis, materials science, and even in medicinal chemistry.

A complex's stability is gauged by the interaction strength between the central metal atom or ion and the ligands. This strength is quantified by the stability constant, representing the equilibrium between the complex and its components. A high stability constant signifies a stable complex, and vice versa.

The creation of a complex can also result in changes in a solution's color. This is because the energy gap between the metal ion's ground state and excited state can be modified by the ligands, leading to the absorption or emission of light of specific wavelengths. This principle forms the foundation of numerous colorimetric analytical methods.

Problem 1: Question: Can you explain the term 'ligand' in the context of a complex ion?

Answer: In the context of a complex ion, a ligand refers to an ion or molecule that contributes a pair of its own electrons to create a coordinate bond with a central metal atom or ion. It could be a single atom, such as Cl-, or a molecule, like NH3.

Problem 2: Question: Given a complex ion with a metal ion at its core, represented by the formula [Fe(H2O)6]3+, what is the metal ion's oxidation state?

Answer: The metal ion's oxidation state is +3. This is calculated from the overall charge of the complex ion. As H2O is a neutral ligand, the oxidation state of Fe matches the charge of the complex ion.

Problem 3: Question: How many ligands are present in the complex ion [Cu(NH3)4]2+?

Answer: The complex ion [Cu(NH3)4]2+ contains four ligands. This is indicated by the subscript 4 following the (NH3), signifying that there are four NH3 molecules serving as ligands.

Problem 4: Question: Why does the complex ion [Ni(CN)4]2- exhibit a color?

Answer: The complex ion [Ni(CN)4]2- displays color due to the inclusion of a transition metal ion (Ni). Transition metals possess partially filled d orbitals. When the complex ion is exposed to light, the d electrons absorb the energy and elevate to higher energy levels. The energy gap between these levels corresponds to a specific light wavelength, which is perceived as color.

Problem 5: Question: What is the coordination number of the metal in the complex ion [Ag(NH3)2]+?

Answer: The coordination number in a complex ion refers to the amount of coordinate bonds the metal ion forms. In this situation, Ag establishes two coordinate bonds with two NH3 molecules, making its coordination number 2.

Transition metal complex structures vary depending on the nature of ligands, coordination number, and the central metal ion's electronic configuration. The geometries observed include linear, square planar, tetrahedral, and octahedral.

Linear complexes, characterized by a coordination number of two, are common in transition metal complexes where the metal ion coordinates with two ligands. The ligands are positioned directly across from each other on either side of the central metal, at a bond angle of 180 degrees. This arrangement minimizes electron repulsion, giving the complex a linear shape.

A coordination number of four characterizes square planar complexes, typical for transition metal complexes with a d8 configuration. The central metal ion is surrounded by four ligands at the corners of a square, with bond angles of 90 degrees, forming a planar structure. This arrangement minimizes electron repulsion, ensuring stability.

Tetrahedral complexes also have a coordination number of four but differ from square planar ones. The central metal ion is in the center of a tetrahedron, with ligands at the corners. The bond angles are approximately 109.5 degrees.

This geometry is common when the central metal ion has a d10 configuration.

Octahedral complexes, with a coordination number of six, have the central metal ion surrounded by six ligands at the corners of an octahedron. The bond angles are 90 degrees between adjacent ligands and 180 degrees between opposite ones. This geometry is common in transition metal complexes with d2 or high spin d6 electronic configurations.

The geometry of transition metal complexes significantly impacts their properties, including color, magnetic properties, reactivity, and stability. The shape also affects the ligand field splitting, essential to the crystal field theory, which describes the electronic structure of transition metal complexes.

In linear complexes, the d-orbitals of the central metal ion have equal energy, resulting in no splitting. In contrast, the d-orbitals in square planar, tetrahedral, and octahedral complexes split into different energy levels based on the geometry.

This splitting is responsible for the vibrant colors seen in many transition metal complexes.

The complex's geometry can also affect its reactivity. For example, in square planar complexes, the unoccupied d-orbitals are often oriented in a specific direction, making them susceptible to nucleophile or electrophile attack. In tetrahedral complexes, the three-dimensional structure shields the ligands, reducing their reactivity. However, this reactivity can be enhanced by the Jahn-Teller distortion, which distorts the geometry and makes certain d-orbitals more accessible for reaction.

In octahedral complexes, the ligand arrangement allows for various types of isomerism, leading to differences in reactivity and properties. These complexes can also exhibit geometric and optical isomerism, creating the possibility for chiral complexes.

Problem 1: What is the geometry and bond angles of a linear transition element complex?

Solution: A linear transition element complex is characterized by a central atom that is flanked by two other atoms or groups of atoms in a straight line. This configuration is linear in geometry with a bond angle of 180 degrees. $[Ag(CN)2]-$, where Ag is the central atom encircled by two cyanide groups, exemplifies this complex.

Problem 2: What is the geometry and bond angles of a square planar transition element complex?

Solution: A square planar transition element complex features a central atom surrounded by four other atoms or groups of atoms that form a square plane. This configuration results in bond angles of 90 degrees. $[Ni(CN)4]2-$, where Ni is the central atom encircled by four cyanide groups, is an example of this complex.

Problem 3: What is the geometry and bond angles of a tetrahedral transition element complex?

Solution: A tetrahedral transition element complex has a central atom that is encircled by four other atoms or groups of atoms configured in a tetrahedral shape. This arrangement leads to bond angles of roughly 109.5 degrees. $[ZnCl4]2-$, where Zn is the central atom encircled by four chloride groups, is an example of this complex.

Problem 4: What is the geometry and bond angles of an octahedral transition element complex?

Solution: An octahedral transition element complex is characterized by a central atom that is encircled by six other atoms or groups of atoms in an octahedral shape. The bond angles in this complex are 90 degrees between adjacent ligands and 180 degrees between opposite ligands. $[Co(NH3)6]3+$, where Co is the central atom encircled by six ammonia groups, exemplifies this complex.

Problem 5: What distinguishes the geometry of a square planar from a tetrahedral transition element complex?

Solution: The primary distinction between a square planar and a tetrahedral transition element complex lies in the arrangement of the atoms or groups of atoms around the central atom. In a square planar complex, the four surrounding atoms form a square plane around the central atom, creating bond angles of 90 degrees. Conversely, in a tetrahedral complex, the four surrounding atoms adopt a tetrahedral shape around the central atom, resulting in bond angles of approximately 109.5 degrees.

The term 'coordination number' is incredibly meaningful in the field of chemistry, specifically in regards to atomic structure, crystals, and complex compounds. Represented by CN, the coordination number refers to the number of neighboring or surrounding atoms an atom or ion has in its immediate vicinity. Essentially, it indicates the amount of bonds a central atom forms within a molecule or a crystal lattice.

The study of transition metals and their compounds heavily relies on the concept of coordination number. These metals frequently create coordination compounds, or complexes, where the coordination number is key in determining the compound's geometry and overall properties. Though the coordination number can range from two to sixteen, it is typically six.

It's important to note that the coordination number doesn't always correspond to the number of bonds an atom can form. It's more closely related to the spatial arrangement of the atoms or ions in a structure. In a complex ion or compound, the coordination number of a central atom is equal to the number of ligands attached to it. Ions or molecules known as ligands attach themselves to a central metal atom, resulting in the formation of a coordination complex. For example, in the compound [Fe(H2O)6]3+, the central Fe3+ ion has a coordination number of 6, indicating it's directly bonded to six water molecules. In another example, [CuCl4]2- has a coordination number of 4, showing that the central copper ion is bonded to four chloride ions.

The coordination number is also crucial in understanding the structure of crystals. In a crystal lattice, the coordination number can provide insight into the packing efficiency, stability, and the lattice type. For instance, a simple cubic lattice has each atom with a coordination number of 6, while in a face-centered cubic lattice, the coordination number is 12.

The study of defects in crystal structures also benefits from understanding the coordination number. Schottky and Frenkel defects, for example, involve changes in the coordination number of atoms or ions, significantly affecting the crystal's properties.

The coordination number can impact the physical and chemical properties of compounds and materials. Compounds with a higher coordination number often have higher boiling and melting points due to the increased number of bonds to be broken.

In biochemistry, the coordination number significantly aids in understanding the structure and function of biological molecules. For instance, in the heme group found in hemoglobin, the iron ion has a coordination number of six.

Keep in mind, the coordination number isn't always fixed and can change under different conditions like pressure and temperature. For example, under high pressure, atoms in a material can be pushed closer together, increasing their coordination number.

Although the concept of coordination number is relatively easy to understand, determining the coordination number in more complex compounds and materials can be difficult. It often necessitates a thorough understanding of the material's structure and bonding.

The coordination number refers to the number of neighboring atoms, ions, or molecules that one central atom or ion can hold in a complex, coordination compound, or crystal.

Problem 1: Q: How many spheres does each sphere in a cubic close-packed structure coordinate with?

Solution: Each sphere in a cubic close-packed structure coordinates with 12 other spheres, hence the coordination number is 12.

Problem 2: Q: What is the coordination number of an atom in a body-centered cubic (BCC) structure?

Solution: In a BCC structure, each atom is situated at the cube's center and is in contact with 8 other atoms at the corners. Therefore, the coordination number is 8.

Problem 3: Q: What is the coordination number of Sodium in Sodium Chloride (NaCl)?

Solution: In Sodium Chloride (NaCl), each sodium ion coordinates with six chloride ions, so the coordination number is 6.

Problem 4: Q: What is the coordination number of Iron (Fe) in the complex [Fe(H2O)6]3+?

Solution: In the [Fe(H2O)6]3+ complex, Fe is surrounded by six H2O molecules, meaning its coordination number is 6.

Problem 5: Q: What is the coordination number of Carbon in Diamond?

Solution: In diamond, each carbon atom is bonded covalently to four other carbon atoms. Therefore, the coordination number is 4.

Understanding the formulation and charge prediction of a complex ion requires knowledge of the elements that contribute to its creation. A complex ion is created when a central metal ion becomes surrounded by ligands, which are usually anions or neutral molecules possessing lone pairs of electrons that can be given to the metal ion. The metal ion accepts these electron pairs and forms coordinate bonds with the ligands. The metal ion can be a transition metal or a part of the actinide or lanthanide series.

The ligand's ability to bond with a metal ion is chiefly influenced by the metal ion's oxidation state and coordination number. The oxidation state refers to the charge the metal ion would carry if all its bonds were ionic. The coordination number is the count of ligand attachment sites. It's crucial to highlight that the coordination number doesn't necessarily match the ligand count, as some ligands can occupy multiple attachment sites.

For instance, a metal ion with a +2 charge or oxidation state and a coordination number of 6 can form a complex ion with a mono-dentate ligand, which can form one bond with the metal ion. Here, the complex ion's formula would consist of one metal ion and six ligand molecules, provided the ligand is neutral. This would result in the complex ion having an overall charge of +2, equivalent to the metal ion, since the ligand doesn't contribute any charge.

However, if the ligand is an anion like chloride (Cl-) or hydroxide (OH-), the complex ion's overall positive charge would decrease. For example, with four chloride ions as ligands for a metal ion with a +2 charge, the complex ion's overall charge would be -2. This is due to the four chloride ions contributing a -4 charge, which when combined with the +2 charge of the metal ion, results in an overall charge of -2.

In the case of polydentate ligands, which can form multiple bonds with the metal ion, fewer ligand molecules are needed to fulfil the coordination number. For example, ethylenediamine (en) is a bidentate ligand that can form two bonds with a metal ion. If the coordination number is 6, then only three ethylenediamine molecules would be needed.

The complex ion's geometry is also influenced by the coordination number and the type of ligands. Common geometries include linear (coordination number 2), trigonal planar (coordination number 3), tetrahedral (coordination number 4), square planar (coordination number 4), and octahedral (coordination number 6).

In many instances, a complex ion's overall charge is not zero. To balance the charge, counter ions are often included in the formula. For example, if the complex ion has a negative charge, positive counter ions (usually a metal ion) are included to balance the charge.

Question 1: The complex contains a Copper (Cu) metal ion with a +2 charge and an Ammonia (NH3) ligand with a coordination number of 4. Can you determine the formula and charge of the complex ion?
Answer: The complex ion's formula is [Cu(NH3)4]2+ and its charge is +2.

Question 2: The complex incorporates an Iron (Fe) metal ion with a +3 charge and a Cyanide (CN-) ligand with a coordination number of 6. Can you calculate the formula and charge of the complex ion?
Answer: The complex ion's formula is [Fe(CN)6]3- and its charge is -3.

Question 3: The complex involves a Chromium (Cr) metal ion with a +3 charge and a Water (H2O) ligand with a coordination number of 6. Can you figure out the formula and charge of the complex ion?
Answer: The complex ion's formula is [Cr(H2O)6]3+ and its charge is +3.

Question 4: The complex has a Nickel (Ni) metal ion with a +2 charge and an Ethylenediamine (en) ligand with a coordination number of 3. Can you determine the formula and charge of the complex ion?
Answer: The complex ion's formula is [Ni(en)3]2+ and its charge is +2.

Question 5: The complex contains a Platinum (Pt) metal ion with a +4 charge and a Chloride (Cl-) ligand with a coordination number of 6.
What would be the charge and formula of the complex ion?
Answer: The complex ion's formula is [Pt(Cl)6]2- and its charge is -2.

The term 'ligand exchange' is used in the field of coordination chemistry, which studies metal complexes, to describe the process where one ligand is replaced by another in a complex ion. Ligands are molecules or ions that can donate an electron pair to form a coordinate bond with the central metal ion. This process can be seen in various metal complexes, including those with copper(II) and cobalt(II), which can undergo ligand exchange with water, ammonia, hydroxide, and chloride ions.

Typically, the process of ligand exchange involves one ligand being displaced by another in a metal complex through various mechanisms - associative, dissociative, and interchange mechanisms. The exact mechanism that takes place is often dependent on the properties of the metal ion and the incoming and outgoing ligands.

For instance, copper(II) ions often exist as the aqua complex $[Cu(H_2O)_6]^{2+}$ in a solution. Here, the copper(II) ion is surrounded by six water molecules, each serving as a ligand and donating an electron pair to the metal ion. However, when ammonia is introduced into the solution, it can lead to ligand exchange due to its stronger affinity with the copper(II) ions as compared to water.

The same process can happen with cobalt(II) ions, where the introduction of ammonia to an aqueous solution can lead to the formation of $[Co(NH_3)_6]^{2+}$ through ligand exchange. Hydroxide and chloride ions can also undergo ligand exchange, resulting in the formation of the complexes $[Cu(OH)_2]$ and $[CoCl_4]^{2-}$, respectively. Ligand exchange reactions play an important role in understanding the behavior of metal ions in solution and are also practically used in the synthesis of various metal complexes. These complexes have applications as catalysts, in medicine, and in various other areas of chemistry.

However, the process of ligand exchange is complex and can be affected by various parameters, such as the nature of the metal ion, the ligands, the concentration of the ligands, temperature, and pH of the solution. Hence, understanding the mechanisms of ligand exchange reactions is crucial for predicting and controlling their outcomes.

Many aspects of ligand exchange reactions, such as their exact mechanisms and the factors influencing their rates, remain unclear. Therefore, further research is needed to fully understand these reactions.

Problem 1: A Copper(II) sulfate solution has a pale blue color. When an excess of aqueous ammonia is introduced, a light blue precipitate initially forms and then dissolves, resulting in a dark blue solution. Could you provide an explanation for this?

Solution: The pale blue color observed in the copper(II) sulfate solution can be attributed to the $[Cu(H_2O)_6]^{2+}$ complex ion. When the aqueous ammonia interacts with this, it replaces the water ligands in the complex, forming a light blue precipitate of copper(II) hydroxide $[Cu(OH)_2]$ in a ligand exchange reaction. The reaction can be represented as:

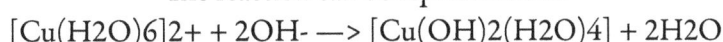

$$[Cu(H_2O)_6]^{2+} + 2OH^- \longrightarrow [Cu(OH)_2(H_2O)_4] + 2H_2O$$

On adding more ammonia, the copper(II) hydroxide dissolves into the excess ammonia, forming a dark blue complex ion $[Cu(NH_3)_4(H_2O)_2]^{2+}$ in another ligand exchange reaction. The equation for this is:

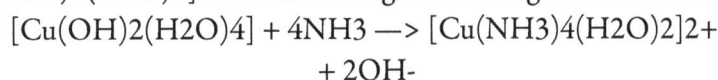

$$[Cu(OH)_2(H_2O)_4] + 4NH_3 \longrightarrow [Cu(NH_3)_4(H_2O)_2]^{2+} + 2OH^-$$

Problem 2: When sodium hydroxide in aqueous form is added to a solution of cobalt(II) chloride, also in aqueous form, a blue precipitate is observed. What are the molecular-level actions causing this?

Solution: Before the addition of sodium hydroxide, the cobalt(II) ions are encircled by water molecules, forming a pink-colored $[Co(H_2O)_6]^{2+}$ complex. The sodium hydroxide then interacts with the water ligands to form a blue precipitate of cobalt(II) hydroxide $[Co(OH)_2]$ in a process known as a ligand exchange reaction. The equation for this reaction is:

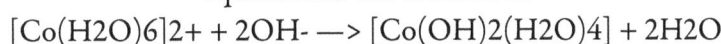

$$[Co(H_2O)_6]^{2+} + 2OH^- \longrightarrow [Co(OH)_2(H_2O)_4] + 2H_2O$$

Problem 3: What causes the color change when concentrated hydrochloric acid is introduced into a solution of copper(II) sulfate?

Solution: Before the addition of hydrochloric acid, the copper(II) ions in the solution are encircled by water molecules forming the blue-colored [Cu(H2O)6]2+ complex. When concentrated hydrochloric acid is added, the chloride ions replace the water molecules around the copper(II) ion, forming the yellow-colored complex [CuCl4]2-. This is another example of a ligand exchange reaction. The equation for this reaction is:

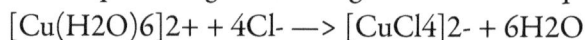

$$[Cu(H_2O)_6]^{2+} + 4Cl^- \longrightarrow [CuCl_4]^{2-} + 6H_2O$$

Redox reactions, also known as oxidation-reduction reactions, are crucial chemical processes that involve changes in the oxidation states of atoms in the reaction. The likelihood of these reactions taking place can be forecasted using standard electrode potentials, denoted as $E^{\ominus}$ values. These values give a quantitative indication of how likely a chemical species is to lose or gain electrons, hence participating in redox reactions.

Transition metals and their ions, which frequently participate in redox reactions, display a broad variety of oxidation states. This is due to the presence of (n-1)d and ns electrons in their outermost shells. These electrons' availability for redox reactions makes transition metals and their ions adaptable redox agents.

The $E^{\ominus}$ value of a transition metal or its ion indicates its reduction potential, or its capability to accept electrons. Transition metals with high $E^{\ominus}$ values are more likely to be reduced, suggesting they are strong oxidizing agents. In contrast, those with lower $E^{\ominus}$ values are more likely to be oxidized, indicating they are powerful reducing agents.

For example, the $E^{\ominus}$ value for the Mn3+/Mn2+ pair is +1.51V and for the Cr3+/Cr2+ pair is -0.41V. This means Mn3+ ions are more prone to accept electrons and reduce to Mn2+, thus being better oxidizing agents than Cr3+ ions.

When predicting the likelihood of redox reactions involving transition metals and their ions, the relative $E^{\ominus}$ values of the reactants is a key factor. A redox reaction is likely if the $E^{\ominus}$ value of the oxidizing agent is higher than that of the reducing agent. This is due to a higher $E^{\ominus}$ value for the oxidizing agent representing a higher likelihood to accept electrons and hence, propel the redox reaction forward.

In a theoretical reaction involving Fe2+ ions ($E^{\ominus}$ = -0.44V) and MnO4- ions ($E^{\ominus}$ = +1.51V) in an acidic medium, the overall $E^{\ominus}$ value for this redox reaction would be positive (+1.95V), indicating a feasible reaction. However, it's important to note that $E^{\ominus}$ values only predict whether a redox reaction can occur, not the speed of the reaction. The rate of a redox reaction is dependent on factors such as temperature, reactant concentration, and the presence of a catalyst.

While $E^{\ominus}$ values are beneficial for predicting redox reaction feasibility, they are not the only considerables. The solution's pH can greatly affect the $E^{\ominus}$ values of certain redox pairs, and, as a result, the likelihood of redox reactions. Transition metal ions often exhibit different $E^{\ominus}$ values under acidic and alkaline conditions, which must be factored in when predicting redox reaction feasibility.

Problem 1: Given the half-reactions below:

$$MnO_4^- + 8H^+ + 5e^- \to Mn^{2+} + 4H_2O \quad E^{\ominus} = 1.51 \text{ V}$$
$$Fe^{3+} + e^- \to Fe^{2+} \quad E^{\ominus} = 0.77 \text{ V}$$

Assess the possibility of the redox reaction between MnO4- and Fe2+.

Solution: A redox reaction requires an oxidation half-reaction (where electrons are released) and a reduction half-reaction (where electrons are absorbed). MnO4- is reduced to Mn2+ in the initial reaction, and Fe3+ is reduced to Fe2+ in the subsequent reaction. We need to invert the latter reaction to obtain an oxidation half-reaction. This modifies the $E^{\ominus}$ sign, so the oxidation half-reaction is now:

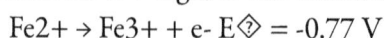

$$Fe^{2+} \to Fe^{3+} + e^- \quad E^{\ominus} = -0.77 \text{ V}$$

The overall $E^{\ominus}$ for the redox reaction equals the sum of the $E^{\ominus}$ values for the oxidation and reduction half-reactions, which works out to be 1.51 V - 0.77 V = 0.74 V. Since the $E^{\ominus}$ for the redox reaction is positive, the reaction is possible.

Problem 2: Given the half-reactions below:

Cu2+ + 2e- → Cu E⬦ = 0.34 V

Zn2+ + 2e- → Zn E⬦ = -0.76 V

Assess the possibility of the redox reaction between Cu2+ and Zn.

Solution: In this case, Cu2+ is reduced to Cu in the first reaction, and Zn2+ is reduced to Zn in the subsequent reaction. We need to invert the latter reaction to obtain an oxidation half-reaction, which changes the sign of E⬦:

Zn → Zn2+ + 2e- E⬦ = 0.76 V

The total E⬦ for the redox reaction equals the sum of the E⬦ values for the oxidation and reduction half-reactions, which is 0.34 V + 0.76 V = 1.10 V. Since the E⬦ for the redox reaction is positive, the reaction is possible.

These problems illustrate how to determine the feasibility of redox reactions using E⬦ values. If the overall E⬦ for the redox reaction is positive, the reaction is feasible. If the value is negative, the reaction is not possible under normal conditions.

The redox reaction between permanganate ions (MnO4-) and oxalate ions (C2O42-) in an acidic environment is a well-known chemical process. In this reaction, permanganate ions serve as a robust oxidizing agent, and oxalate ions function as a reducing agent. This reaction is commonly utilized for analyzing different compounds in complexometry and titrimetry, and for studying reaction kinetics in physical chemistry.

This reaction unfolds in two stages. Firstly, permanganate ions are reduced to manganese(II) ions (Mn2+), and oxalate ions are oxidized to carbon dioxide (CO2). The complete equation for this reaction is:

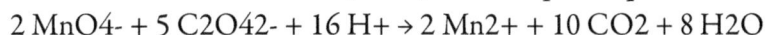

2 MnO4- + 5 C2O42- + 16 H+ → 2 Mn2+ + 10 CO2 + 8 H2O

This equation reveals that the reaction is highly complex and involves the exchange of a considerable number of electrons. The oxidation state of manganese shifts from +7 in the permanganate ion to +2 in the manganese(II) ion, indicating that each permanganate ion receives five electrons. Similarly, the oxidation state of carbon changes from +3 in the oxalate ion to +4 in carbon dioxide, indicating that each oxalate ion gives up two electrons.

To perform computations relating to this reaction, it's essential to understand its stoichiometry, or the ratio in which the reactants combine and the products form. In this reaction, the stoichiometry for permanganate, oxalate, and protons is 2:5:16, respectively. This means that for every two moles of permanganate ions, five moles of oxalate ions and 16 moles of protons are required. Similarly, two moles of manganese(II) ions, ten moles of carbon dioxide, and eight moles of water are produced for every two moles of permanganate ions.

With the knowledge of stoichiometry, one can calculate the quantity of any reactant or product if the amounts of the other reactants or products are known. For instance, if the quantity of permanganate ions is known and one wishes to determine the required amount of oxalate ions or the amount of carbon dioxide produced, one can use the ratio 2:5:10. Similarly, if the quantity of oxalate ions is known and one wishes to determine the required amount of permanganate ions or the amount of manganese(II) ions produced, one can use the ratio 5:2:2.

Besides stoichiometry, other factors such as reactant concentration, temperature, and solution pH can influence the reaction rate. The reaction is quicker at higher concentrations and temperatures, and at lower pH levels. This is because the reaction involves proton transfer, which is promoted by the presence of an acid.

This reaction between permanganate and oxalate ions in an acidic solution is a standard redox reaction example where electron transfer leads to a change in the elements' oxidation state. It serves as a valuable tool for chemists, allowing them to analyze various compounds, study reaction kinetics, and even create new compounds.

Question: What is the balanced equation for the redox reaction between MnO4- and C2O4 2- in an acidic solution?

Answer: The redox balanced equation is:

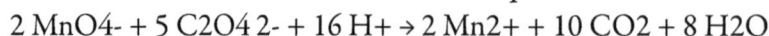

$$2 \text{ MnO4-} + 5 \text{ C2O4 2-} + 16 \text{ H+} \rightarrow 2 \text{ Mn2+} + 10 \text{ CO2} + 8 \text{ H2O}$$

Clarification: The ion MnO4- is reduced to Mn2+, while the ion C2O4 2- is oxidized to CO2. The number of electrons lost in the oxidation half-reaction equals the number of electrons gained in the reduction half-reaction, hence the coefficients in the reaction are balanced.

Question: How many moles of CO2 are produced when 5.0 moles of MnO4- react with 5.0 moles of C2O4 2-?

Answer: According to the balanced equation, 2 moles of MnO4- react with 5 moles of C2O4 2- to produce 10 moles of CO2. So, if 5.0 moles of MnO4- react with 5.0 moles of C2O4 2-, it will yield 25 moles of CO2.

Question: How many moles of Mn2+ are produced when 3.0 moles of MnO4- are added to a solution containing 5.0 moles of C2O4 2-?

Answer: Based on the balanced equation, 2 moles of MnO4- are reduced to 2 moles of Mn2+. Thus, 3.0 moles of MnO4- will yield 3.0 moles of Mn2+.

Question: How many moles of H2O are produced when 10.0 moles of CO2 are formed in the reaction?

Answer: Going by the balanced equation, 10 moles of CO2 correlate to the production of 8 moles of H2O. Hence, 10.0 moles of CO2 will generate 8.0 moles of H2O.

Question: How many moles of H+ are needed for the reaction of 4.0 moles of MnO4- and 10.0 moles of C2O4 2-?

Answer: According to the balanced equation, 16 moles of H+ are needed for the reaction of 2 moles of MnO4- and 5 moles of C2O4 2-. Therefore, for the reaction of 4.0 moles of MnO4- and 10.0 moles of C2O4 2-, 32.0 moles of H+ are needed.

The interaction between the permanganate ion (MnO4-) and the iron (II) ion (Fe2+) in an acidic solution is a quintessential illustration of a redox (reduction-oxidation) reaction in chemistry. This reaction is commonly employed in the lab for titrating iron (II) solutions.

The permanganate ion behaves as an oxidizing agent, and its purple color vanishes when it reduces to the colorless Mn2+ ion. Conversely, the Fe2+ ion gets oxidized to Fe3+ ion. The fading of the permanganate ion's purple color can act as a self-indicator, eliminating the need for an external indicator.

The balanced chemical formula for this acidic medium reaction is: $5\text{Fe2+} + \text{MnO4-} + 8\text{H+} \rightarrow 5\text{Fe3+} + \text{Mn2+} + 4\text{H2O}$. It can be split into two half-reactions. The reduction half-reaction is: $\text{MnO4-} + 8\text{H+} + 5\text{e-} \rightarrow \text{Mn2+} + 4\text{H2O}$. The oxidation half-reaction is: $\text{Fe2+} \rightarrow \text{Fe3+} + \text{e-}$.

The stoichiometry of the reaction reveals that five moles of Fe2+ react with one mole of MnO4-.

For calculations, suppose someone wants to determine the Fe2+ concentration in a solution using titration. The required information would be the volume and concentration of the MnO4- solution used.

The stoichiometric relationship between Fe2+ and MnO4- would be used in the calculation. As one mole of MnO4- reacts with five moles of Fe2+, the number of moles of Fe2+ in the solution would be five times the number of moles of MnO4- used.

The molar concentration of Fe2+ can be calculated by dividing the number of moles of Fe2+ by the volume of the Fe2+ solution (in liters). The same method can be used to find the concentration of an unknown MnO4- solution using a known Fe2+ concentration.

It's crucial to remember that the reaction requires an acidic medium. The H+ ions are needed to balance the charges in the half-reaction equations, and the reaction won't proceed without them.

The reaction rate is temperature-dependent, increasing with temperature. It's also affected by the solution's ionic strength.

Additionally, the reaction is applicable in determining iron in iron ores, assessing water hardness, and analyzing various industrial products.

Problem 1: In an oxidation-reduction reaction, the Permanganate ion (MnO_4^-) interacts with Iron (Fe^{2+}) in an acidic solution, resulting in Manganese (Mn^{2+}) and Iron (III) (Fe^{3+}). If we commence with 0.02 moles of MnO_4^- and 0.06 moles of Fe^{2+}, establish the limiting reactant.

Solution:

First and foremost, jot down the balanced redox reaction:

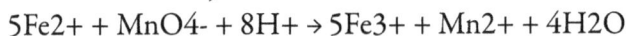

$$5Fe^{2+} + MnO_4^- + 8H^+ \rightarrow 5Fe^{3+} + Mn^{2+} + 4H_2O$$

From the stoichiometry of the balanced chemical equation, it's evident that a single mole of MnO_4^- reacts with five moles of Fe^{2+}.

If we initiate with 0.02 moles of MnO_4^-, we would necessitate $0.02 \times 5 = 0.1$ moles of Fe^{2+} for the reaction to be complete. However, we only possess 0.06 moles of Fe^{2+}.

Hence, Fe^{2+} is the limiting reactant.

Problem 2: Compute the quantity of MnO_4^- ions required to react entirely with 50 mL of 0.1 M Fe^{2+} solution.

Solution:

From the balanced chemical equation, we deduce that 5 moles of Fe^{2+} react with a single mole of MnO_4^-.

Initially, calculate the moles of Fe^{2+}:

Number of moles = molarity volume = 0.1 M 0.05 L = 0.005 moles

The moles of MnO_4^- needed = 0.005 moles Fe^{2+} / 5 = 0.001 moles

Problem 3: If you commence with 0.03 moles of MnO_4^- and 0.1 moles of Fe^{2+}, calculate the quantity of Mn^{2+} produced at the end of the reaction.

Solution: From the balanced chemical equation, we deduce that a single mole of MnO_4^- produces a single mole of Mn^{2+}.

Given that MnO_4^- is less than Fe^{2+} (after considering their stoichiometric ratio), MnO_4^- is the limiting reactant. Therefore, the reaction will cease when all the MnO_4^- is consumed.

So, the quantity of Mn^{2+} produced = number of moles of MnO_4^- = 0.03 moles.

The relationship between copper ions (Cu^{2+}) and iodide ions (I^-) is an intriguing topic within the realm of chemistry. It's vital to study these interactions as they're at the core of many chemical reactions with substantial industrial, biological, and environmental impacts. This subject is especially captivating due to the unique properties of both the copper and iodide ions and the intricate reactions they can partake in.

Copper, a transition metal, is recognized for its ability to produce complex ions. The Cu^{2+} ion, in particular, showcases unique chemistry. Its outermost d orbital is partially filled, enabling it to participate in varying oxidation states and coordination numbers. Cu^{2+} is a potent oxidizing agent, and its electron configuration allows it to create colored complexes, a trait heavily used in analytical chemistry.

On the other hand, the iodide ion, I^-, is a halide ion. It is distinguished by its large size and polarizability compared to other halides. The I^- ion can perform as a nucleophile, a reducing agent, and a ligand, showing its adaptability.

When Cu^{2+} and I^- interact, the Cu^{2+} ion is reduced to Cu^+, and the I^- ion is oxidized to I_2. This redox reaction is common of metal-ligand interactions and can be represented by the equation: $2Cu^{2+} + 4I^- \rightarrow 2CuI + I_2$.

The result of this reaction, copper(I) iodide (CuI), has several applications. It acts as a p-type semiconductor and is utilized in solar cells, LEDs and photocatalysis. CuI also serves as a catalyst in organic synthesis. It's worth noting that CuI doesn't dissolve in water, a trait often used in its production and isolation.

Besides the redox reaction, Cu2+ and I– can produce complex ions under certain circumstances. Complex ions are entities where a central metal ion is surrounded by several ligands. Here, the Cu2+ ion acts as the central ion while the I– ion functions as the ligand. The nature and stability of the resulting complex depend on various factors, including the ion concentration, the solution's pH, and the presence of other ions.

The relationship between Cu2+ and I– also has environmental chemistry implications. Copper is a common water pollutant, derived from industrial discharge and copper pipe corrosion. Conversely, iodide naturally exists in seawater and some mineral deposits. The reaction between these two ions can affect copper's speciation and mobility in the environment.

Furthermore, the redox chemistry of Cu2+ and I– is relevant in biological systems. Copper is a crucial trace element, playing a key role in various physiological processes. Mismanagement of copper homeostasis is linked to several neurodegenerative diseases. Interestingly, iodide is also essential for human health, being a critical component of thyroid hormones.

Problem 1: Given that the standard reduction potentials for Cu2+ to Cu and I2 to I- are +0.34 V and +0.54 V respectively, find out the standard cell potential for the cell reaction Cu2+ + 2I– → Cu + I2.

Solution: Using the formula E°cell = E°cathode – E°anode, and considering that I2 to I- has a higher positive value hence will function as the cathode, the E°cell = (+0.54 V) - (+0.34 V) = +0.20 V.

Problem 2: In a voltaic cell made up of Cu2+ and I– ions, if the standard reduction potential for the Cu2+/Cu pair is +0.34 V and for the I2/I– pair is +0.54 V, identify which will act as the anode and the cathode.

Solution: The half-reaction with the superior reduction potential, in this case, I2, will be reduced at the cathode, leaving Cu2+ to be oxidized at the anode.

Problem 3: Find the equilibrium constant for the reaction Cu2+ + 2I– → Cu + I2 at 25°C given that the standard reduction potentials for Cu2+ to Cu is +0.34 V and I2 to I- is +0.54 V.

Solution: Firstly, calculate the standard cell potential (E°cell) as in Problem 1, which is +0.20 V.

Utilize the Nernst equation to determine the equilibrium constant (K).

$$\Delta G° = -nFE°cell = -RTlnK$$

where n = 2 (number of electrons transferred), F = 96,485 C/mol (Faraday's constant), R = 8.314 J/(mol·K) (gas constant), and T = 298 K (temperature).

Rearranging the Nernst equation gives: $lnK = -nFE°cell / RT$

Substituting these values gives:

$$lnK = -(2)(96485)(0.20) / (8.314)(298) = -47.07$$

So, $K = e^{(-47.07)} \approx 3.4 \times 10^{-21}$.

Problem 4: Find the cell potential for the reaction

Cu2+ + 2I – → Cu + I2 at 25°C if the Cu2+ and

I– concentrations are 0.1 M and 0.01 M respectively.

Solution: Firstly, find the standard cell potential (E°cell) as in Problem 1, which is +0.20 V.

Next, apply the Nernst equation to determine the cell potential (Ecell): Ecell = E°cell - (RT/nF)lnQ

where Q = $[Cu2+]/[I–]^2$ (reaction quotient).

Substituting these values gives:

$$Ecell = 0.20 - ((8.314)(298) / (2)(96485))ln(0.1 / (0.01)^2)$$
$$= 0.20 - 0.0287 = 0.1713 \text{ V}.$$

Redox reactions, also known as reduction-oxidation reactions, are essential in many chemical processes, including the rusting of iron and the metabolism of sugars in our bodies. To successfully perform calculations related to these reactions, an in-depth understanding of the redox process and the associated data is necessary.

The main element of a redox reaction is the transfer of electrons between two chemical substances. The chemical that loses electrons undergoes oxidation, while the one that gains electrons undergoes reduction. It's essential that the number of lost and gained electrons balance, as outlined by the law of conservation of charge.

In order to conduct calculations related to redox reactions, we must first identify the oxidation states of the elements involved in the reactants and products. The oxidation state signifies the level of oxidation of an atom within a compound and can be either positive or negative. The shift in oxidation states allows us to figure out the quantity of electrons transferred during the reaction.

Subsequently, we must list the half-reactions for both the oxidation and reduction processes. The oxidation half-reaction displays the chemical substance being oxidized and the quantity of electrons lost, and the reduction half-reaction shows the chemical substance being reduced and the quantity of electrons gained. Ensuring these half-reactions balance is crucial. If the quantity of electrons lost in the oxidation half-reaction does not match the quantity gained in the reduction half-reaction, we have to adjust the coefficients of the substances involved.

Once the half-reactions are balanced, we can list the balanced overall redox reaction. This displays all the reactants and products, including the electrons, which should equalize on both sides of the equation.

The stoichiometric coefficients in the balanced equation are employed to calculate the quantities of reactants and products. This is where the principle of stoichiometry comes into play, stating that the ratios of reactant and product quantities are directly proportional to their stoichiometric coefficients in the balanced equation.

Another critical factor in redox calculations is determining the standard cell potential, denoted as E°. This is calculated from the standard reduction potentials of the half-reactions, which are listed values. The cell potential provides a measure of the driving force of the redox reaction.

To find the cell potential, the formula E°cell = E°cathode - E°anode is used. Here, E°cathode is the standard reduction potential of the reduction half-reaction (occurring at the cathode), and E°anode is the standard reduction potential of the oxidation half-reaction (occurring at the anode).

For systems where reactant and product concentrations are not standard, the Nernst equation is used to calculate the actual cell potential. This introduces the reaction quotient Q, which is the ratio of product concentrations to reactant concentrations.

The Gibb's free energy change (ΔG) of the reaction can also be calculated using the formula $\Delta G = -nFE°cell$, where n is the electron number transferred, F is the Faraday constant, and E°cell is the standard cell potential. This provides a measure of the reaction's spontaneity. If ΔG is less than 0, the reaction is spontaneous; if ΔG is more than 0, the reaction is non-spontaneous.

Problem 1:

We have a redox reaction as follows:

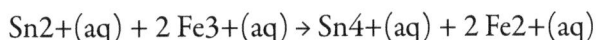

$$Sn^{2+}(aq) + 2\ Fe^{3+}(aq) \rightarrow Sn^{4+}(aq) + 2\ Fe^{2+}(aq)$$

The standard reduction potentials for Sn^{2+}/Sn^{4+} and Fe^{3+}/Fe^{2+} are +0.15V and +0.77V respectively. We need to find the standard electromotive force (E°) for the overall cell reaction.

Solution 1: Recognize the half-reactions first. Sn^{2+} gets oxidized to Sn^{4+} (electron loss) and Fe^{3+} reduces to Fe^{2+} (electron gain). The sum of these two half-reactions gives us the total cell reaction.

We can find the standard electromotive force for the cell reaction using the formula E°cell = E°cathode - E°anode. Here, E°cathode is the reduction potential of the substance undergoing reduction (Fe^{3+}) and E°anode is the reduction potential of the substance undergoing oxidation (Sn^{2+}).

By substituting the values, we get E°cell = 0.77V - 0.15V = 0.62V.

Problem 2:

We are given the redox reaction:

$$2 H2O(l) + 2 e\text{-} \rightarrow H2(g) + 2 OH\text{-}(aq)$$

We need to find the cell potential at 25°C where [OH-] = 1.0M and P(H2) = 1.0 atm. The reduction potential standard for this reaction is -0.83V.

Solution 2: We solve this problem using the Nernst equation which connects the cell potential to the standard cell potential, the reaction quotient, and the temperature. The equation is:

$$E = E° \text{-} (RT/nF)\ln Q$$

Here:

E° = standard cell potential = -0.83V,

R = gas constant = 8.314 J/(mol·K),

T = temperature = 25°C = 298.15K,

n = number of electrons transferred = 2,

F = Faraday's constant = 96485 C/mol,

Q = reaction quotient = [H2][OH-]² = (1.0)(1.0)² = 1.0.

On substituting these values in the equation, we get:

$$E = \text{-}0.83V \text{-} (8.314 J/(mol·K) \ 298.15K / (2 \ 96485 C/mol)) * \ln(1.0)$$

$$E = \text{-}0.83V + 0V = \text{-}0.83V.$$

Therefore, the cell potential under these conditions is -0.83V.

Complex Coloration
Degenerate and Non-Degenerate d Orbitals: An In-depth Analysis
Introduction

In the realm of atomic and molecular physics, degenerate and non-degenerate d orbitals play a crucial role in understanding the electronic structure of atoms and their interactions. These terms are often used to describe the energy levels and spatial orientations of d orbitals, which are subshells that contain five atomic orbitals: dxy, dyz, dzx, dx2-y2, and dz2. By delving into the definitions and applications of these terms, we can unravel their significance in various scientific fields, including chemistry, quantum mechanics, and material science.

Defining Degenerate and Non-Degenerate d Orbitals

First and foremost, let us establish the precise definitions of degenerate and non-degenerate d orbitals. In quantum mechanics, degeneracy refers to the situation where two or more quantum states possess the same energy level. Hence, degenerate d orbitals are those that have an identical energy level. Conversely, non-degenerate d orbitals are those that possess distinct energy levels.

To elaborate further, consider an isolated atom with a partially filled d subshell. In this scenario, each of the five d orbitals will have different energies, and thus, the system exhibits non-degenerate d orbitals. However, when the atom interacts with its environment or within a crystal lattice, the energy levels of the d orbitals can change due to factors such as symmetry breaking or external perturbations. Consequently, the degeneracy of the d orbitals can be lifted, resulting in either partially degenerate or fully non-degenerate d orbitals.

Degeneracy Breaking Mechanisms

Degeneracy breaking can occur through several mechanisms, depending on the nature of the system. One prominent mechanism is the presence of an external magnetic field. When a magnetic field is applied, the orbital angular momentum of the electrons in the d subshell experiences a Zeeman splitting, leading to the lifting of degeneracy. This splitting arises due to the interaction between the magnetic moment associated with the orbital angular momentum and the magnetic field. Consequently, the d orbitals acquire distinct energy levels, and their degeneracy is broken. Another mechanism that can break degeneracy is crystal field splitting. In solid-state systems, such as transition metal complexes or materials with transition metal ions, the d orbitals experience a crystal field generated by the surrounding ligands or crystal lattice. This field can perturb the energy levels of the d orbitals, leading to their splitting. The specific arrangement and symmetry of the ligands determine the extent of degeneracy breaking. For example, in an octahedral coordination environment, the d orbitals split into two sets: t2g (dxz, dyz, and dzx) and eg (dx2-y2 and dz2). This splitting is a result of the electrostatic interaction between the ligand's charges and the d orbitals, causing the degeneracy to be lifted.

Applications in Chemistry

Degenerate and non-degenerate d orbitals find extensive applications in the field of chemistry, particularly in understanding the electronic structure and properties of transition metal complexes. Transition metals, such as iron, copper, and chromium, possess partially filled d orbitals, making them highly versatile for various chemical reactions.

Degenerate d orbitals play a vital role in determining the magnetic properties of coordination compounds. In a high-spin complex, where the ligand field is weak, the degeneracy of the d orbitals remains intact due to minimal splitting. As a result, unpaired electrons can occupy multiple degenerate orbitals, leading to a high magnetic moment. Conversely, in a low-spin complex, strong ligand field splitting occurs, causing the degeneracy of the d orbitals to be lifted. Consequently, the electrons preferentially occupy the lower-energy orbitals, resulting in a low magnetic moment. Moreover, the concept of degenerate and non-degenerate d orbitals is crucial in understanding the spectroscopic properties of transition metal complexes. For instance, the absorption and emission spectra of transition metal ions are highly sensitive to the nature of the d orbitals involved and their degeneracy. The energy differences between these orbitals determine the wavelengths of light that can be absorbed or emitted, providing valuable information about the electronic transitions occurring within the complex.

Quantum Mechanical Perspective

From a quantum mechanical standpoint, degenerate and non-degenerate d orbitals are deeply intertwined with the principles of symmetry and group theory. Symmetry plays a fundamental role in determining the degeneracy of the d orbitals. If a molecular or crystal system possesses a particular symmetry, the d orbitals that transform in the same way under that symmetry operation will be degenerate.

Group theory provides a systematic approach to analyze the symmetry properties of molecular systems. By assigning point group representations to the d orbitals, one can determine their degeneracy or non-degeneracy. For example, in an octahedral coordination complex with an Oh symmetry, the d orbitals can be assigned to the irreducible representations of the Oh group: t2g and eg. The degeneracy of the orbitals is related to the number of orbitals in each irreducible representation.

Material Science Perspective

Degenerate and non-degenerate d orbitals also have significant implications in material science, where they contribute to the understanding of electronic band structures and the properties of materials. In solid-state systems, the interactions between atoms lead to the formation of energy bands, which are ranges of energies that electrons can occupy. The splitting of d orbitals due to crystal field effects can influence the band structure, resulting in variations in material properties.

For instance, the phenomenon of degeneracy breaking in d orbitals plays a crucial role in the emergence of magnetism in certain materials. In a ferromagnetic material, such as iron, the splitting of d orbitals due to crystal field effects leads to a difference in energy levels for spin-up and spin-down electrons. As a result, the number of up-spin electrons in the d orbitals can exceed the number of down-spin electrons, creating a net magnetic moment and giving rise to ferromagnetism. On the other hand, non-degenerate d orbitals can influence the optical properties of materials. The energy difference between non-degenerate d orbitals can be in the range of visible light, allowing for the absorption and emission of photons. This property is exploited in various applications, such as light-emitting diodes (LEDs) and photovoltaics, where d-orbital transitions play a crucial role in energy conversion processes.

Counterarguments and Limitations

While the concepts of degenerate and non-degenerate d orbitals provide valuable insights into the electronic structure and properties of atoms and materials, it is important to acknowledge their limitations and potential counterarguments. Some argue that the terms "degenerate" and "non-degenerate" can be subjective and context-dependent. The exact energy levels and splitting of d orbitals can vary based on the specific system under consideration, making it challenging to generalize these concepts across different cases. Furthermore, the breaking of degeneracy can be influenced by various factors beyond those discussed, such as spin-orbit coupling and relativistic effects. These effects can introduce additional complexities and modify the energy levels of the d orbitals, leading to deviations from the idealized degenerate or non-degenerate scenarios. The splitting of degenerate d orbitals into two non-degenerate sets of d orbitals of higher energy is an important concept in coordination chemistry. This phenomenon, known as the crystal field splitting, occurs when transition metal ions are surrounded by ligands in a coordination complex. The crystal field splitting leads to the formation of different energy levels for the d orbitals, resulting in a split spectrum. Understanding this splitting is crucial for explaining the electronic properties and reactivity of transition metal complexes.

The crystal field splitting in octahedral complexes gives rise to two sets of d orbitals – one higher in energy and the other lower in energy. In this arrangement, the ligands are positioned along the x, y, and z axes, leading to a repulsion between the ligand electrons and the d orbitals. This repulsion causes the d orbitals to split into two different energy levels.

The five d orbitals (dxy, dxz, dyz, dzx, and dz2) are initially degenerate, meaning they have the same energy. However, the presence of the ligands causes this degeneracy to be broken. The dxy, dxz, and dyz orbitals, which lie in between the ligands, experience more repulsion and are raised to a higher energy level. These three orbitals are referred to as the t2g set. On the other hand, the dzx and dz2 orbitals, which point directly towards the ligands, experience less repulsion and are lowered in energy. These two orbitals form the eg set.

The energy difference between the t2g and eg sets is denoted by Δ E. This energy difference is a key factor in determining the electronic and magnetic properties of transition metal complexes. The magnitude of Δ E depends on various factors, such as the nature of the ligands, the metal ion's charge, and the metal-ligand bond strength.

In octahedral complexes, the t2g set is lower in energy, while the eg set is higher. This energy difference is responsible for many observable properties of transition metal complexes. For instance, it affects the color of the complex. When visible light interacts with the complex, it is absorbed if the energy of the incident light matches the energy difference between the t2g and eg orbitals. The absorbed light corresponds to a particular wavelength, and the color we perceive is the complementary color of the absorbed light. For example, if the complex absorbs light in the blue region of the spectrum, it will appear yellow or orange.

Another consequence of the crystal field splitting is the pairing of electrons in the d orbitals. According to Hund's rule, electrons will first occupy the degenerate orbitals singly, with their spins aligned in the same direction. However, due to the energy difference between the t2g and eg sets, it is energetically favorable for the electrons to pair up in the lower energy t2g orbitals before occupying the higher energy eg orbitals. This pairing of electrons has significant implications for the magnetism and reactivity of transition metal complexes.

In tetrahedral complexes, the crystal field splitting is slightly different compared to octahedral complexes. In this arrangement, the ligands are positioned at the corners of a tetrahedron, rather than along the axes. As a result, the splitting pattern is reversed. The eg set is now lower in energy, while the t2g set is higher.

The reversal of the crystal field splitting in tetrahedral complexes can be explained by considering the geometry and symmetry of the arrangement. The ligands in a tetrahedral complex approach the metal ion along the edges of the tetrahedron. This arrangement leads to a different distribution of electron density compared to octahedral complexes. The repulsion between the ligand electrons and the d orbitals is stronger for the eg orbitals, causing them to be raised to a higher energy level. Conversely, the t2g orbitals, which experience less repulsion, are lowered in energy.

The energy difference between the eg and t2g sets in tetrahedral complexes is also denoted by Δ E. This energy difference determines the electronic and magnetic properties of the complex. However, it is important to note that the magnitude of Δ E in tetrahedral complexes is generally smaller than in octahedral complexes. This is because the ligand field in tetrahedral complexes is weaker due to the geometric arrangement of the ligands. The crystal field splitting in tetrahedral complexes leads to three higher energy t2g orbitals and two lower energy eg orbitals. The electron configuration in tetrahedral complexes follows a similar pattern to octahedral complexes, with electrons first occupying the t2g orbitals before pairing up in the eg orbitals. However, due to the reversed splitting pattern, the electronic configuration and magnetic properties of tetrahedral complexes differ from their octahedral counterparts.

The crystal field splitting in both octahedral and tetrahedral complexes has been extensively studied and has found applications in various fields. In spectroscopy, the energy difference between the split d orbitals is used to identify transition metal complexes and study their electronic structure. For example, UV-visible spectroscopy can be used to determine the absorption spectrum of a complex, providing information about the energy levels and electronic transitions.

The crystal field splitting also plays a crucial role in determining the reactivity of transition metal complexes. The energy difference between the t2g and eg sets influences the stability and reactivity of different oxidation states of the metal ion. For instance, a higher energy difference (larger Δ E) generally leads to greater stability of higher oxidation states.

Counterarguments against the crystal field splitting theory do exist. One such argument is the limitations of the crystal field theory in accurately predicting the energy levels of transition metal complexes. The crystal field splitting assumes a purely electrostatic interaction between the ligands and the metal ion, neglecting other factors such as covalency and orbital overlap. These factors can significantly affect the energy levels and may not be adequately accounted for by the crystal field theory.

Another counterargument is the limitations of the crystal field splitting in explaining the magnetic properties of transition metal complexes. While the crystal field splitting provides a qualitative explanation for the pairing of electrons and the resulting magnetic behavior, it fails to account for the observed magnetic properties in some cases. This limitation has led to the development of more advanced theories, such as ligand field theory and molecular orbital theory, which provide a more accurate description of the electronic structure and magnetic properties of transition metal complexes.

Transition Elements and the Formation of Colored Compounds
Introduction

The phenomenon of transition elements forming colored compounds has long fascinated scientists and chemists. Transition elements, also known as transition metals, are a group of elements found in the middle of the periodic table. These elements exhibit unique properties due to the partially filled d orbitals in their electronic configurations. One of the most intriguing characteristics of transition elements is their ability to form compounds that absorb specific frequencies of light, resulting in the perception of color. This essay aims to explore and explain the underlying reasons why transition elements form colored compounds by examining the promotion of electrons between two non-degenerate d orbitals and the subsequent absorption of light.

Understanding Transition Elements

To comprehend why transition elements form colored compounds, it is crucial to first grasp the nature of these elements. Transition elements are characterized by the presence of partially filled d orbitals. Unlike the s and p orbitals, which are completely filled in the ground state, the d orbitals can accommodate a varying number of electrons, leading to diverse electronic configurations. This unique feature allows transition elements to exhibit a wide range of oxidation states and form various compounds with distinct properties.

Electron Promotion and Absorption of Light

The absorption of light by a compound occurs when electrons in the ground state are excited to a higher energy level. In the case of transition elements, the promotion of electrons occurs between two non-degenerate d orbitals. A non-degenerate orbital refers to an orbital with a unique energy level. This promotion of electrons is responsible for the formation of colored compounds.

Electronic Spectra and Absorption of Light

To understand the absorption of light by transition elements, one must examine the concept of electronic spectra. The electronic spectra of transition metal compounds are obtained through a technique known as electronic spectroscopy. This technique involves passing light through a sample and measuring the amount of light absorbed at different wavelengths.

When transition metal compounds absorb light, the energy of the photons matches the energy difference between two non-degenerate d orbitals. The energy difference corresponds to the frequency of light absorbed, and this frequency determines the perceived color. The absorbed light appears as the complementary color to that which was absorbed. For example, if a compound absorbs light in the blue region of the spectrum, it will appear yellow to the human eye.

Crystal Field Theory

To further delve into the promotion of electrons and the absorption of light, crystal field theory provides a valuable framework. Crystal field theory explains the bonding and properties of transition metal complexes by considering the interactions between the transition metal ion and its surrounding ligands.

In crystal field theory, the ligands surrounding the transition metal ion create a crystal field, which splits the d orbitals into two sets of energy levels. The lower energy set of orbitals is called the t2g set, while the higher energy set is referred to as the eg set. The energy difference between these two sets determines the frequency of light that will be absorbed.

d-d Transitions and Color Perception

The promotion of electrons between the t2g and eg sets of orbitals is known as a d-d transition. This transition is responsible for the absorption of light and the resulting color perception.

When a transition element forms a complex with ligands, the crystal field created by the ligands causes the d orbitals to split. The energy difference between the split orbitals corresponds to particular wavelengths of light. If a photon with the corresponding energy is absorbed, it promotes an electron from one orbital to another, resulting in the perception of color.

The specific wavelength of light absorbed depends on the nature and strength of the ligands surrounding the transition metal ion. Different ligands create different crystal fields, leading to unique energy differences between the t2g and eg orbitals. Consequently, each transition metal complex exhibits a distinct absorption spectrum and color.

Examples from Various Fields

The formation of colored compounds by transition elements can be observed in various fields, providing concrete examples of the phenomenon.

Biological Systems: In biological systems, transition metal ions play a pivotal role in various processes, including photosynthesis and oxygen transport. For instance, the heme group in hemoglobin contains an iron ion that forms a complex with porphyrin ligands. This complex absorbs light, enabling hemoglobin to transport oxygen efficiently. The red color of blood is a result of the absorption of light by the iron-porphyrin complex.

Art and Pigments: Transition elements have been widely used in art and pigments throughout history. For instance, copper ions in copper-based pigments, such as verdigris, absorb specific wavelengths of light, resulting in the characteristic green color. Similarly, cobalt-based pigments absorb orange and yellow light, producing a blue appearance. These examples exemplify how the absorption of light by transition metal compounds has been harnessed for artistic purposes.

Photovoltaic Cells: Transition metals also play a crucial role in photovoltaic cells, which convert sunlight into electricity. For instance, the dye-sensitized solar cell (DSSC) utilizes transition metal complexes to absorb light and generate an electrical charge. The absorption of light by these complexes is essential for the efficient conversion of photons into electrical energy.

Counterarguments and Alternate Explanations

While the promotion of electrons between non-degenerate d orbitals provides a comprehensive explanation for the formation of colored compounds by transition elements, alternative theories and counterarguments exist. One such theory is the ligand field theory, which focuses on the interactions between the transition metal ion and the ligands. Ligand field theory provides a different perspective on the absorption of light and color perception. However, it is important to note that both crystal field theory and ligand field theory contribute to our understanding of the phenomenon, and further research is necessary to reconcile and integrate these theories.

The Effects of Different Ligands on ΔE, Frequency of Light Absorbed, and Complementary Colors
Introduction

In the field of chemistry, ligands play a crucial role in the formation of coordination compounds. These compounds consist of a central metal ion surrounded by ligands, which are molecules or ions that donate a pair of electrons to the metal ion. The interaction between the ligand and the metal ion results in the formation of a complex, which often exhibits unique properties, including the ability to absorb specific wavelengths of light. The absorption of light by coordination compounds leads to the observation of complementary colors. This article aims to explore the qualitative effects of different ligands on the energy difference (ΔE), frequency of light absorbed, and the complementary color observed.

Ligands and Coordination Compounds

Before delving into the effects of different ligands, it is important to have a basic understanding of coordination compounds and ligands. Coordination compounds are formed by the reaction of a Lewis acid (the metal ion) with a Lewis base (the ligand). The ligands can be classified into various types, such as monodentate ligands, bidentate ligands, polydentate ligands, and chelating ligands, based on the number of donor atoms they possess.

Energy Difference (ΔE) and Frequency of Light Absorbed

When a coordination compound absorbs light, it undergoes a process known as electronic transition. This transition occurs when an electron in a lower energy level jumps to a higher energy level. The energy difference between these two levels is denoted as ΔE. According to the equation $E = h\nu$, where E is the energy, h is Planck's constant, and ν is the frequency of light, the frequency of light absorbed is directly proportional to the energy difference.

Therefore, ligands can influence the frequency of light absorbed by coordination compounds by altering the energy difference (ΔE). Different ligands have different electronic properties, including their ability to donate electron pairs to the metal ion. This influences the energy levels within the coordination compound and, consequently, the energy difference (ΔE) between them. Ligands with stronger donating abilities tend to result in smaller energy differences, leading to the absorption of higher frequency (shorter wavelength) light.

Complementary Colors

The absorption of light by coordination compounds leads to the observation of complementary colors. Pairs of colors that produce white light when combined are known as complementary colors. Pairs of colors that produce white light when combined are known as complementary colors. Complementary colors can be found on opposite sides of the color wheel. For example, blue and orange are complementary colors, as are red and cyan, and green and magenta.

The complementary color observed in coordination compounds is determined by the color of light that is not absorbed. The absorbed light is subtracted from white light, and the remaining light is what we perceive as the complementary color. For instance, if a coordination compound absorbs light in the blue region of the spectrum, the complementary color observed would be orange.

Effects of Different Ligands on ΔE and Frequency of Light Absorbed

Different ligands have distinct properties that affect the energy difference (ΔE) and, consequently, the frequency of light absorbed by coordination compounds. Ligands can be broadly classified into two categories: high-field ligands and low-field ligands.

High-Field Ligands

High-field ligands are known for their strong donating abilities, leading to a larger splitting of energy levels within the coordination compound. The energy difference (ΔE) between these levels is smaller, resulting in the absorption of higher frequency (shorter wavelength) light. Consequently, coordination compounds with high-field ligands tend to exhibit complementary colors towards the red end of the visible spectrum.

One example of a high-field ligand is cyanide (CN-). Complexes formed with cyanide ligands often absorb light in the red region, resulting in a complementary color towards the green end of the spectrum. This can be observed in compounds such as $[Fe(CN)6]4-$, which appears green due to its absorption of red light.

Low-Field Ligands

In contrast to high-field ligands, low-field ligands have weaker donating abilities, leading to a smaller splitting of energy levels within the coordination compound. The energy difference (ΔE) between these levels is larger, resulting in the absorption of lower frequency (longer wavelength) light. Therefore, coordination compounds with low-field ligands tend to exhibit complementary colors towards the blue end of the visible spectrum.

An example of a low-field ligand is water (H_2O). Complexes formed with water ligands often absorb light in the blue region, resulting in a complementary color towards the orange end of the spectrum. This can be observed in compounds such as $[Cu(H2O)6]2+$, which appears blue due to its absorption of orange light.

Examples from Various Fields

The effects of different ligands on ΔE, the frequency of light absorbed, and complementary colors can be observed in various fields, including biochemistry, medicine, and art.

Biochemistry

In biochemistry, ligands are essential for the functioning of metalloproteins. Metalloproteins are proteins that contain a metal ion within their structure, often coordinated by ligands. One example is hemoglobin, which contains iron (Fe) coordinated by porphyrin ligands. The absorption of light by the iron-porphyrin complex results in the observed red color of oxygenated blood. When oxygen binds to the iron ion, it causes a change in the ligand field, leading to a shift in the absorption spectrum and a change in color from red to blue.

Medicine

In medicine, coordination compounds are used in various applications, including imaging and cancer treatment. For instance, technetium-99m (Tc-99m) complexes are commonly used in nuclear medicine imaging. These complexes, such as Tc-99m sestamibi, absorb light in the green region, leading to a complementary color towards the red end of the spectrum. This property allows for the visualization and detection of specific tissues or organs in the body.

In cancer treatment, cisplatin is a widely used chemotherapy drug that forms coordination complexes with DNA. The absorption of light by cisplatin-DNA complexes can induce DNA damage and cell death. However, the specific ligands in cisplatin and their effects on the observed complementary color are not directly relevant in this context.

Art

The understanding of ligands and their effects on complementary colors is also relevant in the field of art. Artists often use pigments and dyes that contain coordination compounds to achieve specific colors. For example, cobalt blue is a pigment that contains cobalt coordinated by ligands. The absorption of light by the cobalt complex results in the observed blue color.

Counterarguments

While the effects of different ligands on ΔE, frequency of light absorbed, and complementary colors are well-established, it is important to consider some counterarguments. One counterargument is that the ligand field theory, which explains the effects of different ligands, is a simplification of the complex reality. In reality, the interactions between ligands and metal ions are influenced by various factors, such as steric effects and solvent interactions, which can affect the observed complementary colors.

Additionally, the qualitative effects of ligands on complementary colors do not provide precise quantitative predictions. The observed colors are influenced by multiple factors, including the ligands, metal ions, and their respective concentrations. Therefore, predicting the exact complementary color solely based on ligands can be challenging.

The Influence of Ligand Exchange on the Color of Copper (II) and Cobalt (II) Complexes
Introduction

This aims to explore the role of ligand exchange in altering the observed color of complex species involving copper (II) and cobalt (II) ions, with a specific focus on water, ammonia, hydroxide, and chloride ligands.

Copper (II) Complexes

Copper (II) ions possess a d9 electronic configuration, with three unpaired electrons in the d-orbital. This electronic configuration allows copper (II) ions to form a variety of colorful complexes due to the presence of empty d-orbitals that can interact with ligand orbitals. The color of copper (II) complexes arises from the absorption of light in the visible region, corresponding to the energy gap between the d-orbitals of copper (II) and various ligands.

Copper (II) Complexes with Water Ligands

When copper (II) ions are coordinated with water molecules, the resulting complex is often blue in color. The octahedral complex $[Cu(H_2O)_6]^{2+}$ is a classic example. The blue color is attributed to the absorption of light in the orange region of the visible spectrum. This absorption arises from the excitation of an electron from the d-orbital of copper (II) to a higher energy orbital, facilitated by the presence of water ligands. The energy gap between the d-orbital and the ligand orbitals determines the observed color, with larger energy gaps corresponding to shorter wavelengths and colors towards the blue end of the spectrum.

Ligand Exchange: Ammonia Replacing Water

Ligand exchange plays a crucial role in changing the color of copper (II) complexes. For instance, when water is replaced by ammonia ligands in the complex $[Cu(H_2O)_6]^{2+}$, the color changes from blue to deep blue. This change in color is due to the different bonding properties of water and ammonia ligands. While both ligands possess a lone pair of electrons that can interact with the empty d-orbitals of copper (II), ammonia has a stronger field strength than water. This stronger field splits the d-orbitals to a greater extent, resulting in a larger energy gap between the d-orbitals and the ligand orbitals. As a consequence, the absorption of light occurs at shorter wavelengths, shifting the color towards the deep blue region.

Cobalt (II) Complexes

Cobalt (II) ions possess a d7 electronic configuration, with three unpaired electrons in the d-orbital. Similar to copper (II) complexes, cobalt (II) complexes exhibit a wide range of colors due to ligand interactions with the d-orbitals. The observed color of cobalt (II) complexes depends on the nature of the ligands coordinated to the metal ion.

Cobalt (II) Complexes with Water Ligands

Cobalt (II) complexes with water ligands often exhibit a pink color. This color arises from the absorption of light in the green region of the visible spectrum. The energy gap between the d-orbitals of cobalt (II) and the ligand orbitals determines the observed color, with larger energy gaps leading to shorter wavelengths and colors towards the red end of the spectrum.

Ligand Exchange: Hydroxide Replacing Water

When water ligands in cobalt (II) complexes are replaced by hydroxide ligands, the color changes from pink to blue. This change in color is due to the different field strength of hydroxide compared to water. Hydroxide ligands possess a stronger field strength than water, resulting in a larger splitting of the d-orbitals. Consequently, the energy gap between the d-orbitals and the ligand orbitals increases, leading to the absorption of light at shorter wavelengths, which corresponds to the blue region of the visible spectrum.

Counterargument: Ligand Exchange: Chloride Replacing Water

While the replacement of water by chloride ligands in cobalt (II) complexes does result in a change in color, it does not lead to a blue color as observed with hydroxide ligands. Instead, the color changes to green. This discrepancy can be attributed to the different field strength of chloride compared to hydroxide. Chloride ligands possess a weaker field strength than hydroxide, resulting in a smaller splitting of the d-orbitals. Consequently, the energy gap between the d-orbitals and the ligand orbitals decreases, leading to the absorption of light at longer wavelengths, which corresponds to the green region of the visible spectrum.

Stereoisomerism in transition element complexes

Types of Stereoisomerism in Complexes

Stereoisomerism refers to the phenomenon where two or more compounds have the same molecular formula and connectivity of atoms but differ in the spatial arrangement of their atoms. In the field of coordination chemistry, stereoisomerism is commonly observed in metal complexes due to the presence of different ligands around a central metal atom. This article aims to describe the different types of stereoisomerism shown by complexes, with a particular focus on geometrical (cis-trans) isomerism. Furthermore, the discussion will cover complexes involving bidentate ligands, such as [Pt(NH3)2Cl2], [Co(NH3)4(H2O)2]2+, and [Ni(H2NCH2CH2NH2)2(H2O)2]2+.

Geometrical (Cis-Trans) Isomerism

Geometrical isomerism is a type of stereoisomerism that arises when two different groups or ligands are attached to a central atom in a coordination complex, resulting in different spatial arrangements. The most common example of geometrical isomerism is cis-trans isomerism, observed in complexes with a square planar or octahedral geometry.

In square planar complexes, the central metal atom is surrounded by four ligands arranged in a plane. Cis-trans isomerism occurs when two different ligands occupy adjacent positions in the complex. For instance, consider the complex [Pt(NH3)2Cl2]. The cis-isomer depicts the arrangement where the two ammonia (NH3) ligands are positioned next to each other, whereas in the trans-isomer, one NH3 ligand is positioned on the opposite side of the other. The cis and trans isomers can be illustrated in the following manner:

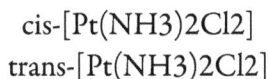

cis-[Pt(NH3)2Cl2]

trans-[Pt(NH3)2Cl2]

On the other hand, octahedral complexes have six ligands arranged around the central metal atom. Cis-trans isomerism in octahedral complexes arises when there are two sets of identical ligands, with each set occupying adjacent positions. One example of an octahedral complex exhibiting cis-trans isomerism is [Co(NH3)4(H2O)2]2+. In the cis-isomer, the two water (H2O) ligands are adjacent to each other, while in the trans-isomer, one H2O ligand is opposite to the other. The cis and trans isomers can be represented as follows:

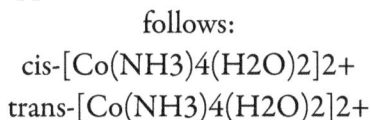

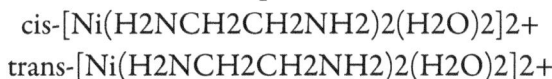

cis-[Co(NH3)4(H2O)2]2+

trans-[Co(NH3)4(H2O)2]2+

Similarly, [Ni(H2NCH2CH2NH2)2(H2O)2]2+ is another example of an octahedral complex exhibiting cis-trans isomerism. In the cis-isomer, the two ethylenediamine (H2NCH2CH2NH2) ligands are adjacent to each other, while in the trans-isomer, one H2NCH2CH2NH2 ligand is opposite to the other. The cis and trans isomers can be represented as follows:

cis-[Ni(H2NCH2CH2NH2)2(H2O)2]2+

trans-[Ni(H2NCH2CH2NH2)2(H2O)2]2+

The presence of cis-trans isomerism in square planar and octahedral complexes can have significant implications on their chemical and physical properties. The trans-isomer of [Pt(NH3)2Cl2] is typically more stable and less reactive compared to its cis-isomer, as an example. This difference in stability arises due to steric effects and electronic factors caused by the proximity of the ligands. In some cases, the cis-isomer may undergo isomerization reactions to convert into the more stable trans-isomer.

Bidentate Ligands and Stereoisomerism

Bidentate ligands are ligands that can form two bonds with a central metal atom, thereby occupying two coordination sites. The presence of bidentate ligands in a complex introduces additional possibilities for stereoisomerism. The most common type of stereoisomerism associated with bidentate ligands is known as chelate isomerism.

Chelate isomerism arises when a bidentate ligand can bind to the central metal atom in two different ways, resulting in distinct spatial arrangements. This phenomenon is particularly significant in complexes with octahedral geometry, as bidentate ligands can occupy two adjacent coordination sites. The two possible arrangements of a bidentate ligand in an octahedral complex can be cis-cis or trans-trans.

Let's consider an example to understand chelate isomerism better. The complex [Ni(H2NCH2CH2NH2)2(H2O)2]2+ contains two bidentate ligands, ethylenediamine (H2NCH2CH2NH2). In the cis-cis isomer, both ethylenediamine ligands are arranged in a cis configuration, occupying adjacent coordination sites. In the trans-trans isomer, both ethylenediamine ligands are arranged in a trans configuration, with one ligand opposite to the other. The cis-cis and trans-trans isomers can be represented as follows:

cis-cis-[Ni(H2NCH2CH2NH2)2(H2O)2]2+

trans-trans-[Ni(H2NCH2CH2NH2)2(H2O)2]2+

Chelate isomerism can have a significant impact on the stability and reactivity of complexes. The different spatial arrangements of bidentate ligands can lead to variations in the steric and electronic properties around the central metal atom, affecting the overall chemical behavior of the complex. Furthermore, chelate isomerism can also influence the coordination geometry of the complex, as the presence of bidentate ligands can introduce distortions or strain in the ligand arrangement.

Types of Steroisomerism in Complexes: Optical Isomerism

We will explore the types of stereoisomerism exhibited by complexes, focusing on optical isomerism, with specific examples involving bidentate ligands.

Optical Isomerism

Optical isomerism, also known as enantiomerism, is a type of stereoisomerism that arises due to the presence of a chiral center in a molecule or complex. A chiral center is an atom that is bonded to four different substituents, resulting in the lack of a plane of symmetry. Optical isomers, also called enantiomers, cannot be superimposed onto their mirror images. Instead, they exhibit distinct left-handed (L) and right-handed (D) forms, named after the Latin terms "levorotatory" and "dextrorotatory" respectively, based on their behavior in plane-polarized light.

In the context of coordination complexes, optical isomerism is commonly observed in complexes containing chiral ligands or complexes with a chiral metal center. The chiral ligands can be monodentate or multidentate, where multidentate ligands, such as bidentate ligands, play a significant role in the formation of optically active complexes.

Example 1: $[Ni(H_2NCH_2CH_2NH_2)_3]^{2+}$

Consider the complex $[Ni(H_2NCH_2CH_2NH_2)_3]^{2+}$. This complex contains a central nickel ion coordinated to three molecules of the bidentate ligand $H_2NCH_2CH_2NH_2$, commonly known as ethylenediamine. Each ethylenediamine ligand coordinates to the nickel ion through two nitrogen atoms, forming a five-membered chelate ring.

The presence of the chiral center in the ethylenediamine ligand gives rise to the possibility of optical isomerism in this complex. The nitrogen atoms in the ethylenediamine ligands can be arranged in either a clockwise or counterclockwise manner around the central nickel ion, resulting in two enantiomers, designated as Λ (lambda) and Δ (delta). These enantiomers cannot be superimposed onto each other and exhibit different optical activities in plane-polarized light.

Example 2: $[Ni(H_2NCH_2CH_2NH_2)_2(H_2O)_2]^{2+}$

Another example of optical isomerism involving bidentate ligands is the complex $[Ni(H_2NCH_2CH_2NH_2)_2(H_2O)_2]^{2+}$. In this case, the nickel ion is coordinated to two molecules of the bidentate ligand ethylenediamine and two water molecules. The coordination geometry around the nickel ion is octahedral, with the ethylenediamine ligands occupying two adjacent coordination sites and the water molecules occupying the remaining two sites.

Similar to the previous example, the ethylenediamine ligands in this complex can exhibit two different arrangements around the central nickel ion, resulting in two enantiomers. However, in this case, the presence of the water molecules introduces additional stereoisomerism possibilities. The two water molecules can be arranged in either a cis or trans configuration with respect to the ethylenediamine ligands. This results in four possible stereoisomers: Λ-cis, Λ-trans, Δ-cis, and Δ-trans.

Importance and Applications

The presence of optical isomerism in coordination complexes has significant implications in various fields of chemistry. One of the most notable applications is in asymmetric catalysis, where chiral complexes play a crucial role in facilitating enantioselective reactions. Enantioselective catalysis allows for the production of single enantiomers of a chiral compound, which is of utmost importance in the pharmaceutical industry. Many drugs and biologically active molecules exist as enantiomeric pairs, where one enantiomer may exhibit therapeutic effects while the other may possess adverse side effects. By utilizing chiral complexes as catalysts, chemists can selectively produce the desired enantiomer, leading to safer and more effective drugs.

Additionally, the study of optical isomerism in coordination complexes contributes to our understanding of stereochemistry and the molecular basis of chirality. It provides insight into the factors governing the stability and reactivity of enantiomers, as well as the interactions between chiral molecules and their surrounding environment. Understanding the behavior of chiral complexes is essential for designing new catalysts, developing efficient separation techniques, and predicting the biological activity of enantiomeric drugs.

Counterarguments and Limitations

While optical isomerism in coordination complexes offers numerous benefits and applications, it is important to acknowledge certain limitations and counterarguments. One counterargument is based on the fact that enantiomers have identical physical and chemical properties in an achiral environment. Therefore, the separation and identification of enantiomers can be challenging, requiring specialized techniques such as chiral chromatography or the use of chiral auxiliaries.

Another limitation is that not all coordination complexes exhibit optical isomerism. For optical isomerism to occur, the complex must possess a chiral center or contain chiral ligands. In many cases, coordination complexes with symmetrical ligands or octahedral coordination geometries lack chiral centers, resulting in the absence of optical isomerism.

Analyzing the Overall Polarity of Complexes

Introduction

Polarity refers to the distribution of electric charge within a molecule or complex, which greatly influences its physical and chemical properties. Complexes, also known as coordination compounds, are formed by the association of a central metal ion with surrounding ligands. Determining the overall polarity of complexes requires an in-depth analysis of the polar nature of both the metal ion and the ligands, as well as their spatial arrangement within the complex.

Factors Influencing Polarity

To deduce the overall polarity of a complex, several factors must be considered. These include the electronegativity of the metal ion and the ligands, the geometry of the complex, and the presence of any dipoles within the ligands or the complex as a whole. The electronegativity of an element refers to its ability to attract electrons towards itself in a chemical bond. The higher the electronegativity difference between the metal ion and the ligands, the more polar the complex is likely to be.

Metal Ion Polarity

The polarity of a complex can be influenced by the nature of the metal ion involved. Transition metal ions, which commonly form complexes, often have partially filled d-orbitals that can interact with ligands to form coordination bonds. The electronegativity of the metal ion can determine the polarity of the complex. For example, in a complex where the metal ion has a high electronegativity, it will attract electrons more strongly, resulting in a polar complex. Conversely, a metal ion with low electronegativity will lead to a less polar complex.

Ligand Polarity

The nature of the ligands surrounding the metal ion in a complex also contributes to its overall polarity. Ligands can be classified as either polar or nonpolar, depending on their electronegativity and molecular structure. Polar ligands possess atoms with high electronegativity, such as oxygen or nitrogen, which create partial charges within the ligand. Nonpolar ligands, on the other hand, have atoms with similar electronegativities, resulting in a symmetrical distribution of charge. The presence of polar ligands in a complex can enhance its overall polarity.

Coordination Geometry

The geometry of a complex greatly affects its polarity. Different coordination geometries, such as octahedral, square planar, or tetrahedral, can give rise to different polar properties. For instance, in an octahedral complex, the ligands are arranged in a symmetrical manner around the central metal ion, resulting in a nonpolar complex. Conversely, a square planar complex, where the ligands lie in the same plane and create a more asymmetrical arrangement, can lead to a polar complex. The geometry of a complex is determined by factors such as the size of the metal ion and the nature of the ligands.

Dipole Moments

Dipole moments can provide valuable insights into the overall polarity of a complex. A dipole moment represents the separation of positive and negative charges within a molecule or complex. If a complex possesses a net dipole moment, it is considered polar. The dipole moment is determined by the vector sum of individual bond dipoles within the complex. For example, if a complex consists of polar ligands and the vectors of their dipole moments do not cancel each other out, the complex will have a net dipole moment and be polar.

Examples from Different Fields

The concept of polarity extends beyond the realm of chemistry and finds applications in various fields. In biology, for instance, the polarity of molecules such as amino acids and nucleotides plays a crucial role in protein folding and DNA structure. The polar nature of these molecules enables the formation of hydrogen bonds, which contribute to the stability and functionality of biological macromolecules. In physics, the concept of polarity is utilized in the study of polarized light and its interaction with matter. By analyzing the orientation and alignment of electromagnetic waves, scientists can deduce information about the polarity and structure of molecules.

Counterarguments

While the factors mentioned above provide a framework for deducing the overall polarity of complexes, it is essential to consider counterarguments. In certain cases, the overall polarity of a complex may not be solely determined by the metal ion, ligands, geometry, or dipole moments. Other factors, such as solvent effects, temperature, and steric hindrance, can influence the overall polarity. Solvent effects, for example, may disrupt the coordination bonds and alter the distribution of charges within the complex, leading to changes in polarity. Additionally, the presence of multiple metal ions or ligands with different polarities can complicate the analysis of overall polarity.

Stability constants, K_{stab}

Stability Constant: Definition and Significance

Introduction

The stability constant, denoted as Kstab, is a fundamental concept in the field of chemistry that measures the equilibrium constant for the formation of a complex ion in a solvent from its constituent ions or molecules. It provides valuable insight into the strength of the bond formed between the constituents and the stability of the resulting complex. This stability constant plays a crucial role in various fields, including inorganic chemistry, biochemistry, and environmental chemistry. In this essay, we will delve into the details of stability constants, exploring their definition, significance, and applications.

Definition of Stability Constant

The stability constant, often referred to as the formation constant, is a quantitative measure of the strength of the bond formed between the constituent ions or molecules in a complex. It is defined as the equilibrium constant for the formation of the complex ion from its constituents in a specific solvent. The stability constant is expressed as the ratio of the concentration of the complex ion to the product of the concentrations of the constituent ions or molecules, each raised to the power of their stoichiometric coefficients.

Mathematically, the stability constant, Kstab, can be represented as:

$$Kstab = [Complex\ ion] / ([Constituent\ ion\ 1]^{n1} \times [Constituent\ ion\ 2]^{n2} \times ...)$$

where [Complex ion] represents the concentration of the complex ion, [Constituent ion 1] and [Constituent ion 2] represent the concentrations of the constituent ions or molecules, and n1 and n2 are the stoichiometric coefficients of the constituent ions or molecules, respectively.

Significance of Stability Constant

The stability constant holds immense significance as it provides valuable information regarding the stability and strength of the bond formed in a complex. It allows chemists to quantify and compare the stability of different complexes, aiding in the prediction of their behavior under various conditions. Some key aspects of the significance of stability constants are discussed below.

Complex Formation and Dissociation

The stability constant plays a crucial role in understanding the formation and dissociation of complexes. A higher stability constant indicates a stronger bond and a more stable complex. Conversely, a lower stability constant suggests a weaker bond and a less stable complex. By comparing the stability constants of different complexes, chemists can determine which complex is more likely to form and persist in a given solvent.

Equilibrium Shifts

Stability constants also influence the equilibrium shifts in chemical reactions involving complex formation. According to Le Chatelier's principle, when a system is subjected to a change, it tends to counteract that change and reach a new equilibrium. In the context of complex formation, an increase in the concentration of one of the constituent ions or molecules can shift the equilibrium towards the formation of the complex if the stability constant is sufficiently high. Conversely, a decrease in the concentration of a constituent ion or molecule can lead to the dissociation of the complex.

Metal Ion Speciation

In inorganic chemistry, stability constants have significant implications for the speciation of metal ions in solution. Metal ions can form complexes with various ligands, and the stability constant determines the likelihood of complex formation. The speciation of metal ions is crucial in understanding their bioavailability, toxicity, and environmental behavior. For example, stability constants play a vital role in predicting the bioavailability of essential metal ions, such as iron and copper, in biological systems.

Coordination Chemistry

The stability constant concept is particularly relevant in coordination chemistry, where metal ions form complexes with ligands. Stability constants provide essential information about the strength of the metal-ligand bond, which determines the stability of the resulting complex. This knowledge is crucial for designing and synthesizing coordination compounds with desired properties, such as catalytic activity and magnetic behavior.

Biochemical Reactions

In biochemistry, stability constants are valuable in understanding and predicting the behavior of biomolecules, such as proteins and nucleic acids. Metal ions often interact with biomolecules, forming complexes that play essential roles in various biological processes. Stability constants help in determining the strength of these interactions, shedding light on the stability and function of the resulting complexes. For example, stability constants are used to study metalloproteins, which are proteins that contain metal ions in their active sites and are involved in crucial biological processes, including enzyme catalysis and electron transfer.

Determination of Stability Constants

The determination of stability constants is a complex and challenging task. Several experimental techniques and theoretical models have been developed to measure and predict stability constants accurately. Some commonly used methods for determining stability constants are discussed below.

Potentiometric Titrations

Potentiometric titrations are widely employed to determine stability constants. In this technique, the pH of a solution containing the constituent ions and the complex ion is measured as a function of the added titrant. The pH changes observed during the titration provide information about the formation and dissociation of the complex, allowing the calculation of the stability constant.

Spectroscopic Methods

Spectroscopic techniques, such as UV-Vis spectroscopy and fluorescence spectroscopy, are often used to determine stability constants. These methods rely on the fact that the formation of a complex results in changes in the absorbance or emission properties of the system. By monitoring these changes, the stability constant can be inferred.

Computational Methods

Computational methods, including molecular dynamics simulations and quantum mechanics calculations, are increasingly employed to predict stability constants. These methods utilize theoretical models and algorithms to calculate the energy of the complex and its constituent ions or molecules. The stability constant can then be obtained from the energy difference between the complex and the constituents.

Applications of Stability Constants

The knowledge of stability constants finds applications in various fields, contributing to a deeper understanding of complex formation and its implications. Some notable applications are discussed below.

Environmental Chemistry

In environmental chemistry, stability constants are used to predict the behavior of metal ions in natural waters and soils. The stability constant of a metal-ligand complex determines its solubility, bioavailability, and potential for transport and accumulation in the environment. These factors are crucial in assessing the environmental impact of pollutants and designing remediation strategies.

Pharmaceutical Chemistry

Stability constants play a critical role in pharmaceutical chemistry, particularly in drug development and formulation. Many drugs interact with metal ions in the body, and stability constants help in understanding these interactions. For instance, stability constants are used to determine the binding affinity of drugs to metal ions, which affects their pharmacokinetics and therapeutic efficacy.

Analytical Chemistry

Stability constants have numerous applications in analytical chemistry. They are utilized in the design and optimization of complexometric titrations, a common analytical technique used to determine metal ion concentrations. Stability constants help in selecting suitable complexing agents and determining the optimal pH conditions for the titration.

Geochemistry

Geochemists employ stability constants to study the behavior of metal ions in geological systems. Stability constants provide insights into the formation and stability of mineral complexes, influencing the distribution and cycling of elements in Earth's crust. This knowledge is crucial in understanding processes such as weathering, mineral dissolution, and ore formation.

Counterarguments and Limitations

While stability constants are highly valuable, it is essential to acknowledge their limitations and potential counterarguments. One limitation is that stability constants are typically determined under specific conditions, such as a particular solvent and temperature. Therefore, they may not accurately reflect the stability of complexes under different environmental conditions. Additionally, stability constants are often measured or predicted for idealized systems, neglecting the impact of factors like ionic strength, pH, and complexation kinetics, which can significantly influence complex stability.

Furthermore, stability constants are influenced by the assumptions and approximations made in the calculations. Different theoretical models and computational methods can yield different stability constants, leading to discrepancies and uncertainties. It is crucial to consider these limitations and carefully interpret stability constant values in the context of the specific system being studied.

Before delving into the specifics of the Kstab expression for a complex, it is essential to establish a solid understanding of the concept of equilibrium and the factors that influence it. Equilibrium refers to a state in which the rates of the forward and reverse reactions are equal, resulting in a constant concentration of reactants and products over time. In other words, the system no longer shows any net change in the concentrations of the species involved.

In a chemical reaction involving a complex formation, one or more reactants combine to form a complex, also known as a coordination compound. A complex consists of a central metal atom or ion surrounded by ligands, which are atoms, ions, or molecules that coordinate with the metal through coordinate covalent bonds. The formation of a complex involves a dynamic process in which ligands bind to the metal center, leading to the establishment of an equilibrium between the free metal ions and the complex.

The equilibrium constant, Kstab, for a complex can be expressed using two different approaches: the concentration-based expression and the formation constant expression. In the concentration-based expression, the concentrations of the species involved in the reaction are used, while in the formation constant expression, the stability constants of the complex and its constituents are employed.

The concentration-based expression for the Kstab of a complex can be written as follows:

Kstab = [Complex]/([Metal]^[a]) [Ligand1]^[b] [Ligand2]^[c] *

In this expression, [Complex] represents the concentration of the complex, [Metal] refers to the concentration of the metal ion, [Ligand1], [Ligand2], and so on, represent the concentrations of the ligands involved, and a, b, c, and so forth denote the stoichiometric coefficients of the respective species in the balanced chemical equation.

It is important to note that the concentrations of the reactants and products are generally expressed in molar units (mol/L) when determining the Kstab. Furthermore, the stoichiometric coefficients present in the expression correspond to the balanced chemical equation, signifying the quantity of molecules participating in the reaction for each species.

To illustrate the application of the concentration-based expression for the Kstab of a complex, let us consider the formation of a simple coordination compound: [Cu(NH3)4]2+. In this case, the reaction can be represented as follows: Cu2+ + 4NH3 ⇌ [Cu(NH3)4]2+

The corresponding Kstab expression can be written as:

Kstab = [Cu(NH3)4]2+/([Cu2+] * [NH3]^4)

Here, [Cu(NH3)4]2+ represents the concentration of the complex, [Cu2+] denotes the concentration of the free copper ions, and [NH3] represents the concentration of ammonia ligands.

Although the concentration-based expression provides a practical approach to determining the Kstab of a complex, it does not take into account the inherent stability of the complex and its constituents. This limitation is overcome by the formation constant expression, which incorporates the stability constants of the complex and its constituents.

The formation constant expression for the Kstab of a complex can be represented as follows:

Kstab = (Kf1 Kf2 ... Kfn)/(Kr1 Kr2 ... Krn)

In this expression, Kf1, Kf2, and so on, represent the formation constants of the complex and its constituents, while Kr1, Kr2, and so forth denote the dissociation constants of the complex and its constituents. The formation constant (Kf) represents the equilibrium constant for the formation of the complex from its constituents, while the dissociation constant (Kr) corresponds to the equilibrium constant for the dissociation of the complex back into its constituents.

To further illustrate the application of the formation constant expression, let us consider the complex formation between iron(III) ion (Fe^{3+}) and the ligand ethylenediamine (en). The reaction can be represented as follows:

$$Fe^{3+} + 2en \rightleftharpoons [Fe(en)2]^{3+}$$

The corresponding formation constant expression can be written as:

$$K_{stab} = ([Fe(en)2]^{3+})/([Fe^{3+}] [en]^2) = K_f/[K_{r1} K_{r2}]$$

Here, K_f represents the formation constant for the complex $[Fe(en)2]^{3+}$, while K_{r1} and K_{r2} denote the dissociation constants for Fe^{3+} and ethylenediamine, respectively.

It is worth noting that the formation constant expression provides a more comprehensive understanding of the stability of the complex, taking into account both the formation and dissociation processes. By considering the individual stability constants of the complex and its constituents, researchers can gain insight into the relative strength of the bonds within the complex, as well as the factors that influence its stability.

Here are five sample problems and their solutions using Kstab expressions in chemistry:

Problem: Calculate the equilibrium concentration of hydrogen gas (H2) in a reaction mixture if the initial concentration of nitrogen gas (N2) is 0.2 M and the equilibrium constant (Kstab) for the reaction $N2 + 3H2 \leftrightarrow 2NH3$ is 0.05 at a certain temperature.

Solution: Since the reaction is in equilibrium, we can use the equilibrium constant expression: $K_{stab} = [NH3]^2 / ([N2] [H2]^3)$. Let's assume the equilibrium concentration of H2 is x M. Plugging in the given values, we get: $0.05 = (x^2) / (0.2 (x^3)^3)$. Solving this equation will give us the equilibrium concentration of H2.

Problem: A reaction has a Kstab value of 2.0×10^4 at a certain temperature. If the initial concentration of reactant A is 0.5 M and the equilibrium concentration of reactant B is 0.1 M, calculate the equilibrium concentration of product C.

Solution: The equilibrium constant expression for the reaction is: $K_{stab} = [C] / ([A] [B])$. Let's assume the equilibrium concentration of C is x M. Plugging in the given values, we get: $2.0 \times 10^4 = x / (0.5\ 0.1)$. Solving this equation will give us the equilibrium concentration of C.

Problem: A reaction has a Kstab value of 0.01 at a certain temperature. If the initial concentration of reactant X is 0.2 M and the equilibrium concentration of reactant Y is 0.05 M, calculate the equilibrium concentration of reactant Z.

Solution: The equilibrium constant expression for the reaction is: $K_{stab} = [Z] / ([X] [Y])$. Let's assume the equilibrium concentration of Z is x M. Plugging in the given values, we get: $0.01 = x / (0.2\ 0.05)$. Solving this equation will give us the equilibrium concentration of Z.

Problem: A reaction has a Kstab value of 5.0×10^6 at a certain temperature. If the initial concentration of reactant P is 0.1 M and the equilibrium concentration of product Q is 0.02 M, calculate the equilibrium concentration of reactant R.

Solution: The equilibrium constant expression for the reaction is: $K_{stab} = ([Q]^2) / ([P] [R])$. Let's assume the equilibrium concentration of R is x M. Plugging in the given values, we get: $5.0 \times 10^6 = (0.02^2) / (0.1\ x)$. Solving this equation will give us the equilibrium concentration of R.

Problem: A reaction has a Kstab value of 0.05 at a certain temperature. If the initial concentration of reactant M is 0.2 M and the equilibrium concentration of reactant N is 0.1 M, calculate the equilibrium concentration of product O.

Solution: The equilibrium constant expression for the reaction is: $K_{stab} = [O] / ([M] [N])$. Let's assume the equilibrium concentration of O is x M. Plugging in the given values, we get: $0.05 = x / (0.2\ 0.1)$. Solving this equation will give us the equilibrium concentration of O.

Ligand Exchanges and the Stability of Complex Ions
Introduction

In chemistry, ligand exchange reactions play a crucial role in the formation and transformation of complex ions. These reactions involve the substitution of one ligand (a molecule or ion) for another in a coordination complex. The stability of the resulting complex ion is determined by the equilibrium constant, known as the stability constant (Kstab), which quantifies the extent of the ligand exchange reaction. A large Kstab value signifies the formation of a stable complex ion, indicating a favorable ligand exchange process. This essay aims to describe and explain ligand exchanges in terms of Kstab values, emphasizing the correlation between a large Kstab and the formation of stable complex ions.

Ligand Exchange Reactions and Stability Constants

Ligand exchange reactions occur when a ligand in a coordination complex is replaced by another ligand. These reactions can be either associative or dissociative, depending on the mechanism by which the ligand substitution takes place. In the associative mechanism, the incoming ligand binds to the metal center before the leaving ligand dissociates, while in the dissociative mechanism, the leaving ligand dissociates before the incoming ligand binds. Regardless of the mechanism, the stability of the resulting complex ion is determined by the stability constant, Kstab, which is a measure of the equilibrium position of the ligand exchange reaction.

Kstab, the stability constant, is calculated by dividing the concentration of the complex ion by the product of the concentrations of the unbound ligands, with each concentration raised to the power of its respective stoichiometric coefficient in the equilibrium equation. Mathematically, it can be expressed as:

$$Kstab = [M(Ln)] / ([M] [L1] [L2] ... [Ln])$$

Where $[M(Ln)]$ represents the concentration of the complex ion, $[M]$ represents the concentration of the metal ion, and $[L1]$, $[L2]$, ..., $[Ln]$ represent the concentrations of the ligands involved in the ligand exchange reaction.

The Role of Stability Constants in Ligand Exchanges

Stability constants provide valuable information about the thermodynamics and kinetics of ligand exchange reactions. A large Kstab value indicates a high stability of the complex ion, suggesting a low energy barrier for the ligand substitution process. On the other hand, a small Kstab value signifies a less stable complex ion and a higher energy barrier for ligand exchange.

Thermodynamic Aspects

From a thermodynamic perspective, the stability constant represents the equilibrium position of the ligand exchange reaction. Higher values of Kstab indicate that the complex ion is more stable, as the concentration of the complex ion in the numerator is greater compared to the product of the concentrations of the free ligands in the denominator. This stability is attributed to the formation of strong bonds between the metal ion and the ligands involved in the complex.

For instance, in the field of bioinorganic chemistry, metalloproteins often contain transition metal ions coordinated by specific ligands. The stability of these metalloproteins is crucial for their biological functions. In the case of hemoglobin, the oxygen transport protein in red blood cells, the iron ion is coordinated by six ligands, including four nitrogen atoms from the porphyrin ring and two histidine residues from the protein. The stability constant of this complex is exceptionally high, leading to a stable oxygen-binding site.

Kinetic Aspects

The stability constant also influences the kinetics of ligand exchange reactions. A large Kstab value indicates a rapid ligand substitution process, as the complex ion is highly stable and readily allows ligands to exchange. On the other hand, a small Kstab value implies a slower ligand exchange process due to the lower stability of the complex ion.

The kinetics of ligand exchange reactions are crucial in various fields, such as catalysis and drug design. For example, in homogeneous catalysis, transition metal complexes act as catalysts, and ligand exchange reactions play a pivotal role in their reactivity. The stability of the complex ion determines the rate at which ligands can be exchanged, impacting the efficiency of the catalytic process.

Factors Influencing Stability Constants

Several factors influence the stability constants of ligand exchange reactions. Understanding these factors is essential for predicting and controlling the stability of complex ions.

Nature of the Metal Ion

The nature of the metal ion strongly influences the stability of the resulting complex ion. Transition metal ions with partially filled d-orbitals exhibit a high affinity for ligands, forming stable complexes. The size and charge of the metal ion also play a role in determining the stability of the complex. Generally, metal ions with higher positive charges form more stable complexes due to the increased electrostatic attraction between the metal ion and the ligands.

Nature of the Ligands

The nature of the ligands involved in the ligand exchange reaction significantly affects the stability of the resulting complex ion. Strong ligands, such as those with multiple donor atoms or high charge densities, tend to form more stable complexes. For example, the ligand ethylenediamine (en), which contains two nitrogen atoms, is a strong donor and forms stable complexes with metal ions.

Chelating Effect

The chelating effect refers to the enhanced stability of a complex due to the formation of multiple bonds between a ligand and a metal ion. Chelating ligands, often referred to as chelators, possess two or more donor atoms capable of coordinating to the metal ion simultaneously. The formation of chelate rings increases the stability of the complex, as the multiple bonds provide additional binding energy. The chelating effect is commonly observed in biological systems, where chelators play a crucial role in metal ion transport and storage.

pH and Redox Conditions

The pH and redox conditions of the reaction medium can significantly influence the stability constants of ligand exchange reactions. Changes in pH alter the protonation states of the ligands and may affect their coordination ability. In the field of bioinorganic chemistry, enzymes often utilize pH and redox gradients to regulate ligand exchange reactions, allowing for precise control of biological processes.

Counterarguments and Limitations

While stability constants provide valuable insights into the stability of complex ions, they have some limitations and counterarguments that need to be considered.

Equilibrium-Based Approach

The stability constants are determined based on the assumption of equilibrium conditions. However, in many practical scenarios, ligand exchange reactions may not reach complete equilibrium due to kinetic factors or other factors, such as solubility limitations or steric hindrance. Therefore, the stability constants obtained under idealized conditions may not accurately represent the ligand exchange process in real systems.

Solvent Effects

The stability constants of ligand exchange reactions can be influenced by the nature of the solvent. Solvent molecules may compete with ligands for coordination sites on the metal ion, affecting the stability of the complex. Additionally, the polarity and dielectric constant of the solvent can influence the strength of the metal-ligand bonds. Therefore, the stability constants determined in a specific solvent may not be directly applicable to other solvent systems.

Characteristic organic reactions

Electrophilic Substitution in Organic Mechanisms

In organic chemistry, electrophilic substitution is a crucial reaction wherein a molecule's functional group or atom is replaced by an electrophile. It is a crucial reaction that allows for the synthesis of a wide range of organic compounds, including pharmaceuticals, agrochemicals, and dyes. In this article, we will delve into the concept of electrophilic substitution, its mechanism, and provide examples of its application in various fields.

Introduction to Electrophilic Substitution

Electrophilic substitution reactions involve the substitution of a functional group or atom in a molecule with an electrophile. An electrophile is a species that is electron-deficient and seeks to gain electrons. It can be a positively charged ion or a neutral molecule with an electron-deficient center. The electrophile attacks the electron-rich region of the molecule, leading to the displacement of the original functional group or atom.

The electrophilic substitution reaction is often depicted as a two-step process: the attack of the electrophile on the substrate and the elimination of the leaving group. The overall reaction can be summarized as follows:

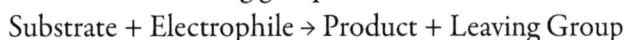

Substrate + Electrophile → Product + Leaving Group

The leaving group is a functional group or atom that is displaced from the substrate during the reaction. It typically carries away a pair of electrons, allowing the electrophile to form a new bond with the substrate.

Mechanism of Electrophilic Substitution

The mechanism of electrophilic substitution involves several steps, including the generation of the electrophile, attack on the substrate, and regeneration of the aromaticity.

Let's explore each step in detail:

1. Generation of the Electrophile

In the first step of the mechanism, the electrophile is generated from a precursor molecule. This precursor molecule can be a Lewis acid, a strong acid, or a radical initiator. The electrophile is formed by accepting an electron pair from the precursor, leading to the formation of a positive charge or an electron-deficient center.

2. Attack on the Substrate

Once the electrophile is generated, it attacks the substrate molecule. The attack occurs at a region of high electron density in the substrate, such as a lone pair of electrons or a π-bond. This region is often associated with a functional group, such as an aromatic ring or a double bond.

The electrophile forms a bond with the substrate, displacing the original functional group or atom. The leaving group carries away a pair of electrons, resulting in the formation of a new bond between the electrophile and the substrate.

3. Regeneration of the Aromaticity

In many electrophilic substitution reactions, the substrate contains an aromatic ring. After the attack of the electrophile, the aromaticity of the ring is disrupted. However, the aromaticity is restored in the final step of the mechanism.

The regeneration of aromaticity involves the loss of a proton or another group from the intermediate formed after the electrophile attack. This step is often facilitated by the presence of a strong base or an acid, which abstracts the proton or group, leading to the formation of the final product with restored aromaticity.

Examples of Electrophilic Substitution in Various Fields

Electrophilic substitution reactions find broad applications in various fields of organic chemistry.

Let's explore a few examples to understand their significance:

1. Pharmaceuticals: Electrophilic substitution reactions play a crucial role in the synthesis of pharmaceutical compounds. One example is the synthesis of salicylic acid, which is a precursor for the widely used pain-relieving drug, aspirin.

Salicylic acid is synthesized through the electrophilic substitution of phenol with acetic anhydride. The electrophile in this reaction is the acylium ion generated from the acetic anhydride. The attack of the electrophile on the phenol leads to the formation of a new bond, displacing the hydroxyl group. The final product, salicylic acid, is obtained after the regeneration of aromaticity.

2. Agrochemicals: Electrophilic substitution reactions are also employed in the synthesis of agrochemicals, such as herbicides and insecticides. For example, the synthesis of the herbicide 2,4-Dichlorophenoxyacetic acid (2,4-D) involves an electrophilic substitution reaction.

In this reaction, a chloroacetic acid derivative acts as the electrophile, which replaces one of the chlorine atoms in the starting material, 2,4-dichlorophenol. The attack of the electrophile results in the formation of the desired product, 2,4-D, which exhibits herbicidal properties.

3. Dyes: Electrophilic substitution reactions are extensively utilized in the synthesis of dyes. A notable example is the synthesis of azo dyes, which are widely used in the textile industry. Azo dyes are characterized by the presence of one or more azo groups
(-N=N-) in their structure.

The synthesis of azo dyes involves the electrophilic substitution of an aromatic amine with a diazonium salt. The diazonium salt acts as the electrophile, and it attacks the amino group of the aromatic amine. This reaction leads to the formation of a new bond, resulting in the incorporation of the azo group into the dye molecule.

Counterarguments and Limitations

While electrophilic substitution reactions are widely utilized in organic chemistry, they do have certain limitations and challenges. One limitation is the regioselectivity of the reaction, which refers to the preference of the electrophile to attack a specific position in the substrate molecule.

In some cases, multiple positions in the substrate can be susceptible to attack by the electrophile. This can result in the formation of multiple products with different substitution patterns. Controlling the regioselectivity of the reaction is a significant challenge, and various strategies, such as directing groups and steric effects, are employed to achieve the desired selectivity.

Another limitation is the potential for side reactions, such as elimination or rearrangement. In some cases, the attack of the electrophile can lead to the formation of unstable intermediates, which may undergo undesired reactions. Careful optimization of reaction conditions and choice of reagents is necessary to minimize these side reactions and enhance the selectivity of the electrophilic substitution reaction.

Addition-Elimination Mechanism in Organic Chemistry
Introduction:

The addition-elimination mechanism is a fundamental concept in organic chemistry that describes a two-step reaction process involving the addition of a molecule to a substrate followed by the elimination of a leaving group. This mechanism plays a crucial role in the synthesis of a wide range of organic compounds and is a key concept for understanding various reactions involving nucleophiles, electrophiles, and leaving groups.

Definition and Explanation:

The addition-elimination mechanism, also known as the nucleophilic substitution-elimination mechanism, involves two distinct steps: addition and elimination. In the addition step, a nucleophile attacks an electrophilic center, resulting in the formation of a new chemical bond. The species that donates a pair of electrons to create a new bond is referred to as the nucleophile, whereas the species that accepts a pair of electrons to create a new bond is known as the electrophile. Once the addition step occurs, the intermediate formed undergoes elimination in the second step. During elimination, a leaving group, which is a substituent or atom that can depart with the electron pair it shared with the rest of the molecule, is expelled. This results in the formation of a new double bond or the regeneration of an aromatic system.

The addition-elimination mechanism is commonly encountered in various organic reactions, including nucleophilic substitutions, such as SN1 and SN2 reactions, and elimination reactions, such as E1 and E2 reactions. These reactions are essential for the synthesis of complex organic molecules and are widely utilized in the pharmaceutical, agrochemical, and materials industries.

Nucleophilic Substitution Reactions:

Nucleophilic substitution reactions are a class of reactions in which a nucleophile replaces a leaving group in a molecule. These reactions can proceed via either an SN1 or an SN2 mechanism, both of which involve an addition-elimination sequence.

In an SN1 reaction, the addition step involves the attack of a nucleophile on an electrophilic center, resulting in the formation of a carbocation intermediate. This carbocation is stabilized by adjacent electron-withdrawing groups or resonance effects. In the elimination step, a leaving group is expelled, and a new bond is formed. An example of an SN1 reaction is the hydrolysis of tert-butyl chloride:

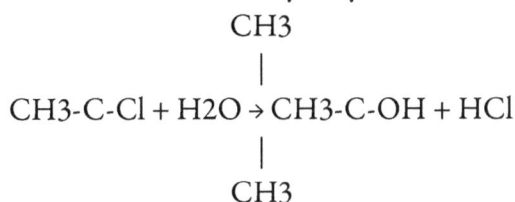

$$CH3$$
$$|$$
$$CH3\text{-}C\text{-}Cl + H2O \rightarrow CH3\text{-}C\text{-}OH + HCl$$
$$|$$
$$CH3$$

In contrast, an SN2 reaction proceeds via a concerted mechanism, where the addition and elimination steps occur simultaneously. The nucleophile directly displaces the leaving group, resulting in the formation of a new bond and the expulsion of the leaving group. An example of an SN2 reaction is the reaction between ethyl bromide and hydroxide ion:

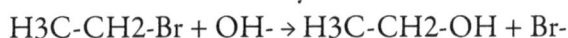

$$H3C\text{-}CH2\text{-}Br + OH\text{-} \rightarrow H3C\text{-}CH2\text{-}OH + Br\text{-}$$

Elimination Reactions:

Elimination reactions involve the removal of a leaving group and the formation of a double bond or the regeneration of an aromatic system. These reactions can proceed via either an E1 or an E2 mechanism, both of which follow an addition-elimination process.

In an E1 reaction, the elimination step occurs after the formation of a carbocation intermediate through the addition step. The leaving group departs, and a new bond is formed, resulting in the formation of an alkene or an aromatic system. An example of an E1 reaction is the dehydration of 2-butanol:

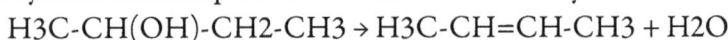

$$H3C\text{-}CH(OH)\text{-}CH2\text{-}CH3 \rightarrow H3C\text{-}CH=CH\text{-}CH3 + H2O$$

On the other hand, an E2 reaction proceeds via a concerted mechanism, similar to the SN2 reaction. The nucleophile, acting as a base, abstracts a proton from the substrate, resulting in the formation of a new bond and the expulsion of the leaving group. An example of an E2 reaction is the dehydrohalogenation of 2-bromobutane:

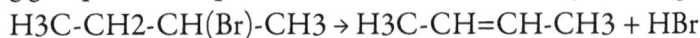

$$H3C\text{-}CH2\text{-}CH(Br)\text{-}CH3 \rightarrow H3C\text{-}CH=CH\text{-}CH3 + HBr$$

Applications and Importance:

The addition-elimination mechanism is of paramount importance in organic chemistry due to its wide applicability and versatility. It is used extensively in the synthesis of organic compounds, including pharmaceuticals, agrochemicals, and materials.

In the pharmaceutical industry, the addition-elimination mechanism is employed in the synthesis of various drugs. For example, nucleophilic substitutions with addition-elimination steps are crucial in the production of antibiotics like penicillin. By selectively replacing a leaving group with a nucleophile, chemists can modify the structure of the antibiotic, enhancing its efficacy or reducing side effects.

Similarly, the addition-elimination mechanism plays a vital role in the synthesis of agrochemicals, such as pesticides and herbicides. By using nucleophilic substitutions or elimination reactions, chemists can develop compounds that selectively target pests or unwanted plants, minimizing environmental impact. Furthermore, the addition-elimination mechanism is utilized in the production of materials with specific properties. For instance, in polymer chemistry, the addition-elimination mechanism is involved in the synthesis of various polymers, such as polyesters and polyamides. By controlling the addition and elimination steps, chemists can tailor the properties of the resulting polymer, such as its mechanical strength, flexibility, and thermal stability.

Counterarguments:

Despite the importance and utility of the addition-elimination mechanism, there are certain limitations and challenges associated with its application. One limitation is the regioselectivity and stereoselectivity of the reaction. Depending on the nature of the nucleophile, leaving group, and substrate, different regioisomers or stereoisomers may be formed. Achieving the desired selectivity can be challenging and requires careful optimization of reaction conditions and choice of reagents. Another challenge is the presence of competing reactions. In some cases, other reaction pathways may compete with the addition-elimination mechanism, leading to undesired side products. For example, in nucleophilic substitutions, the competing elimination pathway can result in the formation of an alkene instead of the desired substitution product.

Shapes of aromatic organic molecules; σ and π bonds
The Shape of Benzene and Other Aromatic Molecules: A Comprehensive Analysis
Introduction

Benzene, a widely known aromatic hydrocarbon, exhibits a unique and captivating structure. Its shape, characterized by a hexagonal ring of carbon atoms with alternating double bonds, has intrigued scientists for centuries. In this essay, we will explore the shape of benzene and other aromatic molecules, examining the concept of sp2 hybridization, σ bonds, and the presence of a delocalized π system. By thoroughly analyzing these factors, we will gain a deeper understanding of the structural properties and stability of these compounds.

Sp^2 Hybridization and σ Bonds

To comprehend the shape of benzene and aromatic molecules, it is crucial to first examine the concept of sp^2 hybridization. Hybridization refers to the mixing of atomic orbitals to form new hybrid orbitals, allowing for the maximum overlap of electron densities and the formation of stronger bonds. In the case of benzene, each carbon atom is sp^2 hybridized, meaning that it utilizes one s orbital and two p orbitals to form three sp^2 hybrid orbitals.

These sp^2 hybrid orbitals arrange themselves in a trigonal planar geometry around each carbon atom. With an angle of 120 degrees between each hybrid orbital, the carbon atoms in benzene are optimally positioned to form σ bonds, which are formed by the overlapping of sp^2 hybrid orbitals. The hexagonal structure of benzene is formed by each carbon atom bonding with its neighboring carbon atoms through σ bonds. The formation of σ bonds between carbon atoms in benzene provides the foundation for its shape. These strong covalent bonds contribute to the stability and rigidity of the molecule, making it a fundamental building block in organic chemistry.

Delocalized π System

While the σ bonds in benzene play a critical role in its structure, the presence of a delocalized π system further defines its shape and aromaticity. The π system arises from the unhybridized p orbitals that are perpendicular to the plane of the carbon atoms. In benzene, each carbon atom contributes one unhybridized p orbital, resulting in a ring of six overlapping p orbitals above and below the plane of the molecule.

The overlapping of these p orbitals forms a delocalized π system, which is responsible for the unique stability and reactivity of aromatic compounds. This delocalization allows for the distribution of electron density across the entire ring, creating a resonance structure where the π electrons are shared between all six carbon atoms. This resonance energy contributes significantly to the overall stability of benzene and other aromatic molecules. The delocalized π system also influences the shape of benzene. Due to the presence of overlapping p orbitals, the π electrons are not restricted to localized bonds between specific carbon atoms. Instead, they occupy a region above and below the plane of the molecule. This leads to a flat, planar structure for benzene, with all carbon atoms lying in the same plane.

Stability of Aromatic Molecules

The unique shape of benzene and other aromatic molecules is closely tied to their exceptional stability. Aromatic compounds exhibit a lower reactivity compared to non-aromatic compounds, making them highly valuable in various fields, including pharmaceuticals, materials science, and organic synthesis. The stability of aromatic molecules can be attributed to several factors. Firstly, the delocalized π system in benzene provides resonance energy, which lowers the overall energy of the molecule. This energy stabilization is a result of the cyclic overlap of p orbitals, allowing for the formation of a large electron cloud. The increased stability arising from this delocalization contributes to the characteristic unreactivity of aromatic compounds. Furthermore, the planar geometry of benzene, resulting from the sp2 hybridization and π system, minimizes the steric strain between adjacent atoms. Steric strain refers to the repulsion between atoms due to their spatial arrangement. In benzene, the optimal arrangement of carbon atoms in a flat plane minimizes these repulsive forces, enhancing its stability.

Counterarguments and Alternative Perspectives

While the explanation presented above provides a solid understanding of the shape of benzene and aromatic molecules, it is essential to acknowledge alternative perspectives and counterarguments. One such counterargument is the concept of non-planar aromatic compounds. Non-planar aromatic compounds, such as cyclooctatetraene (C_8H_8), deviate from the traditional flat structure of benzene. These compounds exhibit a puckered shape, resulting from a phenomenon known as "banana bonding." In banana bonding, the electron density is primarily localized above and below the carbon ring, rather than being delocalized across the entire structure. The presence of non-planar aromatic compounds challenges the notion that all aromatic molecules possess a flat, planar shape. However, it is important to note that these non-planar structures are exceptions to the general trend exhibited by most aromatic compounds. The vast majority of aromatic molecules, including benzene, adopt a planar geometry due to the favorable energetic and electronic properties of the planar structure.

Isomerism: Optical

Introduction

Enantiomers are a fascinating class of molecules that share identical physical and chemical properties, except for their ability to rotate plane-polarized light and their potential biological activity. The study of enantiomers is crucial in various scientific fields, including chemistry, pharmacology, and biochemistry. In this essay, we will explore the concept of enantiomers, their physical and chemical properties, the phenomenon of optical activity, and their significance in biological systems.

Enantiomers: Definition and Characteristics

Enantiomers are a type of stereoisomers, which are molecules that have the same chemical formula and connectivity but differ in their spatial arrangement. Enantiomers're non-superimposable mirror images of one another. This means that although enantiomers possess identical physical and chemical properties, they cannot be aligned by rotation or translation. To illustrate this concept, consider your hands – they are mirror images of each other but cannot be superimposed.

Enantiomers share several essential characteristics. Firstly, they have the same boiling point, melting point, density, solubility, and other physical properties. This is due to the fact that these properties are primarily determined by intermolecular forces and molecular weight, which remain unchanged between enantiomers. Secondly, enantiomers exhibit the same chemical reactivity, as their functional groups and bonding patterns are identical. Therefore, their ability to undergo chemical reactions and form new compounds is indistinguishable.

Optical Activity: Chirality and Polarized Light

The key difference between enantiomers lies in their ability to rotate plane-polarized light. This property is known as optical activity and is a result of their unique structural arrangement. Enantiomers are chiral molecules, meaning they lack an internal plane of symmetry. This lack of symmetry causes the interaction of enantiomers with plane-polarized light to be different, resulting in the rotation of the plane of polarization.

To understand this phenomenon, let's consider an example. Imagine two enantiomers, (+)-carvone and (-)-carvone, which are found in spearmint and caraway seeds, respectively. When a beam of plane-polarized light passes through a solution of
(+)-carvone, the plane of polarization will rotate clockwise. Conversely, when the same light passes through a solution of
(-)-carvone, the plane of polarization will rotate counterclockwise. The specific rotation angle depends on factors such as temperature, solvent, and concentration.

Importance in Biological Systems

The significance of enantiomers in biological systems cannot be overstated. Many biological processes are chiral in nature, and the interaction between enantiomers and biological receptors can lead to vastly different outcomes. The best-known example of this is the thalidomide tragedy. Thalidomide was a drug prescribed to pregnant women in the 1950s to alleviate morning sickness. However, it was later discovered that one enantiomer caused severe birth defects, while the other enantiomer had the desired therapeutic effects.

This example highlights the importance of considering enantiomeric purity in drug design and development. Pharmaceutical companies must carefully analyze and separate enantiomers to ensure the efficacy and safety of their products. The field of enantioselective synthesis has emerged to address this challenge, aiming to selectively produce one enantiomer over the other.

Enantioselective Reactions and Resolution

Enantioselective reactions are chemical transformations that yield predominantly one enantiomer over the other. These reactions are essential in the synthesis of chiral compounds, including pharmaceuticals, agricultural chemicals, and flavors. One widely used method to obtain enantiomerically pure compounds is through asymmetric catalysis. By employing chiral catalysts, which are themselves enantiomers, chemists can influence the reaction pathway to favor the formation of a specific enantiomer. Another approach to obtain enantiopure compounds is through resolution. Resolution involves the physical separation of enantiomers, typically by exploiting their different interactions with a chiral resolving agent. For example, a racemic mixture of enantiomers can be reacted with a chiral acid or base, forming diastereomers that can be separated using conventional techniques. Once separated, the enantiomers can be converted back to their pure form.

Analytical Techniques for Enantiomeric Analysis

Accurate analysis and characterization of enantiomers are crucial in both research and industry. Several analytical techniques are available to determine the enantiomeric composition of a sample. One of the most widely used methods is chiral chromatography, which employs chiral stationary phases to separate enantiomers based on their affinity for the stationary phase.

Other techniques include circular dichroism spectroscopy and nuclear magnetic resonance (NMR) spectroscopy. Circular dichroism measures the differential absorption of left and right circularly polarized light, providing information about the chiral properties of a molecule. NMR spectroscopy can also be employed, as the chemical shifts of enantiomers in a chiral environment differ due to their distinct interactions with the surrounding molecules.

Counterarguments and Limitations

While enantiomers generally exhibit identical physical and chemical properties, there are exceptions to this rule. In some cases, enantiomers may display slight differences in their properties due to their interactions with other chiral molecules or environments. These differences, often referred to as weak enantiomeric discrimination, can be observed in certain physical properties such as specific rotation, melting point, or solubility. Additionally, it is important to note that biological systems can sometimes differentiate between enantiomers, resulting in different biological activities. This can occur when enantiomers interact with chiral biomolecules like enzymes or receptors. In such cases, one enantiomer may bind more tightly or have a different pharmacological effect compared to its mirror image. Therefore, it is crucial to evaluate the biological activity of each enantiomer separately.

Optically Active and Racemic Mixture: Unveiling the World of Chirality

Introduction

Optically active substances and racemic mixtures play a significant role in various scientific fields, including chemistry, biology, pharmacology, and material science. Understanding the concepts of chirality, enantiomers, and racemization is crucial for comprehending the behavior and properties of these compounds. This article aims to explain and explore the terms "optically active" and "racemic mixture" in detail, providing examples from different disciplines and addressing counterarguments objectively.

Chirality: The Origin of Optical Activity

Chirality is a fundamental concept in chemistry that arises from the presence of a central atom or a molecular arrangement that cannot be superimposed on its mirror image. Such molecules are called chiral, and their mirror images are referred to as enantiomers. Enantiomers share identical physical and chemical properties, except for their interaction with plane-polarized light. This phenomenon is known as optical activity.

Optically Active Substances

Optically active substances rotate the plane of polarization of plane-polarized light as it passes through them. These compounds can be divided into two categories: dextrorotatory (d-) and levorotatory (l-), based on the direction of rotation. If the light is rotated clockwise (to the right) when looking towards the light source, it is referred to as dextrorotation. Conversely, if the light is rotated counterclockwise (to the left), it is considered levorotation.

To determine the extent of optical activity, a polarimeter is used. This device measures the angle of rotation, termed specific rotation (α), in degrees. The specific rotation is defined as: $\alpha = \alpha\_observed / (c \times l)$

Where $\alpha\_observed$ is the observed rotation, c is the concentration of the substance in g/mL, and l is the path length of the polarimeter tube in decimeters.

Enantiomers: A Mirror Image Duality

Enantiomers exhibit identical physical properties, such as boiling point, melting point, and solubility, but they differ in their interaction with other chiral compounds (e.g., enzymes) and the rotation of plane-polarized light. Enantiomers have an equal but opposite specific rotation (α) and are often denoted as (+) and (-) or (d-) and (l-), respectively.

A prominent example of optically active substances can be found in the field of pharmacology. Many drugs exist as enantiomers, and their biological activity can vary significantly depending on the enantiomeric form. For instance, the drug Ibuprofen is available as a racemic mixture of both (+) and (-) enantiomers. However, only the (-) enantiomer exhibits anti-inflammatory activity, while the (+) enantiomer is inactive and may even cause adverse effects.

Racemic Mixture: A Balance of Opposites

A racemic mixture, also known as a racemate or racemic compound, is a 50:50 mixture of enantiomers. It contains equal amounts of both (+) and (-) enantiomers, resulting in a net optical rotation of zero. The term "racemic" originates from the Latin word "racemus," meaning "bunch of grapes," which symbolizes the equal distribution of enantiomers.

In nature, many organic compounds are chiral and exist as enantiomers. However, the synthesis of these compounds often results in the formation of a racemic mixture due to the lack of chiral control during chemical reactions. This racemization process occurs because chemical reactions do not distinguish between the two enantiomers, leading to the formation of an equal mixture.

Racemization: A Battle of Enantiomers

Racemization is a process where a chiral compound interconverts between its enantiomers, resulting in the formation of a racemic mixture. This process can occur spontaneously through various mechanisms, such as thermal isomerization, acid/base catalysis, or enzymatic reactions. Racemization can significantly impact the properties and behavior of chiral compounds, making it an essential aspect to consider in various scientific disciplines.

One example of racemization is the conversion of L-amino acids to their D-enantiomers in living organisms. This process occurs due to the action of enzymes called racemases, which catalyze the interconversion of enantiomers. The racemization of amino acids plays a crucial role in protein synthesis, as both L- and D-amino acids are involved in the formation of peptides and proteins.

Analyzing Counterarguments

Although the concepts of optical activity and racemic mixtures are well-established, it is important to consider counterarguments and alternative viewpoints to foster a comprehensive understanding of the subject matter.

One counterargument against the significance of chirality and enantiomers is the idea that their different properties are merely a result of physical orientation and lack true functional differences. While it is true that enantiomers exhibit identical physical properties, such as boiling point and solubility, their interactions with other chiral compounds can have profound consequences. For instance, the interaction between chiral drugs and enzymes can vary greatly depending on the enantiomeric form. This can lead to differences in biological activity, potency, and potential side effects, as exemplified by the case of Ibuprofen mentioned earlier.

Another counterargument suggests that racemic mixtures are simply a product of chemical synthesis and do not have significant implications. However, racemization processes can often occur spontaneously, even in natural systems. Understanding racemization is crucial in fields such as drug development, where the biological activity of enantiomers can differ significantly. Separation of enantiomers from racemic mixtures is an important step towards improving drug efficacy and reducing potential side effects.

The Effect of Optical Isomers on Plane Polarised Light

Introduction

Plane polarised light refers to light waves that vibrate in a single plane, usually achieved by passing unpolarised light through a polarising filter. Optical isomers, also known as enantiomers, are molecular compounds that have the same chemical formula but differ in their spatial arrangement, resulting in mirror-image structures.

These isomers have profound effects on plane polarised light due to their ability to rotate the plane of polarization. In this essay, we will explore the concept of optical isomers, their impact on plane polarised light, and the applications of this phenomenon in various scientific disciplines.

Optical Isomers: A Brief Overview

Optical isomers arise when a molecule possesses a chiral center, which is an atom that is bonded to four different substituents. This property leads to two non-superimposable mirror-image structures, known as enantiomers.

The term "optical isomer" comes from the fact that these compounds exhibit different interactions with plane polarised light. One enantiomer will cause the plane of polarisation to rotate in a clockwise direction, whereas the other enantiomer will cause it to rotate in a counterclockwise direction. This effect is known as optical activity and is measured using a polarimeter.

Optical Activity and Plane Polarised Light

The Interaction of Plane Polarised Light with Optical Isomers

When plane polarised light interacts with an optically active substance, the angle of rotation depends on various factors such as the concentration of the compound, the path length, and the wavelength of light. This rotation occurs due to the interaction between the electric field vector of the polarised light and the chiral molecules present in the substance. The rotation can be clockwise or counterclockwise, depending on the enantiomer present.

Specific Rotation and Enantiomeric Excess

The extent of rotation caused by an optical isomer is quantified using the concept of specific rotation. Specific rotation (α) is defined as the angle of rotation (θ) divided by the concentration © of the compound and the path length (l) traveled by the light: $\alpha = (\theta / l) * (1 / c)$

The specific rotation is a characteristic property of a compound and is often reported at a specific wavelength of light. Enantiomeric excess (ee) is another important parameter that describes the purity of an enantiomeric mixture. It represents the difference in the concentrations of the two enantiomers: $ee = (c(S) - c(R)) / (c(S) + c(R))$

Here, c(S) and c(R) represent the concentrations of the sinister (left-handed) and rectus (right-handed) enantiomers, respectively. A pure enantiomer will have an enantiomeric excess of $\pm 100\%$, while a racemic mixture will have an enantiomeric excess of 0%.

Polarimetry and Chiral Analysis

Polarimetry is a technique used to measure the rotation of plane polarised light by an optically active substance. It involves passing a beam of plane polarised light through a sample, measuring the angle of rotation, and then calculating the specific rotation. This technique is widely used in chiral analysis to determine the enantiomeric purity of a sample. By comparing the measured specific rotation with known values, scientists can identify the presence of a specific enantiomer or determine the quality of a pharmaceutical product.

Applications of Optical Isomers and Plane Polarised Light

Pharmaceutical Industry

The impact of optical isomers on plane polarised light has significant implications in the pharmaceutical industry. Many drugs are chiral compounds, and the biological activity of enantiomers can differ greatly. For example, one enantiomer may be effective in treating a particular condition, while the other may cause adverse side effects or have no therapeutic effect at all. Therefore, the ability to separate and analyze enantiomers is crucial for drug development and quality control. Optical rotation measurements help ensure that pharmaceutical products contain the desired enantiomer in the correct concentration.

Chemistry and Synthesis

Optical isomers also play a crucial role in chemical synthesis. When synthesizing chiral compounds, it is often necessary to control the stereochemistry of the product. Reactions that produce a racemic mixture of enantiomers are termed "non-stereoselective," while those that yield a single enantiomer are called "stereoselective." By carefully designing and controlling reaction conditions, chemists can manipulate the formation of enantiomers and increase the yield of the desired enantiomer. Chiral catalysts and auxiliaries are commonly used in stereoselective synthesis to achieve this goal.

Analytical Chemistry and Forensic Science

The ability of optical isomers to interact differently with plane polarised light has also found applications in analytical chemistry and forensic science. Optical rotation measurements are used to identify unknown compounds by comparing their specific rotations to known values in databases. This technique is particularly useful for identifying illicit drugs, where slight differences in enantiomeric composition can significantly affect the physiological effects and legal classification of the substance.

Biological Systems

Optical isomers are prevalent in biological systems, where they can exhibit different physiological effects. For example, the two enantiomers of a drug may bind differently to receptors or enzymes in the body, leading to variations in efficacy and side effects. Understanding the interactions of optical isomers with biological systems is crucial for developing safe and effective pharmaceuticals. It also has implications in fields such as biochemistry, pharmacology, and toxicology.

Counterarguments and Limitations

While optical isomers have broad applications and profound effects on plane polarised light, it is essential to consider some counterarguments and limitations. Firstly, the specific rotation of a compound can vary with temperature, solvent, and wavelength of light. Therefore, it is crucial to ensure consistent experimental conditions when comparing specific rotations. Additionally, some substances may exhibit no optical activity due to the absence of a chiral center or the presence of an internal plane of symmetry. Furthermore, the separation of enantiomers can be a challenging task, often involving complex and time-consuming techniques such as chiral chromatography or asymmetric synthesis. The high cost and difficulty associated with achieving enantiomerically pure compounds can limit their widespread use in certain applications. Finally, it is important to note that optical rotation measurements alone do not provide information about the absolute configuration of a compound. Additional analytical techniques, such as X-ray crystallography or nuclear magnetic resonance spectroscopy, are often required for complete structural determination.

Introduction

The concept of chirality plays a crucial role in the synthetic preparation of drug molecules. Chirality refers to the property of an object or a molecule that is non-superimposable on its mirror image. In the context of drug molecules, chirality arises due to the presence of a carbon atom bonded to four different groups, creating two mirror-image structures known as enantiomers. Chirality is of utmost importance in drug synthesis because the two enantiomers of a drug molecule can exhibit different biological activities, pharmacokinetics, and toxicities. This essay aims to explore the relevance of chirality in the synthetic preparation of drug molecules and delve into the potential different biological activities of enantiomers.

Chirality and Drug Synthesis

Chiral Centers: A chiral center in a molecule is a carbon atom bonded to four different groups. The presence of a chiral center leads to the formation of two enantiomers, designated as R (rectus) and S (sinister) based on the sequence rules of the Cahn-Ingold-Prelog system. The enantiomers possess the same physical properties, such as melting point, boiling point, and solubility, but they differ in their interaction with chiral biological receptors. Stereochemistry: The stereochemistry of a drug molecule is determined by the spatial arrangement of its atoms. This arrangement influences its interaction with biological receptors, enzymes, and proteins. The synthetic preparation of drug molecules involves controlling the stereochemistry to ensure the desired biological activity. Stereochemistry can be controlled by selecting appropriate chiral starting materials, using chiral catalysts, or employing highly selective asymmetric reactions.

Enantiomer Separation: It is often necessary to separate enantiomers to avoid undesirable effects or to utilize the specific activity of a single enantiomer. Various techniques like chiral chromatography, chiral HPLC (High-Performance Liquid Chromatography), and chiral resolution methods can be employed to achieve enantiomer separation. Chiral separation is important because the biological activity of enantiomers can be significantly different.

Different Biological Activities of Enantiomers

Pharmacodynamics: Enantiomers can exhibit different pharmacological activities, i.e., they can bind to different receptors or exhibit different affinities for the same receptor. This phenomenon is known as enantioselectivity. Several examples illustrate the relevance of chirality in pharmacodynamics:

Thalidomide: Thalidomide is a notorious example where the different activities of the enantiomers led to tragic consequences. The (+)-enantiomer acted as a sedative, while the (-)-enantiomer caused severe birth defects. This highlights the importance of enantiomer-specific testing and regulatory measures in drug development.

Beta-blockers: Drugs like propranolol and atenolol, used in the treatment of hypertension and cardiac diseases, exhibit different activities based on their enantiomers. For example, the (-)-enantiomer of propranolol is a potent beta-blocker, whereas the (+)-enantiomer is less active or inactive.

Antihistamines: Enantiomers of antihistamines, such as cetirizine and loratadine, have different affinities for histamine receptors and hence exhibit different therapeutic effects. For example, the (-)-enantiomer of cetirizine possesses antihistaminic activity, while the (+)-enantiomer is inactive.

Pharmacokinetics: Enantiomers can also display differences in their pharmacokinetic properties, including absorption, distribution, metabolism, and excretion. These differences can be attributed to the varying stereochemical interactions of enantiomers with transporters, enzymes, and other biological components:

Ibuprofen: The (+)-enantiomer of ibuprofen undergoes a different metabolic pathway than the (-)-enantiomer. The (-)-enantiomer is metabolized primarily by the liver, while the (+)-enantiomer undergoes rapid renal clearance. This difference in metabolism influences the duration and intensity of the drug's action.

Antidepressants: Selective serotonin reuptake inhibitors (SSRIs) like fluoxetine and citalopram exhibit different pharmacokinetic properties due to their enantiomers. The (-)-enantiomer of fluoxetine is responsible for the therapeutic activity, while the (+)-enantiomer has a longer half-life and contributes to the overall pharmacokinetics. Similarly, citalopram consists of both (+)- and (-)-enantiomers, but only the (-)-enantiomer is responsible for the antidepressant effect.

Toxicity and Side Effects: The different biological activities of enantiomers can also manifest in varying toxicities and side effects. Chiral drug molecules may interact with different biological targets, leading to distinct adverse effects:

Terbutaline: Terbutaline, a bronchodilator used in asthma treatment, consists of two enantiomers. The (-)-enantiomer shows bronchodilatory activity, while the (+)-enantiomer can cause tachycardia and other cardiovascular side effects.

Methamphetamine: Methamphetamine is a central nervous system stimulant and a controlled substance. The (+)-enantiomer exhibits potent psychostimulant effects, while the (-)-enantiomer is less active and has fewer side effects. This example illustrates the importance of chirality in controlling the abuse potential and toxicity of drugs.

Counterarguments and Limitations

While the potential different biological activities of enantiomers highlight the importance of chirality in drug synthesis, it is worth noting some counterarguments and limitations:

Racemates: Racemates, also known as racemic mixtures, contain equal amounts of both enantiomers. In some cases, racemates can exhibit similar or even superior therapeutic effects compared to individual enantiomers. A famous example is the antidepressant drug fluoxetine, where the racemate exhibits better efficacy than the individual (-)-enantiomer.

Stereochemistry and Toxicity: While chirality can influence the toxicity of drug molecules, it is not the sole determinant. Other factors such as drug dosage, metabolism, drug-drug interactions, and individual patient characteristics can also contribute to toxicity. Therefore, the analysis of chirality should be complemented with a comprehensive evaluation of other factors.

Cost and Complexity: The synthesis and separation of enantiomers can be challenging, time-consuming, and expensive. The need to produce single enantiomers can significantly increase the cost of drug synthesis and limit their availability. In some cases, the cost-benefit analysis may favor the use of racemates or a single enantiomer with acceptable side effects.

The Relevance of Chirality in Synthetic Drug Preparation

Introduction

It will explore the significance of chirality in synthetic drug preparation, focusing on the necessity of separating racemic mixtures and the challenges associated with it.

Background on Chirality

To understand the relevance of chirality in drug synthesis, it is important to grasp the concept of chirality itself. Chirality is a geometric property of a molecule resulting from the presence of an asymmetric carbon atom, also known as a chiral center. A carbon atom is chiral when it is bonded to four different substituents, creating two non-superimposable mirror images. These mirror images are enantiomers, and their distinct spatial arrangements can lead to different physiological effects.

Drug Development and Enantiomers

In drug development, chiral molecules are commonly encountered due to the prevalence of chiral centers in bioactive compounds. It is estimated that around 90% of pharmaceuticals contain at least one chiral center. The biological activity of a drug is often determined by the interaction of the drug molecule with specific receptors or enzymes in the body. Enantiomers can exhibit different affinities for these targets, resulting in variations in pharmacological activity and side effects.

For example, the drug Thalidomide serves as a poignant reminder of the importance of chirality in drug synthesis. In the late 1950s, Thalidomide was prescribed as a sedative to pregnant women but tragically resulted in severe birth defects. It was later discovered that while one enantiomer of Thalidomide caused the desired sedative effects, the other enantiomer was responsible for the devastating teratogenic effects. This case highlighted the need to separate enantiomers and led to stricter regulations in drug development.

Racemic Mixtures and its Challenges

A racemic mixture, also known as a racemate or a racemic compound, is a 50:50 mixture of two enantiomers. In many cases, drug synthesis results in the formation of racemic mixtures due to the inherent difficulty in controlling the stereochemistry of chemical reactions. However, the biological activity of the enantiomers within a racemic mixture can vary significantly. Therefore, it becomes imperative to separate the racemic mixture into its pure enantiomers to ensure the desired therapeutic effect and minimize potential side effects.

The separation of enantiomers is a challenging task due to their identical physicochemical properties. Enantiomers have the same boiling point, melting point, and solubility, making conventional separation techniques ineffective. Traditional separation methods such as distillation, crystallization, and extraction are not suitable for enantiomeric resolution. As a result, specialized techniques and strategies have been developed to address this issue.

Enantiomeric Separation Techniques

Several techniques have been developed for the separation of enantiomers, each utilizing the subtle differences in their interactions with other molecules. One of the most common techniques is chromatography, which involves the separation of enantiomers based on their differential affinity for a chiral stationary phase. Chiral chromatographic columns, such as those containing a chiral selector like a cyclodextrin or a chiral ligand, can selectively retain and separate enantiomers.

Another widely used technique is the formation of diastereomers. Diastereomers are stereoisomers that are not mirror images of each other, and their separation properties differ from enantiomers. By selectively reacting the enantiomers with a chiral resolving agent, diastereomers are formed and can be separated using conventional techniques. The diastereomeric separation can then be reversed, yielding the pure enantiomers. Other techniques include enzymatic resolution, where enzymes selectively react with one enantiomer, and kinetic resolution, which exploits the different reaction rates of enantiomers. Each technique has its advantages and disadvantages, and the choice depends on the specific compound and its practical considerations.

Importance of Separating Racemic Mixtures

The separation of racemic mixtures is crucial in drug synthesis for several reasons. Firstly, the pharmacological activity of enantiomers can differ significantly. One enantiomer may exhibit the desired therapeutic effect, while the other could be inactive or even possess adverse effects. Therefore, the administration of a racemic mixture can lead to inefficiency in treatment or potentially harmful side effects. By separating the racemic mixture, it is possible to obtain the desired enantiomer with the desired pharmacological properties.

Secondly, regulatory agencies such as the U.S. Food and Drug Administration (FDA) often require the evaluation of individual enantiomers during the drug approval process. This is especially true for chiral drugs, where the safety and efficacy of each enantiomer must be demonstrated. Therefore, separating and characterizing the individual enantiomers is essential for obtaining regulatory approval.

Thirdly, the synthesis of enantiopure drugs allows for better control over the pharmacokinetics and pharmacodynamics of the drug. Enantiomers can exhibit different rates of absorption, metabolism, and elimination within the body. By utilizing a single enantiomer, the dosing and therapeutic response can be precisely tailored. This provides greater control over optimizing drug efficacy while minimizing potential adverse effects.

Counterarguments and Limitations

While the separation of racemic mixtures is often necessary, there are instances where racemic drugs are preferred. In some cases, the desired therapeutic effect may arise from the combined action of both enantiomers. For example, the antihistamine drug cetirizine is administered as a racemic mixture because the enantiomers exhibit complementary effects on histamine receptors. Separating the enantiomers in such cases may not offer significant therapeutic advantages.

Additionally, the separation of racemic mixtures can be a time-consuming and costly process. Specialized techniques and equipment are often required, which may increase production costs. The development of new chiral separation methods and the optimization of existing techniques are ongoing areas of research to address these limitations.

❦

The Relevance of Chirality in the Synthetic Preparation of Drug Molecules

Introduction

We will explore the relevance of chirality to the synthetic preparation of drug molecules, with a specific focus on the use of chiral catalysts to produce a single pure optical isomer.

Chirality and Drug Development

Chirality plays a crucial role in drug development due to its effect on the interaction between drugs and biological systems. Many biological processes involve the interaction of drugs with specific target molecules, such as receptors or enzymes. These target molecules often exhibit chirality, and therefore, the chirality of the drug molecule can significantly influence its binding affinity and activity.

Enantiomers and Drug Activity

The mirror-image relationship between enantiomers, which are chiral molecules, leads to differences in their three-dimensional structures. These differences can result in distinct interactions with biological target molecules. As a result, enantiomers can exhibit different pharmacological effects, including variations in potency, efficacy, and side effects.

An illustrative example is the drug Thalidomide, which was developed in the 1950s as a sedative and anti-nausea medication. However, it was later discovered that while one enantiomer was therapeutically effective, the other enantiomer caused severe birth defects. This tragic incident highlighted the importance of considering chirality in drug development and led to the establishment of regulations requiring the evaluation of chirality in pharmaceutical compounds.

Chiral Catalysts in Drug Synthesis

The synthesis of chiral drug molecules often requires the production of a single pure enantiomer to ensure optimal pharmacological properties. Chiral catalysts are instrumental in achieving this goal. A chiral catalyst is a molecule or compound that can selectively catalyze a chemical reaction to yield a single enantiomer. This is achieved by the presence of a chiral center within the catalyst molecule, which imparts its own chirality onto the product.

Asymmetric Synthesis and Chiral Catalysts

Asymmetric synthesis is the process of selectively producing one enantiomer over the other during a chemical reaction. Chiral catalysts play a central role in asymmetric synthesis by guiding the reaction towards the desired enantiomer. They achieve this by interacting with the reactants in a stereospecific manner, leading to the formation of a specific enantiomer as the major product. There are various types of chiral catalysts employed in drug synthesis, including transition metal complexes, enzymes, and organocatalysts. Transition metal complexes, such as those based on rhodium or ruthenium, are commonly used in asymmetric hydrogenation reactions. Enzymes, on the other hand, are employed in biocatalysis, where they can catalyze a wide range of reactions with high selectivity. Organocatalysts, such as proline derivatives, have found applications in asymmetric aldol reactions.

Advantages of Chiral Catalysts

The use of chiral catalysts offers several advantages in drug synthesis. One key advantage is the ability to obtain high enantioselectivity, producing a single pure enantiomer. This is crucial as it ensures that the pharmacological properties and safety profile of the drug are consistent and predictable. Additionally, chiral catalysts often enable the use of more efficient and environmentally friendly reaction conditions, reducing waste and minimizing the need for complex purification processes.

Chiral Catalyst Examples

To illustrate the impact of chiral catalysts in drug synthesis, let us consider a few examples. One prominent example is the synthesis of the antidepressant drug Escitalopram. Escitalopram is the S-enantiomer of the racemic mixture citalopram and exhibits higher potency and selectivity as a selective serotonin reuptake inhibitor (SSRI). The synthesis of Escitalopram utilizes a chiral catalyst to selectively convert the racemic mixture into the desired enantiomer.

Another example is the synthesis of the antiviral drug Oseltamivir, commonly known as Tamiflu. Oseltamivir is an inhibitor of the neuraminidase enzyme, which plays a crucial role in the replication of the influenza virus. The synthesis of Oseltamivir involves several steps, including the use of a chiral catalyst to generate the key chiral center in the molecule. The chiral catalyst ensures the formation of the desired enantiomer, which exhibits potent antiviral activity.

Counterarguments and Limitations

While the use of chiral catalysts in drug synthesis offers numerous advantages, there are also some limitations and counterarguments to consider. One counterargument is the potential cost associated with the use of chiral catalysts. Chiral catalysts are often expensive to synthesize or obtain, which can impact the overall cost of drug production. Additionally, the development of new and efficient chiral catalysts for specific reactions can be challenging and time-consuming.

Another limitation is the potential formation of by-products or impurities during asymmetric synthesis, even when using chiral catalysts. These impurities can be difficult to remove and may impact the purity and safety of the final drug product. Therefore, rigorous purification methods are necessary to ensure the removal of any residual enantiomers or impurities.

Hydrocarbons

Part 3: Organic Chemistry
Arenes
Introduction

Arenes are a class of organic compounds that contain a benzene ring as their core structure. The chemistry of arenes, particularly benzene and its derivatives, is of great importance due to their wide range of applications in various fields such as pharmaceuticals, agrochemicals, dyes, and polymers. One of the key reactions of benzene and its derivatives is substitution reactions with halogens (chlorine and bromine) in the presence of catalysts such as aluminum chloride ($AlCl_3$) or aluminum bromide ($AlBr_3$). This reaction leads to the formation of halogenoarenes, also known as aryl halides. In this essay, we will explore the chemistry of arenes, focusing on the substitution reactions of benzene and methylbenzene with chlorine and bromine, and the role of catalysts in these reactions.

Substitution Reactions of Benzene and Methylbenzene

Benzene

Benzene, C_6H_6, is a highly stable and aromatic compound due to its delocalized electron system. The delocalization of electrons in the benzene ring makes it resistant to addition reactions, in which two reactants combine to form a single product. However, benzene readily undergoes substitution reactions, in which one or more atoms or groups are replaced by another atom or group.

Substitution with Chlorine

When benzene reacts with chlorine in the presence of a catalyst such as aluminum chloride ($AlCl_3$), the substitution reaction takes place. The following equation can represent the reaction: $C_6H_6 + Cl_2 \rightarrow C_6H_5Cl + HCl$

In this reaction, one hydrogen atom of benzene is replaced by a chlorine atom, resulting in the formation of chlorobenzene. The catalyst, $AlCl_3$, plays a crucial role in this reaction by activating the chlorine molecule and facilitating the substitution process. The mechanism of this reaction involves the formation of a complex between the benzene molecule and the catalyst. The chlorine molecule then attacks this complex, leading to the formation of a cyclic intermediate known as a arenium ion. The arenium ion is highly unstable and quickly undergoes rearrangement to form the final product, chlorobenzene.

Substitution with Bromine

Similar to the reaction with chlorine, benzene can also undergo substitution with bromine in the presence of a catalyst such as aluminum bromide ($AlBr_3$). The representation of the reaction is given by the following equation: $C_6H_6 + Br_2 \rightarrow C_6H_5Br + HBr$. In this reaction, one hydrogen atom of benzene is replaced by a bromine atom, resulting in the formation of bromobenzene. The reaction mechanism of this process is similar to the reaction mechanism involving chlorine. It begins with the creation of a complex between the benzene molecule and the catalyst, leading to the formation of an arenium ion. This ion then undergoes rearrangement to produce the ultimate product.

Methylbenzene (Toluene)

Methylbenzene, also known as toluene, is a derivative of benzene in which a methyl group (CH3) is attached to the benzene ring. The presence of the methyl group in toluene affects its reactivity towards substitution reactions compared to benzene.

Substitution with Chlorine

When toluene reacts with chlorine in the presence of a catalyst such as aluminum chloride (AlCl3), the substitution reaction takes place. The following equation represents the reaction. C6H5CH3 + Cl2 → C6H5CH2Cl + HCl

Chloromethylbenzene is formed when a chlorine atom replaces one hydrogen atom in the methyl group during this reaction. The presence of the methyl group in toluene makes it more reactive towards substitution reactions compared to benzene. This is due to the electron-donating nature of the methyl group, which increases the electron density on the benzene ring and makes it more susceptible to attack by electrophiles such as chlorine.

Substitution with Bromine

Similar to the reaction with chlorine, toluene can also undergo substitution with bromine in the presence of a catalyst such as aluminum bromide (AlBr3). The reaction can be represented by the following equation:

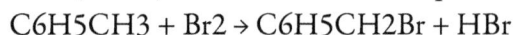

C6H5CH3 + Br2 → C6H5CH2Br + HBr

In this reaction, one hydrogen atom of the methyl group is replaced by a bromine atom, resulting in the formation of bromomethylbenzene. The mechanism of this reaction is analogous to the mechanism of the reaction with chlorine, involving the formation of a complex between the toluene molecule and the catalyst, followed by the formation of an arenium ion and rearrangement to form the final product.

Role of Catalysts in Substitution Reactions

The role of catalysts such as aluminum chloride (AlCl3) or aluminum bromide (AlBr3) in the substitution reactions of benzene and methylbenzene with chlorine and bromine is crucial. By offering a different reaction pathway with reduced activation energy, catalysts enhance the speed of a chemical reaction. In the case of substitution reactions of benzene and methylbenzene, the catalysts act as Lewis acids, which are electron pair acceptors. They coordinate with the electron-rich benzene or methylbenzene molecule, facilitating the attack of the electrophilic chlorine or bromine molecule. The coordination of the catalyst with the reactant molecule leads to the formation of a complex, which is more reactive towards substitution reactions. The catalysts also stabilize the arenium ion intermediate formed during the reaction. The arenium ion is a cyclic intermediate in which the positive charge is delocalized over the benzene ring. This intermediate is highly unstable and tends to undergo rearrangement to form a more stable product. The catalysts help in stabilizing the arenium ion by coordinating with it and reducing the positive charge density on the benzene ring.

Applications of Halogenoarenes

Halogenoarenes, also known as aryl halides, have a wide range of applications in various fields such as pharmaceuticals, agrochemicals, dyes, and polymers. Some of the key applications of halogenoarenes are discussed below:

Pharmaceuticals

Halogenoarenes are widely used in the synthesis of pharmaceutical compounds. The presence of halogen atoms in the aryl ring can significantly affect the biological activity and pharmacokinetic properties of the compounds. For example, chlorobenzene derivatives are used as antiseptics and disinfectants, while bromobenzene derivatives have antiviral and antifungal properties. The introduction of halogen atoms into the aryl ring can also improve the metabolic stability and lipophilicity of the compounds, making them more suitable for drug development.

Agrochemicals

Halogenoarenes are also used in the synthesis of agrochemicals, including herbicides, fungicides, and insecticides. The presence of halogen atoms in the aryl ring can enhance the biological activity and selectivity of these compounds towards specific pests and diseases. For example, chlorobenzene derivatives are used as herbicides to control weed growth, while bromobenzene derivatives are used as fungicides to prevent fungal infections in crops. The introduction of halogen atoms into the aryl ring can also improve the stability and persistence of these compounds in the environment.

Dyes

Halogenoarenes are widely used in the synthesis of dyes and pigments. The presence of halogen atoms in the aryl ring can affect the color and stability of the dyes. For example, chlorobenzene derivatives are used to produce yellow and orange dyes, while bromobenzene derivatives are used to produce red and violet dyes. The introduction of halogen atoms into the aryl ring can also improve the lightfastness and washfastness of the dyes, making them more suitable for textile applications.

Polymers

Halogenoarenes are also used in the synthesis of polymers, including plastics and elastomers. The presence of halogen atoms in the aryl ring can improve the thermal stability, flame retardancy, and electrical properties of the polymers. For example, chlorobenzene derivatives are used in the production of PVC (polyvinyl chloride) plastics, while bromobenzene derivatives are used in the production of flame retardant polymers. The introduction of halogen atoms into the aryl ring can also affect the mechanical properties and processability of the polymers, making them more suitable for specific applications.

Counterarguments

While substitution reactions of benzene and methylbenzene with chlorine and bromine in the presence of catalysts have numerous applications, there are some concerns associated with the use of halogenoarenes. Some of the counterarguments against the use of halogenoarenes are discussed below:

Environmental Impact

Halogenoarenes, particularly chlorinated and brominated compounds, can have a negative impact on the environment. These compounds are persistent and can accumulate in the environment, leading to long-term ecological effects. They can also undergo transformation reactions in the environment, leading to the formation of toxic byproducts such as dioxins and furans. The release of halogenoarenes into the environment should be carefully managed to minimize their impact on ecosystems and human health.

Health Effects

Halogenoarenes, especially chlorinated and brominated compounds, can have adverse health effects on humans and animals. Some of these compounds have been identified as carcinogens or mutagens, meaning they can cause cancer or genetic mutations. The use and handling of halogenoarenes should be done with proper safety precautions to minimize exposure and potential health risks.

Alternatives

There is a growing interest in developing alternative methods for the synthesis of organic compounds, including halogenoarenes, that are more sustainable and environmentally friendly. Green chemistry principles advocate for the use of renewable resources, catalytic reactions, and safer solvents to reduce the environmental impact of chemical processes. Researchers are exploring new catalysts and reaction conditions to minimize or eliminate the use of toxic or hazardous substances in the synthesis of organic compounds.

Nitration of Benzene and Methylbenzene

Introduction

One important reaction of arenes is nitration, where a nitro group (-NO2) is introduced onto the benzene ring. In this article, we will explore the chemistry of arenes, focusing specifically on the nitration of benzene and methylbenzene.

Nitration of Benzene

Nitration of benzene involves the substitution of a hydrogen atom on the benzene ring with a nitro group. This reaction is typically carried out using a mixture of concentrated nitric acid (HNO3) and concentrated sulfuric acid (H2SO4) as the nitrating agent. The reaction is exothermic and proceeds at a temperature between 25 °C and 60 °C. The following equation represents the reaction:C6H6 + HNO3 → C6H5NO2 + H2O

Several steps are involved in the mechanism of the nitration reaction. Initially, the nitric acid molecule protonates the sulfuric acid, forming a stronger acid:

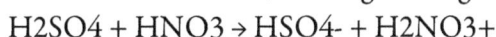

H2SO4 + HNO3 → HSO4- + H2NO3+

The nitronium ion (NO2+) is generated by the transfer of a proton from the nitric acid to the sulfuric acid:

H2NO3+ → HNO3 + H+

The nitronium ion is the active electrophile that attacks the benzene ring. It forms an intermediate complex with the benzene molecule: C6H6 + NO2+ → C6H6NO2

The intermediate complex then loses a proton to form the final product, nitrobenzene, and a hydronium ion:

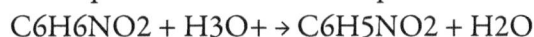

C6H6NO2 + H3O+ → C6H5NO2 + H2O

Factors Influencing the Nitration of Benzene

Several factors influence the nitration of benzene. One important factor is the temperature at which the reaction is carried out. The reaction proceeds faster at higher temperatures due to the increased kinetic energy of the reactant molecules. However, temperatures above 60 °C can lead to undesirable side reactions, such as the oxidation of the nitro group to a nitroso group (-NO) or the formation of dinitro compounds. Therefore, a temperature range between 25 °C and 60 °C is chosen to optimize the yield of nitrobenzene.

The concentration of nitric acid and sulfuric acid also plays a crucial role in the nitration reaction. Higher concentrations of these acids increase the concentration of the nitronium ion, leading to a higher reaction rate. However, excessive concentrations can result in the formation of undesired by-products. Therefore, a balanced ratio of the two acids is necessary to achieve the desired reaction outcome.

Nitration of Methylbenzene (Toluene)

Methylbenzene, commonly known as toluene, is a methyl-substituted benzene compound. The nitration of toluene follows a similar mechanism to that of benzene. However, due to the presence of the electron-donating methyl group (-CH3), the reaction proceeds at a faster rate compared to benzene. Additionally, the methyl group directs the substitution of the nitro group predominantly at the ortho and para positions relative to the methyl group. This phenomenon is known as the ortho-para directing effect.

The nitration of toluene can yield three possible mononitro products: ortho-nitrotoluene, meta-nitrotoluene, and para-nitrotoluene. The distribution of these isomers depends on the reaction conditions, such as the temperature and concentration of the nitrating agent. At low temperatures (below 10 °C), the major product is para-nitrotoluene, whereas at higher temperatures (around 100 °C), the major product is ortho-nitrotoluene. The meta-isomer is usually obtained in smaller quantities. The ortho-para directing effect can be explained by the resonance stabilization of the intermediate formed during the reaction. The methyl group donates electron density to the benzene ring, increasing the electron density at the ortho and para positions. This makes these positions more susceptible to electrophilic attack by the nitronium ion.

Applications of Nitration

Nitration reactions have significant industrial applications. Nitrobenzene, which is produced by the nitration of benzene, is an important intermediate in the synthesis of various chemicals. It is primarily used as a precursor in the production of aniline, which is further used in the manufacture of dyes, pharmaceuticals, and rubber chemicals.

The nitration of toluene is a key step in the production of TNT (trinitrotoluene), a powerful explosive. Toluene is first nitrated to give a mixture of mono- and dinitro isomers, which are then further nitrated to obtain trinitrotoluene. The explosive properties of TNT make it useful in military applications, mining, and demolition.

Counterarguments and Limitations

While the nitration of benzene and methylbenzene is a widely used reaction, it does have some limitations and drawbacks. One major limitation is the potential for the formation of multiple substitution products. Subsequent nitration reactions can occur, leading to the formation of dinitro-compounds. These by-products are often unwanted and can reduce the yield of the desired mononitro products.

Furthermore, the nitration reaction can also result in the formation of side products, such as the oxidation of the nitro group to a nitroso group (-NO), or the formation of polynitro compounds. These side reactions reduce the efficiency of the reaction and can complicate the purification of the desired product.

Friedel-Crafts Alkylation of Benzene and Methylbenzene

Introduction

In this analysis, we will explore the chemistry of arenes by focusing on the Friedel-Crafts alkylation reaction. Specifically, we will discuss the alkylation of benzene and methylbenzene using CH3Cl (methyl chloride) and AlCl3 (aluminum chloride) as the catalyst. This reaction, known as Friedel-Crafts alkylation, allows for the introduction of alkyl groups onto the aromatic ring, resulting in the formation of alkylated arenes.

Friedel-Crafts Alkylation Reaction

The Friedel-Crafts alkylation reaction is a classic example of electrophilic aromatic substitution (EAS), a type of reaction that involves the substitution of a hydrogen atom on an aromatic ring with an electrophile. The electrophile in this case is the alkyl group, which is introduced onto the aromatic ring through the reaction with an alkyl halide in the presence of a Lewis acid catalyst, such as aluminum chloride (AlCl3).

The reaction mechanism of Friedel-Crafts alkylation involves several steps. Firstly, the Lewis acid catalyst (AlCl3) coordinates with the alkyl halide (CH3Cl), forming a complex with a positive charge on the carbon atom of the alkyl group. This complex is highly reactive and acts as an electrophile, attacking the aromatic ring of benzene or methylbenzene.

Benzene as a Substrate

Let's first consider the alkylation of benzene using CH3Cl and AlCl3 as the catalyst. The following equation represents the reaction.

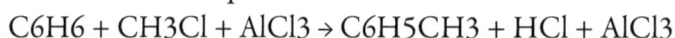

$$C_6H_6 + CH_3Cl + AlCl_3 \rightarrow C_6H_5CH_3 + HCl + AlCl_3$$

The formation of a complex between CH3Cl and AlCl3 is facilitated by the Lewis acid catalyst AlCl3 in this reaction. The coordination of AlCl3 to CH3Cl enhances the electrophilicity of the carbon atom in CH3Cl, making it more prone to attack by the π electrons of benzene.

The electrophilic attack of the complex on benzene initiates the reaction. One of the π electrons from the benzene ring forms a bond with the carbon atom of the alkyl group, resulting in the formation of a carbocation intermediate. The positive charge on the carbon atom is stabilized by the delocalization of the electrons in the π system of the benzene ring. However, the positive charge on the carbocation intermediate makes it highly reactive and prone to rearrangement. Rearrangement of the carbocation can occur through the migration of a hydrogen atom or an alkyl group to a neighboring carbon atom, resulting in the formation of different alkylated products. This phenomenon, known as carbocation rearrangement, is a common occurrence in Friedel-Crafts alkylation reactions.

To control the regioselectivity of the reaction and minimize carbocation rearrangement, a bulky substituent is often introduced onto the aromatic ring to block one of the possible reaction sites. For example, in the case of benzene, the introduction of a methyl group (C6H5CH3) onto the ring would result in the formation of toluene as the major product, with the methyl group acting as a directing group.

Methylbenzene as a Substrate

Now, let's consider the alkylation of methylbenzene (toluene) using CH3Cl and AlCl3 as the catalyst. The reaction can be represented by the following equation:

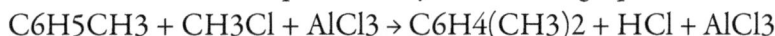

C6H5CH3 + CH3Cl + AlCl3 → C6H4(CH3)2 + HCl + AlCl3

The regioselectivity of the reaction is influenced by the presence of the methyl group on the aromatic ring in this particular scenario. The methyl group acts as a directing group, directing the incoming alkyl group to the ortho and para positions relative to itself. The mechanism of the reaction is similar to that of benzene alkylation. The Lewis acid catalyst AlCl3 coordinates with CH3Cl, forming a complex with enhanced electrophilicity. The complex then attacks the aromatic ring of methylbenzene, resulting in the formation of a carbocation intermediate. However, due to the presence of the methyl group, the carbocation intermediate is more stable than in the case of benzene alkylation. The delocalization of the positive charge through resonance with the methyl group and the aromatic ring provides additional stability to the intermediate. As a result, the likelihood of carbocation rearrangement is reduced, and the ortho and para positions relative to the methyl group are favored for the alkylation reaction.

Advantages and Limitations of Friedel-Crafts Alkylation

The Friedel-Crafts alkylation reaction offers several advantages for the synthesis of alkylated arenes. Firstly, it allows for the introduction of a wide range of alkyl groups onto the aromatic ring, providing access to a diverse array of compounds with different physical and chemical properties. This versatility makes the reaction valuable in the synthesis of pharmaceuticals, agrochemicals, and fine chemicals.

Secondly, the reaction is relatively straightforward and can be carried out under mild conditions. The use of inexpensive and readily available reagents, such as CH3Cl and AlCl3, makes the reaction cost-effective and practical for large-scale applications. Additionally, the reaction can be performed on a wide range of substrates, including benzene and various substituted benzenes, enabling the synthesis of complex molecules with tailored functionalities.

Despite its advantages, Friedel-Crafts alkylation also has some limitations. One major limitation is the susceptibility of carbocation intermediates to rearrangement. The formation of different alkylated products through carbocation rearrangement can reduce the selectivity and efficiency of the reaction. Furthermore, the reaction can be limited by the formation of undesired byproducts, such as polyalkylated compounds, which can be difficult to separate and purify.

To address these limitations, several modifications and alternative methods have been developed. For example, the use of protecting groups on the aromatic ring can prevent unwanted reactions and increase the selectivity of the alkylation. Additionally, different catalysts and reaction conditions, such as the use of superacids or solid-supported catalysts, have been explored to improve the efficiency and selectivity of the reaction.

Friedel-Crafts Acylation of Benzene and Methylbenzene

Introduction

Arenes, also known as aromatic hydrocarbons, are a class of organic compounds that contain one or more benzene rings. These compounds exhibit unique chemical reactivity due to the delocalization of pi electrons within the aromatic ring. One of the most important reactions of arenes is the Friedel-Crafts acylation, which involves the introduction of an acyl group onto the aromatic ring. In this article, we will explore the chemistry of arenes, focusing on the Friedel-Crafts acylation reactions of benzene and methylbenzene.

The Structure and Stability of Arenes

Arenes are characterized by the presence of a benzene ring, which consists of six carbon atoms arranged in a cyclic structure with alternating single and double bonds. The delocalization of pi electrons over the entire ring gives rise to aromaticity, a unique property that imparts exceptional stability to arenes. This stability is a result of the resonance stabilization provided by the delocalized pi electrons.

The stability of benzene and other arenes can be further explained using molecular orbital theory. According to this theory, the pi electrons in the benzene ring occupy a set of six molecular orbitals, three bonding orbitals (π bonds) and three antibonding orbitals (π^* bonds). The delocalization of the pi electrons results in a lower energy for the bonding orbitals and a higher energy for the antibonding orbitals, leading to increased stability.

Friedel-Crafts Acylation

The Friedel-Crafts acylation is a classic reaction in organic chemistry that allows for the introduction of an acyl group onto an aromatic ring. It involves the reaction of an arene with an acyl chloride (RCOCl) in the presence of a Lewis acid catalyst, typically aluminum chloride (AlCl3). The reaction proceeds through the formation of a carbocation intermediate, followed by nucleophilic attack of the arene on the carbocation and subsequent loss of a proton.

The mechanism of the Friedel-Crafts acylation can be divided into three main steps: activation of the acyl chloride, generation of the electrophile, and reaction with the arene. In the first step, the acyl chloride is activated by coordination with the Lewis acid catalyst. This coordination increases the electrophilicity of the acyl chloride, facilitating the subsequent reaction with the arene.

In the second step, the electrophile is generated through the formation of a complex between the activated acyl chloride and the Lewis acid catalyst. This complexation leads to the formation of a carbocation intermediate, which is stabilized by resonance with the aromatic ring.

Finally, in the third step, the nucleophilic attack of the arene on the carbocation occurs, resulting in the formation of a new carbon-carbon bond. The Lewis acid catalyst facilitates the departure of a proton from the intermediate, regenerating the aromaticity of the ring.

Friedel-Crafts Acylation of Benzene

The Friedel-Crafts acylation of benzene is a fundamental reaction in organic synthesis. It allows for the introduction of various acyl groups onto the benzene ring, leading to the formation of a wide range of substituted aromatic compounds. The reaction is typically carried out using benzene as the arene and acetyl chloride (CH3COCl) as the acylating agent.

The reaction proceeds in the presence of a Lewis acid catalyst, such as aluminum chloride (AlCl3), which activates the acetyl chloride and facilitates the reaction with benzene. The reaction is exothermic and is usually performed under reflux conditions to ensure the completion of the reaction.

The following equation represents the overall reaction. CH3COCl + C6H6 → CH3C6H5 + HCl

In this reaction, a hydrogen atom from the benzene ring is replaced by the acetyl group, resulting in the formation of acetophenone. The acetyl group is added to the ring in a position that maintains the aromaticity of the benzene ring.

Friedel-Crafts Acylation of Methylbenzene (Toluene)

The Friedel-Crafts acylation reaction can also be applied to methylbenzene, commonly known as toluene. Toluene is a derivative of benzene in which one of the hydrogen atoms is replaced by a methyl group (-CH3). The presence of the methyl group in toluene affects the reactivity of the aromatic ring and the regioselectivity of the reaction.

When toluene is subjected to Friedel-Crafts acylation, the acyl group can be introduced at either the ortho, meta, or para position relative to the methyl group. The regioselectivity of the reaction is influenced by steric effects and electronic factors. The steric hindrance caused by the methyl group favors ortho and para substitution, while the electron-donating effect of the methyl group enhances the reactivity of the ortho and para positions. For example, when toluene is reacted with acetyl chloride in the presence of aluminum chloride, the major product obtained is o-tolylacetophenone. The reaction can be represented as follows:

$$CH3COCl + CH3C6H5 \rightarrow CH3C6H4COCH3 + HCl$$

The ortho position is preferred due to the steric hindrance caused by the methyl group, which makes the meta position less accessible. Additionally, the electron-donating effect of the methyl group enhances the reactivity of the ortho and para positions, leading to a higher percentage of ortho substitution.

Limitations and Considerations

While the Friedel-Crafts acylation is a versatile method for the synthesis of aromatic compounds, it has certain limitations and considerations. One of the main limitations is the susceptibility of the acylating agent to side reactions, such as self-condensation and rearrangement. These side reactions can lead to the formation of undesired byproducts and reduce the yield of the desired product. Moreover, the reaction is typically limited to the use of arenes that are not significantly deactivated or substituents that are not strongly activating. Strongly deactivated arenes, such as nitrobenzene, do not undergo the Friedel-Crafts acylation reaction efficiently. Similarly, strongly activating substituents can lead to multiple acylations and complex mixtures of products. Furthermore, the choice of the Lewis acid catalyst is crucial for the success of the reaction. While aluminum chloride is commonly used, it may react with the acylating agent and form complex mixtures of products. Other Lewis acid catalysts, such as ferric chloride (FeCl3) or boron trifluoride (BF3), can be employed to overcome these issues and improve the selectivity of the reaction.

Applications of Friedel-Crafts Acylation

The Friedel-Crafts acylation reaction has found extensive applications in the synthesis of various aromatic compounds. It is widely used in the pharmaceutical and agrochemical industries for the preparation of drugs, dyes, and flavors.

For example, the synthesis of aspirin, a widely used pain reliever, involves the Friedel-Crafts acylation of salicylic acid with acetic anhydride. The reaction leads to the formation of acetylsalicylic acid, which is then hydrolyzed to produce aspirin. Similarly, the synthesis of ibuprofen, another common nonsteroidal anti-inflammatory drug, involves the Friedel-Crafts acylation of isobutylbenzene.

The Friedel-Crafts acylation reaction is also utilized in the synthesis of fragrances and flavors. For instance, the compound vanillin, which imparts the characteristic flavor and aroma of vanilla, can be synthesized through the Friedel-Crafts acylation of guaiacol with acetic anhydride.

Introduction

In this discussion, we will explore the chemistry of arenes through the reactions of benzene and methylbenzene (toluene). Specifically, we will focus on the complete oxidation of the side-chain of methylbenzene using hot alkaline KMnO4 followed by dilute acid, resulting in the formation of benzoic acid.

The delocalization of π electrons is a result of the resonance structure of the benzene ring. Each carbon atom in the ring is sp2 hybridized, forming sigma bonds with its adjacent carbon atoms. The remaining p-orbital on each carbon atom overlaps with the p-orbitals of the neighboring carbon atoms, creating a continuous network of π bonds above and below the plane of the ring. This delocalized π electron cloud results in a more stable system compared to localized double bonds in alkenes.

Due to this stability, arenes exhibit lower reactivity towards addition reactions, making them less prone to undergo electrophilic addition or hydrogenation. Instead, arenes predominantly engage in substitution reactions, either electrophilic or nucleophilic, which occur at the reactive positions on the benzene ring.

Oxidation of Methylbenzene

The oxidation of methylbenzene, or toluene, involves the introduction of an oxygen atom into the side-chain of the molecule. This reaction can be achieved by utilizing a strong oxidizing agent such as hot alkaline potassium permanganate ($KMnO_4$). The overall reaction can be represented as follows:

$$CH_3C_6H_5 + 3O_2 \rightarrow CH_3CO_2H + H_2O$$

Here, toluene ($CH_3C_6H_5$) is oxidized to benzoic acid (CH_3CO_2H) in the presence of oxygen (O_2). The reaction proceeds through several steps, involving the formation of various intermediates.

Step 1: Formation of Methylbenzyl Alcohol

The initial step in the oxidation of methylbenzene involves the reaction of toluene with $KMnO_4$ in alkaline conditions. The strong oxidizing agent, $KMnO_4$, is reduced to MnO_2, while toluene is converted to methylbenzyl alcohol ($CH_3C_6H_4CH_2OH$). The reaction can be represented as follows:

$$CH_3C_6H_5 + KMnO_4 \rightarrow CH_3C_6H_4CH_2OH + MnO_2$$

In this step, the alkyl side-chain of methylbenzene is oxidized to an alcohol functional group, resulting in the formation of methylbenzyl alcohol.

Step 2: Formation of Benzaldehyde

The next step involves the oxidation of methylbenzyl alcohol to benzaldehyde (C_6H_5CHO). This reaction occurs in the presence of a mild oxidizing agent, such as chromic acid (H_2CrO_4) or potassium dichromate ($K_2Cr_2O_7$). The overall reaction can be represented as follows:

$$CH_3C_6H_4CH_2OH + [O] \rightarrow C_6H_5CHO + H_2O$$

In this step, the alcohol group of methylbenzyl alcohol is oxidized to a carbonyl group, resulting in the formation of benzaldehyde. The oxidizing agent, denoted as [O], provides the necessary oxygen atom for the oxidation process.

Step 3: Formation of Benzoic Acid

The final step in the complete oxidation of the side-chain of methylbenzene involves the conversion of benzaldehyde to benzoic acid. This reaction is achieved by treating benzaldehyde with dilute acid, typically hydrochloric acid (HCl), in the presence of an oxidizing agent. The overall reaction can be represented as follows:

$$C_6H_5CHO + [O] \rightarrow CH_3CO_2H$$

In this step, the aldehyde group of benzaldehyde is further oxidized to a carboxylic acid group, resulting in the formation of benzoic acid. The oxidizing agent, denoted as [O], provides the necessary oxygen atom for the oxidation process.

Mechanism of the Oxidation Reaction

The oxidation of methylbenzene proceeds through a series of intermediate steps before forming the final product, benzoic acid. Let's delve into the detailed mechanism of this reaction.

Step 1: Formation of Methylbenzyl Alcohol

The initial step involves the oxidation of the methyl group of methylbenzene to an alcohol group. It occurs through the following mechanism:

The strong oxidizing agent, $KMnO_4$, is reduced to MnO_2. This reduction is facilitated by the basic medium, typically achieved by adding a strong base such as sodium hydroxide (NaOH) to the reaction mixture. The oxygen atom from $KMnO_4$ attacks the methyl group of methylbenzene, leading to the formation of a carbon-oxygen bond. Simultaneously, one of the MnO_2 species gets reduced to MnO_4^- ion. The resulting intermediate, called a carbocation, undergoes deprotonation by the hydroxide ion (OH^-) present in the reaction mixture, leading to the formation of methylbenzyl alcohol.

Step 2: Formation of Benzaldehyde

The second step involves the oxidation of methylbenzyl alcohol to benzaldehyde. This reaction occurs in the presence of a mild oxidizing agent, such as chromic acid (H_2CrO_4) or potassium dichromate ($K_2Cr_2O_7$). The mechanism of this step is as follows: The oxidizing agent, denoted as [O], is reduced during this step. The alcohol group of methylbenzyl alcohol is attacked by the [O] species, resulting in the formation of a carbonyl group.

Simultaneously, the [O] species gets reduced to water, completing the oxidation process. The resulting product is benzaldehyde, which contains a carbonyl group.

Step 3: Formation of Benzoic Acid

The final step involves the oxidation of benzaldehyde to benzoic acid. This reaction occurs in the presence of dilute acid, typically hydrochloric acid (HCl), and an oxidizing agent. The mechanism of this step is as follows: The hydrochloric acid provides a proton (H+) to benzaldehyde, leading to the formation of a resonance-stabilized carbocation. The oxidizing agent, denoted as [O], attacks the carbocation, resulting in the formation of a carboxylic acid group. Simultaneously, the [O] species gets reduced to water, completing the oxidation process.

The resulting product is benzoic acid, which contains a carboxylic acid group.

Hydrogenation of Benzene and Methylbenzene

Introduction

In this article, we will explore the chemistry of arenes, focusing on the hydrogenation of benzene and methylbenzene to form a cyclohexane ring.

The hydrogenation of benzene is a highly exothermic reaction that releases a large amount of energy. The reaction proceeds in a step-wise manner, with each step involving the addition of one hydrogen atom to the benzene ring. The final product of the reaction is cyclohexane, a saturated hydrocarbon with a six-membered ring.

Mechanism of Hydrogenation

The mechanism of hydrogenation involves several steps, including adsorption of hydrogen on the catalyst surface, diffusion of hydrogen atoms across the catalyst, and their subsequent addition to the aromatic ring. The overall reaction can be divided into three main stages: adsorption, hydrogenation, and desorption.

Adsorption

In the first step of the hydrogenation process, hydrogen molecules (H_2) are adsorbed onto the surface of the catalyst, which is usually platinum or nickel. This adsorption occurs due to the weak Van der Waals forces between the hydrogen molecules and the metal surface. The presence of the catalyst provides a favorable environment for the reaction to occur by reducing the activation energy barrier.

Hydrogenation

Once adsorbed, the hydrogen molecules dissociate into individual hydrogen atoms. These atoms then diffuse across the catalyst surface until they encounter a benzene molecule. The interaction between the hydrogen atom and the benzene ring activates the aromatic system, making it more susceptible to reaction. The addition of hydrogen to the benzene ring occurs through a series of steps known as cyclohexadiene intermediates. In the first step, a hydrogen atom attacks the benzene ring, breaking one of the carbon-carbon bonds. This results in the formation of a cyclohexadiene intermediate, which is a transient species. In the subsequent steps, additional hydrogen atoms are added to the cyclohexadiene intermediate, resulting in the formation of cyclohexene and cyclohexane intermediates. These intermediates can further react with hydrogen to produce fully hydrogenated cyclohexane.

Desorption

After the hydrogenation process is complete, the cyclohexane product desorbs from the catalyst surface, allowing the catalyst to be reused for subsequent reactions. The desorption step involves the weakening of the interaction between the catalyst and the product, leading to the release of the cyclohexane molecule.

Hydrogenation of Benzene

The hydrogenation of benzene is a widely studied and industrially important reaction. It is typically carried out using a heterogeneous catalyst, such as platinum or nickel, supported on a solid material like alumina. The reaction is conducted at elevated temperatures and pressures to improve the reaction rate and yield.

The hydrogenation of benzene to cyclohexane can be represented by the following chemical equation:

$$C_6H_6 + 3H_2 \rightarrow C_6H_{12}$$

A substantial amount of energy is released during the highly exothermic reaction. This energy is harnessed for various industrial applications, such as the production of cyclohexane, which is used as a solvent in the manufacture of nylon and other polymers.

Hydrogenation of Methylbenzene

Methylbenzene, also known as toluene, is a common aromatic compound that contains a methyl group attached to the benzene ring. Its hydrogenation is similar to that of benzene but involves an additional step due to the presence of the methyl group.

The hydrogenation of methylbenzene to cyclohexylmethane can be represented by the following chemical equation:

$$C_6H_5CH_3 + 4H_2 \rightarrow C_6H_{11}CH_3$$

The presence of the methyl group introduces steric hindrance, which affects the reaction rate and selectivity. Steric hindrance refers to the repulsion between atoms or groups in a molecule, which restricts their movement and influences the reactivity of the molecule. In the case of methylbenzene, the steric hindrance caused by the methyl group can affect the adsorption of hydrogen on the catalyst surface and the subsequent addition of hydrogen atoms to the benzene ring.

Industrial Applications

The hydrogenation of benzene and methylbenzene has numerous industrial applications. One of the major applications is the production of cyclohexane, which is used as a solvent in the chemical industry. Cyclohexane is also a precursor for the production of adipic acid, a key raw material for the synthesis of nylon. In addition to cyclohexane production, the hydrogenation of arenes is used in the synthesis of various pharmaceuticals and fine chemicals. The ability to selectively hydrogenate specific functional groups in an aromatic compound is crucial for the synthesis of complex molecules.

Limitations and Challenges

While hydrogenation reactions have proven to be highly versatile and useful, they are not without limitations and challenges. One of the main challenges is achieving selective hydrogenation, particularly in the presence of multiple functional groups. The reactivity of different functional groups can vary significantly, making it difficult to control the reaction and avoid over-hydrogenation.

Another challenge is the high energy requirements of the reaction. A large amount of energy is released during the exothermic process of the hydrogenation of arenes. This energy needs to be carefully managed to prevent runaway reactions and ensure the safety of the process. Furthermore, the use of transition metal catalysts, such as platinum and nickel, can be expensive and environmentally unfriendly. Efforts are being made to develop more sustainable and cost-effective catalysts for hydrogenation reactions.

Introduction

Electrophilic substitution is a fundamental reaction in organic chemistry that involves the replacement of a hydrogen atom in an aromatic ring by an electrophile. This mechanism is widely observed in the formation of various aromatic compounds, including nitrobenzene and bromobenzene. By understanding the mechanism of electrophilic substitution in arenes, we can gain insights into the reactivity and functionalization of aromatic compounds.

Electrophilic Substitution in Arenes

Mechanism

The mechanism of electrophilic substitution in arenes involves a series of steps including activation, attack, and deactivation. Let's explore this mechanism using the examples of nitrobenzene and bromobenzene.

Activation

The first step in electrophilic substitution is the activation of the electrophile. This is achieved by combining the electrophile with a Lewis acid catalyst, such as $AlCl_3$ or $FeCl_3$. The Lewis acid coordinates with the electrophile, making it more electrophilic and facilitating its attack on the aromatic ring.

Attack

Once the electrophile is activated, it can attack the π electron cloud of the aromatic ring. The attack occurs at one of the carbon atoms in the ring, resulting in the formation of a sigma complex. The sigma complex is an intermediate species in which the electrophile is bonded to the aromatic ring through a carbon atom. This intermediate is also known as an arenium ion or Wheland intermediate.

Deactivation

After the attack, the sigma complex undergoes deactivation to restore the aromaticity of the ring. Deactivation involves the loss of a proton from the sigma complex, leading to the formation of the substituted aromatic compound. The loss of a proton is facilitated by a base, such as H_2O or a halide ion. In the case of nitrobenzene, the base can be water, while in the case of bromobenzene, a halide ion (e.g., Br-) can act as the base.

Nitrobenzene Formation

Let's now examine the formation of nitrobenzene through electrophilic substitution. Nitrobenzene is formed by the reaction of benzene (C_6H_6) with nitronium ion (NO_2^+). The overall reaction can be represented as follows:

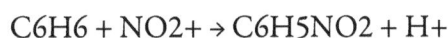

$$C_6H_6 + NO_2^+ \rightarrow C_6H_5NO_2 + H^+$$

In the first step of the mechanism, the nitronium ion is activated by a Lewis acid catalyst, typically sulfuric acid (H_2SO_4). The activation involves the coordination of the nitronium ion with the Lewis acid, resulting in the formation of a stronger electrophile. The activated nitronium ion then attacks the π electron cloud of the benzene ring, leading to the formation of a sigma complex. The sigma complex is stabilized by resonance, as the positive charge can be delocalized over the ring. This resonance stabilization is a key factor that makes the formation of nitrobenzene favorable. In the final step of the mechanism, a proton is abstracted from the sigma complex by water, resulting in the formation of nitrobenzene. The overall reaction is exothermic, and the released energy contributes to the stability of the final product.

Bromobenzene Formation

Now, let's explore the formation of bromobenzene through electrophilic substitution. The reaction of benzene with bromine (Br_2) in the presence of a Lewis acid catalyst, like iron(III) chloride ($FeCl_3$), results in the formation of bromobenzene. The overall reaction can be represented as follows: $C_6H_6 + Br_2 \rightarrow C_6H_5Br + HBr$

In the first step of the mechanism, bromine is activated by coordination with the Lewis acid catalyst. The coordination increases the electrophilicity of bromine, making it capable of attacking the aromatic ring. The activated bromine attacks the π electron cloud of the benzene ring, forming a sigma complex. The sigma complex is stabilized by resonance, similar to the case of nitrobenzene formation. The resonance stabilization contributes to the high reactivity of bromine towards aromatic substitution. In the final step of the mechanism, a proton is abstracted from the sigma complex by a bromide ion (Br-) acting as a base. This deprotonation leads to the formation of bromobenzene. The released proton combines with another bromide ion to form hydrogen bromide (HBr), which is often observed as a byproduct of this reaction.

Factors Affecting Electrophilic Substitution

The mechanism of electrophilic substitution in arenes is influenced by various factors, including the nature of the electrophile, the substituents on the aromatic ring, and the reaction conditions. Let's discuss some of these factors in more detail.

Nature of the Electrophile

The reactivity of electrophiles towards aromatic substitution depends on their ability to accept electron density. Strongly electron-withdrawing groups, such as the nitro group (NO2), enhance the electrophilicity of the attacking species. On the other hand, electron-donating groups, such as alkyl groups, decrease the electrophilicity of the attacking species.

Substituents on the Aromatic Ring

The presence of substituents on the aromatic ring can influence the reactivity of electrophilic substitution. Electron-donating substituents, such as alkyl groups, activate the ring towards electrophilic substitution by increasing the electron density on the ring. Conversely, electron-withdrawing substituents, such as nitro or halogen groups, deactivate the ring by withdrawing electron density, making it less reactive towards electrophiles.

Reaction Conditions

The choice of reaction conditions, including the solvent and temperature, can affect the rate and selectivity of electrophilic substitution. Polar solvents, such as sulfuric acid, enhance the solubility and reactivity of the reactants. Higher temperatures generally increase the rate of reaction, but excessive heat can lead to side reactions or decomposition of the reactants.

Applications and Significance

Electrophilic substitution in arenes is a versatile tool for the synthesis of a wide range of aromatic compounds. This reaction allows the introduction of various functional groups onto the benzene ring, enabling the synthesis of pharmaceuticals, dyes, polymers, and other important organic compounds. For example, nitrobenzene is a crucial intermediate in the production of aniline, which is used for the synthesis of numerous dyes and pharmaceuticals. Bromobenzene finds applications in organic synthesis as a versatile starting material for the introduction of other functional groups. Understanding the mechanism of electrophilic substitution in arenes also provides insights into the reactivity and stability of aromatic compounds. The resonance stabilization of the sigma complex and the restoration of aromaticity through deprotonation contribute to the high stability of substituted aromatic compounds. This stability is essential for the functionalization and manipulation of aromatic compounds in various chemical reactions.

Counterarguments and Limitations

While electrophilic substitution in arenes is a widely studied and utilized reaction, it is not without limitations. One limitation is the regioselectivity of the reaction, which refers to the preference for substitution at a specific position of the aromatic ring. In some cases, multiple substitution products can be formed, making the reaction challenging to control. The regioselectivity can be influenced by the nature of the electrophile and the substituents on the aromatic ring.

Another limitation is the compatibility of the reaction conditions with the functional groups present in the reactants. Strongly acidic conditions, such as those used in the activation step, can cause unwanted side reactions or decomposition of sensitive functional groups. In such cases, milder reaction conditions or alternative synthetic routes may be required.

The Dominance of Substitution over Addition

Introduction

We will explore the mechanism of electrophilic substitution in arenes and delve into the concept of delocalization, its role in aromatic stabilization, and how these factors contribute to the predominance of substitution reactions over addition reactions.

Electrophilic Substitution Reactions

Electrophilic substitution reactions involve the replacement of an atom or a group in a molecule by an electrophile—an electron-deficient species. In the case of arenes, which are cyclic compounds containing a conjugated system of double bonds, electrophilic substitution reactions can occur at the aromatic ring. The most common electrophilic substitution reactions in arenes include nitration, halogenation, sulfonation, and Friedel-Crafts acylation/alkylation.

Delocalization and Aromatic Stabilization

Delocalization refers to the spread of electrons over a larger area than a single bond or atom. In the case of arenes, the delocalization of π (pi) electrons occurs within the conjugated system of double bonds, resulting in the formation of a stable, resonance hybrid structure. Aromatic compounds exhibit unique stability and reactivity due to the delocalization of electrons. The concept of aromatic stabilization can be understood by considering the example of benzene, the simplest aromatic compound. Benzene consists of a planar ring of six carbon atoms, each bonded to a hydrogen atom. The resonance hybrid structure of benzene can be represented by two contributing structures, often denoted as Kekulé structures.

$$H - C = C - C = C - C = C - C = C - H$$

In the Kekulé structures, each carbon atom is alternatively single or double bonded to its neighboring carbon atom. However, experimental evidence shows that the actual structure of benzene is a resonance hybrid of these two Kekulé structures, with delocalized π electrons spread evenly around the ring.

Explanation of Substitution Dominance over Addition

The predominance of electrophilic substitution over addition in arenes can be explained by the aromatic stabilization resulting from delocalization of electrons. This stabilization is lost in the case of addition reactions, leading to a higher activation energy and a less favorable reaction pathway.

Activation Energy and Reaction Kinetics

In an addition reaction, the electrophile attacks the π electron cloud of the aromatic ring, resulting in the formation of a new σ (sigma) bond. However, this disrupts the delocalization of π electrons, leading to the loss of aromatic stabilization. As a consequence, addition reactions typically require higher activation energies compared to substitution reactions.

The higher activation energy in addition reactions can be rationalized by considering the reorganization of electron density during the reaction. In the case of a substitution reaction, the electrophile interacts with the π electron cloud, which is already relatively concentrated above and below the plane of the aromatic ring. This results in a relatively smooth transition state, facilitating the reaction with a lower energy barrier. On the other hand, in an addition reaction, the electrophile needs to interact with the π electron cloud and form a new σ bond simultaneously. This requires significant reorganization of electron density, leading to a higher activation energy. Consequently, substitution reactions tend to be kinetically favored over addition reactions.

Aromatic Stabilization and Reaction Thermodynamics

Apart from the kinetic considerations, the thermodynamics of electrophilic substitution reactions also favor substitution over addition in arenes. The aromatic stabilization resulting from the delocalization of electrons provides a significant energetic advantage to substitution reactions. The aromatic stabilization energy arises from the resonance hybrid structure of the aromatic compound. The delocalized π electrons are spread over the entire ring, minimizing electron-electron repulsion and stabilizing the compound. As a result, the resonance hybrid structure has lower energy compared to the hypothetical Kekulé structures. When an electrophile attacks the aromatic ring in a substitution reaction, the resulting product still retains the delocalized π electron cloud to a significant extent. Therefore, the product remains stabilized by aromatic resonance, and the overall energy change in the reaction is relatively small. In contrast, in an addition reaction, the resulting product no longer possesses the delocalized π electron cloud, leading to the loss of aromatic stabilization. This results in a relatively large energy change in the reaction, making addition reactions less thermodynamically favorable compared to substitution reactions.

Examples from Various Fields

The dominance of substitution over addition in electrophilic aromatic substitution reactions is evident across various fields of chemistry and has numerous practical applications.

Pharmaceutical Chemistry

Pharmaceutical chemists often utilize electrophilic substitution reactions to introduce specific functional groups into aromatic compounds, thereby modulating their biological activity. For instance, the introduction of halogen atoms or other substituents in drug molecules can enhance their lipophilicity, solubility, and receptor affinity. These modifications are typically achieved through substitution reactions rather than addition reactions, ensuring the retention of aromatic stabilization.

Materials Science

In materials science, the synthesis of polymers often involves electrophilic substitution reactions in aromatic monomers. These reactions allow for the incorporation of various substituents that can influence the properties of the resulting polymer, such as mechanical strength, thermal stability, and optical properties. Again, substitution reactions are preferred over addition reactions to maintain the aromatic stabilization and obtain the desired properties.

Synthetic Organic Chemistry

Electrophilic substitution reactions play a crucial role in synthetic organic chemistry, enabling the construction of complex organic molecules. The ability to selectively substitute one group for another in an aromatic compound allows chemists to control the regioselectivity and stereochemistry of the reaction, leading to the desired product. Addition reactions, on the other hand, would disrupt the aromaticity and hinder the synthesis of specific compounds.

Counterarguments

While the predominance of substitution over addition in electrophilic aromatic substitution reactions is well-established, it is important to acknowledge counterarguments and limitations.

Addition Reactions in Non-Aromatic Systems

In non-aromatic systems, addition reactions can be more prevalent compared to substitution reactions. For example, in alkenes, electrophilic addition reactions are highly favorable due to the presence of localized π bonds. In these cases, the electron density is not delocalized, and the activation barriers for addition reactions are significantly lower compared to substitution reactions.

Substitution in Non-Aromatic Systems

Substitution reactions can also occur in non-aromatic systems, where they are not influenced by aromatic stabilization. For instance, in aliphatic compounds, nucleophilic substitution reactions are common. These reactions involve the replacement of one atom or group by a nucleophile, rather than an electrophile as in electrophilic substitution reactions. The prevalence of substitution reactions in non-aromatic systems is primarily dictated by the electronic and steric factors specific to those systems.

Predicting the Location of Halogenation in Arene Compounds

Introduction

In this analysis, we will explore the factors that influence the location of halogenation in arenes and discuss how reaction conditions can be manipulated to control the outcome of the halogenation reaction.

Factors Influencing the Location of Halogenation

The location of halogenation in arenes is primarily governed by two factors: the reactivity of the aromatic ring and the reactivity of the side-chain.

Let's explore each of these factors more extensively.

Reactivity of the Aromatic Ring

The reactivity of the aromatic ring plays a crucial role in determining the location of halogenation. The electron density distribution in the ring is affected by the presence of substituents and the delocalized π-electron system. Halogens, being highly electronegative, tend to preferentially attack electron-rich sites in the aromatic ring.

The electron density of an aromatic ring can be influenced by various factors, including the presence of electron-donating or electron-withdrawing substituents. Electron-donating groups, such as alkyl groups, increase the electron density in the ring, making it more susceptible to attack by electrophiles like halogens. On the other hand, electron-withdrawing groups, such as nitro or carbonyl groups, decrease the electron density, making the ring less reactive towards halogenation.

For example, in the reaction between benzene (C6H6) and chlorine (Cl2), if there are no substituents on the ring, halogenation can occur at any position. However, when a substituent like methyl (-CH3) is present, it donates electron density to the ring, making it more reactive. As a result, the halogenation reaction is more likely to occur at the ortho or para positions relative to the methyl group.

Reactivity of the Side-Chain

In addition to the reactivity of the aromatic ring, the nature of the side-chain also influences the location of halogenation. Side-chains can be classified into two categories: electron-rich and electron-deficient. The reactivity of the side-chain is determined by its ability to stabilize positive charges and its susceptibility to nucleophilic attack.

Electron-rich side-chains, such as alkyl groups, stabilize positive charges through the donation of electron density. As a result, they are less likely to undergo halogenation. On the other hand, electron-deficient side-chains, such as carbonyl or nitro groups, are more susceptible to halogenation due to their ability to stabilize positive charges and their vulnerability to nucleophilic attack.

For instance, in the case of toluene (C6H5CH3), the side-chain consists of a methyl group (-CH3). The electron-rich nature of the methyl group stabilizes positive charges, making it less reactive towards halogenation. Hence, the halogenation reaction is more likely to occur on the aromatic ring rather than the side-chain.

Reaction Conditions and their Impact on Halogenation

The choice of reaction conditions can significantly influence the location of halogenation in arenes. The most commonly employed halogenation methods include electrophilic aromatic substitution (EAS) and free radical halogenation. Let's explore how these reaction conditions affect the location of halogenation:

Electrophilic Aromatic Substitution (EAS)

EAS is a common method for halogenating arenes. It involves the attack of an electrophilic species, such as a halogen cation, on the electron-rich aromatic ring. The reaction conditions, including the choice of reagents, solvent, and temperature, can be adjusted to control the outcome of the halogenation reaction.

Choice of Reagent

The choice of halogenating reagent can have a significant impact on the location of halogenation. Different halogens have different electron-withdrawing abilities, which affect their reactivity towards the electron-rich aromatic ring. For example, chlorine is less electron-withdrawing compared to bromine or iodine. Hence, chlorine is more likely to undergo halogenation at the ortho or para positions, where electron density is higher.

Solvent Effects

The choice of solvent can also influence the location of halogenation in EAS reactions. Polar solvents, such as nitrobenzene or acetic acid, stabilize the intermediate carbocation formed during the reaction. This stabilization can lead to an increased likelihood of halogenation occurring at the ortho or para positions. Non-polar solvents, on the other hand, may favor halogenation at the meta position.

Temperature

The temperature at which the EAS reaction is carried out can impact the location of halogenation. Lower temperatures tend to favor ortho and para halogenation, as they slow down the reaction rate, allowing more time for the electrophile to approach the electron-rich positions. Higher temperatures, on the other hand, may favor meta halogenation due to the increased mobility of the aromatic ring.

Free Radical Halogenation

Free radical halogenation involves the use of halogen radicals, generated from compounds like N-halosuccinimide (NBS), to halogenate arenes. This method typically leads to substitution in the side-chain rather than the aromatic ring. The selectivity of this reaction can be influenced by various factors.

Reactivity of the Side-Chain

As mentioned earlier, the reactivity of the side-chain plays a crucial role in determining the location of halogenation in free radical reactions. Electron-deficient side-chains, such as carbonyl or nitro groups, are more susceptible to halogenation due to their ability to stabilize positive charges. Hence, they are more likely to undergo halogenation in the side-chain rather than the aromatic ring.

Reaction Conditions

The reaction conditions, including the choice of solvent and temperature, can also influence the location of halogenation in free radical reactions. Non-polar solvents, such as carbon tetrachloride or chloroform, tend to favor side-chain halogenation. Lower temperatures may also increase the likelihood of side-chain halogenation due to slower radical formation and reduced mobility of the aromatic ring.

Examples from Various Fields

The concepts discussed above can be illustrated with examples from different fields of chemistry:

Medicinal Chemistry

In the field of medicinal chemistry, the location of halogenation in drug molecules can significantly impact their biological activity. For instance, the introduction of halogen atoms at specific positions in a drug molecule can enhance its lipophilicity, metabolic stability, and binding affinity to target proteins. By carefully selecting the reaction conditions, medicinal chemists can control the location of halogenation to optimize the desired properties of the drug.

Material Science

Halogenation plays a crucial role in modifying the properties of materials. For example, the introduction of halogen atoms into polymers can enhance their thermal stability, flame retardancy, and chemical resistance. The location of halogenation in polymers can be controlled by adjusting the reaction conditions, allowing the synthesis of tailored materials with specific properties.

Environmental Chemistry

Halogenation reactions can occur naturally in the environment, leading to the formation of harmful byproducts. For instance, the halogenation of organic compounds in drinking water sources can result in the formation of disinfection byproducts, such as trihalomethanes, which are known to be carcinogenic. Understanding the factors influencing the location of halogenation can aid in the development of strategies to minimize the formation of such byproducts.

Counterarguments and Limitations

While the factors discussed above provide a general framework for predicting the location of halogenation in arenes, it is important to note that there can be exceptions and limitations to these predictions. The presence of multiple substituents on the aromatic ring can complicate the reactivity patterns, making it challenging to predict the exact location of halogenation. Additionally, steric hindrance and electronic interactions between substituents can also influence the outcome of the halogenation reaction.

Furthermore, the choice of reaction conditions and reagents may not always result in the desired location of halogenation. Other factors, such as competing reactions or side reactions, can occur under certain conditions, leading to unexpected products. Hence, careful experimentation and optimization are necessary to achieve the desired outcome in halogenation reactions.

Directing Effects of Different Substituents

Introduction

The electrophilic substitution of arenes is a fundamental concept in organic chemistry, where a substituent is added to an aromatic ring through the attack of an electrophilic species. In this process, the substituents on the ring can have a significant impact on the regioselectivity of the reaction, directing the electrophile to specific positions on the aromatic ring. This phenomenon is commonly referred to as the "directing effects" of substituents. In this essay, we will explore the directing effects of some commonly encountered substituents, namely –NH2, –OH, –R, –NO2, –COOH, and –COR, and analyze how they influence the regiochemistry of electrophilic substitution reactions on arenes.

Overview of Electrophilic Substitution Reactions

Before delving into the directing effects of substituents, it is essential to have a clear understanding of electrophilic substitution reactions themselves. Electrophilic aromatic substitution reactions involve the attack of an electrophile on an aromatic ring, resulting in the substitution of a hydrogen atom with a substituent. The electrophile is often generated by the addition of a Lewis acid or through an electrophilic addition reaction.

The regioselectivity of electrophilic substitution reactions can vary based on the nature of the substituent present on the aromatic ring. The directing effects of substituents can be broadly classified into two categories: ortho/para-directing and meta-directing.

Ortho/Para-Directing Substituents

Ortho/para-directing substituents direct the incoming electrophile to the ortho (o) and para (p) positions on the aromatic ring. These positions are adjacent or opposite, respectively, to the substituent itself. Let us examine the directing effects of –NH2, –OH, and –R substituents.

–NH2 Substituent

The –NH2 substituent exhibits a strong ortho/para-directing effect due to its electron-donating nature. The lone pair of electrons on the nitrogen atom is capable of delocalizing into the π system of the aromatic ring, increasing the electron density at the ortho and para positions. This increased electron density makes these positions more nucleophilic and, therefore, more susceptible to attack by electrophiles.

For instance, when an electrophile such as a nitronium ion (NO2+) is introduced to an aromatic compound with an –NH2 substituent, it will preferentially attack the ortho or para positions. This is exemplified in the nitration of aniline, where the nitronium ion adds to the ortho and para positions, resulting in the formation of ortho-nitroaniline and para-nitroaniline, respectively.

–OH Substituent

Similar to the –NH2 substituent, the –OH substituent is also an ortho/para-directing group due to its electron-donating nature. The lone pair of electrons on the oxygen atom can delocalize into the π system of the aromatic ring, increasing the electron density at the ortho and para positions. This increased electron density makes these positions more nucleophilic and, thus, more susceptible to electrophilic attack.

For example, in the nitration of phenol, the –OH group directs the nitronium ion to the ortho and para positions, leading to the formation of ortho-nitrophenol and para-nitrophenol, respectively.

–R Substituent

The –R substituent represents a general alkyl group that is neither an electron-donating nor an electron-withdrawing group. Alkyl groups are considered weakly activating, which means they exhibit a moderate ortho/para-directing effect. The presence of alkyl groups increases the electron density at the ortho and para positions but to a lesser extent compared to strongly activating groups like –NH2 and –OH.

For instance, in the Friedel-Crafts alkylation of toluene with an alkyl halide, the –R group directs the incoming electrophile to the ortho and para positions, resulting in the formation of ortho-substituted and para-substituted alkylated products.

Meta-Directing Substituents

Unlike ortho/para-directing substituents, meta-directing substituents direct the incoming electrophile to the meta (m) position on the aromatic ring, which is one carbon away from the substituent itself. This occurs due to the electron-withdrawing or deactivating nature of these substituents, which reduces the electron density at the ortho and para positions, making them less nucleophilic and less favorable for electrophilic attack. Let us examine the directing effects of –NO2, –COOH, and –COR substituents.

–NO2 Substituent

The –NO2 substituent is a classic example of a meta-directing group. It is highly electron-withdrawing, primarily due to the presence of the nitro group, which contains both an electron-withdrawing nitroso (–NO) group and an electron-withdrawing nitro (–NO2) group. The strong electron-withdrawing nature of –NO2 decreases the electron density at the ortho and para positions, thereby making these positions less nucleophilic and unfavorable for electrophilic attack.

For instance, in the nitration of nitrobenzene, the –NO2 group directs the nitronium ion to the meta position, leading to the formation of meta-nitrobenzene.

–COOH Substituent

The –COOH substituent, also known as a carboxylic acid group, is another example of a meta-directing group. It is highly electron-withdrawing due to the presence of the carbonyl (C=O) group and the electron-withdrawing nature of the oxygen atom. The electron-withdrawing nature of –COOH reduces the electron density at the ortho and para positions, leading to meta-directing effects.

For example, in the nitration of benzoic acid, the –COOH group directs the nitronium ion to the meta position, resulting in the formation of meta-nitrobenzoic acid.

–COR Substituent

The –COR substituent represents a general acyl group. Similar to the –COOH group, –COR is also a meta-directing group due to the presence of the carbonyl (C=O) group, which exhibits electron-withdrawing characteristics. The electron-withdrawing nature of the carbonyl group reduces the electron density at the ortho and para positions, favoring meta-substitution.

Halogen Compounds

Introduction

We will discuss the reactions by which halogenoarenes can be produced, focusing on the substitution of arenes with chlorine or bromine. We will explore the mechanism of this reaction, exemplified by the formation of chlorobenzene from benzene and 2-chloromethylbenzene and 4-chloromethylbenzene from methylbenzene. We will also discuss the role of the catalyst and the conditions required for the reaction to occur.

Substitution of Arene with Chlorine or Bromine

The substitution of an arene molecule with a halogen atom, such as chlorine or bromine, is a common reaction in organic chemistry. This reaction involves the replacement of a hydrogen atom on the aromatic ring with a halogen atom. The resulting product is a halogenoarene.

The substitution reaction can be represented by the general equation:

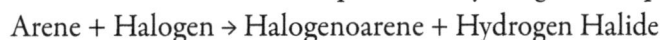

Arene + Halogen → Halogenoarene + Hydrogen Halide

For example, the reaction between benzene and chlorine can be represented as: $C_6H_6 + Cl_2 \rightarrow C_6H_5Cl + HCl$

Similarly, the reaction between methylbenzene (toluene) and chlorine can be represented as:

$$C_6H_5CH_3 + Cl_2 \rightarrow C_6H_4ClCH_3 + HCl$$

In both cases, a hydrogen atom on the aromatic ring is replaced by a chlorine atom, resulting in the formation of chlorobenzene and 2-chloromethylbenzene, respectively.

Mechanism of the Reaction

The substitution of an arene molecule with chlorine or bromine occurs through an electrophilic aromatic substitution mechanism. This mechanism involves the generation of an electrophile, which attacks the electron-rich aromatic ring, leading to the substitution of a hydrogen atom with a halogen atom.

The reaction begins with the generation of a halogen cation, either Cl^+ or Br^+, from the halogen molecule. This step is facilitated by the presence of a Lewis acid catalyst, such as aluminum chloride ($AlCl_3$) or aluminum bromide ($AlBr_3$). The Lewis acid coordinates with the halogen molecule, facilitating the formation of the halogen cation.

Once the halogen cation is formed, it acts as an electrophile and attacks the electron-rich aromatic ring. The electrophilic attack disrupts the aromaticity of the ring, leading to the formation of a sigma complex. This sigma complex is an intermediate species that contains a halogen atom bonded to the aromatic ring.

The sigma complex is not stable and undergoes a rearrangement to restore the aromaticity of the ring. The rearrangement involves the migration of a hydrogen atom from a neighboring carbon atom to the carbon atom that is bonded to the halogen atom. This migration is facilitated by the presence of the Lewis acid catalyst.

The final step of the reaction involves the elimination of the Lewis acid catalyst and the formation of the halogenoarene product. The Lewis acid catalyst is regenerated and can participate in further substitution reactions.

Example 1: Formation of Chlorobenzene from Benzene

Let's consider the reaction between benzene and chlorine to form chlorobenzene. The reaction can be represented as:

$$C_6H_6 + Cl_2 \rightarrow C_6H_5Cl + HCl$$

The reaction begins with the generation of a chlorine cation (Cl+) from the chlorine molecule. This step is facilitated by the presence of aluminum chloride (AlCl3) as a Lewis acid catalyst. The aluminum chloride coordinates with the chlorine molecule, facilitating the formation of the chlorine cation.

Once the chlorine cation is formed, it acts as an electrophile and attacks the electron-rich aromatic ring of benzene. The electrophilic attack disrupts the aromaticity of the ring, leading to the formation of a sigma complex.

The sigma complex is not stable and undergoes a rearrangement to restore the aromaticity of the ring. In this case, the migration of a hydrogen atom occurs from a neighboring carbon atom to the carbon atom that is bonded to the chlorine atom. This migration is facilitated by the presence of aluminum chloride.

The final step of the reaction involves the elimination of aluminum chloride and the formation of chlorobenzene as the product. The aluminum chloride is regenerated and can participate in further substitution reactions.

Example 2: Formation of 2-Chloromethylbenzene and 4-Chloromethylbenzene from Methylbenzene

Now, let's consider the reaction between methylbenzene (toluene) and chlorine to form 2-chloromethylbenzene and 4-chloromethylbenzene. The reaction can be represented as:

$$C6H5CH3 + Cl2 \rightarrow C6H4ClCH3 + HCl$$

Similar to the previous example, the reaction begins with the generation of a chlorine cation (Cl+) from the chlorine molecule. Aluminum chloride (AlCl3) is used as a catalyst to facilitate this step.

The chlorine cation acts as an electrophile and attacks the electron-rich aromatic ring of methylbenzene. The electrophilic attack disrupts the aromaticity of the ring, leading to the formation of a sigma complex.

The sigma complex undergoes a rearrangement to restore the aromaticity of the ring. In this case, two different migration pathways are possible. The migration of a hydrogen atom can occur either to the carbon atom ortho (2-position) or meta (4-position) to the methyl group.

As a result, two different products are formed: 2-chloromethylbenzene and 4-chloromethylbenzene. In the case of 2-chloromethylbenzene, the hydrogen atom migrates to the carbon atom ortho to the methyl group. In the case of 4-chloromethylbenzene, the hydrogen atom migrates to the carbon atom meta to the methyl group.

The final step of the reaction involves the elimination of aluminum chloride and the formation of 2-chloromethylbenzene and 4-chloromethylbenzene as the products. The aluminum chloride is regenerated and can participate in further substitution reactions.

Role of the Catalyst

The Lewis acid catalyst, such as aluminum chloride or aluminum bromide, plays a crucial role in the substitution of arenes with chlorine or bromine. The catalyst facilitates the generation of the halogen cation and enhances the electrophilicity of the halogen atom.

The Lewis acid catalyst coordinates with the halogen molecule, facilitating the formation of the halogen cation. This coordination increases the electrophilicity of the halogen atom, making it more reactive towards the electron-rich aromatic ring. In addition to facilitating the formation of the halogen cation, the Lewis acid catalyst also stabilizes the sigma complex and promotes the rearrangement of the sigma complex to restore the aromaticity of the ring. The presence of the catalyst lowers the activation energy of these steps, making the reaction more efficient. Furthermore, the Lewis acid catalyst can participate in multiple substitution reactions. After the formation of the halogenoarene product, the catalyst is regenerated and can react with another molecule of arene and halogen to initiate another substitution reaction.

The Difference in Reactivity between Halogenoalkanes and Halogenoarenes

Introduction

It will explore and analyze the differences in reactivity between chloroethane (a halogenoalkane) and chlorobenzene (a halogenoarene), providing a comprehensive understanding of their distinct properties.

Halogenoalkanes: Chloroethane

Chloroethane (C_2H_5Cl) is an organic compound with a halogen (chlorine) atom attached to an alkane (ethane) molecule. Halogenoalkanes are generally more reactive than halogenoarenes due to the presence of a highly polarized carbon-halogen bond. In the case of chloroethane, the carbon-halogen bond is polarized, with the chlorine atom being more electronegative than the carbon atom. This polarity leads to differences in reactivity compared to halogenoarenes.

Nucleophilic Substitution Reactions

One of the most important reactions involving halogenoalkanes is nucleophilic substitution. In nucleophilic substitution reactions, a nucleophile replaces the halogen atom, resulting in the formation of a new compound. Due to the polarized carbon-halogen bond, halogenoalkanes are susceptible to nucleophilic attack. For example, chloroethane can undergo nucleophilic substitution with hydroxide ions (OH-) to form ethanol:

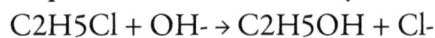

$$C_2H_5Cl + OH^- \rightarrow C_2H_5OH + Cl^-$$

This reaction occurs in the presence of a strong base, such as sodium hydroxide (NaOH), which provides the hydroxide ions necessary for substitution. The nucleophile attacks the carbon atom, leading to the displacement of the chlorine atom and the formation of the alcohol product.

Reactivity of Halogenoalkanes

Halogenoalkanes exhibit higher reactivity than halogenoarenes due to the nature of the carbon-halogen bond. The polarized carbon-halogen bond makes the carbon atom partially positive, rendering it more susceptible to nucleophilic attack. The reactivity of halogenoalkanes can be influenced by factors such as the nature of the halogen, the substitution pattern, and steric hindrance. For instance, primary halogenoalkanes (where the carbon atom attached to the halogen is bonded to only one other carbon) tend to be more reactive than secondary or tertiary halogenoalkanes. This is because primary halogenoalkanes have less steric hindrance, allowing nucleophiles to approach the carbon atom more easily.

Halogenoarenes: Chlorobenzene

Chlorobenzene (C_6H_5Cl) is an aromatic compound with a halogen (chlorine) atom attached to a benzene ring. Unlike halogenoalkanes, halogenoarenes are less reactive due to the aromatic nature of the benzene ring. The presence of the conjugated π-electron system in the benzene ring affects the reactivity of the carbon-halogen bond and influences the behavior of halogenoarenes in chemical reactions.

Electrophilic Aromatic Substitution Reactions

Halogenoarenes are best known for their reactivity in electrophilic aromatic substitution (EAS) reactions. In EAS, an electrophile replaces a hydrogen atom on the aromatic ring, resulting in the formation of a substituted aromatic compound. This reaction is characteristic of aromatic compounds and is facilitated by the delocalized π-electron system of the benzene ring. For example, chlorobenzene can undergo EAS with a strong electrophile, such as nitronium ion (NO_2^+), to form nitrobenzene: $C_6H_5Cl + NO_2^+ \rightarrow C_6H_5NO_2 + HCl$

In this reaction, the delocalized π-electron system of the benzene ring attracts the electrophile, leading to the substitution of the chlorine atom and the formation of the nitro group.

Reactivity of Halogenoarenes

Halogenoarenes exhibit lower reactivity compared to halogenoalkanes due to the stabilization provided by the aromatic system. The π-electron delocalization in the benzene ring reduces the electrophilic character of the carbon-halogen bond, making halogenoarenes less prone to nucleophilic attack. This is evident in the limited substitution reactions that occur with halogenoarenes, primarily through EAS.

In addition, the reactivity of halogenoarenes can be influenced by various factors, such as the nature of the halogen, the substitution pattern on the ring, and electronic effects from substituents. For instance, electron-withdrawing groups, such as nitro (-NO2) or carbonyl (-C=O) groups, increase the reactivity of halogenoarenes by activating the aromatic ring towards EAS.

Comparison and Analysis

The differences in reactivity between chloroethane (a halogenoalkane) and chlorobenzene (a halogenoarene) can be attributed to the polarized carbon-halogen bond in halogenoalkanes and the conjugated π-electron system in halogenoarenes. The polarized carbon-halogen bond in chloroethane makes it more susceptible to nucleophilic attack, leading to various substitution reactions. On the other hand, the conjugated π-electron system in chlorobenzene stabilizes the carbon-halogen bond, reducing its electrophilic character and making it less reactive towards nucleophiles.

While both halogenoalkanes and halogenoarenes exhibit distinct reactivity patterns, it is important to note that these are general trends, and specific reactions may deviate from these patterns. For example, halogenoarenes can undergo nucleophilic aromatic substitution (NAS) reactions under certain conditions, where the nucleophile replaces the halogen atom on the aromatic ring.

Counterarguments

One counterargument could be that the reactivity of halogenoarenes can be enhanced by using strong nucleophiles or harsh reaction conditions. While this is true to some extent, the overall reactivity of halogenoarenes towards nucleophilic substitution is still lower compared to halogenoalkanes due to the stabilization provided by the aromatic system.

Another counterargument might suggest that the reactivity of halogenoalkanes is solely dependent on the nature of the halogen atom. Although the halogen atom does influence the reactivity, factors such as steric hindrance and substitution pattern also play significant roles. For instance, a primary halogenoalkane with a bulky halogen atom may exhibit lower reactivity compared to a secondary halogenoalkane with a smaller halogen atom.

Hydroxy Compounds

Alcohols

Reaction of Acyl Chlorides with Ethyl Ethanoate to Form Esters

Introduction

We will focus on the reaction between acyl chlorides and ethyl ethanoate to form esters, exploring the mechanism, factors influencing the reaction, and various applications in different fields.

Mechanism of the Reaction

The reaction between acyl chlorides and ethyl ethanoate proceeds via a nucleophilic acyl substitution mechanism. It involves the attack of the alcohol oxygen on the carbonyl carbon of the acyl chloride, followed by the elimination of the chloride ion. The resulting intermediate reacts with ethyl ethanoate, leading to the formation of an ester and HCl as a byproduct. The overall reaction can be represented as follows:

Acyl Chloride + Ethyl Ethanoate → Ester + HCl

The reaction mechanism can be further explained using the example of the reaction between ethanoyl chloride and ethyl ethanoate. Initially, the oxygen atom of the ethyl ethanoate acts as a nucleophile and attacks the carbonyl carbon of the ethanoyl chloride, forming a tetrahedral intermediate. This intermediate collapses, leading to the formation of the ester and the expulsion of chloride ion. The reaction can be represented as follows:

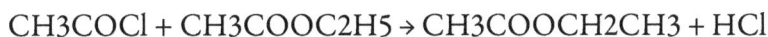

$$CH_3COCl + CH_3COOC_2H_5 \rightarrow CH_3COOCH_2CH_3 + HCl$$

Factors Influencing the Reaction

Several factors can influence the reaction between acyl chlorides and ethyl ethanoate. Some of the key factors that affect the reaction include the nature of the acyl chloride, the nature of the alcohol, the reaction conditions, and the presence of catalysts.

Nature of the Acyl Chloride

The reactivity of acyl chlorides varies depending on the nature of the substituents attached to the carbonyl carbon. Electron-withdrawing groups, such as halogens or nitro groups, increase the electrophilicity of the carbonyl carbon, making the acyl chloride more reactive. On the other hand, electron-donating groups decrease the electrophilicity, resulting in a slower reaction. For example, an acyl chloride containing a halogen substituent will react more readily with ethyl ethanoate than an acyl chloride with an alkyl substituent.

Nature of the Alcohol

The nature of the alcohol used in the reaction also affects the rate of ester formation. Primary alcohols are generally more reactive than secondary or tertiary alcohols due to the increased availability of the alcohol oxygen for nucleophilic attack. Additionally, the presence of electron-donating or electron-withdrawing groups in the alcohol molecule can influence the reaction rate by altering the electron density on the oxygen atom. For example, an alcohol with a hydroxyl group adjacent to an electron-withdrawing group will react more readily with acyl chlorides.

Reaction Conditions

The reaction conditions, including temperature and solvent choice, play a crucial role in the reaction between acyl chlorides and ethyl ethanoate. Higher temperatures generally increase the reaction rate by providing more kinetic energy to the reactant molecules. However, excessively high temperatures can also lead to side reactions or degradation of the reactants. The choice of solvent is important as it affects the solubility and reactivity of the reactants. Common solvents used in esterification reactions include dichloromethane, chloroform, or diethyl ether.

Catalysts

The use of catalysts can significantly enhance the rate of ester formation. Acidic catalysts, such as sulfuric acid or hydrochloric acid, can be employed to increase the acidity of the reaction medium, facilitating the nucleophilic attack of the alcohol oxygen on the acyl chloride. These catalysts can also act as proton donors, promoting the elimination of the chloride ion. Additionally, Lewis acids, such as aluminum chloride or zinc chloride, can be used to activate the carbonyl carbon, making it more susceptible to nucleophilic attack.

Applications in Various Fields

The reaction between acyl chlorides and ethyl ethanoate to form esters finds extensive applications in various fields, including organic synthesis, pharmaceuticals, fragrances, and food industry.

Organic Synthesis

Esterification reactions are fundamental in organic synthesis as they allow the synthesis of a wide range of esters, which serve as important building blocks for the production of various organic compounds. These esters can be further functionalized to obtain complex molecules with desired properties. For example, the reaction of acyl chlorides with ethyl ethanoate can be used to synthesize esters with specific functional groups, such as carboxylic acid esters or aromatic esters, which are widely employed in the production of pharmaceuticals, agrochemicals, and materials.

Pharmaceuticals

Esters play a crucial role in the pharmaceutical industry, both as active pharmaceutical ingredients and as functional groups in drug molecules. The reaction between acyl chlorides and ethyl ethanoate provides a versatile method for the synthesis of ester-based drugs. By selecting appropriate acyl chlorides and alcohols, chemists can tailor the ester structure to achieve desired pharmacological properties, such as improved bioavailability or controlled release. Additionally, esters can serve as prodrugs, which are inactive compounds that are converted into active drugs upon metabolic activation in the body.

Fragrances

Esters are key components of many fragrances and perfumes due to their pleasant smell and volatility. The reaction between acyl chlorides and ethyl ethanoate offers a straightforward route for the synthesis of ester-based fragrances. By varying the acyl chloride or the alcohol, chemists can create a wide variety of esters with different aromatic profiles. For example, the reaction of ethanoyl chloride with ethyl ethanoate results in the formation of ethyl ethanoate, a fruity-smelling ester often used in the production of artificial fruit flavors.

Food Industry

Esters play an essential role in the food industry, contributing to the taste and aroma of various food products. The reaction between acyl chlorides and ethyl ethanoate can be utilized to synthesize esters that mimic natural flavors. These esters can be used as food additives to enhance the sensory experience of consumers. For instance, the reaction between butanoyl chloride and ethyl ethanoate yields ethyl butanoate, a compound responsible for the characteristic smell of pineapple. Ethyl butanoate is commonly used as a flavoring agent in the production of pineapple-flavored beverages and confectioneries.

Counterarguments and Limitations

While the reaction between acyl chlorides and ethyl ethanoate offers numerous advantages, it also has some limitations and potential drawbacks. One limitation is the potential for side reactions, such as the formation of acid chlorides, which can reduce the yield of the desired ester. Additionally, the use of acyl chlorides can be challenging due to their hazardous nature and corrosive properties. Thus, appropriate safety precautions and handling procedures should be followed when working with acyl chlorides. Furthermore, the reaction conditions and choice of reactants must be carefully optimized to avoid undesired byproducts or impurities.

Phenol

Introduction

Phenol is an important organic compound that is widely used in various industries. It is commonly used as a starting material for the synthesis of many other organic compounds. There are several methods to produce phenol, and one of the most common methods is the reaction of phenylamine with HNO_2 or $NaNO_2$ and dilute acid, followed by further warming of the resulting diazonium salt with water. In this essay, we will explore the reactions, reagents, and conditions involved in this process in detail. Reaction of phenylamine with HNO_2 or $NaNO_2$ and dilute acid

The first step in the production of phenol is the reaction of phenylamine (also known as aniline) with HNO_2 or $NaNO_2$ and dilute acid. A diazonium salt is formed as a result of this reaction. The diazonium salt is an intermediate compound that is formed by the replacement of the amino group (NH_2) in phenylamine with a diazonium group (N_2^+). The reaction is typically carried out at temperatures below 10 °C to prevent unwanted side reactions.

The reagents and conditions involved in this step are as follows:

Phenylamine ($C_6H_5NH_2$): It is an aromatic amine that serves as the starting material for the reaction. Phenylamine can be synthesized by reducing nitrobenzene ($C_6H_5NO_2$) with a reducing agent such as tin and hydrochloric acid.

HNO_2 or $NaNO_2$: These are the sources of nitrous acid (HNO_2), which is required for the formation of the diazonium salt. HNO_2 can be generated in situ by the reaction of $NaNO_2$ with dilute acid.

Dilute acid: It is used to provide the acidic conditions required for the reaction. Typically, hydrochloric acid (HCl) or sulfuric acid (H_2SO_4) is used as the dilute acid.

The reaction proceeds as follows:

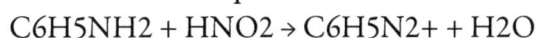

$$C_6H_5NH_2 + HNO_2 \rightarrow C_6H_5N_2^+ + H_2O$$

The diazonium salt that is formed in this step is unstable and highly reactive. It is important to carry out this reaction at low temperatures to prevent its decomposition or reaction with other compounds.

Warming of the diazonium salt with water

The second step in the production of phenol involves the further warming of the diazonium salt with water. This step leads to the conversion of the diazonium salt into phenol.

The reagents and conditions involved in this step are as follows:

Diazonium salt: It is the intermediate compound formed in the previous step. The diazonium salt is highly reactive and can undergo a variety of reactions.

Water (H_2O): It is used to provide the hydroxyl group (OH) required for the formation of phenol. Water is a common and readily available solvent that is used in many chemical reactions.

The reaction proceeds as follows:

$$C_6H_5N_2^+ + H_2O \rightarrow C_6H_5OH + N_2$$

The diazonium group (N_2^+) is replaced by a hydroxyl group (OH) in this step, resulting in the formation of phenol. The nitrogen gas (N_2) that is evolved during the reaction is a byproduct.

Mechanism of the reaction

The reaction of phenylamine with HNO_2 or $NaNO_2$ and dilute acid can be understood in terms of its mechanism. The mechanism involves the following steps:

Formation of the diazonium salt: The reaction begins with the protonation of the amino group in phenylamine by the dilute acid. This is followed by the formation of the nitrous acid (HNO2) from NaNO2 and dilute acid. The nitrous acid then reacts with the protonated phenylamine to form the diazonium salt.

Rearrangement of the diazonium salt: The diazonium salt undergoes a rearrangement reaction in which the nitrogen atom migrates from the diazonium group to the adjacent carbon atom. This rearrangement leads to the formation of a new compound called a tautomeric form. The tautomeric form is stabilized by resonance, which involves the delocalization of the positive charge on the nitrogen atom.

Formation of phenol: The tautomeric form of the diazonium salt is further heated with water, leading to the formation of phenol. The hydroxyl group (OH) is transferred from water to the tautomeric form, resulting in the formation of phenol. The nitrogen gas (N2) is evolved as a byproduct of the reaction.

Significance of the reaction

The reaction of phenylamine with HNO2 or NaNO2 and dilute acid to produce phenol is of great significance in organic synthesis. It provides a convenient and efficient method for the production of phenol, which is an important compound in various industries.

Phenol is widely used as a starting material for the synthesis of many other organic compounds. It is used in the production of plastics, resins, detergents, pharmaceuticals, and dyes, among other things. The ability to produce phenol from readily available starting materials such as phenylamine and dilute acid makes it a cost-effective and environmentally friendly process.

Counterarguments

While the reaction of phenylamine with HNO2 or NaNO2 and dilute acid is a common method for the production of phenol, it has certain limitations and drawbacks. One of the main drawbacks is the formation of unwanted byproducts and side reactions. During the reaction, there is a possibility of the formation of unwanted products such as nitrobenzene and azo compounds. These byproducts not only reduce the yield of phenol but also pose environmental and health hazards. Therefore, it is important to optimize the reaction conditions and control the reaction parameters to minimize the formation of byproducts. Another limitation of this method is the requirement of low temperatures for the reaction to proceed. Carrying out the reaction at low temperatures requires additional cooling equipment and increases the overall cost of the process. Moreover, the reaction rate is slower at low temperatures, which can lead to longer reaction times and lower productivity.

Reactivity with Bases

Introduction

We will explore the chemistry behind the reaction of phenol with bases, focusing on the example of NaOH. We will discuss the mechanism, factors influencing the reaction, and the role of phenoxide in various fields like pharmaceuticals and polymers. Additionally, we will present counterarguments and alternative reactions to provide a comprehensive understanding of this topic.

Reaction Mechanism

The reaction of phenol with NaOH proceeds via a nucleophilic substitution mechanism known as the "Aromatic Nucleophilic Substitution" or "SNAr" mechanism. This mechanism involves the attack of the hydroxide ion (OH-) on the phenol ring, leading to the substitution of the hydroxyl group with the negatively charged phenoxide ion.

The first step of the reaction involves the generation of the phenoxide ion ($C_6H_5O^-$) by the deprotonation of phenol (C_6H_5OH). This deprotonation occurs due to the presence of a strong base, such as NaOH, which abstracts the acidic hydrogen from the hydroxyl group. The resulting phenoxide ion is stabilized by the resonance delocalization of the negative charge across the aromatic ring.

Once the phenoxide ion is formed, it can undergo various reactions depending on the reaction conditions and the nature of the nucleophile present. In the case of the reaction with NaOH, the phenoxide ion reacts with sodium cations (Na^+) to form sodium phenoxide (NaC_6H_5O):

$$C_6H_5O^- + Na^+ \rightarrow NaC_6H_5O$$

This reaction is driven by the strong electrostatic attraction between the negatively charged phenoxide ion and the positively charged sodium ion.

Factors Influencing the Reaction

Several factors can influence the reaction of phenol with bases like NaOH. These factors include the concentration of the reagents, temperature, reaction time, and the presence of catalysts.

Concentration of Reagents

The concentration of phenol and NaOH affects the rate of the reaction. Higher concentrations of phenol or NaOH result in a faster reaction rate due to the increased collision frequency between the reactant molecules. However, excessively high concentrations can lead to side reactions or the formation of unwanted by-products.

Temperature

Temperature plays a crucial role in the reaction kinetics. Generally, increasing the temperature enhances the reaction rate by providing more energy to the reacting species, leading to more frequent and energetic collisions. However, excessively high temperatures can also lead to the decomposition of phenol or the formation of undesirable by-products.

Reaction Time

The reaction time determines the extent of the reaction. Extending the reaction time allows for a higher conversion of phenol to sodium phenoxide. However, excessively long reaction times can result in side reactions or degradation of the desired product.

Catalysts

The addition of catalysts can significantly influence the reaction kinetics and improve the yield of sodium phenoxide. For example, transition metal catalysts like copper or iron salts can accelerate the reaction by facilitating the deprotonation step or stabilizing the phenoxide intermediate. Catalysts can also promote the selectivity of the reaction, leading to a higher yield of the desired product.

Applications of Phenoxide in Various Fields

The formation of sodium phenoxide through the reaction of phenol with NaOH has several applications in various fields, including pharmaceuticals and polymers.

Pharmaceuticals

Phenol derivatives, such as salicylic acid and aspirin, are widely used in the pharmaceutical industry. The synthesis of these compounds often involves the use of sodium phenoxide as an intermediate. For example, the synthesis of salicylic acid from phenol includes the reaction of phenol with NaOH to form sodium phenoxide, which is then treated with carbon dioxide to yield salicylate:

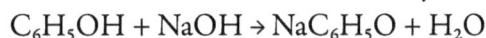

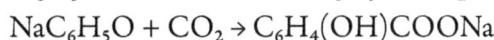

$$C_6H_5OH + NaOH \rightarrow NaC_6H_5O + H_2O$$
$$NaC_6H_5O + CO_2 \rightarrow C_6H_4(OH)COONa$$

Salicylic acid is a key precursor for the production of nonsteroidal anti-inflammatory drugs (NSAIDs) like aspirin. Thus, the reaction between phenol and NaOH plays a crucial role in the pharmaceutical industry.

Polymers

Phenol is also used in the synthesis of various polymers, such as phenolic resins and polycarbonates. The reaction of phenol with formaldehyde (HCHO) in the presence of a base, often NaOH, leads to the formation of phenolic resins. These resins exhibit excellent heat resistance and mechanical properties, making them suitable for applications in adhesives, coatings, and electrical insulation. Similarly, the reaction of phenol with phosgene

($COCl_2$) in the presence of a base results in the formation of polycarbonates. Polycarbonates are widely utilized in the production of transparent plastics, optical lenses, and electronic components due to their high impact strength and heat resistance. In both cases, the reaction of phenol with a base, such as NaOH, serves as a key step in the synthesis of these valuable polymers.

Counterarguments and Alternative Reactions

While the reaction of phenol with NaOH to produce sodium phenoxide is a widely used method, alternative reactions and counterarguments exist.

Reactions with Other Bases

Phenol can react with bases other than NaOH. For example, potassium hydroxide (KOH) can be used instead of NaOH to form potassium phenoxide. The choice of base depends on factors such as the desired product, reaction conditions, and availability of reagents. Different bases may provide slightly different reaction rates or selectivities.

Alternative Deprotonation Methods

Apart from the reaction with bases, phenol can also be deprotonated using other methods. For instance, strong acids like sulfuric acid (H_2SO_4) or Lewis acids like aluminum chloride ($AlCl_3$) can be employed to generate phenoxide ions. These acid-catalyzed reactions proceed through different mechanisms, offering alternative pathways for the synthesis of phenoxide.

Other Applications of Phenol

Phenol finds various other applications beyond the formation of sodium phenoxide. It is used as a disinfectant, antiseptic, and preservative due to its antimicrobial properties. Phenol is also utilized in the production of explosives, herbicides, and antioxidants. Therefore, the chemistry of phenol extends beyond its reaction with bases.

Recall the Chemistry of Phenol

We will delve into the chemistry of phenol and focus on one specific reaction: the reaction of phenol with sodium (Na) to produce sodium phenoxide ($NaC6H5O$) and hydrogen gas ($H2$). This reaction is of particular interest due to its importance in the synthesis of pharmaceuticals, dyes, and other organic compounds.

Structure and Properties of Phenol

Before diving into the reaction itself, it is essential to understand the structure and properties of phenol. Phenol ($C6H6O$) consists of a benzene ring with a hydroxyl (-OH) group attached to it. The hydroxyl group imparts certain characteristics to phenol, such as its acidic nature and enhanced reactivity compared to benzene.

The hydroxyl group in phenol makes it a weak acid, capable of donating a proton ($H+$) to a base. This acidity arises from the partial positive charge on the hydrogen atom, which is in resonance with the aromatic ring and contributes to the stabilization of the phenoxide ion formed after deprotonation.

Phenol is a colorless crystalline solid at room temperature, but it can also exist as a liquid. It has a distinctive odor and is slightly soluble in water, readily forming hydrogen bonds with water molecules. This solubility is due to the presence of the hydroxyl group, which allows for the formation of hydrogen bonds with water.

The Reaction of Phenol with Sodium

Phenol reacts with sodium (Na) in a highly exothermic reaction, resulting in the formation of sodium phenoxide and hydrogen gas. The reaction can be represented by the following equation:

$$C6H5OH + Na \rightarrow NaC6H5O + \frac{1}{2}H2$$

This reaction is a type of acid-base reaction, where phenol acts as the acid and sodium acts as the base. The hydroxyl group in phenol donates a proton to sodium, resulting in the formation of sodium phenoxide. Simultaneously, hydrogen gas is liberated as a byproduct. The reaction between phenol and sodium is often carried out in an inert solvent, such as diethyl ether or tetrahydrofuran (THF). These solvents prevent the reaction from being hindered by atmospheric moisture or oxygen. The reaction is typically conducted under reflux conditions to ensure efficient conversion of phenol to sodium phenoxide.

Mechanism of the Reaction

To gain a deeper understanding of the reaction, let us examine its mechanism. The reaction proceeds through a series of steps, involving the initial formation of a sodium phenoxide intermediate.

Formation of Sodium Phenoxide:

The reaction begins with the donation of a proton from the hydroxyl group of phenol to the sodium metal. This transfer of a proton leads to the formation of sodium phenoxide and the release of a hydrogen ion ($H+$). The reaction can be represented as: $C_6H_5OH + Na \rightarrow NaC_6H_5O + H+$

The formation of sodium phenoxide is facilitated by the high reactivity of sodium, which readily donates its electron to the hydroxyl group of phenol, resulting in the formation of a sodium cation and a phenoxide anion.

Release of Hydrogen Gas:

Simultaneously, the hydrogen ion ($H+$) released during the formation of sodium phenoxide reacts with another phenol molecule to regenerate phenol and liberate hydrogen gas. This reaction proceeds as follows: $C_6H_5OH + H+ \rightarrow C_6H_6 + H_2$

The liberation of hydrogen gas contributes to the exothermic nature of the overall reaction.

The overall reaction, combining both steps, can be summarized as: $2C_6H_5OH + 2Na \rightarrow 2NaC_6H_5O + H_2$

Significance of the Reaction

The reaction of phenol with sodium to produce sodium phenoxide and hydrogen gas holds immense significance in various fields. Let us explore some of its applications and implications.

Pharmaceutical Synthesis

Phenol and its derivatives are widely used in the synthesis of pharmaceutical compounds. The reaction of phenol with sodium provides a crucial step in many pharmaceutical syntheses. Sodium phenoxide, obtained through this reaction, serves as a valuable intermediate in the production of numerous drugs, including analgesics, antiseptics, and antipyretics. For example, the synthesis of salicylic acid, a key precursor for the production of aspirin, involves the reaction of phenol with sodium hydroxide to yield sodium phenoxide. This intermediate is subsequently acidified to obtain salicylic acid, which is then converted to acetylsalicylic acid (aspirin).

Dye Synthesis

Phenol and its derivatives find extensive use in the production of dyes and pigments. The reaction of phenol with sodium is employed in the synthesis of various dyes, including the popular phenolphthalein, which changes color depending on the pH of the solution. Sodium phenoxide is a crucial intermediate in the synthesis of phenolphthalein.

Furthermore, the reaction of phenol with sodium can lead to the formation of aryl ethers, which are essential components of many dyes. These aryl ethers can be further modified to create a wide range of dye molecules, each with its distinct color and properties.

Polymer Production

Phenol is a key ingredient in the production of certain types of polymers, such as phenolic resins. These resins have excellent heat resistance and are widely used in the manufacturing of various products, including adhesives, coatings, and molded objects. The reaction of phenol with formaldehyde leads to the formation of a phenolic resin, a process known as the phenol-formaldehyde resin synthesis. The reaction of phenol with sodium plays a vital role in the synthesis of phenolic resins. Sodium phenoxide, obtained through this reaction, serves as a precursor for the formation of the resin. The resulting phenolic resins possess desirable properties, such as high mechanical strength, thermal stability, and resistance to chemical degradation.

Counterarguments and Limitations

While the reaction of phenol with sodium offers numerous advantages and applications, it is essential to address some of the counterarguments and limitations associated with this process. One key concern is the safety hazards associated with the use of sodium, which is a highly reactive metal. Sodium reacts violently with water, releasing hydrogen gas and generating heat. Handling sodium requires caution and expertise to prevent accidents. Additionally, the use of an inert solvent and the exclusion of moisture and oxygen are crucial to ensure a safe and efficient reaction. Another limitation of the reaction is its sensitivity to impurities and side reactions. Phenol can undergo various side reactions, such as oxidation, during the reaction with sodium. These side reactions can reduce the yield of the desired product and complicate the purification process. Therefore, careful control of reaction conditions and the use of high-purity reagents are necessary to achieve optimal results.

A Comprehensive Analysis of Reactions with Diazonium Salts to Form Azo Compounds

Introduction

One important aspect of phenol chemistry is its reaction with diazonium salts, resulting in the formation of azo compounds. This process, known as diazotization, plays a crucial role in the synthesis of azo dyes, which have vibrant colors and find extensive use in the textile industry. In this comprehensive analysis, we will delve into the chemistry of phenol, exploring the mechanisms and applications of its reactions with diazonium salts.

I. Phenol: Structure and Properties

Phenol, with the molecular formula C_6H_6O, consists of a benzene ring (C_6H_6) attached to a hydroxyl group (-OH). This unique combination of aromatic and alcohol functionalities imparts distinct chemical properties to phenol. The presence of the hydroxyl group makes phenol more acidic compared to other aromatic compounds, allowing it to undergo various reactions not typically seen in simple benzene derivatives.

II. Diazonium Salts: Structure and Preparation

Diazonium salts, also known as diazonium compounds, are organic compounds containing the diazonium functional group ($-N_2^+$). These salts are prepared by the diazotization of primary aromatic amines, typically in the presence of an acid such as hydrochloric acid (HCl). The general reaction involved in the synthesis of diazonium salts is as follows:

$$Ar-NH_2 + HNO_2 + HCl \rightarrow Ar-N_2^+Cl^- + 2H_2O$$

Here, Ar represents an aromatic ring, and the diazonium salt is formed by the replacement of the amino group ($-NH_2$) with a diazonium group ($-N_2^+$).

III. Diazotization of Phenol

The reaction between phenol and diazonium salts is a fundamental process in organic chemistry. When phenol reacts with a diazonium salt in the presence of a strong base, such as sodium hydroxide (NaOH), an electrophilic aromatic substitution reaction occurs. This reaction is commonly known as Sandmeyer's reaction and involves the formation of azo compounds.

IV. Mechanism of Diazotization of Phenol

The mechanism of the diazotization of phenol can be divided into three main steps: diazotization, coupling, and oxidation.

A. Diazotization

The diazotization step involves the conversion of phenol into a diazonium salt. Initially, the phenol reacts with nitrous acid (HNO2), which is formed in situ by the reaction of sodium nitrite (NaNO2) with hydrochloric acid (HCl). The formation of a diazonium ion is the outcome of this reaction.

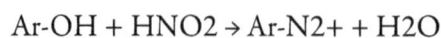

$$Ar-OH + HNO_2 \rightarrow Ar-N_2^+ + H_2O$$

The diazonium ion is highly reactive and unstable due to the presence of a positive charge on the nitrogen atom. Hence, it needs to be used immediately in the subsequent steps.

B. Coupling

After the diazotization step, the diazonium ion undergoes a coupling reaction with an aromatic compound or its derivatives. In the case of phenol, the diazonium ion reacts with the phenol itself, resulting in the formation of an azo compound:

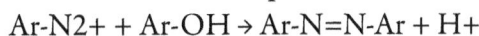

$$Ar-N_2^+ + Ar-OH \rightarrow Ar-N=N-Ar + H^+$$

This reaction, known as electrophilic aromatic substitution, occurs due to the electron-deficient nature of the diazonium ion, which acts as an electrophile. The phenol, on the other hand, donates electrons to the ring, making it nucleophilic. Consequently, a nucleophilic attack by the phenol on the diazonium ion leads to the formation of an azo compound.

C. Oxidation

The final step in the diazotization of phenol involves the oxidation of the azo compound to a more stable product. This oxidation is typically achieved by treating the azo compound with an oxidizing agent, such as nitric acid (HNO_3). The oxidation converts the azo compound into a more stable aromatic compound:

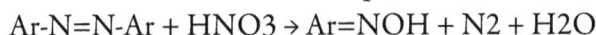

$$Ar-N=N-Ar + HNO_3 \rightarrow Ar=NOH + N_2 + H_2O$$

The resulting product is called a nitroso compound and is often used as an intermediate in the synthesis of various organic compounds.

V. Applications of Diazotization of Phenol: Azo Compounds and Azo Dyes

The diazotization of phenol plays a vital role in the synthesis of azo compounds and azo dyes. Azo compounds are organic compounds containing the functional group R-N=N-R', where R and R' represent aromatic or aliphatic groups. These compounds exhibit bright colors due to the presence of conjugated double bonds between the nitrogen atoms. Azo dyes, derived from azo compounds, are widely used in the textile industry to impart vibrant and durable colors to fabrics.

A. Synthesis of Azo Compounds

The diazotization of phenol is an essential step in the synthesis of azo compounds. By coupling the diazonium salt derived from phenol with various aromatic compounds or their derivatives, a wide range of azo compounds can be obtained. For example, coupling phenol diazonium salt with aniline ($C_6H_5NH_2$) yields azobenzene ($C_6H_5N=N-C_6H_5$):

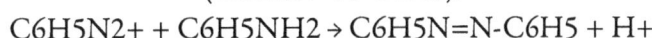

$$C_6H_5N_2^+ + C_6H_5NH_2 \rightarrow C_6H_5N=N-C_6H_5 + H^+$$

Similarly, by coupling phenol diazonium salt with naphthalene ($C_{10}H_8$), azonaphthalene ($C_{10}H_7N=N-C_{10}H_7$) can be synthesized.

B. Synthesis of Azo Dyes

Azo dyes are extensively used in the textile industry to achieve a wide range of shades and colors. The synthesis of azo dyes often involves the diazotization of phenol followed by coupling with specific aromatic compounds or their derivatives. The resulting azo compounds exhibit excellent color fastness and are resistant to fading, making them ideal for dyeing textiles.

VI. Counterarguments and Limitations

While the diazotization of phenol offers numerous advantages in the synthesis of azo compounds and azo dyes, there are certain limitations and counterarguments to consider.

A. Toxicity and Environmental Concerns

Some azo compounds and azo dyes have been found to be toxic and pose environmental hazards. Certain azo dyes, when released into water bodies during textile dyeing processes, can undergo degradation and produce harmful aromatic amines. These amines are known to possess carcinogenic properties and can pose a significant risk to human health and the environment. Therefore, it is essential to ensure proper waste treatment and disposal methods to mitigate these risks.

B. Stability and Reactivity

Azo compounds, including those derived from the diazotization of phenol, can exhibit varying degrees of stability and reactivity. Certain azo compounds are prone to decomposition under specific conditions, limiting their practical applications. Additionally, the reactivity of diazonium salts can pose challenges in controlling the selectivity of the coupling reaction, leading to the formation of unwanted byproducts. Therefore, careful optimization and control of reaction conditions are necessary to achieve the desired azo compounds selectively.

C. Synthesis Complexity and Cost

The synthesis of azo compounds and azo dyes often involves multiple steps, including the diazotization of phenol, coupling reactions, and subsequent oxidations. These multistep processes can be time-consuming, require specific reagents, and increase the overall cost of production. Furthermore, the availability and cost of starting materials, such as phenol and aromatic compounds, may also impact the feasibility and commercial viability of large-scale synthesis.

We will explore the chemistry of phenol, focusing on the nitration of the aromatic ring with dilute $HNO_3(aq)$ at room temperature. This reaction leads to the formation of a mixture of 2-nitrophenol and 4-nitrophenol. We will delve into the mechanism of this reaction, discuss the factors that influence the selectivity of the products, and explore the significance of this reaction in different applications.

Nitration of Phenol

Nitration is a common chemical reaction that involves the introduction of a nitro group ($-NO_2$) into an organic compound. It is widely used in the synthesis of various chemicals, including explosives, dyes, and pharmaceuticals. When phenol undergoes nitration, it reacts with dilute nitric acid (HNO_3) to produce a mixture of 2-nitrophenol and 4-nitrophenol. The following equation represents the reaction: $C_6H_5OH + HNO_3 \rightarrow C_6H_4(NO_2)(OH) + H_2O$

A new compound is formed through the replacement of the hydroxyl group in phenol with the nitro group in this reaction. The nitro group is attached to either the ortho (2-position) or para (4-position) of the benzene ring, giving rise to the two possible isomers - 2-nitrophenol and 4-nitrophenol.

Mechanism of the Reaction

The nitration of phenol follows a mechanism similar to other electrophilic aromatic substitution reactions. It involves the generation of a highly reactive species, the nitronium ion (NO_2^+), which acts as the electrophile. The overall mechanism can be divided into three main steps: generation of the electrophile, attack of the electrophile on the aromatic ring, and regeneration of the aromatic system.

Step 1: Generation of the Electrophile

In the presence of dilute nitric acid, a proton from the nitric acid transfers to the phenol molecule, creating a resonance-stabilized phenoxide ion. This ion is more reactive than the phenol itself and is necessary for the subsequent steps of the reaction. The reaction can be represented as:

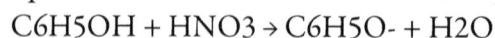

$$C_6H_5OH + HNO_3 \rightarrow C_6H_5O^- + H_2O$$

Step 2: Electrophile attacking the aromatic ring.

The nitronium ion (NO_2^+) is generated by the reaction between nitric acid and a strong acid, such as sulfuric acid (H_2SO_4). The nitronium ion is an electrophile that can attack the electron-rich aromatic ring of the phenoxide ion. The oxygen atom of the phenoxide ion donates its lone pair of electrons to the nitronium ion, resulting in the formation of a sigma bond between the carbon atom of the aromatic ring and the nitro group.

This step can be represented as:

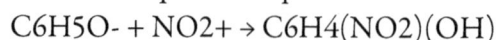

$$C_6H_5O^- + NO_2^+ \rightarrow C_6H_4(NO_2)(OH)$$

Step 3: Regeneration of the Aromatic System
The final step involves the regeneration of the aromatic system by the removal of a proton from the nitrophenol molecule. This deprotonation process is catalyzed by the presence of water or a base. The resulting product is a mixture of 2-nitrophenol and 4-nitrophenol, as shown in the equation below:

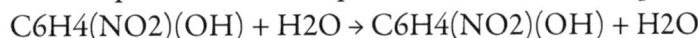

$$C6H4(NO2)(OH) + H2O \rightarrow C6H4(NO2)(OH) + H2O$$

It is important to note that the nitration reaction can be influenced by various factors, such as temperature, concentration of reagents, and the presence of catalysts. These factors can affect the selectivity of the reaction and the relative amounts of the isomers formed.

Factors Influencing Selectivity

The selectivity of the nitration reaction is determined by the relative reactivity of the ortho and para positions of the aromatic ring. The ortho position refers to the carbon atom adjacent to the hydroxyl group, while the para position refers to the carbon atom opposite to the hydroxyl group.

The reactivity of these positions can be explained by considering the electron density distribution in the aromatic ring. The presence of the hydroxyl group in phenol leads to an increase in electron density at the ortho and para positions, as it functions as an electron-donating group. This higher electron density makes these positions more susceptible to attack by the electrophile and, therefore, more reactive. However, the steric hindrance caused by the bulky hydroxyl group makes the ortho position less accessible to the nitronium ion. As a result, the para position is favored in the nitration of phenol, leading to the formation of predominantly 4-nitrophenol. This can be attributed to the lower steric hindrance at the para position, which allows for a more favorable transition state and a faster reaction rate.

The selectivity of the nitration reaction can be further influenced by the reaction conditions. For example, increasing the temperature or the concentration of the reagents can enhance the rate of the reaction. However, it can also lead to increased side reactions, such as the formation of dinitrophenols or higher nitro-substituted products, which can reduce the selectivity of the desired products.

Significance and Applications

The nitration of phenol and the subsequent formation of 2-nitrophenol and 4-nitrophenol have significant implications in various fields. These compounds find applications as intermediates in the synthesis of numerous chemicals, including pharmaceuticals, dyes, and agricultural chemicals.

In the field of pharmaceuticals, nitrophenols have been utilized as starting materials for the synthesis of analgesics, antipyretics, and antiseptics. For example, 2-nitrophenol is a key intermediate in the synthesis of paracetamol, a widely used painkiller. In the dye industry, nitrophenols are employed as precursors for the synthesis of azo dyes. These dyes are widely used to color textiles, plastics, and other materials. The nitro group in the nitrophenols can be easily reduced to an amino group, which is essential for the formation of azo dyes. Furthermore, 2-nitrophenol and 4-nitrophenol have also been studied for their antimicrobial properties. These compounds have shown activity against various bacteria and fungi, making them potential candidates for the development of new antimicrobial agents.

Counterarguments

While the nitration of phenol is a useful reaction for the synthesis of 2-nitrophenol and 4-nitrophenol, it is not without limitations. One limitation is the formation of unwanted byproducts, such as dinitrophenols or higher nitro-substituted products. These byproducts can reduce the selectivity of the desired products and complicate the purification process.

Another limitation is the potential for side reactions, such as oxidation or polymerization of the phenol. These side reactions can lead to the formation of undesired compounds and decrease the overall yield of the desired products.

To overcome these limitations, various strategies have been developed, such as the optimization of reaction conditions, the use of catalysts, and the development of selective nitration methods. These strategies aim to improve the selectivity and efficiency of the reaction, minimize the formation of byproducts, and enhance the overall yield of the desired products.

Bromination of the Aromatic Ring with Br2(aq) to Form 2,4,6-Tribromophenol

Introduction

We will focus on the bromination of the aromatic ring of phenol using bromine (Br2) in an aqueous solution. This reaction results in the formation of 2,4,6-tribromophenol, a compound with three bromine atoms attached to the phenol ring. We will explore the mechanism, reaction conditions, and applications of this reaction, providing a comprehensive analysis of the subject matter.

The Bromination of Phenol: Mechanism

The bromination of phenol with Br2(aq) involves the substitution of a hydrogen atom on the aromatic ring with a bromine atom. This reaction is typically carried out in the presence of a catalyst, which enhances the reaction rate. One commonly used catalyst is iron(III) bromide (FeBr3), which acts as a Lewis acid, facilitating the formation of a reactive electrophile.

The mechanism of the bromination reaction can be divided into several steps:

Generation of the electrophile: Iron(III) bromide coordinates with bromine, forming a complex that activates the bromine molecule. This complex, known as the bromonium ion, acts as an electrophile, seeking an electron-rich site on the phenol ring.

Electrophilic attack: The bromonium ion attacks the aromatic ring of phenol, specifically targeting the electron-rich ortho- and para- positions. This attack results in the formation of a sigma complex, where the bromine atom is temporarily bonded to the ring.

Generation of the phenoxide ion: The sigma complex undergoes proton transfer, leading to the formation of a phenoxide ion. The phenoxide ion is stabilized by resonance, as the negative charge is delocalized throughout the aromatic ring.

Substitution of the bromine atom: The phenoxide ion acts as a nucleophile, attacking the bromine atom in the sigma complex. This substitution reaction results in the replacement of a hydrogen atom on the phenol ring with a bromine atom.

Regeneration of the catalyst: The catalyst, iron(III) bromide, is regenerated in the final step of the reaction. It abstracts a proton from the newly formed tribromophenol, restoring its original form and allowing it to participate in subsequent bromination reactions.

Reaction Conditions

The bromination of phenol with Br2(aq) requires specific reaction conditions to ensure the desired product is obtained efficiently. These conditions include temperature, concentration, and the presence of a catalyst.

Temperature: The bromination reaction is typically carried out at room temperature or slightly above. Higher temperatures can lead to side reactions or undesired products. Additionally, low temperatures may slow down the reaction rate significantly.

Concentration: The concentration of the reactants influences the reaction rate. Higher concentrations of phenol and bromine can increase the likelihood of successful collisions between the molecules, leading to a faster reaction. However, excessively high concentrations can also result in competing reactions or increased side products.

Catalyst: As mentioned earlier, iron(III) bromide is commonly used as a catalyst in the bromination of phenol. The catalyst enhances the electrophilicity of bromine, promoting the formation of the bromonium ion and facilitating the substitution reaction. The amount of catalyst used can vary depending on the specific reaction conditions and desired yield.

Applications of the Bromination Reaction

The bromination of phenol plays a crucial role in various fields, including pharmaceuticals, dyes, and agricultural chemicals. Some notable applications of 2,4,6-tribromophenol, the product of this reaction, are discussed below:

Flame retardants: 2,4,6-tribromophenol is widely used as a flame retardant in the production of plastics, textiles, and electronic devices. It functions by inhibiting the combustion process, preventing the spread of fire and reducing the release of toxic gases.

Wood preservatives: Due to its antimicrobial properties, 2,4,6-tribromophenol is utilized as a wood preservative to protect against decay, fungi, and insect infestations. It enhances the durability and longevity of wood products, such as furniture, decking, and utility poles.

Pharmaceuticals: The bromination of phenol is an important step in the synthesis of various pharmaceutical compounds. For example, 2,4,6-tribromophenol derivatives have been investigated for their potential as antiviral, antitumor, and antibacterial agents. The presence of bromine atoms can alter the biological activity and pharmacokinetics of the compounds.

Dye intermediates: 2,4,6-tribromophenol serves as a precursor for the synthesis of dyes and pigments. It undergoes further reactions, such as coupling with aromatic amines or diazonium salts, to produce colored compounds used in the textile, printing, and coloring industries.

Counterarguments and Limitations

While the bromination of phenol to form 2,4,6-tribromophenol offers numerous advantages and applications, there are also some limitations and counterarguments to consider.

Environmental concerns: The use of brominated compounds, including 2,4,6-tribromophenol, has raised environmental concerns due to their persistence and potential toxicity. These compounds can accumulate in ecosystems and pose risks to human health and the environment. Therefore, alternative methods and greener approaches are being explored to reduce the reliance on bromine-based reactions.

Selectivity and regiochemistry: The bromination reaction can lead to multiple regioisomers, depending on the position of bromine substitution on the phenol ring. Achieving high selectivity for a specific isomer can be challenging, and the presence of multiple isomers may affect the desired properties and applications of the final product.

Side reactions and byproducts: The bromination of phenol can result in side reactions and the formation of unwanted byproducts. These side reactions may reduce the overall yield and purity of the desired product. Careful optimization of reaction conditions, catalyst selection, and purification methods is necessary to minimize these issues.

Health and safety considerations: Bromine and phenol are both hazardous substances, and their handling requires proper safety precautions. Protective equipment, such as gloves, goggles, and fume hoods, should be used to minimize exposure and ensure safe laboratory practices.

The Acidity of Phenol: A Comprehensive Analysis

Introduction

We will delve into the intricacies of phenol's acidity, exploring its unique characteristics, factors influencing acidity, and its applications in different areas. Our approach will involve maintaining a formal and scholarly style, employing sophisticated language and grammar to conduct a comprehensive examination of the topic. We will explain complex concepts clearly, incorporating examples from diverse fields, and objectively present counterarguments.

Phenol's Acidic Nature

Phenol exhibits acidity due to the presence of the hydroxyl group, which can donate a proton (H+) to a suitable acceptor. The hydroxyl group in phenol is stabilized by the aromatic ring, making the proton more easily donated compared to aliphatic alcohols. This stabilization arises from the resonance effect, where the negative charge resulting from the loss of a proton can delocalize within the aromatic ring. As a result, phenol possesses a higher acidity than aliphatic alcohols such as ethanol.

Factors Influencing Phenol's Acidity

Several factors influence the acidity of phenol, including the presence of substituents on the aromatic ring, the electron-donating or electron-withdrawing nature of these substituents, and the resonance effects within the molecule.

Substituents on the Aromatic Ring

The presence of substituents on the aromatic ring can significantly affect the acidity of phenol. For instance, electron-donating substituents, such as alkyl groups (-CH3), increase the electron density on the aromatic ring. This increased electron density results in a decreased acidity compared to pure phenol. Conversely, electron-withdrawing substituents, such as nitro groups (-NO2), decrease the electron density on the aromatic ring, leading to enhanced acidity. This phenomenon can be explained by the electron-donating or electron-withdrawing nature of the substituents, which influences the stability of the resulting negative charge.

Resonance Effects

The resonance effects within phenol play a crucial role in determining its acidity. The aromatic ring in phenol exhibits resonance, resulting in the delocalization of electrons. This delocalization allows the negative charge resulting from the loss of a proton to be distributed across the entire ring. Consequently, the stability of the negative charge increases, making the proton more easily donated. This resonance effect significantly contributes to phenol's acidity, differentiating it from aliphatic alcohols without an aromatic ring.

Acid-Base Equilibrium and Phenol

To understand the acidity of phenol, it is essential to examine its acid-base equilibrium. Phenol can donate a proton to an appropriate acceptor, such as a base, resulting in the formation of a conjugate base and a corresponding acid. For instance, when phenol donates a proton to water, it forms phenoxide ion (C6H5O-) and hydronium ion (H3O+), representing the acid-base equilibrium:

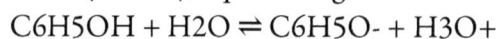

$$C6H5OH + H2O \rightleftharpoons C6H5O- + H3O+$$

The strength of the acid can be quantified using the acid dissociation constant (Ka), which is the ratio of the concentration of the conjugate base to the concentration of the acid. For phenol, the Ka value is approximately 1.3×10^{-10} M, indicating its weak acidic nature.

Comparison with Other Acids

To comprehend the acidity of phenol, it is enlightening to compare it with other well-known acids. Phenol's acidity lies between that of strong mineral acids like sulfuric acid (H2SO4) and weak organic acids like acetic acid (CH3COOH). Sulfuric acid is a strong acid, fully dissociating into its ions in water, resulting in a low pH. In contrast, phenol only partially dissociates, resulting in a pH closer to neutral. Acetic acid, on the other hand, is a weak acid, but still more acidic than phenol due to the presence of a carboxyl group, which exhibits greater electron-withdrawing effects compared to the hydroxyl group in phenol.

Applications of Phenol's Acidity

The acidity of phenol finds applications in various fields, including medicine, industry, and organic synthesis.

Medicine

Phenol has been widely used in medicine as an antiseptic and disinfectant due to its acidic properties. It denatures proteins, making it effective against bacteria and other microorganisms. Moreover, phenol's acidity allows it to penetrate cell membranes, enhancing its antimicrobial activity. However, its corrosive nature limits its direct use on living tissues.

Industry

In the industrial sector, phenol's acidity is utilized in the production of plastics, resins, and dyes. Phenol is a key precursor in the synthesis of materials like Bakelite, a thermosetting plastic. The acidity of phenol enables it to react with formaldehyde to form a cross-linked polymer, providing the desired properties to the resulting plastic.

Organic Synthesis

Phenol's acidity plays a crucial role in organic synthesis. It can act as a catalyst in various reactions, such as esterification and acylation. Phenol's acidity facilitates the formation of esters by donating a proton to an alcohol, resulting in the formation of water and the desired ester. Additionally, phenol can undergo acylation reactions, wherein it donates a proton to an acylating agent, such as an acid chloride, leading to the formation of an aromatic ketone.

Counterarguments and Perspectives

While phenol's acidity is undoubtedly significant, alternative viewpoints exist regarding its relevance and applications. Some argue that the corrosive nature of phenol limits its utility in certain areas, such as medicine. The availability of less harmful antiseptics and disinfectants has diminished the use of phenol in medical practices. Moreover, the emergence of antibiotic resistance has shifted the focus towards alternative antimicrobial agents. Furthermore, the high acidity of phenol can pose challenges in specific reactions. Its reactivity towards nucleophiles can lead to side reactions, necessitating careful optimization in organic synthesis. Scientists are continually exploring alternative methods that offer milder reaction conditions and improved selectivity.

Relative Acidities of Water, Phenol, and Ethanol

Introduction

We will explore the factors that contribute to the acidity of these compounds and provide a thorough analysis of their relative acidities.

Acid-Base Equilibrium

To comprehend the acidities of water, phenol, and ethanol, it is essential to have a fundamental understanding of acid-base equilibrium. An acid is a substance that donates a proton (H+) in a chemical reaction, while a base accepts a proton. The strength of an acid is determined by its ability to donate a proton, which is quantified by its dissociation constant (Ka).

A higher Ka value indicates a stronger acid.

Water: The Prototypical Acid-Base Compound

Water, the universal solvent, is a fundamental compound in chemical reactions and plays a crucial role in acid-base chemistry. Although water is considered neutral, it can act as both an acid and a base due to the self-ionization process. In this process, a water molecule donates a proton to another water molecule, forming a hydronium ion (H3O+) and a hydroxide ion (OH-):

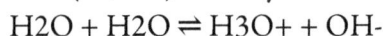

$$H2O + H2O \rightleftharpoons H3O+ + OH-$$

The equilibrium constant for this reaction, known as the ion product of water (Kw), is 1.0×10^{-14} at 25°C. The concentration of hydronium ions and hydroxide ions in pure water is equal, each being 1.0×10^{-7} M. Thus, the pH of pure water is neutral, around 7. In terms of acidity, water can function as a base when it accepts a proton, forming a hydroxide ion. For example, when water reacts with a strong acid such as hydrochloric acid (HCl), it accepts a proton to form a hydronium ion: $H2O + HCl \rightarrow H3O+ + Cl-$

On the other hand, water can also act as an acid when it donates a proton to a base. For example, when water reacts with a strong base such as sodium hydroxide (NaOH), it donates a proton to form a hydroxide ion:

$$H2O + NaOH \rightarrow Na+ + OH-$$

The acidity of water, in comparison to phenol and ethanol, is relatively low due to its limited ability to donate or accept protons. The Ka value for water is 1.8×10^{-16}, indicating its weak acid strength. This weak acidity is primarily attributed to the high stability of the water molecule, resulting from the strong bonding between oxygen and hydrogen.

Phenol: An Aromatic Acid

Phenol, an aromatic compound, exhibits stronger acidity compared to water. It consists of a hydroxyl group (-OH) attached to a benzene ring. The presence of the benzene ring introduces resonance stabilization, which significantly influences the acidity of phenol. The hydroxyl group in phenol can donate a proton, forming a phenoxide ion. This donation is facilitated by the resonance stabilization of the negative charge within the benzene ring. The conjugate base of phenol, known as phenoxide ion, is more stable than the conjugate base of water (hydroxide ion). As a result, phenol has a higher Ka value, indicating stronger acidity. Furthermore, the electronic effects of the benzene ring contribute to the acidity of phenol. The benzene ring is electron-rich, and it donates electron density towards the oxygen atom in the hydroxyl group. This electron-donating effect enhances the ability of the hydroxyl group to donate a proton, making phenol a stronger acid. For instance, when phenol reacts with a strong base such as sodium hydroxide, it donates a proton to form a phenoxide ion:

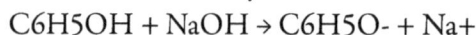

$$C6H5OH + NaOH \rightarrow C6H5O- + Na+$$

The Ka value for phenol is approximately 1.6×10^{-10}, indicating its moderate acid strength. This moderate acidity, influenced by both resonance stabilization and electron-donating effects, allows phenol to participate in various chemical reactions, such as electrophilic aromatic substitution.

Ethanol: A Weak Acid

Ethanol, commonly known as alcohol, is a compound with a hydroxyl group (-OH) attached to a carbon chain. Unlike water and phenol, ethanol exhibits weak acidity due to its distinct molecular structure. The hydroxyl group in ethanol can donate a proton, forming an ethoxide ion. However, the acidity of ethanol is significantly lower than that of water and phenol. This is primarily attributed to the electron-donating effect of the alkyl group (carbon chain) attached to the hydroxyl group. The alkyl group destabilizes the negative charge on the conjugate base, reducing the acidity of ethanol. For example, when ethanol reacts with a strong base such as sodium hydroxide, it donates a proton to form an ethoxide ion:

$$CH3CH2OH + NaOH \rightarrow CH3CH2O- + Na+$$

The Ka value for ethanol is approximately 1.4×10^{-16}, indicating its weak acid strength. Although ethanol can participate in acid-base reactions, its limited acidity restricts its reactivity compared to water and phenol.

Comparative Analysis

To provide a thorough analysis of the relative acidities of water, phenol, and ethanol, it is crucial to compare their Ka values and consider the factors influencing their acidity.

As mentioned earlier, water has a Ka value of 1.8×10^{-16}, indicating its weak acid strength. The high stability of the water molecule, resulting from the strong bonding between oxygen and hydrogen, hinders its ability to donate or accept protons. Thus, water is considered a neutral compound in terms of acidity. Phenol, on the other hand, has a Ka value of approximately 1.6×10^{-10}, indicating its moderate acid strength. This increased acidity is primarily attributed to the resonance stabilization of the phenoxide ion, facilitated by the presence of the benzene ring. Additionally, the electron-donating effect of the benzene ring enhances the ability of the hydroxyl group to donate a proton.

Ethanol exhibits the weakest acidity among the three compounds, with a Ka value of approximately 1.4×10^{-16}. This low acidity is primarily due to the destabilizing effect of the alkyl group attached to the hydroxyl group. The alkyl group reduces the electron density on the oxygen atom, decreasing the ability of ethanol to donate a proton.

In summary, the relative acidities of water, phenol, and ethanol can be ranked as follows: phenol > water > ethanol. Phenol exhibits the highest acidity due to the resonance stabilization and electron-donating effects of the benzene ring. Water, although it can act as both an acid and a base, has a weak acidity due to the high stability of the water molecule. Ethanol displays the weakest acidity among the three compounds, primarily due to the destabilizing effect of the alkyl group.

Introduction

The nitration and bromination reactions are important processes in organic chemistry that involve the substitution of a hydrogen atom with a nitro group (NO2) or a bromine atom, respectively. While both reactions can be carried out on both phenol and benzene, the reagents and conditions used for each reaction differ significantly. This essay will delve into the reasons behind these differences, considering various factors such as the reactivity of the substrates, the stability of the reaction intermediates, and the selectivity of the reactions. By examining these factors, we will gain a deeper understanding of the underlying principles that dictate the reagents and conditions for nitration and bromination in phenol and benzene.

Reactivity of Phenol and Benzene

One of the primary reasons for the differences in the reagents and conditions between the nitration and bromination of phenol and benzene lies in their varying reactivities. Phenol, as a derivative of benzene, contains an -OH group attached to one of its carbon atoms. This functional group imparts distinct reactivity to phenol compared to benzene.

The presence of the hydroxyl group makes phenol more reactive than benzene due to the electron-donating nature of -OH. The lone pairs of electrons on the oxygen atom of the -OH group donate electron density to the benzene ring, increasing its electron density and making it more susceptible to electrophilic attack. Consequently, phenol is more susceptible to electrophilic aromatic substitution reactions compared to benzene.

Nitration

Nitration of Benzene

The mixture of concentrated nitric acid (HNO3) and concentrated sulfuric acid (H2SO4) is employed as the reagent in the process of benzene nitration. The sulfuric acid acts as a catalyst, protonating the nitric acid to generate a highly electrophilic species, the nitronium ion (NO2+). The nitronium ion is the actual electrophile responsible for the substitution reaction. The reaction proceeds through an electrophilic aromatic substitution mechanism, where the nitronium ion attacks the benzene ring, forming a sigma complex. This sigma complex then rearranges to eliminate a proton, regenerating the aromaticity of the benzene ring. The resulting product is nitrobenzene. The use of concentrated sulfuric acid is crucial in this reaction as it contributes to the formation of the nitronium ion. Additionally, the presence of sulfuric acid helps to remove water, which is a byproduct of the reaction, thus driving the equilibrium towards the formation of the nitrobenzene product.

Nitration of Phenol

In the nitration of phenol, the reagents and conditions differ from those used for benzene. Due to the presence of the hydroxyl group in phenol, the reaction conditions need to be altered to prevent unwanted side reactions. The presence of the hydroxyl group in phenol makes it more susceptible to electrophilic attack by the nitronium ion. The lone pairs of electrons on the oxygen atom donate electron density to the benzene ring, further increasing its reactivity. Consequently, nitration of phenol under the same conditions as benzene would result in the formation of multiple products due to the electrophilic attack of the nitronium ion at both the ortho and para positions to the -OH group. To avoid the formation of multiple products, a milder condition is used for the nitration of phenol. Instead of using concentrated nitric acid, a dilute mixture of nitric acid (HNO3) and sulfuric acid (H2SO4) is employed. The dilution of nitric acid reduces the concentration of the nitronium ion, thereby minimizing the extent of electrophilic attack at the ortho and para positions. This selective nitration of phenol predominantly occurs at the meta position to the -OH group, yielding meta-nitrophenol as the major product. The milder conditions for the nitration of phenol also prevent the oxidation of the -OH group to a carbonyl group, which would occur under harsher conditions. The presence of the hydroxyl group makes phenol more susceptible to oxidation, and hence, reducing the concentration of the nitronium ion helps to prevent this side reaction.

Bromination

Bromination of Benzene

In the bromination of benzene, the reagent used is bromine (Br2) in the presence of a Lewis acid catalyst, usually iron(III) bromide (FeBr3). The Lewis acid catalyst facilitates the formation of the electrophilic bromonium ion (Br+), which is the active electrophile in the substitution reaction.

Similar to the nitration of benzene, the bromination reaction proceeds through an electrophilic aromatic substitution mechanism. The bromonium ion attacks the benzene ring, forming a sigma complex, which then rearranges to regenerate the aromaticity of the benzene ring. The resulting product is bromobenzene. It is worth noting that bromination is less reactive than nitration due to the relatively lower electron-withdrawing nature of bromine compared to the nitro group. The bromonium ion is less electrophilic than the nitronium ion, thereby requiring a Lewis acid catalyst to enhance its reactivity.

Bromination of Phenol

Unlike the nitration of phenol, the bromination of phenol can be carried out under the same conditions as benzene. This is because the hydroxyl group in phenol does not significantly affect the reactivity of the benzene ring towards electrophilic attack by the bromonium ion. The hydroxyl group in phenol does not donate electron density to the benzene ring to the same extent as it does in the case of nitration. Consequently, the reactivity of phenol towards bromination is not significantly altered compared to benzene. The presence of the hydroxyl group does not lead to multiple products or side reactions, as observed in the nitration of phenol. Therefore, the bromination of phenol can be carried out using bromine and a Lewis acid catalyst, such as iron(III) bromide, under the same conditions as the bromination of benzene.

The Directing Effect of the Hydroxyl Group in Phenols

Introduction

We will explore the directing effect of the hydroxyl group in phenols, focusing on its directing influence towards the 2-, 4-, and 6-positions of the aromatic ring. We will analyze the reasons behind this directing effect, provide examples from various fields, and present counterarguments objectively.

The Directing Effect of the Hydroxyl Group

Phenols exhibit a unique directing effect due to the electron-donating nature of the hydroxyl group. This effect is a consequence of the resonance stabilization and inductive effects caused by the presence of the hydroxyl group. To understand this directing effect, it is essential to consider the electronic structure and reactivity of phenols.

Electronic Structure and Reactivity of Phenols

The aromatic ring in phenols consists of a conjugated π-electron system, wherein the delocalized electrons provide stability. The presence of the hydroxyl group introduces a significant change in the electronic structure of the molecule. The oxygen atom in the hydroxyl group is more electronegative than carbon, resulting in a polar covalent bond. This polarity induces a partial positive charge on the carbon atom directly bonded to the hydroxyl group. Consequently, the electron density in the aromatic ring is increased at the ortho and para positions relative to the hydroxyl group

(2- and 4-positions, respectively).

Resonance Stabilization

The hydroxyl group in phenols also exhibits resonance stabilization, further enhancing the directing effect. The lone pair of electrons on oxygen can delocalize into the aromatic ring, resulting in resonance structures. This delocalization allows the negative charge to be distributed across the π-electron system, reducing electron density on the oxygen atom and increasing electron density on the ortho and para positions. As a result, electrophilic reagents are more likely to attack these positions due to the higher electron density.

Inductive Effect

The inductive effect is another contributing factor to the directing effect of the hydroxyl group. The electronegativity difference between oxygen and carbon causes a partial positive charge on the carbon atom bonded to the hydroxyl group. This positive charge is transmitted through sigma bonds, affecting the electron density at other positions on the aromatic ring. Consequently, the ortho and para positions experience a greater positive charge, making them more susceptible to nucleophilic attack.

Examples from Various Fields

The directing effect of the hydroxyl group in phenols can be observed in various chemical reactions and synthetic applications. For instance, in electrophilic aromatic substitution reactions, the hydroxyl group directs incoming electrophiles to the ortho and para positions. This effect is exploited in the synthesis of numerous pharmaceutical compounds, natural products, and dyes. Additionally, the directing effect of the hydroxyl group in phenols is crucial in the biosynthesis of various plant secondary metabolites, including flavonoids and lignin.

Counterarguments and Limitations

While the directing effect of the hydroxyl group in phenols is generally observed towards the ortho and para positions, there are exceptions and limitations to this rule. For instance, steric hindrance caused by bulky substituents adjacent to the hydroxyl group can alter the directing effect. In such cases, the directing effect may shift towards the meta position. Additionally, the presence of other functional groups, such as nitro (-NO2) or carbonyl (C=O), can override the directing effect of the hydroxyl group, leading to different regioselectivity.

Reactions of Phenol and Naphthol

Introduction

We will explore the reactions of phenol in comparison to those of naphthol, a compound with two aromatic rings fused together. By analyzing the similarities and differences in their reactions, we can gain a deeper understanding of the chemical behavior of these compounds and their applications in various fields.

Structure and Reactivity

Phenol, also known as carbolic acid, is the simplest member of the phenolic compound family. Its molecular formula is C6H6O, and it consists of a benzene ring with a hydroxyl group attached to it. Naphthol, on the other hand, has a molecular formula of C10H8O and is composed of two benzene rings fused together, with a hydroxyl group attached to one of the rings. The difference in structure between phenol and naphthol gives rise to different reactivity patterns.

The presence of the hydroxyl group in both phenol and naphthol allows them to participate in several types of reactions, including acid-base reactions, electrophilic substitution reactions, and oxidation reactions.

Acid-Base Reactions

Phenol and naphthol both exhibit weak acidic properties due to the presence of the hydroxyl group, which can donate a proton (H+) to a base. However, phenol is a stronger acid compared to naphthol due to the electron-donating effect of the second ring in naphthol, which destabilizes the conjugate base formed after deprotonation. Consequently, phenol readily donates a proton to a base, whereas naphthol requires stronger bases to undergo deprotonation. The acidic nature of phenol and naphthol makes them useful in various applications. For example, they can be used as intermediates in the synthesis of pharmaceuticals, dyes, and polymers. Additionally, their acidic properties enable them to act as preservatives and disinfectants in certain products.

Electrophilic Substitution Reactions

Phenol and naphthol both undergo electrophilic substitution reactions, in which an electrophile replaces a hydrogen atom on the aromatic ring. However, the presence of the second ring in naphthol affects the reactivity of the compound.

In the case of phenol, the hydroxyl group activates the ring towards electrophilic substitution reactions, making it more reactive compared to benzene. This activation occurs through resonance effects, where the lone pair of electrons on the oxygen atom delocalizes into the ring, increasing its electron density. As a result, phenol undergoes reactions with electrophiles such as nitric acid (HNO_3) and bromine (Br_2) more readily than benzene.

Naphthol, on the other hand, exhibits similar reactivity to phenol but with some differences due to the presence of the second ring. The second ring can further delocalize the electron density, making naphthol even more reactive than phenol towards electrophilic substitution reactions. This enhanced reactivity is particularly evident when naphthol is compared to compounds with a single ring, such as phenol or benzene. The ability of phenol and naphthol to undergo electrophilic substitution reactions plays a crucial role in their applications. For instance, these compounds are commonly employed as intermediates in the synthesis of dyes, pharmaceuticals, and fragrances. Their reactivity towards electrophiles allows for the introduction of functional groups onto the aromatic ring, enabling the modification of their chemical and physical properties.

Oxidation Reactions

Both phenol and naphthol can undergo oxidation reactions, where the hydroxyl group is converted into a carbonyl group. Phenol can be oxidized to form benzoquinone, while naphthol can be oxidized to form naphthoquinone. These reactions occur through the removal of hydrogen atoms from the hydroxyl group, followed by the rearrangement of electrons to form a double bond between the carbon and oxygen atoms. The oxidation reactions of phenol and naphthol are often catalyzed by oxidizing agents such as potassium permanganate ($KMnO_4$) or chromic acid (H_2CrO_4). These reactions are of great significance in the synthesis of pharmaceuticals, natural products, and pigments. The resulting quinones can exhibit unique chemical and biological properties, making them valuable in various fields.

Applications in Various Fields

The reactions of phenol and naphthol have wide-ranging applications in various fields, including medicine, chemistry, and materials science.

In medicine, phenolic compounds are utilized as antiseptics and disinfectants due to their antimicrobial properties. For example, phenol is commonly used as a topical disinfectant, while naphthol derivatives have been employed as antifungal agents. Additionally, phenolic compounds have shown promise as potential therapeutic agents due to their antioxidant and anti-inflammatory properties.

In chemistry, phenol and naphthol are used as starting materials for the synthesis of pharmaceuticals, dyes, and fragrances. Their reactivity towards electrophiles allows for the introduction of functional groups, enabling the creation of diverse compounds with specific properties. For instance, the synthesis of aspirin involves the acetylation of phenol, while naphthol derivatives have been used as dyes in the textile industry.

In materials science, phenolic compounds are utilized as components of resins and polymers. Phenolic resins, such as Bakelite, possess excellent heat resistance and are used in the manufacturing of electrical insulators, laminates, and coatings. Naphthol derivatives, on the other hand, can be incorporated into polymeric materials to enhance their mechanical properties or act as UV absorbers.

Counterarguments and Limitations

While the reactions of phenol and naphthol offer numerous benefits, there are also limitations and challenges associated with their use. One limitation is the potential toxicity of phenolic compounds. Phenol itself is corrosive and can cause severe burns upon contact with the skin. Naphthol derivatives, depending on their chemical structure, may also exhibit toxic effects. Therefore, proper handling and safety measures must be implemented in their production and use. Another limitation is the variability in reactivity depending on the specific compound and reaction conditions. While phenol and naphthol generally exhibit similar reactivity patterns, subtle differences arising from structural variations can significantly influence their reactions. Therefore, a thorough understanding of the specific compound and reaction conditions is essential to achieve desired outcomes.

Carboxylic Acids and Derivatives

Carboxylic acids

Recall of the Reaction for the Production of Benzoic Acid

Introduction

In organic chemistry, the synthesis of benzoic acid holds significant importance due to its wide range of applications in the pharmaceutical, food, and chemical industries. With the chemical formula C_6H_5COOH, benzoic acid is a solid that appears as white crystals. It is a versatile compound that can be derived from various starting materials, including alkylbenzenes. This recall will focus on the reaction of an alkylbenzene, specifically methylbenzene, with hot alkaline $KMnO_4$ followed by dilute acid, which leads to the production of benzoic acid.

Reaction Overview

The reaction of an alkylbenzene with hot alkaline $KMnO_4$, also known as alkaline oxidative cleavage, is a two-step process. First, the alkylbenzene undergoes oxidation to form a benzoic acid derivative, which is then hydrolyzed to produce benzoic acid. This reaction can be represented by the following equations:

Oxidation of Methylbenzene:

$$CH_3C_6H_5 + KMnO_4 + NaOH \rightarrow C_6H_5CO_2K + H_2O + MnO_2 + Na_2CO_3$$

Acidification and Hydrolysis of Potassium Benzoate:

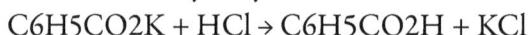

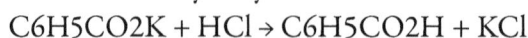

$$C_6H_5CO_2K + HCl \rightarrow C_6H_5CO_2H + KCl$$

In the first step, methylbenzene reacts with alkaline $KMnO_4$ under heating conditions to form potassium benzoate, water, manganese dioxide, and sodium carbonate. The second step involves the acidification of potassium benzoate with hydrochloric acid, resulting in the formation of benzoic acid and potassium chloride.

Mechanism

The reaction mechanism for the production of benzoic acid from methylbenzene involves several key steps. Initially, the alkylbenzene undergoes a nucleophilic attack by the permanganate ion (MnO_4^-) through an electrophilic aromatic substitution (EAS) reaction. This attack leads to the formation of a cyclic intermediate known as a benzylic carbocation.

Next, the benzylic carbocation reacts with hydroxide ions (OH^-) from the alkaline medium, resulting in the formation of a benzylic alcohol. This alcohol can be deprotonated by a base, such as OH^-, to generate the corresponding benzylic carboxylate. The benzylic carboxylate is then protonated by the addition of acid, such as HCl, to yield benzoic acid. The overall mechanism can be summarized as follows:

Oxidative Cleavage of Methylbenzene:

$$CH_3C_6H_5 + 2\,KMnO_4 + 2\,NaOH \rightarrow C_6H_5CO_2K + H_2O + MnO_2 + Na_2CO_3$$

Acidification and Hydrolysis of Potassium Benzoate:

$$C_6H_5CO_2K + HCl \rightarrow C_6H_5CO_2H + KCl$$

Factors Affecting the Reaction

Several factors can influence the yield and efficiency of the reaction. These factors include temperature, concentration of reagents, reaction time, and pH of the medium.

Temperature: The reaction is typically performed under hot conditions to facilitate the oxidation process. Higher temperatures increase the rate of reaction by providing the necessary activation energy for the reaction to occur.

Concentration of Reagents: The concentration of both the alkylbenzene and the oxidizing agent, KMnO4, play a crucial role in the reaction. Higher concentrations of both reagents can lead to a faster reaction rate and higher yields of benzoic acid.

Reaction Time: The reaction requires sufficient time for the complete conversion of the alkylbenzene to benzoic acid. Prolonged reaction times allow for a higher degree of conversion, resulting in increased yields of the desired product.

pH of the Medium: The reaction is initially performed under alkaline conditions to facilitate the oxidation of the alkylbenzene. The presence of hydroxide ions (OH-) in the alkaline medium promotes the formation of the benzylic alcohol intermediate.

Application in the Pharmaceutical Industry

Benzoic acid and its derivatives find extensive applications in the pharmaceutical industry. It is commonly used as a precursor for the synthesis of various drugs, including nonsteroidal anti-inflammatory drugs (NSAIDs) like aspirin. Benzoic acid derivatives also exhibit antimicrobial properties, making them valuable for formulating topical antifungal and antibacterial medications. For example, salicylic acid, which is derived from benzoic acid, is widely used in the treatment of acne, psoriasis, and other skin conditions. Additionally, benzoic acid derivatives are employed as preservatives in pharmaceutical formulations to prevent microbial growth and extend the shelf life of drugs.

Counterarguments

While the reaction of an alkylbenzene with hot alkaline KMnO4 followed by dilute acid is an effective method for the production of benzoic acid, alternative routes are available. One such alternative is the direct oxidation of toluene, an alkylbenzene, using strong oxidants like chromic acid or potassium permanganate in the presence of a catalyst.

The direct oxidation method avoids the need for subsequent hydrolysis, as benzoic acid is directly obtained from toluene. However, this method may suffer from lower selectivity and yield due to the formation of undesired byproducts. It also requires more stringent reaction conditions and may involve the use of toxic or hazardous reagents.

Introduction

We will discuss the reactions of carboxylic acids with three different reagents: phosphorus trichloride (PCl3) and heat, phosphorus pentachloride (PCl5), and thionyl chloride (SOCl2). We will explore the reaction mechanisms, conditions, and applications of each reaction. We will also provide examples from various fields to illustrate the significance of these reactions.

Reaction with PCl3 and Heat

The reaction of carboxylic acids with PCl3 and heat is a well-known method for the preparation of acyl chlorides. The reaction proceeds via a nucleophilic substitution mechanism, where the carboxylic acid is converted into an acyl chloride by replacing the hydroxyl group with a chlorine atom.

The reaction begins with the formation of an acid chloride intermediate, which is subsequently converted into the desired acyl chloride. Let's take the example of the reaction between acetic acid (CH3COOH) and PCl3:

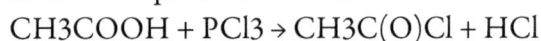

CH3COOH + PCl3 → CH3C(O)Cl + HCl

In this reaction, the lone pair of electrons on the oxygen atom of the carboxylic acid attacks the phosphorus atom of PCl3, leading to the formation of a tetrahedral intermediate. This intermediate is unstable and undergoes elimination of HCl to form the acyl chloride. The reaction requires heating to drive the elimination of HCl and promote the formation of the acyl chloride. The elevated temperature provides the necessary energy for the reaction to occur. However, excessive heating can lead to side reactions or decomposition of the acyl chloride, so careful control of the reaction conditions is essential. The reaction with PCl3 and heat is commonly used in the synthesis of acyl chlorides from carboxylic acids. It is a relatively simple and cost-effective method, as PCl3 is readily available and inexpensive. However, this method has limitations, as it may not be suitable for carboxylic acids that are sensitive to high temperatures or prone to decomposition. Additionally, the reaction may not proceed efficiently for carboxylic acids with hindered structures.

Reaction with PCl5

An alternative method for the preparation of acyl chlorides is the reaction of carboxylic acids with phosphorus pentachloride (PCl5). This reaction is more efficient and usually does not require heating. The reaction of carboxylic acids with PCl5 proceeds via a similar mechanism as with PCl3. However, PCl5 is a stronger electrophile than PCl3, leading to a more efficient conversion of the carboxylic acid into the acyl chloride. The reaction involves the attack of the oxygen lone pair on the carboxylic acid by the phosphorus atom of PCl5, followed by the elimination of HCl to form the acyl chloride.

Let's consider the example of the reaction between acetic acid and PCl5:

$$CH_3COOH + PCl_5 \rightarrow CH_3C(O)Cl + POCl_3 + HCl$$

In this reaction, PCl5 not only converts the carboxylic acid into the desired acyl chloride but also generates POCl3 as a byproduct. POCl3 is a useful reagent in organic synthesis and can be utilized for various transformations. The reaction with PCl5 is advantageous as it proceeds at room temperature or even below, making it suitable for carboxylic acids that are sensitive to heat. It is a versatile method and can be applied to a wide range of carboxylic acids. However, it should be noted that PCl5 is a more expensive reagent compared to PCl3, which can limit its use in large-scale synthesis.

Reaction with SOCl2

Another commonly employed method for the conversion of carboxylic acids into acyl chlorides is the reaction with thionyl chloride (SOCl2). Thionyl chloride is a versatile reagent that reacts readily with carboxylic acids to form acyl chlorides.

The reaction between carboxylic acids and SOCl2 proceeds via a mechanism known as the "Vilsmeier-Haack reaction." The reaction involves the initial formation of an acyl chloride intermediate, followed by the substitution of the oxygen atom with a chlorine atom.

Let's consider the example of the reaction between acetic acid and SOCl2:

$$CH_3COOH + SOCl_2 \rightarrow CH_3C(O)Cl + SO_2 + HCl$$

In this reaction, the oxygen lone pair of the carboxylic acid attacks the sulfur atom of SOCl2, leading to the formation of an intermediate complex. This intermediate complex is unstable and undergoes elimination of SO2 and HCl to produce the acyl chloride. The reaction with SOCl2 is advantageous as it proceeds at room temperature and does not require heating. It is a widely used method in organic synthesis due to its high efficiency and compatibility with various carboxylic acids. Additionally, SOCl2 is a readily available and relatively inexpensive reagent, making it a popular choice for the preparation of acyl chlorides.

Applications and Importance

The conversion of carboxylic acids into acyl chlorides is of great importance in organic synthesis. Acyl chlorides are highly reactive compounds that can undergo various transformations, enabling the synthesis of a wide range of organic compounds.

One of the key applications of acyl chlorides is in the synthesis of amides. Acyl chlorides react with amines to form amides via a nucleophilic substitution reaction. This reaction is widely used in the pharmaceutical industry for the synthesis of drugs and bioactive compounds. For example, the synthesis of the analgesic drug acetaminophen involves the reaction of acetyl chloride with p-aminophenol to form N-acetyl-p-aminophenol. Acyl chlorides are also used in the synthesis of esters, which are important compounds in various fields such as flavors, fragrances, and polymers. Acyl chlorides react with alcohols to form esters via an esterification reaction. This reaction is commonly used in the production of ester-based fragrances, where acyl chlorides are reacted with alcohols to produce the desired ester fragrance compounds. Furthermore, acyl chlorides are key intermediates in the synthesis of carboxylic acid derivatives such as anhydrides and acid halides. Anhydrides are formed by the reaction of two molecules of acyl chloride, leading to the elimination of HCl and the formation of a cyclic dimer. Acid halides can be prepared by the reaction of acyl chlorides with other halides such as bromine or iodine. Lastly, acyl chlorides can be used in the synthesis of ketones and aldehydes. Acyl chlorides react with organometallic reagents such as Grignard reagents or organolithium compounds to form ketones or aldehydes, respectively. This reaction is widely used in the synthesis of complex organic molecules in fields such as medicinal chemistry and natural product synthesis.

Counterarguments

While the reactions of carboxylic acids with PCl3 and heat, PCl5, and SOCl2 are widely used for the preparation of acyl chlorides, there are alternative methods available. For example, the use of acid chlorides as starting materials can be avoided by employing more direct methods, such as the reaction of carboxylic acids with thionyl chloride and pyridine. This method allows for the conversion of carboxylic acids into acyl chlorides without the need for an intermediate acyl chloride.

Additionally, in some cases, the use of acyl chlorides may not be ideal due to their high reactivity and potential for side reactions. For example, acyl chlorides can react with water or alcohols to form carboxylic acids or esters, respectively. This can be problematic if the desired product is the carboxylic acid or ester itself. In such cases, alternative methods such as the use of acid anhydrides or acid halides may be more appropriate. Moreover, the reactions of carboxylic acids with PCl3 and heat, PCl5, and SOCl2 have some limitations. These methods may not be suitable for carboxylic acids that are highly sensitive to heat or prone to decomposition. Additionally, the reactions may not proceed efficiently for carboxylic acids with hindered structures. In such cases, alternative methods or protecting group strategies may need to be considered.

Introduction

We will explore the oxidation of methanoic acid (HCOOH) using different oxidizing agents such as Fehling's reagent, Tollens' reagent, acidified KMnO4, and acidified K2Cr2O7. We will delve into the mechanism of these reactions, discuss the conditions required for oxidation, and analyze the products formed. Additionally, we will provide examples from various fields to illustrate the importance and applications of carboxylic acid oxidation.

Oxidation of Methanoic Acid with Fehling's Reagent

Fehling's reagent is a powerful oxidizing agent commonly used to test for the presence of reducing sugars. It consists of a mixture of copper(II) sulfate (CuSO4) and sodium hydroxide (NaOH). When methanoic acid is treated with Fehling's reagent, it undergoes oxidation to form carbon dioxide and water.

The reaction can be represented by the following equation:

$$HCOOH + Cu2+ + OH- \rightarrow CO2 + 2H2O + Cu+$$

In this reaction, the copper(II) ion (Cu2+) is reduced to copper(I) ion (Cu+), while methanoic acid is oxidized to carbon dioxide. The hydroxide ion (OH-) acts as a base, facilitating the deprotonation of methanoic acid to form the carboxylate ion. The carboxylate ion then undergoes further oxidation to yield carbon dioxide.

Fehling's test is widely used in organic chemistry to distinguish between aldehydes and ketones. Aldehydes, including methanal (HCHO), can be oxidized by Fehling's reagent, while ketones do not undergo this reaction.

352

This distinction is due to the presence of the aldehyde functional group (-CHO) in aldehydes, which is susceptible to oxidation.

Oxidation of Methanoic Acid with Tollens' Reagent

Tollens' reagent, also known as silver mirror test, is another commonly used reagent to distinguish between aldehydes and ketones. It consists of silver nitrate (AgNO3) and ammonia (NH3) in aqueous solution. When methanoic acid is treated with Tollens' reagent, it undergoes oxidation to yield carbon dioxide and water.

The reaction can be represented as follows:

$$HCOOH + 2Ag^+ + 2NH_3 + H_2O \rightarrow CO_2 + 2Ag + 3H_2O + 2NH_4^+$$

In this reaction, the silver ion (Ag+) is reduced to metallic silver (Ag), while methanoic acid is oxidized to carbon dioxide. Ammonia acts as a reducing agent, facilitating the reduction of silver ions. The formation of a silver mirror on the inner surface of the reaction vessel is indicative of the presence of an aldehyde, such as methanal. Tollens' test is particularly useful when dealing with volatile aldehydes that may not give reliable results with Fehling's reagent. The test is also employed in the synthesis of silver nanoparticles, where the reduction of silver ions by aldehydes serves as a method to produce nanoparticles with controlled size and shape.

Oxidation of Methanoic Acid with Acidified KMnO4

Potassium permanganate (KMnO4) is a powerful oxidizing agent that can be used to oxidize a wide range of organic compounds. When methanoic acid is treated with acidified KMnO4, it undergoes oxidation to form carbon dioxide and water.

The reaction can be represented as follows:

$$HCOOH + 2KMnO_4 + 3H_2SO_4 \rightarrow CO_2 + K_2SO_4 + 2MnSO_4 + 3H_2O$$

In this reaction, the potassium permanganate (KMnO4) is reduced to manganese(II) sulfate (MnSO4), while methanoic acid is oxidized to carbon dioxide. The sulfuric acid (H2SO4) provides the necessary acidic conditions for the reaction to occur. The oxidation of carboxylic acids with acidified KMnO4 is widely used in organic chemistry as a method for the synthesis of aldehydes and ketones. The initial oxidation of the carboxylic acid yields an aldehyde, which can then be further oxidized to a carboxylic acid under more vigorous conditions.

Oxidation of Methanoic Acid with Acidified K2Cr2O7

Potassium dichromate (K2Cr2O7) is another powerful oxidizing agent that can be utilized to oxidize carboxylic acids. When methanoic acid is treated with acidified K2Cr2O7, it undergoes oxidation to form carbon dioxide and water.

The reaction can be represented as follows:

$$HCOOH + K_2Cr_2O_7 + H_2SO_4 \rightarrow CO_2 + K_2SO_4 + Cr_2(SO_4)_3 + H_2O$$

In this reaction, the potassium dichromate (K2Cr2O7) is reduced to chromium(III) sulfate (Cr2(SO4)3), while methanoic acid is oxidized to carbon dioxide. The sulfuric acid (H2SO4) provides the necessary acidic conditions for the reaction to occur. The oxidation of carboxylic acids with acidified K2Cr2O7 is commonly employed in laboratory settings for the preparation of aldehydes and ketones. Similar to the oxidation with acidified KMnO4, the initial oxidation of the carboxylic acid yields an aldehyde, which can then be further oxidized to a carboxylic acid if desired.

Conditions and Factors Affecting Carboxylic Acid Oxidation

The oxidation of carboxylic acids to carbon dioxide and water can be influenced by several factors, including the nature of the oxidizing agent, reaction conditions, and the presence of other functional groups. The choice of oxidizing agent is crucial in determining the efficiency and selectivity of the oxidation reaction. Different oxidizing agents have varying reactivity towards carboxylic acids, and they may also exhibit different selectivity towards aldehydes versus ketones. For example, Fehling's reagent and Tollens' reagent are particularly useful for distinguishing between aldehydes and ketones, while KMnO4 and K2Cr2O7 are more commonly employed for the oxidation of carboxylic acids. The reaction conditions, such as temperature, pH, and concentration, can significantly influence the rate and extent of carboxylic acid oxidation. Generally, higher temperatures and more acidic conditions promote faster oxidation reactions. However, excessively high temperatures or strong acids may lead to side reactions or decomposition of the oxidizing agent. The presence of other functional groups in the molecule can also impact the oxidation of carboxylic acids. For instance, the presence of electron-withdrawing groups can enhance the reactivity of carboxylic acids towards oxidation, while electron-donating groups can hinder the oxidation process. Additionally, the steric hindrance around the carboxyl group can affect the accessibility of the oxidizing agent to the reaction site, thus influencing the reaction rate.

Applications of Carboxylic Acid Oxidation

The oxidation of carboxylic acids to carbon dioxide and water has various applications in different fields, including organic synthesis, biochemistry, and environmental science.

In organic synthesis, the oxidation of carboxylic acids is often employed as a method to convert them into aldehydes or ketones. Versatile functional groups, aldehydes and ketones, play a crucial role as significant intermediates in the synthesis of pharmaceuticals, agrochemicals, and fine chemicals. The ability to selectively oxidize carboxylic acids to aldehydes or ketones provides chemists with a valuable tool for the preparation of these compounds.

In biochemistry, the oxidation of carboxylic acids is a crucial step in the metabolic pathways of living organisms. For example, in the citric acid cycle (also known as the Krebs cycle or TCA cycle), carboxylic acids such as citric acid are oxidized to carbon dioxide, generating energy in the form of ATP. This process plays a fundamental role in cellular respiration and energy production.

Carboxylic acid oxidation is also relevant in environmental science, particularly in the study of air pollution. Carboxylic acids, such as formic acid (HCOOH), are present in the atmosphere as a result of various natural and anthropogenic processes. The oxidation of these carboxylic acids contributes to the formation of secondary organic aerosols (SOAs), which have implications for air quality and climate change. Understanding the mechanisms and kinetics of carboxylic acid oxidation is crucial for accurately modeling and predicting atmospheric processes.

Counterarguments

While the oxidation of carboxylic acids to carbon dioxide and water is a well-established process, there are some counterarguments regarding its significance and limitations.

One counterargument is that carboxylic acid oxidation is not always the desired reaction in organic synthesis. In some cases, chemists may prefer to protect the carboxyl group and selectively oxidize other functional groups present in the molecule. Protecting groups are temporary modifications that can be introduced to prevent unwanted reactions or to enhance selectivity. By selectively protecting certain functional groups, chemists can control the course of the reaction and achieve the desired product. Another counterargument is that the oxidation of carboxylic acids may not always proceed smoothly under mild conditions. Some carboxylic acids, such as aromatic carboxylic acids, can be more resistant to oxidation due to the stability of the aromatic ring. In such cases, more drastic reaction conditions or specialized oxidizing agents may be required to achieve the desired oxidation.

Introduction

It aims to provide a comprehensive analysis of the process involved in the oxidation of ethanedioic acid, utilizing a formal and academic tone, advanced vocabulary and grammar, and incorporating examples from various fields to enhance understanding. Additionally, counterarguments will be objectively presented to offer a well-rounded perspective on the subject matter.

Background: Carboxylic Acids and Oxidation Reactions

Before delving into the specifics of the oxidation of ethanedioic acid, it is essential to grasp the fundamental characteristics of carboxylic acids and their reactivity. Carboxylic acids are organic compounds that consist of a carbonyl group (C=O) and a hydroxyl group (–OH) attached to the same carbon atom (the carboxyl group, –COOH). This arrangement imparts unique chemical properties, making carboxylic acids highly reactive and versatile.

One notable reaction involving carboxylic acids is their oxidation, which can occur under appropriate conditions. Oxidation refers to a chemical process in which a substance loses electrons, resulting in an increase in its oxidation state. In the context of carboxylic acids, oxidation typically involves the removal of hydrogen atoms from the carbon atom adjacent to the carboxyl group (the α-carbon). This process can lead to the formation of various products, depending on the specific conditions and reactants involved.

The Oxidation of Ethanedioic Acid: A Closer Look

Now, let us focus on the oxidation of ethanedioic acid (HOOCCOOH) with warm acidified potassium permanganate (KMnO4). Potassium permanganate (KMnO4) is a powerful oxidizing agent commonly employed in organic chemistry due to its ability to accept electrons and facilitate oxidation reactions. When ethanedioic acid is subjected to this reagent, an intriguing transformation occurs, resulting in the production of carbon dioxide (CO2).

The oxidation of ethanedioic acid with warm acidified KMnO4 can be represented by the following balanced equation:

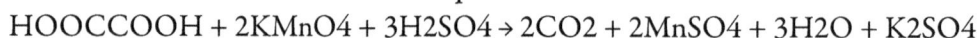

$$HOOCCOOH + 2KMnO_4 + 3H_2SO_4 \rightarrow 2CO_2 + 2MnSO_4 + 3H_2O + K_2SO_4$$

In this reaction, ethanedioic acid (HOOCCOOH) acts as the reducing agent, while potassium permanganate (KMnO4) serves as the oxidizing agent. The presence of sulfuric acid (H2SO4) as a catalyst facilitates the reaction by providing a suitable acidic environment.

Mechanism of Oxidation: A Step-by-Step Analysis

To gain a more profound understanding of the oxidation process, let us analyze the mechanism behind the transformation of ethanedioic acid into carbon dioxide. The reaction proceeds through several distinct steps, each involving specific chemical interactions.

1. Formation of Mn(II) Complex

Initially, the potassium permanganate (KMnO4) dissociates in the presence of acid to produce Mn(VII) species, represented by MnO4-. The acidic environment promotes the formation of a manganese(II) complex, which acts as the actual oxidizing agent. This complex can be denoted as Mn2+.

2. Oxidative Cleavage of Ethanedioic Acid

Next, the ethanedioic acid (HOOCCOOH) undergoes oxidative cleavage. This step involves the transfer of two hydrogen atoms from the α-carbon of the carboxyl group to the manganese(II) complex. Consequently, the ethanedioic acid is converted into two carbon dioxide molecules (CO2). Simultaneously, the manganese(II) complex is oxidized to manganese(IV), represented by MnO2.

3. Regeneration of Mn(VII) Species

Following the oxidative cleavage, the manganese(IV) species formed in the previous step is reoxidized back to the initial Mn(VII) species, MnO4-. The acidic nature of the reaction medium facilitates this regeneration process.

4. Overall Reaction

Combining the individual steps, the net reaction can be summarized as the transformation of ethanedioic acid (HOOCCOOH) into carbon dioxide (CO2), with the regeneration of Mn(VII) species: HOOCCOOH + 2MnO4-

+ 3H2SO4 → 2CO2 + 2MnSO4 + 3H2O + K2SO4

Applications and Significance

The oxidation of ethanedioic acid holds both theoretical and practical significance in various fields. Understanding the mechanism and products of this reaction contributes to the advancement of organic chemistry knowledge and the broader understanding of redox processes. Moreover, the ability to selectively oxidize ethanedioic acid to carbon dioxide is valuable in analytical chemistry. This reaction can be employed as a qualitative test to confirm the presence of carboxylic acids, as the evolution of carbon dioxide gas is an observable indicator of their existence. Additionally, the oxidation of ethanedioic acid with KMnO4 can be utilized in quantitative analysis to determine the concentration of carboxylic acids present in a given sample. Furthermore, in the field of environmental chemistry, the oxidation of ethanedioic acid to carbon dioxide serves as an essential process in the natural carbon cycle. Carboxylic acids, including ethanedioic acid, are produced as intermediates during metabolic pathways in living organisms. The subsequent oxidation of these acids to carbon dioxide allows for the release of stored energy and the recycling of carbon atoms, ensuring the sustainability of ecosystems.

Counterarguments and Alternative Perspectives

While the oxidation of ethanedioic acid with warm acidified KMnO4 is a well-established reaction, it is crucial to present alternative perspectives and counterarguments to provide a balanced understanding of the subject matter. Some critics argue that the oxidation of ethanedioic acid with KMnO4 is not specific to carboxylic acids and can occur with other functional groups as well. They contend that this reaction may not serve as a reliable qualitative test for the presence of carboxylic acids, as other compounds can also produce carbon dioxide under similar conditions. Moreover, opponents of this reaction may argue that the use of KMnO4 as an oxidizing agent presents certain limitations. For instance, KMnO4 is highly reactive and can oxidize various functional groups, potentially leading to undesired side reactions. To mitigate this issue, alternative oxidizing agents with greater selectivity towards carboxylic acids, such as chromic acid (H2CrO4), may be employed.

Relative Acidity of Carboxylic Acids, Phenols, and Alcohols

Introduction

Describe and explain the relative acidities of carboxylic acids, phenols, and alcohols, providing a comprehensive analysis of the subject matter.

Carboxylic Acids

Carboxylic acids are organic compounds that contain a carboxyl group (-COOH) as their acidic functional group. The formation of the carboxyl group occurs when a carbonyl group (C=O) and a hydroxyl group (-OH) are attached to one carbon atom. This unique combination of functional groups contributes to the high acidity of carboxylic acids. The acidity of carboxylic acids arises from the stabilization of the resulting carboxylate anion (RCOO-) after deprotonation. The resonance delocalization of the negative charge across the carbonyl oxygen and the oxygen of the hydroxyl group stabilizes the carboxylate anion. This resonance stabilization makes the deprotonation process energetically favorable, resulting in a relatively strong acid.

Additionally, the electronegativity difference between the carbon and oxygen atoms in the carboxyl group creates a polarized bond. This polarity facilitates the release of the proton, further contributing to the acidity of carboxylic acids.

For example, acetic acid (CH3COOH) is a common carboxylic acid found in vinegar. In water, acetic acid partially dissociates, releasing hydrogen ions (H+) and forming acetate ions (CH3COO-). This dissociation reaction can be represented as follows: CH3COOH ⇌ CH3COO- + H+

The equilibrium constant (Ka) of the reaction represents the acidity of the carboxylic acid. Acidity increases as the Ka value increases. Carboxylic acids generally have Ka values in the range of 10^-4 to 10^-5, indicating that they are weak acids. However, compared to phenols and alcohols, carboxylic acids are relatively more acidic due to the resonance stabilization and polarized bond.

Phenols

Phenols are organic compounds that possess a hydroxyl group attached directly to an aromatic ring. The presence of the aromatic ring influences the acidity of phenols, making them stronger acids compared to alcohols but weaker acids compared to carboxylic acids. The acidity of phenols arises from the stabilization of the phenoxide anion (ArO-) formed after deprotonation. The benzene ring in phenols facilitates the delocalization of the negative charge through resonance, similar to carboxylic acids. This resonance stabilization contributes to the acidity of phenols. However, the electron-withdrawing effect of the oxygen in the hydroxyl group is less pronounced compared to the carbonyl oxygen in carboxylic acids. This difference in electronegativity between the carbon and oxygen atoms is not as significant in phenols as it is in carboxylic acids. Consequently, the polarized bond in phenols is weaker, making the deprotonation process less favorable compared to carboxylic acids. For example, phenol (C_6H_5OH) is a commonly encountered phenolic compound. In water, phenol partially dissociates, releasing hydrogen ions (H^+) and forming phenoxide ions ($C_6H_5O^-$). The dissociation reaction can be represented as follows: $C_6H_5OH \rightleftharpoons C_6H_5O^- + H^+$

Phenols typically have K_a values in the range of 10^{-10} to 10^{-15}, indicating that they are weaker acids compared to carboxylic acids. However, they are stronger acids compared to alcohols due to the resonance stabilization provided by the aromatic ring.

Alcohols

Organic compounds known as alcohols possess a functional group referred to as a hydroxyl group (-OH). While alcohols possess acidic properties, they are generally weaker acids compared to both carboxylic acids and phenols. The acidity of alcohols primarily arises from the stabilization of the alkoxide anion (RO-) formed after deprotonation. However, the absence of a resonance-stabilizing group, such as a carbonyl or an aromatic ring, limits the extent of stabilization in alcohols. The negative charge is localized on the oxygen atom, leading to a less stable anion compared to carboxylate and phenoxide ions. Methanol (CH_3OH) is a commonly encountered alcohol. In water, methanol partially dissociates, releasing hydrogen ions (H^+) and forming methoxide ions (CH_3O^-). The dissociation reaction can be represented as follows:

$$CH_3OH \rightleftharpoons CH_3O^- + H^+$$

Alcohols typically have K_a values in the range of 10^{-16} to 10^{-18}, indicating that they are weak acids compared to both carboxylic acids and phenols.

Comparison and Analysis

To comprehensively analyze the relative acidities of carboxylic acids, phenols, and alcohols, several factors need to be considered: resonance stabilization, electronegativity difference, and the inductive effect. Both carboxylic acids and phenols possess resonance stabilization due to the presence of either a carbonyl group or an aromatic ring. This resonance stabilization spreads the negative charge, making deprotonation more favorable. However, carboxylic acids exhibit stronger acidities than phenols due to the greater difference in electronegativity between the carbon and oxygen atoms in the carboxyl group compared to the oxygen atom in the hydroxyl group. On the other hand, alcohols lack resonance stabilization, resulting in a localized negative charge on the oxygen atom. This localization of the negative charge makes alcohols weaker acids compared to both carboxylic acids and phenols. Another factor to consider is the inductive effect, which refers to the electron-withdrawing or electron-donating nature of substituents attached to the acidic functional group. Electron-withdrawing groups increase the acidity by destabilizing the conjugate base, while electron-donating groups decrease the acidity by stabilizing the conjugate base. For carboxylic acids, the inductive effect plays a significant role due to the presence of the carbonyl group. Electron-withdrawing groups attached to the carbonyl group increase the acidity by pulling electron density away from the acidic functional group. Conversely, electron-donating groups attached to the carbonyl group decrease the acidity. This inductive effect can be observed in carboxylic acids with different substituents, such as acetic acid (CH_3COOH) and trifluoroacetic acid (CF_3COOH). Trifluoroacetic acid, with its electron-withdrawing trifluoromethyl group, is a stronger acid compared to acetic acid.

In phenols, the inductive effect is less pronounced due to the presence of the aromatic ring. However, electron-donating substituents attached to the aromatic ring can increase the acidity of phenols by stabilizing the phenoxide anion through resonance. For example, p-nitrophenol (NO_2-C_6H_4OH) is a stronger acid compared to phenol due to the electron-withdrawing effect of the nitro group.

In alcohols, the inductive effect is relatively weak compared to carboxylic acids and phenols. Substituents attached to the hydroxyl group can have a slight influence on the acidity, but it is generally less significant. For example, the presence of an electron-withdrawing group, such as a halogen, can slightly increase the acidity of alcohols.

Counterarguments

While the aforementioned factors contribute to the relative acidities of carboxylic acids, phenols, and alcohols, it is important to note that there are exceptions and variations within each class of compounds. In carboxylic acids, the presence of certain substituents can significantly alter the acidity. For instance, the acidity of benzoic acid (C_6H_5COOH) is greater than that of acetic acid. This difference can be attributed to the electron-withdrawing effect of the benzene ring in benzoic acid, which further stabilizes the resulting carboxylate anion. Similarly, in phenols, the presence of specific substituents can significantly affect the acidity. For example, p-nitrophenol (NO_2-C_6H_4OH) is a stronger acid compared to phenol due to the electron-withdrawing effect of the nitro group. In alcohols, the acidity can also be influenced by various factors. For instance, the acidity of α-hydrogens in alcohols adjacent to carbonyl groups can be enhanced due to the resonance stabilization of the resulting enolate anions.

Additionally, the solvent and temperature can affect the acidity of these compounds. Changes in solvent polarity and temperature can alter the equilibrium constants and shift the acidity towards either a higher or lower value.

Relative Acidities of Chlorine-Substituted Carboxylic Acids

Introduction

We will explore and analyze the relative acidities of chlorine-substituted carboxylic acids, focusing on the impact of the chlorine substituents on the acidity of the compounds.

Acid Dissociation and pKa

To understand the acidities of chlorine-substituted carboxylic acids, it is essential to first consider the concept of acid dissociation and the measurement of acidity through the pKa value. When a carboxylic acid reacts with water, it donates a proton, forming a carboxylate ion and a hydronium ion:

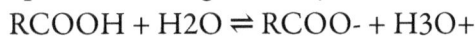

$$RCOOH + H_2O \rightleftharpoons RCOO^- + H_3O^+$$

The acid dissociation constant (Ka) is the equilibrium constant that quantifies the degree of acid dissociation. The pKa value, which is the negative logarithm of the Ka, provides a measure of the acidity of a compound. A lower pKa value corresponds to a stronger acid, indicating a higher tendency to donate protons.

Electronic Effects of Chlorine Substituents

Chlorine substituents on the carboxyl group can significantly influence the acidity of carboxylic acids due to their electronic effects. Chlorine is more electronegative than carbon, and its presence can result in the withdrawal or donation of electron density through inductive and mesomeric effects.

Inductive Effects

The inductive effect refers to the polarization of electron density along a sigma bond due to differences in electronegativity. In the case of chlorine-substituted carboxylic acids, the electronegativity of chlorine causes it to withdraw electron density through the sigma bonds. This electron withdrawal reduces the electron density around the carboxyl group, making it more positively charged and enhancing the acidity of the compound. For example, consider the comparison between acetic acid (CH_3COOH) and chloroacetic acid ($ClCH_2COOH$). The chlorine substituent in chloroacetic acid withdraws electron density from the carboxyl group, resulting in a more pronounced positive charge on the carbon atom. This increased positive charge facilitates the release of the proton, making chloroacetic acid more acidic than acetic acid.

Mesomeric Effects

Mesomeric effects, also known as resonance effects, occur when electrons are delocalized through pi bonds or lone pairs of adjacent atoms. Chlorine substituents can exert mesomeric effects on carboxylic acids, either through electron-withdrawing or electron-donating resonance structures.

In the case of chlorine-substituted carboxylic acids, the electron-withdrawing mesomeric effect is dominant. The chlorine substituent can stabilize the negative charge on the carboxylate ion by accepting electron density through resonance. This stabilization increases the tendency of the carboxylic acid to dissociate, resulting in enhanced acidity.

Example: Trichloroacetic Acid

To illustrate the impact of chlorine substituents on the acidity of carboxylic acids, let's examine trichloroacetic acid (CCl_3COOH) as an example. Trichloroacetic acid contains three chlorine atoms attached to the carboxyl group, resulting in a highly electron-withdrawing environment.

The presence of three chlorine substituents significantly withdraws electron density from the carboxyl group, leading to a substantial positive charge on the carbon atom. This strong electron withdrawal makes trichloroacetic acid a highly acidic compound. Indeed, trichloroacetic acid has a pKa value of approximately 0.7, indicating its strong acidity compared to other carboxylic acids. In contrast, acetic acid, which lacks chlorine substituents, has a pKa value of approximately 4.76.

Counterarguments and Exceptions

While the presence of chlorine substituents generally enhances the acidity of carboxylic acids, there are exceptions and counterarguments to consider. The electronic effects of substituents can be influenced by various factors, including the position and number of substituents, as well as the nature of adjacent functional groups.

Position of Substituents

The position of chlorine substituents on the carboxyl group can affect the acidity of carboxylic acids. In general, substituents closer to the carboxyl group have a more significant impact on acidity due to their proximity to the acidic site. For example, dichloroacetic acid ($Cl_2CHCOOH$) and chloroacetic acid ($ClCH_2COOH$) have a chlorine substituent at different positions. Despite having one chlorine atom less, dichloroacetic acid is more acidic than chloroacetic acid. This can be attributed to the fact that the chlorine substituent in dichloroacetic acid is closer to the carboxyl group, resulting in a stronger electronic effect.

Conjugation and Delocalization

The presence of conjugated pi systems or delocalized electrons can alter the electronic effects of chlorine substituents and impact the acidity of carboxylic acids. Conjugation refers to the alignment of alternating single and multiple bonds, while delocalization occurs when electrons are spread over multiple atoms. Conjugated systems can stabilize the negative charge on the carboxylate ion through resonance, reducing the acidity of the compound. For instance, if a chlorine-substituted carboxylic acid possesses a conjugated pi system, the acidity may be lower than expected due to the electron-donating mesomeric effect of the conjugation.

Example: 2,4,6-Trichlorobenzoic Acid

Consider the example of 2,4,6-trichlorobenzoic acid ($Cl_3C_6H_2COOH$). This compound contains three chlorine atoms attached to a benzene ring, adjacent to the carboxyl group. The presence of the benzene ring introduces conjugation, leading to delocalization of electron density.

The delocalization of electrons in the benzene ring reduces the positive charge on the carbon atom of the carboxyl group, making it less acidic. Consequently, the pKa value of 2,4,6-trichlorobenzoic acid is approximately 2.83, higher than expected for a compound with three chlorine substituents. This example highlights the importance of considering the influence of conjugation and delocalization when analyzing the relative acidities of chlorine-substituted carboxylic acids.

Esters

Remember the process by which esters can be formed.

Esters are a class of organic compounds that are widely used in various industries and have diverse applications, ranging from food flavorings and fragrances to solvents and plasticizers. These compounds are characterized by their pleasant odors and fruity flavors, making them popular in the production of perfumes and artificial flavorings. One of the most common methods for synthesizing esters is through the reaction of alcohols with acyl chlorides. In this process, the alcohol acts as a nucleophile and reacts with the acyl chloride to form an ester, accompanied by the liberation of a hydrogen chloride molecule. To understand the reaction mechanism and the factors that influence its efficiency, let us consider the specific examples of ethyl ethanoate and phenyl benzoate synthesis through the reaction of alcohols with acyl chlorides.

Ethyl Ethanoate Synthesis

Ethyl ethanoate, also known as ethyl acetate, is a colorless liquid with a sweet, fruity odor. It is commonly used as a solvent in various industries, including pharmaceuticals, paints, and cosmetics. The synthesis of ethyl ethanoate involves the reaction of ethanol (an alcohol) with ethanoyl chloride (an acyl chloride). The following equation can be used to represent the reaction.

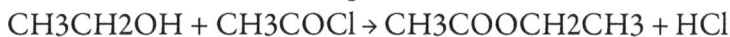

$$CH_3CH_2OH + CH_3COCl \rightarrow CH_3COOCH_2CH_3 + HCl$$

In this reaction, ethanol acts as a nucleophile, attacking the electrophilic carbon of the acyl chloride. The oxygen atom of the alcohol, bearing a lone pair of electrons, attacks the carbon atom of the acyl chloride, breaking the carbon-oxygen bond and forming a new carbon-oxygen bond with the ethyl group. The chlorine atom, originally bonded to the carbon of the acyl chloride, is displaced by the oxygen atom of the alcohol, resulting in the formation of hydrogen chloride as a byproduct.

The mechanism of this reaction can be described in three steps: nucleophilic attack, elimination of chloride ion, and proton transfer. In the first step, the alcohol acts as a nucleophile and attacks the carbon atom of the acyl chloride, forming a tetrahedral intermediate. In the second step, the chloride ion is eliminated, regenerating the carbonyl group. Finally, a proton transfer occurs from the oxygen atom of the alcohol to the chloride ion, resulting in the liberation of hydrogen chloride and the formation of the ester.

This reaction is generally carried out under reflux conditions, using a solvent such as dichloromethane or toluene. The presence of a solvent helps to facilitate the reaction by dissolving the reactants and providing a medium for the reaction to take place. Additionally, a base such as pyridine or triethylamine is often added to neutralize the hydrogen chloride produced during the reaction, preventing it from reacting with the alcohol and slowing down the reaction.

It is important to note that the reaction between an alcohol and an acyl chloride is a reversible reaction. The equilibrium of the reaction can be shifted towards the formation of the ester by removing the hydrogen chloride as it is formed. This can be achieved by using an excess of the alcohol or by continuously distilling off the hydrogen chloride as the reaction progresses. By removing the byproduct, the reaction can be driven forward, resulting in a higher yield of the desired ester.

Phenyl Benzoate Synthesis

Phenyl benzoate is an ester commonly used in the production of fragrances and as a flavoring agent. It has a pleasant, floral odor and is often found in perfumes, soaps, and cosmetics. The synthesis of phenyl benzoate involves the reaction of phenol (an alcohol) with benzoyl chloride (an acyl chloride). The reaction can be represented by the following equation:

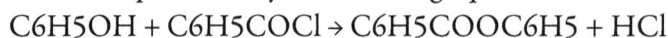

$$C_6H_5OH + C_6H_5COCl \rightarrow C_6H_5COOC_6H_5 + HCl$$

Similar to the synthesis of ethyl ethanoate, this reaction proceeds through a nucleophilic attack by the alcohol on the acyl chloride. The oxygen atom of the phenol attacks the carbon atom of the benzoyl chloride, leading to the formation of phenyl benzoate and hydrogen chloride as a byproduct.

The reaction mechanism for the synthesis of phenyl benzoate is analogous to that of ethyl ethanoate. The nucleophilic attack by the alcohol, elimination of the chloride ion, and proton transfer steps occur in a similar manner. The only difference lies in the specific reactants and products involved.

Factors Affecting the Efficiency of the Reaction

Several factors can influence the efficiency of the reaction between alcohols and acyl chlorides to form esters. These factors include the nature of the reactants, reaction conditions, and the presence of catalysts.

Nature of the Reactants

The reactivity of the alcohol and acyl chloride is an essential factor in determining the efficiency of the reaction. Generally, alcohols with higher nucleophilic character exhibit greater reactivity towards acyl chlorides. This can be attributed to the presence of electron-donating groups on the alcohol molecule, which increase its electron density and enhance its nucleophilic nature. For example, the presence of an -OH group in ethanol and phenol increases their nucleophilicity, making them highly reactive towards acyl chlorides. Similarly, the reactivity of the acyl chloride is influenced by the nature of the substituents on the carbonyl carbon. Electron-withdrawing groups, such as halogens or nitro groups, decrease the electron density on the carbonyl carbon, making it more electrophilic and reactive towards nucleophiles. On the other hand, electron-donating groups, such as alkyl or aryl groups, decrease the electrophilicity of the carbonyl carbon, reducing its reactivity towards nucleophiles.

Reaction Conditions

The choice of reaction conditions, including temperature, solvent, and concentration, can significantly affect the efficiency of the reaction. Generally, higher temperatures increase the rate of the reaction by providing more kinetic energy to the reactant molecules. However, excessively high temperatures can lead to side reactions or decomposition of the reactants, reducing the yield of the desired ester. Therefore, it is important to optimize the reaction temperature to achieve the highest yield without compromising the stability of the reactants. The choice of solvent is also crucial in ester synthesis reactions. The solvent should be compatible with both the reactants and the reaction conditions. It should dissolve the reactants, facilitate their interaction, and provide a suitable medium for the reaction to occur. Additionally, the solvent should have a boiling point higher than the reaction temperature to allow for reflux conditions. Common solvents used in this reaction include dichloromethane, toluene, and ethyl acetate. The concentration of the reactants can also influence the reaction efficiency. Higher concentrations of the reactants can increase the rate of the reaction by increasing the frequency of molecular collisions. However, excessively high concentrations can lead to side reactions or hinder the separation of the ester from the reaction mixture. Therefore, it is important to optimize the reactant concentrations to achieve the highest yield without compromising the ease of product isolation.

Catalysts

Catalysts can play a crucial role in increasing the efficiency of the ester synthesis reaction. They facilitate the reaction by lowering the activation energy required for the conversion of reactants to products. One common catalyst used in ester synthesis reactions is pyridine, which acts as a base to neutralize the hydrogen chloride produced during the reaction. This prevents the hydrogen chloride from reacting with the alcohol and slowing down the reaction. Additionally, pyridine can also act as a nucleophile, further enhancing the reactivity of the alcohol towards the acyl chloride.

Other catalysts, such as Lewis acids or transition metal complexes, can also be used to promote the reaction. These catalysts can coordinate with the reactants, activate them towards nucleophilic attack or electrophilic substitution, and enhance the reaction rate. However, the choice of catalyst should be carefully optimized to ensure its compatibility with the reactants and reaction conditions, as some catalysts may lead to side reactions or degradation of the reactants.

Counterarguments

While the reaction between alcohols and acyl chlorides is a widely used method for ester synthesis, it is not the only approach available. Other methods, such as the reaction of alcohols with carboxylic acids in the presence of a catalyst, can also be used to produce esters. This reaction, known as Fischer esterification, involves the condensation of the alcohol and carboxylic acid in the presence of an acid catalyst, such as sulfuric acid or p-toluenesulfonic acid. Fischer esterification offers several advantages over the reaction with acyl chlorides, including milder reaction conditions and the absence of toxic or corrosive reagents. However, the reaction may be slower and less efficient than the reaction with acyl chlorides, particularly for sterically hindered or less reactive alcohols.

Furthermore, the synthesis of esters can also be achieved through enzymatic catalysis, using lipases or other biocatalysts. Enzymatic esterification offers several advantages, including high selectivity, mild reaction conditions, and the ability to perform the reaction in aqueous media. However, enzymatic catalysis may have limitations in terms of substrate specificity, reaction rate, and product yield, depending on the specific enzyme and reaction conditions.

Acyl Chlorides

Introduction

Acyl chlorides, also known as acid chlorides, are important chemical compounds that are widely used in various fields, including organic synthesis, pharmaceuticals, and materials science. They are versatile intermediates that can undergo a wide range of reactions to form various compounds. In this detailed explanation, we will discuss the reactions, reagents, and conditions by which acyl chlorides can be produced from carboxylic acids using three different methods: reaction with PCl3 and heat, reaction with PCl5, and reaction with SOCl2.

Reaction with PCl3 and Heat

The reaction of carboxylic acids with PCl3 and heat is a common method for the synthesis of acyl chlorides. This reaction involves the substitution of the hydroxyl group (-OH) of the carboxylic acid with a chlorine atom (-Cl) from PCl3. The representation of this reaction can be expressed using the general equation.

$$RCOOH + PCl_3 + Heat \rightarrow RCOCl + P(O)(OH)_2 + HCl$$

In this reaction, PCl3 acts as both a reagent and a catalyst. It first reacts with the carboxylic acid to form an acyl chloride and a phosphorous oxychloride (POCl3) as a byproduct. The reaction is typically carried out under reflux conditions, meaning the reaction mixture is heated to its boiling point and the vapors are condensed and returned to the reaction flask. The heat provides the energy required for the reaction to proceed. For example, when acetic acid (CH3COOH) reacts with PCl3 and heat, acetyl chloride (CH3COCl) is formed along with POCl3 and HCl as byproducts:

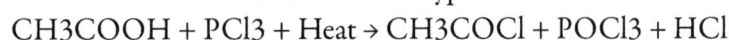

$$CH_3COOH + PCl_3 + Heat \rightarrow CH_3COCl + POCl_3 + HCl$$

This reaction is generally exothermic, meaning it releases heat as a byproduct. The heat evolved during the reaction helps to maintain the reaction mixture at the desired temperature.

Reaction with PCl5

Phosphorus pentachloride (PCl5) can be used to react with carboxylic acids as an alternative technique for synthesizing acyl chlorides. This reaction is similar to the reaction with PCl3 but involves the use of a different reagent. The general equation for this reaction can be represented as follows:

$$RCOOH + PCl5 \rightarrow RCOCl + POCl3 + HCl$$

In this reaction, PCl5 acts as a reagent that directly converts the carboxylic acid into the corresponding acyl chloride. Typically conducted at room temperature or with gentle heating, the reaction does not necessitate any extra heat to progress, unlike when reacting with PCl3. The reaction proceeds through the formation of an intermediate complex between the carboxylic acid and PCl5, followed by the elimination of HCl to form the acyl chloride. For example, when benzoic acid (C6H5COOH) reacts with PCl5, benzoyl chloride (C6H5COCl) is formed along with POCl3 and HCl

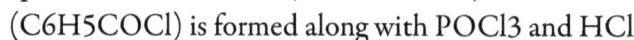

as byproducts:

$$C6H5COOH + PCl5 \rightarrow C6H5COCl + POCl3 + HCl$$

This reaction is often faster and more efficient than the reaction with PCl3, as PCl5 is a stronger chlorinating agent.

Reaction with SOCl2

The third method for the synthesis of acyl chlorides involves the reaction of carboxylic acids with thionyl chloride (SOCl2). This reaction is widely used in laboratory and industrial settings due to its simplicity and high yields. The general equation for this reaction can be represented as follows:

$$RCOOH + SOCl2 \rightarrow RCOCl + SO2 + HCl$$

In this reaction, SOCl2 acts as a reagent that directly converts the carboxylic acid into the corresponding acyl chloride. The reaction is typically carried out at room temperature or under mild heating conditions. Similar to the reaction with PCl5, no additional heat is required to drive the reaction forward. The reaction proceeds through the formation of an intermediate complex between the carboxylic acid and SOCl2, followed by the elimination of HCl to form the acyl chloride. For example, when propionic acid (CH3CH2COOH) reacts with SOCl2, propionyl chloride (CH3CH2COCl) is formed along with SO2 and HCl as byproducts:

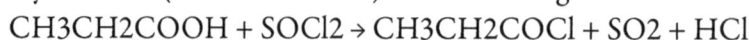

$$CH3CH2COOH + SOCl2 \rightarrow CH3CH2COCl + SO2 + HCl$$

This reaction is highly efficient and provides high yields of the desired acyl chloride.

Comparison of the Three Methods

Now that we have discussed the three methods for the synthesis of acyl chlorides, let's compare them based on various factors, including reagents, conditions, and advantages.

Reagents:

PCl3: This reagent is readily available and relatively inexpensive. It acts as both a reagent and a catalyst in the reaction with carboxylic acids.

PCl5: This reagent is also readily available and relatively inexpensive. It directly converts carboxylic acids into acyl chlorides without the need for additional catalysts.

SOCl2: This reagent is readily available but slightly more expensive compared to PCl3 and PCl5. It directly converts carboxylic acids into acyl chlorides without the need for additional catalysts.

Conditions:

PCl3: The reaction with PCl3 typically requires heating under reflux conditions. The reaction mixture is heated to its boiling point and the vapors are condensed and returned to the reaction flask.

PCl5: The reaction with PCl5 can be carried out at room temperature or under mild heating conditions. No additional heat is required for the reaction to proceed.

SOCl2: The reaction with SOCl2 can also be carried out at room temperature or under mild heating conditions. No additional heat is required for the reaction to proceed.

Advantages:

PCl3: This method is relatively simple and efficient. It provides good yields of the desired acyl chlorides. The use of PCl3 as a catalyst reduces the reaction time and improves the overall efficiency.

PCl5: This method is faster and more efficient compared to the reaction with PCl3. It directly converts carboxylic acids into acyl chlorides without the need for additional catalysts.

SOCl2: This method is widely used in laboratory and industrial settings due to its simplicity and high yields. It provides high yields of the desired acyl chlorides.

Application in Various Fields

The synthesis of acyl chlorides using the three methods described above finds extensive applications in various fields, including organic synthesis, pharmaceuticals, and materials science. Let's explore some examples of how acyl chlorides are used in these fields.

Organic Synthesis:

Acyl chlorides are versatile intermediates that can undergo numerous reactions to form various compounds. They can be used for the preparation of esters, amides, anhydrides, and ketones through nucleophilic substitution, addition, or condensation reactions. For example, acyl chlorides can react with alcohols to form esters. This reaction, known as the Fischer esterification, is a widely used method for the synthesis of esters in organic chemistry.

Acyl chlorides can also react with primary or secondary amines to form amides. This reaction, known as the Schotten-Baumann reaction, is commonly used for the synthesis of amides in both laboratory and industrial settings.

Pharmaceuticals:

Acyl chlorides play a crucial role in the synthesis of pharmaceutical compounds. They are often used as key intermediates for the introduction of functional groups or the modification of existing functional groups in drug molecules.

For example, acyl chlorides can be used to introduce an amide group into a drug molecule, thereby enhancing its stability and bioavailability. The amide bond is a common structural motif in many pharmaceutical compounds. Acyl chlorides can also be used to introduce an ester group into a drug molecule, which can influence its pharmacokinetic properties, such as solubility and metabolic stability.

Materials Science:

Acyl chlorides find applications in materials science for the synthesis of polymers and functional materials. For example, acyl chlorides can be used as monomers for the synthesis of polyesters or polyamides through polymerization reactions. These polymers have diverse properties and can be tailored for specific applications, such as fibers, films, or coatings. Acyl chlorides can also be used to functionalize surfaces or modify the properties of materials. For instance, they can react with hydroxyl groups present on the surface of materials to form covalent bonds, leading to the attachment of desired functional groups.

Counterarguments

While the methods described above are widely used for the synthesis of acyl chlorides, there are alternative methods available as well. For example, acyl chlorides can also be synthesized using oxalyl chloride (COCl)2 or thionyl chloride (SOCl2) in the presence of a tertiary amine as a catalyst. These methods provide an alternative route for the synthesis of acyl chlorides, especially in cases where PCl3 or PCl5 are not readily available. The choice of method depends on various factors, including the availability of reagents, reaction conditions, and desired yields.

Hydrolysis of Acyl Chlorides

Hydrolysis is a chemical reaction in which a compound reacts with water to produce new compounds. Acyl chlorides, also known as acid chlorides, are highly reactive compounds that readily undergo hydrolysis. This reaction involves the substitution of the chlorine atom in the acyl chloride with a hydroxyl group from water, resulting in the formation of a carboxylic acid and hydrochloric acid (HCl). In this article, we will discuss the hydrolysis of acyl chlorides in detail, including the mechanism, factors affecting the reaction, and the importance of this reaction in various fields.

Mechanism of Hydrolysis

The hydrolysis of acyl chlorides occurs through a nucleophilic substitution reaction, where a nucleophile attacks the electrophilic carbon atom of the acyl chloride, leading to the displacement of the chloride ion. Water acts as a nucleophile in this reaction and attacks the carbonyl carbon of the acyl chloride. The reaction proceeds in two steps: the initial attack of water and the subsequent elimination of chloride ion.

Initial Attack of Water

In the first step of the reaction, water, which acts as a nucleophile, attacks the carbonyl carbon of the acyl chloride. The oxygen atom of water carries a partial negative charge due to its higher electronegativity compared to hydrogen. This partial negative charge on the oxygen atom makes it a nucleophile, i.e., an electron-rich species that can donate its electrons to an electron-deficient species.

The carbonyl carbon of the acyl chloride, on the other hand, carries a partial positive charge due to the highly polarized nature of the C=O bond. This partial positive charge makes the carbonyl carbon an electrophile, i.e., an electron-deficient species that can accept electrons from a nucleophile. Hence, water, acting as a nucleophile, attacks the carbonyl carbon of the acyl chloride.

The attack of water on the carbonyl carbon results in the formation of a tetrahedral intermediate. In this intermediate, the chlorine atom is displaced by the hydroxyl group of water. This step is usually the rate-determining step of the reaction, as it involves the breaking of a relatively strong C-Cl bond.

Elimination of Chloride Ion

In the second step of the hydrolysis reaction, the chloride ion, which is now displaced from the acyl chloride, is eliminated. The elimination of the chloride ion occurs through the protonation of the oxygen atom in the tetrahedral intermediate. The transfer of a proton from the oxygen atom to a water molecule leads to the formation of the carboxylic acid and a hydronium ion (H3O+).

The carboxylic acid produced in this reaction is the result of the hydrolysis of the acyl chloride. The carboxyl group (-COOH) is made up of a carbonyl group (C=O) and a hydroxyl group (OH) bonded to the carbon atom. The hydrochloric acid (HCl) produced as a byproduct of the reaction is a strong acid that dissociates in water to produce hydrogen ions (H+) and chloride ions (Cl-).

Factors Affecting Hydrolysis

Several factors can influence the rate of hydrolysis of acyl chlorides. These factors include the nature of the acyl chloride, the concentration of water, the presence of catalysts, and the reaction temperature.

Nature of the Acyl Chloride

The nature of the acyl chloride plays a significant role in determining the rate of hydrolysis. Acyl chlorides with electron-withdrawing groups attached to the carbonyl carbon are more susceptible to hydrolysis. Electron-withdrawing groups, such as halogens (e.g., fluorine, chlorine), nitro groups (-NO2), and cyano groups (-CN), increase the electrophilic character of the carbonyl carbon, making it more prone to attack by nucleophiles like water.

On the other hand, acyl chlorides with electron-donating groups attached to the carbonyl carbon are less reactive toward hydrolysis. Electron-donating groups, such as alkyl groups (-R), amino groups (-NH2), and alkoxy groups (-OR), decrease the electrophilic character of the carbonyl carbon, making it less susceptible to nucleophilic attack.

Concentration of Water

The concentration of water in the reaction mixture also affects the rate of hydrolysis. An increase in the concentration of water leads to an increase in the rate of hydrolysis. This is because a higher concentration of water provides more nucleophiles for attacking the acyl chloride, resulting in a higher reaction rate.

Presence of Catalysts

Catalysts can significantly accelerate the hydrolysis of acyl chlorides. A catalyst is a substance that increases the rate of a chemical reaction without being consumed in the process. In the case of hydrolysis of acyl chlorides, catalysts such as Lewis acids, bases, or transition metal complexes can enhance the reaction rate.

Lewis acids, such as aluminum chloride ($AlCl_3$) or iron(III) chloride ($FeCl_3$), can coordinate with the carbonyl oxygen of the acyl chloride, making it more susceptible to nucleophilic attack. This coordination activates the carbonyl carbon and facilitates the hydrolysis reaction.

Bases, such as pyridine or triethylamine, can also act as catalysts by deprotonating the hydroxyl group of water, making it a better nucleophile. The deprotonation of water increases its nucleophilicity, allowing it to attack the acyl chloride more effectively.

Transition metal complexes, such as palladium(II) chloride ($PdCl_2$) or rhodium(III) chloride ($RhCl_3$), can catalyze the hydrolysis of acyl chlorides through coordination with the carbonyl oxygen. These metal complexes stabilize the carbonyl carbon, making it more susceptible to nucleophilic attack.

Reaction Temperature

The reaction temperature is another crucial factor that affects the rate of hydrolysis. Generally, an increase in temperature leads to an increase in the reaction rate. This is because higher temperatures provide more thermal energy to the reactant molecules, increasing their kinetic energy and collision frequency. As a result, more effective collisions occur between the acyl chloride and water molecules, leading to a higher rate of hydrolysis. However, extremely high temperatures can also lead to side reactions or decomposition of the reactants. Therefore, the reaction temperature should be optimized to achieve the desired rate of hydrolysis without causing unwanted side reactions.

Importance of Hydrolysis of Acyl Chlorides

The hydrolysis of acyl chlorides is a fundamental reaction in organic chemistry with significant importance in various fields. Some of the key applications and implications of this reaction are discussed below.

Synthesis of Carboxylic Acids

Hydrolysis of acyl chlorides is one of the most common methods for the synthesis of carboxylic acids. Carboxylic acids are versatile compounds used in various industries, including pharmaceuticals, polymers, and food additives. By selectively hydrolyzing specific acyl chlorides, chemists can access a wide range of carboxylic acids with diverse chemical properties.

For example, the hydrolysis of acetyl chloride (CH_3COCl) yields acetic acid (CH_3COOH), a widely used carboxylic acid in the production of vinegar, solvents, and pharmaceuticals. Similarly, the hydrolysis of benzoic acid (C_6H_5COCl) produces benzoic acid (C_6H_5COOH), a compound used as a food preservative and in the production of pharmaceuticals, dyes, and fragrances.

Formation of Acid Chloride Derivatives

The hydrolysis of acyl chlorides can also be used to synthesize acid chloride derivatives. Acid chlorides are reactive compounds that can undergo a variety of reactions to form new chemical compounds. By selectively hydrolyzing acyl chlorides and subsequently modifying the resulting carboxylic acid, chemists can introduce different functional groups or substituents into the molecule, expanding its chemical diversity. For instance, the hydrolysis of acetyl chloride, followed by the reaction with an alcohol in the presence of a base, leads to the formation of an ester. Esters are widely used in the fragrance and flavor industry, as well as in the production of plastics, solvents, and pharmaceuticals.

Similarly, the hydrolysis of acyl chlorides, followed by the reaction with an amine, results in the formation of an amide. Amides have diverse applications, ranging from pharmaceuticals and agrochemicals to polymers and dyes.

Importance in Pharmaceutical Chemistry

Hydrolysis of acyl chlorides is of significant importance in pharmaceutical chemistry. Many pharmaceutical drugs contain carboxylic acid functional groups, which can be introduced by the hydrolysis of acyl chlorides. The ability to selectively hydrolyze specific acyl chlorides allows chemists to synthesize a wide range of carboxylic acid derivatives, which can then be further modified to yield specific drugs.

For example, the synthesis of acetylsalicylic acid (aspirin) involves the hydrolysis of acetyl chloride to produce acetic acid, followed by the reaction with salicylic acid. Acetylsalicylic acid is a commonly utilized drug for alleviating pain, reducing fever, and combating inflammation.

Limitations and Challenges

While the hydrolysis of acyl chlorides offers numerous possibilities for the synthesis of carboxylic acids and their derivatives, there are certain limitations and challenges associated with this reaction.

One limitation is the reactivity of acyl chlorides. Acyl chlorides are highly reactive compounds that can undergo unwanted side reactions or decompose under certain conditions. Therefore, careful control of reaction conditions, such as temperature, pH, and reaction time, is crucial to ensure the desired hydrolysis reaction occurs.

Another challenge is the selectivity of the hydrolysis reaction. Acyl chlorides can have multiple reactive sites, leading to the formation of multiple products. Controlling the selectivity of the reaction requires the use of specific catalysts, controlling the reaction conditions, or employing protecting groups to selectively hydrolyze a particular acyl chloride.

Introduction

We will explore the mechanism of this reaction, its conditions, and its applications in various fields.

Reaction Mechanism

The reaction between an acyl chloride and an alcohol involves the substitution of the chlorine atom in the acyl chloride with the alkyl group of the alcohol. This substitution occurs through a nucleophilic acyl substitution mechanism. The reaction can be divided into three main steps: activation of the acyl chloride, nucleophilic attack of the alcohol, and proton transfer.

In the first step, the acyl chloride is activated by the addition of a Lewis base, such as pyridine or triethylamine. This Lewis base reacts with the acyl chloride to form a complex, stabilizing the reactive acyl chloride and facilitating the subsequent reaction.

In the second step, the activated acyl chloride undergoes nucleophilic attack by the alcohol. The lone pair of electrons on the oxygen atom of the alcohol attacks the electrophilic carbonyl carbon atom of the acyl chloride, leading to the formation of a tetrahedral intermediate. Simultaneously, the chloride ion is displaced from the acyl chloride, resulting in the formation of hydrogen chloride (HCl).

Finally, in the third step, a proton transfer occurs from the hydroxyl group of the alcohol to the chloride ion, resulting in the formation of the ester and regeneration of the Lewis base catalyst. This step is crucial for the completion of the reaction and the formation of the desired ester product.

Overall, the reaction between an acyl chloride and an alcohol can be represented by the following equation:

Acyl chloride + Alcohol → Ester + HCl

Reaction Conditions

The reaction between an acyl chloride and an alcohol to produce an ester and HCl typically occurs at room temperature. This is advantageous as it allows for a more convenient and energy-efficient process compared to reactions that require high temperatures. However, the reaction rate can be accelerated by heating, especially for less reactive acyl chlorides or alcohols with hindered nucleophilic centers.

The choice of reagents and catalysts is also important for the success of the reaction. Common catalysts used in this reaction include pyridine, triethylamine, and dimethylaminopyridine (DMAP). These catalysts not only activate the acyl chloride but also serve as proton acceptors during the final proton transfer step, facilitating the formation of the ester product. The choice of alcohol is also crucial, as primary alcohols are more reactive than secondary or tertiary alcohols due to the strength of their nucleophilic centers.

The reaction can be carried out in various solvents, including polar aprotic solvents such as dichloromethane, ethyl acetate, or tetrahydrofuran. These solvents help to dissolve both the acyl chloride and the alcohol, promoting their interaction and facilitating the reaction. Additionally, the reaction mixture is often stirred or refluxed to ensure thorough mixing and maximize the contact between the reactants.

Applications

The reaction of acyl chlorides with alcohols to form esters has numerous applications in various fields. One of the most common applications is in the synthesis of pharmaceuticals and natural products. Ester functional groups are present in many biologically active compounds, and the ability to selectively introduce ester groups using acyl chlorides allows for the efficient production of these compounds. For example, the synthesis of aspirin involves the reaction of acetyl chloride with salicylic acid to form acetylsalicylic acid, the active ingredient in aspirin.

This reaction also finds applications in the fragrance and flavor industry. Many natural fragrances and essential oils contain ester functional groups, which contribute to their pleasant aromas. By using acyl chlorides and alcohols with specific structures, chemists can create synthetic esters that mimic the natural fragrances and flavors of fruits, flowers, and other sources. This is particularly useful in the production of perfumes, cosmetics, and food additives.

In addition to the synthesis of specific compounds, the reaction between acyl chlorides and alcohols can also be used for functional group transformations. For example, the reaction can be employed to convert carboxylic acids into their corresponding esters. By first converting the carboxylic acid into an acyl chloride, followed by reaction with an alcohol, the desired ester can be obtained. This transformation is valuable for the modification of natural products or the preparation of compounds with improved properties, such as increased stability or bioavailability.

Counterarguments and Critiques

While the reaction of acyl chlorides with alcohols to produce esters is a well-established and widely used method, there are certain limitations and challenges associated with it. One limitation is the reactivity of specific acyl chlorides. Some acyl chlorides may be less reactive due to the presence of electron-withdrawing groups or steric hindrance around the carbonyl carbon. In such cases, additional reagents or modifications to the reaction conditions may be necessary to achieve the desired ester product. Another challenge is the potential for side reactions or competing reactions. For example, the alcohol used in the reaction may act as a nucleophile towards other functional groups present in the acyl chloride or in the reaction mixture. This can lead to the formation of undesired byproducts or the consumption of reagents without the desired ester product being formed. Careful selection of reactants, catalysts, and reaction conditions can help minimize these side reactions and improve the selectivity of the desired ester formation.

Introduction

We will explore the mechanism and factors affecting this reaction, as well as its applications in various fields.

Reaction Mechanism

The reaction between acyl chlorides and phenols involves the substitution of the chlorine atom in the acyl chloride with the phenolic hydroxyl group. This reaction is generally known as an acylation reaction, where the acyl group of the acyl chloride is transferred to the phenol, resulting in the formation of an ester. The reaction can be represented by the following general equation:

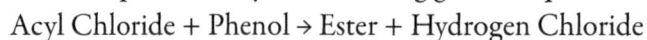

Acyl Chloride + Phenol → Ester + Hydrogen Chloride

To understand the reaction mechanism in more detail, let's consider the reaction between acetyl chloride (CH_3COCl) and phenol (C_6H_5OH) as an example:

$$CH_3COCl + C_6H_5OH \rightarrow CH_3COOC_6H_5 + HCl$$

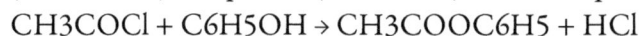

This reaction proceeds in two steps: nucleophilic attack and elimination.

Nucleophilic Attack: In the first step, the lone pair of electrons on the oxygen atom of the phenol attacks the electrophilic carbon atom of the acyl chloride, resulting in the formation of a tetrahedral intermediate. This process is facilitated by the Lewis acid-base interaction between the electron-deficient carbon atom and the electron-rich oxygen atom.

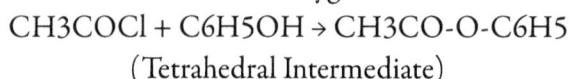

$$CH_3COCl + C_6H_5OH \rightarrow CH_3CO\text{-}O\text{-}C_6H_5$$

(Tetrahedral Intermediate)

Elimination: In the second step, the tetrahedral intermediate formed in the previous step undergoes elimination of the chloride ion (Cl^-) to regenerate the aromaticity of the phenol ring. This elimination process leads to the formation of the ester and hydrogen chloride.

$$CH_3CO\text{-}O\text{-}C_6H_5 \rightarrow CH_3COOC_6H_5 + HCl$$

The overall reaction can be seen as the replacement of the chlorine atom in the acyl chloride with the phenolic hydroxyl group, resulting in the formation of an ester. The hydrogen chloride is released as a byproduct of the reaction.

Factors Affecting the Reaction

Several factors can influence the reaction between acyl chlorides and phenols. These factors include the nature of the acyl chloride and phenol, reaction conditions, and the presence of catalysts.

Nature of Acyl Chloride: The reactivity of acyl chlorides depends on the nature of the acyl group attached to the carbonyl carbon. Electron-withdrawing groups attached to the acyl group increase the electrophilicity of the carbonyl carbon, making the reaction more favorable. For example, acyl chlorides with halogen substituents, such as chloroacetyl chloride ($ClCH_2COCl$), are more reactive than simple acetyl chloride (CH_3COCl).

Nature of Phenol: The reactivity of phenols depends on the presence of electron-donating or electron-withdrawing groups on the aromatic ring. Electron-donating groups, such as alkyl or methoxy groups, increase the nucleophilicity of the phenol and enhance the reaction rate. Conversely, electron-withdrawing groups, such as nitro or cyano groups, decrease the nucleophilicity of the phenol and slow down the reaction.

Reaction Conditions: The reaction between acyl chlorides and phenols generally occurs at room temperature or slightly elevated temperatures. Higher temperatures may increase the reaction rate but can also lead to side reactions or decomposition of the reactants. The reaction is typically carried out in an inert solvent, such as dichloromethane or benzene, to facilitate mixing and provide a medium for the reaction.

Catalysts: The reaction between acyl chlorides and phenols can be catalyzed by various acids or base catalysts. Acid catalysts, such as sulfuric acid or hydrochloric acid, can enhance the electrophilicity of the carbonyl carbon and promote the reaction. Base catalysts, such as pyridine or triethylamine, can facilitate the deprotonation of the phenol and increase its nucleophilicity. The choice of catalyst depends on the specific reaction conditions and desired reaction rate.

Applications

The reaction between acyl chlorides and phenols to produce esters and hydrogen chloride has numerous applications in various fields. Some of the notable applications are discussed below:

Organic Synthesis: The acylation reaction between acyl chlorides and phenols is a fundamental transformation in organic synthesis. It provides a versatile method for the introduction of acyl groups into aromatic compounds, allowing the synthesis of a wide range of esters. These esters can serve as important building blocks for the synthesis of pharmaceuticals, agrochemicals, and other fine chemicals.

Polymer Chemistry: The reaction between acyl chlorides and phenols is also used in the synthesis of polyesters. By using bifunctional phenols and acyl chlorides, it is possible to form ester linkages between the phenolic hydroxyl groups, resulting in the formation of linear or cross-linked polyester polymers. These polyester polymers find applications in various industries, including coatings, adhesives, and textiles.

Drug Discovery: The acylation reaction between acyl chlorides and phenols has been utilized in drug discovery programs to modify the properties of pharmacologically active compounds. By introducing an acyl group onto a phenolic moiety, the lipophilicity, solubility, and stability of the compound can be altered, leading to improved drug-like properties. This strategy has been successfully employed in the development of several drugs, including anti-inflammatory agents and antiviral drugs.

Natural Product Synthesis: The acylation reaction between acyl chlorides and phenols has found applications in the synthesis of natural products. Many natural products contain ester functionalities that can be formed through acylation reactions. By using acyl chlorides and phenols as starting materials, chemists can access complex natural products through a series of selective acylation reactions. This approach has been instrumental in the total synthesis of various natural products, including polyketides, flavonoids, and alkaloids.

Counterarguments

While the reaction between acyl chlorides and phenols is a valuable tool in organic synthesis, it is not without limitations. Some of the potential drawbacks and counterarguments to consider include:

Side Reactions: The reaction between acyl chlorides and phenols can sometimes lead to side reactions, such as Friedel-Crafts acylation or O-acylation of the phenol. These side reactions can result in the formation of undesired byproducts and reduce the overall yield of the desired ester. Careful reaction optimization and selection of appropriate reaction conditions are necessary to minimize these side reactions.

Reactivity and Selectivity: The reactivity of acyl chlorides can vary depending on the nature of the acyl group. Highly reactive acyl chlorides may undergo undesired side reactions or react with other nucleophiles present in the reaction mixture. Additionally, the selectivity of the reaction may be influenced by the presence of multiple hydroxyl groups in the phenol or steric hindrance around the reaction site. These factors can complicate the reaction and require careful control of reaction conditions.

Safety Considerations: Acyl chlorides are highly reactive and can release toxic hydrogen chloride gas during the reaction. Proper safety precautions, such as working in a well-ventilated area and using appropriate personal protective equipment, should be followed to ensure the safety of the researcher. Additionally, acyl chlorides should be handled and stored with caution due to their corrosive nature.

Introduction

We will explore the reaction of acyl chlorides with ammonia at room temperature, focusing on the mechanism, factors influencing the reaction, and potential applications.

Reaction Mechanism

The reaction between an acyl chloride and ammonia involves a nucleophilic substitution reaction, where the ammonia molecule acts as a nucleophile attacking the acyl chloride. This reaction proceeds in several steps:

Nucleophilic Attack: The ammonia molecule, which has a lone pair of electrons on the nitrogen atom, attacks the carbonyl carbon of the acyl chloride. This attack results in the formation of a tetrahedral intermediate, with the chlorine atom of the acyl chloride becoming a leaving group.

Chloride Ion Formation: The tetrahedral intermediate is highly unstable due to the partial positive charge on the carbonyl carbon and the repulsion between the negatively charged chloride ion and the leaving group. As a result, the tetrahedral intermediate collapses, leading to the formation of a chloride ion.

Amide Formation: The chloride ion then reacts with another molecule of ammonia, resulting in the formation of an ammonium chloride salt. This salt can be easily separated from the desired product, which is the amide. The amide is formed when the remaining ammonia molecule attacks the carbonyl carbon, displacing the chloride ion.

The overall reaction can be represented as follows:

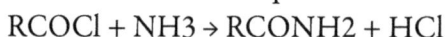

$$RCOCl + NH_3 \rightarrow RCONH_2 + HCl$$

where R represents an organic group attached to the carbonyl carbon.

Factors Influencing the Reaction

Several factors can influence the reaction between acyl chlorides and ammonia. Understanding these factors is crucial for optimizing the reaction conditions and obtaining the desired product. Some of the key factors include:

Steric Hindrance: The presence of bulky substituents on the acyl chloride can hinder the nucleophilic attack by ammonia. This is because the steric hindrance can prevent the ammonia molecule from approaching the carbonyl carbon. Therefore, acyl chlorides with less steric hindrance tend to react more efficiently with ammonia.

Electron-Withdrawing Groups: The presence of electron-withdrawing groups on the acyl chloride can increase the reactivity of the carbonyl carbon. These groups, such as halogens or nitro groups, can enhance the polarization of the carbonyl bond, making it more susceptible to nucleophilic attack. Consequently, acyl chlorides with electron-withdrawing groups generally react more rapidly with ammonia.

Temperature: The reaction between acyl chlorides and ammonia can be influenced by the reaction temperature. At room temperature, the reaction proceeds relatively slowly, requiring a longer reaction time for completion. However, higher temperatures can accelerate the reaction, allowing for faster conversion of acyl chlorides to amides. It is important to note that excessively high temperatures can also lead to side reactions or decomposition of the starting materials.

Solvent Choice: The choice of solvent can significantly impact the reaction between acyl chlorides and ammonia. Polar solvents, such as dichloromethane or tetrahydrofuran, are commonly used as reaction media. These solvents facilitate the dissolution of both the acyl chloride and ammonia, promoting their interaction. Additionally, the solvent can affect the solubility of the products, making it easier to separate the desired amide from any byproducts or unreacted starting materials.

Applications and Examples

The reaction of acyl chlorides with ammonia to produce amides has widespread applications in various fields, including pharmaceuticals, polymers, and materials science. Some notable examples include:

Pharmaceutical Industry: Amides are an essential class of compounds in the field of drug discovery and development. They serve as building blocks for numerous drugs, including antibiotics, analgesics, and anticonvulsants. The reaction between acyl chlorides and ammonia provides a straightforward route for the synthesis of amides, enabling the production of a wide range of pharmaceutical compounds.

Polymer Chemistry: Amides are commonly found in polymers, where they contribute to the mechanical properties and stability of the materials. The reaction of acyl chlorides with ammonia can be used to introduce amide linkages in polymer chains, allowing for the synthesis of polyamides. Polyamides, such as nylon, exhibit excellent tensile strength, durability, and resistance to wear, making them suitable for applications in textiles, engineering plastics, and composites.

Catalysis: The reaction between acyl chlorides and ammonia can also be utilized in catalytic processes. For example, palladium-catalyzed aminocarbonylation reactions involve the coupling of acyl chlorides with amines and carbon monoxide to form amides. This catalytic approach offers a more sustainable and atom-economical alternative to traditional methods, reducing the need for stoichiometric amounts of reagents and minimizing waste generation.

While the reaction of acyl chlorides with ammonia to produce amides is widely used, it is important to acknowledge some potential limitations and challenges. One such limitation is the selectivity of the reaction. In some cases, competing side reactions may occur, leading to the formation of undesired byproducts. Additionally, the use of ammonia as a nucleophile can result in the formation of mixtures of primary, secondary, and tertiary amides, depending on the nature of the acyl chloride and reaction conditions. Achieving high selectivity and controlling the reaction outcome often requires careful optimization of reaction parameters and the use of appropriate catalysts.

Reactions of Acyl Chlorides with Amines: Formation of Amides and HCl

Introduction

We will explore the mechanism and various aspects of the reaction between acyl chlorides and amines, focusing on the formation of amides and HCl. We will delve into the key concepts and theories behind this reaction, providing clear explanations and utilizing examples from different fields to enhance our understanding. Additionally, we will consider counterarguments and alternative perspectives to present a comprehensive analysis of the subject matter.

Reaction Mechanism

The reaction between an acyl chloride and a primary or secondary amine occurs through a nucleophilic substitution mechanism, with the amine acting as the nucleophile. This mechanism involves several steps, including nucleophilic attack, formation of a tetrahedral intermediate, and subsequent elimination of HCl to yield the amide product.

Nucleophilic Attack: The reaction begins with the nucleophilic attack of the amine on the electrophilic carbonyl carbon of the acyl chloride. The amine donates a pair of electrons to the carbon, breaking the carbon-oxygen double bond and forming a tetrahedral intermediate. This step is facilitated by the lone pair of electrons on the nitrogen atom of the amine, which acts as the nucleophile.

Tetrahedral Intermediate: The nucleophilic attack results in the formation of a tetrahedral intermediate, where the chlorine atom from the acyl chloride is replaced by the nitrogen atom of the amine. This intermediate is stabilized by various factors, such as resonance and inductive effects. The resonance stabilization arises from the delocalization of the electrons in the carbonyl group, while the inductive effects arise from the electronegativity difference between the chlorine and nitrogen atoms.

Elimination of HCl: In the final step of the reaction, elimination of HCl occurs, leading to the formation of the amide product. This elimination is facilitated by the presence of a strong base or acid, which can abstract the proton from the amine, generating a chloride ion. The chloride ion then combines with the hydrogen atom from the adjacent carbon to form HCl, which is released as a byproduct.

The overall reaction can be represented by the following equation:

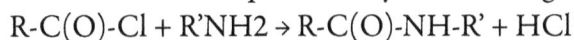

$$R\text{-}C(O)\text{-}Cl + R'NH_2 \rightarrow R\text{-}C(O)\text{-}NH\text{-}R' + HCl$$

where R and R' represent different organic groups.

Factors Affecting the Reaction

Several factors influence the reaction between acyl chlorides and amines, impacting the rate and selectivity of the reaction. Understanding these factors is crucial in optimizing the reaction conditions and obtaining the desired products.

Steric Hindrance: The presence of bulky groups around the carbonyl carbon of the acyl chloride can hinder the nucleophilic attack of the amine, slowing down the reaction. This steric hindrance arises from the spatial arrangement of the substituents around the carbon atom, which can restrict the approach of the amine and reduce the reactivity.

Electronic Effects: The electronic nature of the substituents attached to the carbonyl carbon also influences the reaction. Electron-withdrawing groups, such as halogens or nitro groups, increase the electrophilicity of the carbon, enhancing the rate of the reaction. Conversely, electron-donating groups, such as alkyl or aryl groups, decrease the electrophilicity and may require more vigorous reaction conditions.

Solvent Choice: The choice of solvent plays a vital role in the reaction between acyl chlorides and amines. Polar solvents, such as dichloromethane or acetonitrile, are commonly used to dissolve both the acyl chloride and the amine, facilitating their interaction. These solvents also provide stability to the resulting tetrahedral intermediate through solvation effects.

Temperature: The reaction between acyl chlorides and amines typically proceeds at room temperature, providing a balance between reactivity and control. Higher temperatures can increase the reaction rate but may also promote side reactions or decomposition of the products. Lower temperatures, on the other hand, can slow down the reaction or even inhibit it.

Importance and Applications

The formation of amides from acyl chlorides and amines is a fundamental reaction in organic synthesis, finding widespread applications in various fields. The versatility of amides as chemical intermediates contributes to their significance in the pharmaceutical, polymer, and agricultural industries.

Pharmaceuticals: Amides are essential components in the synthesis of many pharmaceuticals. They serve as building blocks for drug molecules, providing structural stability and influencing their biological activity. For example, the well-known painkiller acetaminophen contains an amide functional group.

Polymers: Amides are key constituents in the production of polyamides, which are widely used as engineering plastics and fibers. The reaction between acyl chlorides and amines is a crucial step in the synthesis of these polymeric materials, allowing the formation of strong and durable structures.

Agrochemicals: Many agrochemicals, such as herbicides and insecticides, contain amide moieties. The reaction between acyl chlorides and amines enables the synthesis of these compounds, which play a vital role in modern agriculture by protecting crops and controlling pests.

Bioconjugation: Amides are also employed in bioconjugation reactions, which involve attaching biomolecules, such as proteins or peptides, to other compounds. This enables the creation of conjugates with enhanced properties, such as improved stability or targeted drug delivery.

Counterarguments and Alternative Perspectives

While the reaction between acyl chlorides and amines is widely utilized and well-established, alternative methods for amide synthesis have also been developed. One such alternative is the reaction between carboxylic acids and amines, which proceeds under milder conditions and avoids the use of toxic and corrosive acyl chlorides. This method, known as the amidation reaction, involves the activation of the carboxylic acid through the use of coupling agents or catalysts.

The choice between using acyl chlorides or carboxylic acids depends on several factors, such as the desired reaction conditions, the availability of starting materials, and the specific requirements of the synthesis. Acyl chlorides offer high reactivity and are well-suited for reactions requiring fast and efficient amide formation. On the other hand, carboxylic acids provide a more sustainable and environmentally friendly approach, making them attractive in certain applications.

We will explore the addition-elimination mechanism in the context of hydrolysis, esterification with alcohol and phenol, and amide formation with ammonia and primary or secondary amines.

Hydrolysis is the process of breaking down a compound by the addition of water. When an acyl chloride is subjected to hydrolysis, water acts as a nucleophile and attacks the carbonyl carbon of the acyl chloride. This nucleophilic attack results in the formation of a tetrahedral intermediate, where the carbonyl oxygen is bonded to the water molecule. Subsequently, a chloride ion is eliminated as a leaving group, regenerating the carbonyl functionality. The resulting compound is a carboxylic acid. For example, let's consider the hydrolysis of acetyl chloride (CH_3COCl):

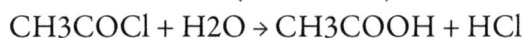

$$CH_3COCl + H_2O \rightarrow CH_3COOH + HCl$$

In this reaction, water acts as a nucleophile and attacks the carbonyl carbon of acetyl chloride. The chloride ion is then eliminated as a leaving group, and acetic acid is formed.

In addition to hydrolysis, acyl chlorides can also undergo esterification reactions with alcohols and phenols. In these reactions, the alcohol or phenol acts as a nucleophile and attacks the carbonyl carbon of the acyl chloride. Similar to hydrolysis, a tetrahedral intermediate is formed, followed by the elimination of a chloride ion as a leaving group.

The resulting compound is an ester.

Let's consider the reaction between acetyl chloride and methanol:

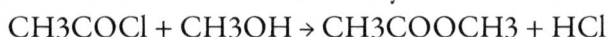

$$CH_3COCl + CH_3OH \rightarrow CH_3COOCH_3 + HCl$$

In this reaction, methanol acts as a nucleophile and attacks the carbonyl carbon of acetyl chloride. The chloride ion is then eliminated, and methyl acetate is formed.

Similarly, when acyl chlorides react with phenols, esters are formed. The phenol molecule acts as a nucleophile, attacking the carbonyl carbon of the acyl chloride. Again, a tetrahedral intermediate is formed, followed by the elimination of a chloride ion as a leaving group. For example, let's consider the reaction between acetyl chloride and phenol:

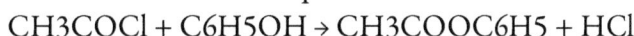

$$CH_3COCl + C_6H_5OH \rightarrow CH_3COOC_6H_5 + HCl$$

In this reaction, phenol acts as a nucleophile and attacks the carbonyl carbon of acetyl chloride. The chloride ion is then eliminated, and phenyl acetate is formed.

In addition to esterification, acyl chlorides can also undergo amide formation reactions with ammonia and primary or secondary amines. In these reactions, the ammonia or amine molecule acts as a nucleophile and attacks the carbonyl carbon of the acyl chloride. As before, a tetrahedral intermediate is formed, followed by the elimination of a chloride ion as a leaving group. The resulting compound is an amide.

Let's consider the reaction between acetyl chloride and ammonia:

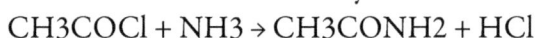

$$CH_3COCl + NH_3 \rightarrow CH_3CONH_2 + HCl$$

In this reaction, ammonia acts as a nucleophile and attacks the carbonyl carbon of acetyl chloride. The chloride ion is then eliminated, and acetamide is formed.

Similarly, when acyl chlorides react with primary or secondary amines, amides are formed. The amine molecule acts as a nucleophile, attacking the carbonyl carbon of the acyl chloride. Again, a tetrahedral intermediate is formed, followed by the elimination of a chloride ion as a leaving group.

For example, let's consider the reaction between acetyl chloride and ethylamine:

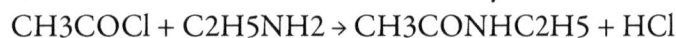

$$CH_3COCl + C_2H_5NH_2 \rightarrow CH_3CONHC_2H_5 + HCl$$

In this reaction, ethylamine acts as a nucleophile and attacks the carbonyl carbon of acetyl chloride. The chloride ion is then eliminated, and N-ethylacetamide is formed.

Overall, the addition-elimination mechanism plays a crucial role in the reactions of acyl chlorides. Through nucleophilic attack, followed by the elimination of a leaving group, various compounds such as carboxylic acids, esters, and amides can be formed. These reactions have significant applications in organic synthesis, pharmaceuticals, and the production of polymers.

It is worth mentioning that the addition-elimination mechanism is not the only mechanism observed in acyl chloride reactions. In some cases, other mechanisms such as nucleophilic acyl substitution or addition-addition can also occur. However, the addition-elimination mechanism is the most common and well-studied mechanism in the context of acyl chloride reactions.

The Relative Ease of Hydrolysis of Acyl Chlorides, Alkyl Chlorides, and Halogenoarenes

Introduction

Hydrolysis is a fundamental chemical reaction in which a compound reacts with water, resulting in the breaking of chemical bonds and the formation of new compounds. The ease of hydrolysis of different compounds can vary significantly depending on their chemical structure and functional groups. In this analysis, we will examine the hydrolysis of acyl chlorides, alkyl chlorides, and halogenoarenes varies in terms of relative ease. These compounds belong to different classes of organic compounds and exhibit distinct reactivity patterns towards hydrolysis. By understanding the factors that influence the hydrolysis rates of these compounds, we can gain insights into their chemical behavior and potential applications.

Acyl Chlorides

Acyl chlorides, also known as acid chlorides, are organic compounds with the general formula RCOCl, where R represents an alkyl or aryl group. They are highly reactive compounds due to the presence of the electrophilic carbonyl carbon and the leaving group chlorine. The formation of a tetrahedral intermediate occurs during the hydrolysis of acyl chlorides when water attacks the carbonyl carbon through nucleophilic attack. This intermediate subsequently collapses, resulting in the formation of a carboxylic acid and hydrogen chloride (HCl). The overall reaction can be represented as follows:

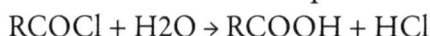

$$RCOCl + H_2O \rightarrow RCOOH + HCl$$

The hydrolysis of acyl chlorides is generally rapid and exothermic, making it a useful reaction in various chemical processes. The reactivity of acyl chlorides towards hydrolysis can be attributed to several factors, including the nature of the acyl group and the stability of the resulting carboxylic acid.

One key factor that influences the hydrolysis rate is the electron-withdrawing effect of the acyl group. Electron-withdrawing groups, such as halogens or carbonyl groups, increase the electrophilicity of the carbonyl carbon, facilitating the attack by the nucleophilic water molecule. For example, acetyl chloride (CH3COCl) hydrolyzes rapidly due to the strong electron-withdrawing effect of the carbonyl group. On the other hand, acyl chlorides with electron-donating groups, such as alkyl groups, are less reactive towards hydrolysis. This is because the electron-donating groups decrease the electrophilicity of the carbonyl carbon, making it less susceptible to nucleophilic attack.

Furthermore, the stability of the resulting carboxylic acid also influences the hydrolysis rate of acyl chlorides. Carboxylic acids with resonance-stabilized structures, such as benzoic acid, are more stable and thus hydrolyze at a slower rate compared to acyl chlorides that form less stable carboxylic acids. This stability arises from the delocalization of the negative charge through the conjugated system of the aromatic ring, which reduces the reactivity of the carboxylic acid towards hydrolysis.

Alkyl Chlorides

Alkyl chlorides, also known as haloalkanes, are organic compounds that contain a halogen atom (usually chlorine) bonded to an alkyl group. The general formula for alkyl chlorides is R-Cl, where R represents an alkyl group. Compared to acyl chlorides, alkyl chlorides are less reactive towards hydrolysis. This reduced reactivity can be attributed to the weaker electrophilicity of the carbon-halogen bond as compared to the carbonyl carbon in acyl chlorides.

The hydrolysis of alkyl chlorides can occur through two different mechanisms: nucleophilic substitution and elimination. Nucleophilic substitution involves the attack of a nucleophile, such as water, on the carbon bonded to the halogen atom, resulting in the displacement of the halogen and the formation of an alcohol. On the other hand, elimination reactions involve the removal of the halogen and a proton from an adjacent carbon, leading to the formation of an alkene. The outcome of the hydrolysis reaction depends on the reaction conditions, including the presence of a nucleophile or a base.

The reactivity of alkyl chlorides towards hydrolysis depends on various factors, such as the nature of the alkyl group, the type of halogen atom, and the reaction conditions. In general, primary alkyl chlorides exhibit higher reactivity compared to secondary alkyl chlorides, and secondary alkyl chlorides are more reactive than tertiary alkyl chlorides. This trend can be explained by the increasing stability of the carbocation intermediate formed during the hydrolysis reaction. Primary alkyl chlorides readily form a primary carbocation, which is relatively unstable and therefore reacts more rapidly with water. In contrast, tertiary alkyl chlorides form a more stable tertiary carbocation, leading to slower hydrolysis rates.

The type of halogen atom in alkyl chlorides also affects the hydrolysis rate. Alkyl chlorides are generally more reactive than alkyl bromides or alkyl iodides towards hydrolysis. This is because chlorine is a better leaving group than bromine or iodine. The weaker bond between carbon and chlorine allows for easier nucleophilic attack and subsequent substitution or elimination reactions.

Halogenoarenes (Aryl Chlorides)

Halogenoarenes, commonly referred to as aryl halides, are organic compounds in which a halogen atom is directly bonded to an aromatic ring. In the case of aryl chlorides, the halogen atom is chlorine. Aryl chlorides exhibit significantly lower reactivity towards hydrolysis compared to both acyl chlorides and alkyl chlorides. This reduced reactivity can be attributed to the stability of the aromatic ring and the weaker electrophilicity of the carbon-halogen bond.

The hydrolysis of aryl chlorides can occur through nucleophilic aromatic substitution (SNAr) reactions. In these reactions, a nucleophile attacks the carbon atom of the aromatic ring, leading to the substitution of the chlorine atom. However, the electron-rich nature of the aromatic ring, due to its conjugated π system, makes it less susceptible to nucleophilic attack. This is because the electron density of the ring is already high, making it difficult for a nucleophile to approach and attack the carbon atom.

Additionally, the carbon-chlorine bond in aryl chlorides is less polarized compared to alkyl chlorides or acyl chlorides, rendering it less electrophilic. As a result, the rate of hydrolysis of aryl chlorides is significantly slower compared to other chlorinated organic compounds.

Nitrogen Compounds

Primary and secondary amines

Reaction of Halogenoalkanes with NH3 in Ethanol

The reaction of halogenoalkanes with NH3 in ethanol heated under pressure is a nucleophilic substitution reaction. It involves the substitution of the halogen atom in the halogenoalkane with an amino group (-NH2), resulting in the formation of a primary or secondary amine, depending on the structure of the halogenoalkane.

Primary Amine Formation

When a halogenoalkane reacts with NH3, the nucleophilic substitution occurs, leading to the formation of a primary amine. The general reaction can be represented as follows:

$$R\text{-}X + NH3 \rightarrow R\text{-}NH2 + HX$$

Here, R represents the alkyl group attached to the halogen atom, X represents the halogen atom, and HX represents the hydrogen halide formed as a byproduct. The reaction is typically carried out in a closed vessel under pressure to ensure the progress of the reaction and to increase the yield of the desired primary amine.

Secondary Amine Formation

In the synthesis of secondary amines, the reaction of halogenoalkanes with NH3 follows a two-step process. The initial nucleophilic substitution reaction produces a primary amine, followed by a subsequent reaction of the primary amine with another halogenoalkane molecule, resulting in the formation of a secondary amine. The overall reaction can be represented as follows:

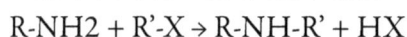

$$R\text{-}X + NH3 \rightarrow R\text{-}NH2 + HX$$

$$R\text{-}NH2 + R'\text{-}X \rightarrow R\text{-}NH\text{-}R' + HX$$

In this series of reactions, R represents an alkyl group attached to the halogen atom in the first step, X represents the halogen atom, R' represents a different alkyl group in the second step, and HX represents the hydrogen halide formed as a byproduct. The reaction conditions, including temperature, pressure, and the molar ratio of reactants, play a crucial role in determining the yield and selectivity of the secondary amine formation.

Reagents and Conditions

Halogenoalkanes

Halogenoalkanes are the key starting materials in this reaction. They are typically organic compounds containing a halogen atom, such as chlorine (Cl), bromine (Br), or iodine (I), attached to a carbon atom in an alkyl group. The choice of halogenoalkane depends on the desired primary or secondary amine product. Different halogenoalkanes have varying reactivity, and the nature of the halogen atom affects the rate of the nucleophilic substitution reaction.

NH3 (Ammonia)

Ammonia (NH3) is a colorless gas that acts as the nucleophile in the reaction. It donates a lone pair of electrons to the carbon atom bonded to the halogen atom, facilitating the displacement of the halogen atom and the formation of a new bond with the carbon atom. The reaction between NH3 and the halogenoalkane occurs due to the high electronegativity of the halogen atom, making it susceptible to nucleophilic attack.

Ethanol

Ethanol (C2H5OH), also known as alcohol, serves as the solvent in this reaction. It helps in dissolving both the halogenoalkane and NH3, facilitating their reactions. Ethanol also acts as a nucleophile, enhancing the nucleophilic substitution process by stabilizing the transition state and solvating the reactants and intermediates.

Heating under Pressure

The reaction of halogenoalkanes with NH3 in ethanol is typically performed under elevated temperatures and pressure. Heating the reaction mixture increases the rate of reaction by providing the necessary energy to break the existing bonds and form new ones. The pressure, maintained within a closed vessel, promotes the equilibrium shift towards the desired product, resulting in higher yields of primary and secondary amines.

Examples and Applications

The reaction of halogenoalkanes with NH3 in ethanol heated under pressure has several practical applications across various fields.

Pharmaceuticals

Primary and secondary amines are vital building blocks in the synthesis of pharmaceutical compounds. They serve as precursors for the production of drugs, such as antibiotics, antihistamines, and anti-cancer agents. The ability to selectively produce specific primary or secondary amines through this reaction provides a valuable tool in medicinal chemistry for designing and synthesizing new therapeutic agents.

Agrochemicals

Amines find extensive use in the development of agrochemicals, including herbicides, insecticides, and fungicides. The reaction of halogenoalkanes with NH3 enables the synthesis of amine-based active ingredients that exhibit potent biological activities against pests, weeds, and diseases. The versatility of this reaction allows for the modification of amine structures, leading to the development of more effective and environmentally friendly agrochemicals.

Materials Science

Amines play a crucial role in materials science, particularly in the production of polymers and resins. Primary and secondary amines can be incorporated into polymer chains, enhancing their properties, such as strength, flexibility, and adhesion. The reaction of halogenoalkanes with NH3 provides a means to introduce amino groups into the polymer backbone, leading to the synthesis of amine-modified polymers with tailored functionalities.

Counterarguments

While the reaction of halogenoalkanes with NH3 in ethanol heated under pressure offers significant advantages in the synthesis of primary and secondary amines, there are some limitations and potential drawbacks to consider.

Competing Reactions

In some cases, competing reactions may occur, leading to the formation of undesired byproducts. For example, elimination reactions, such as the E2 mechanism, could compete with the nucleophilic substitution reaction, resulting in the formation of alkenes or other elimination products. The choice of reaction conditions, including temperature and concentration, must be carefully controlled to minimize such side reactions and maximize the yield of the desired amines.

Steric Hindrance

In reactions involving secondary amines, steric hindrance can pose a challenge. The bulkiness of the alkyl groups attached to the nitrogen atom can hinder the approach of the halogenoalkane in the second step of the reaction, reducing the reaction rate and selectivity. It is crucial to consider the steric effects and choose appropriate halogenoalkanes and reaction conditions to overcome this limitation.

Reactions for the Production of Primary and Secondary Amines

Reaction Mechanism

The reaction between halogenoalkanes and primary amines involves a nucleophilic substitution reaction. Nucleophilic substitution reactions occur when a nucleophile replaces a leaving group in an organic compound. In the case of the reaction between halogenoalkanes and primary amines, the nucleophile is the primary amine, and the leaving group is the halogen atom.

The general reaction can be represented as follows:

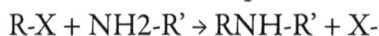

R-X + NH2-R' → RNH-R' + X-

Where R represents an alkyl group, X represents a halogen atom, and R' represents an alkyl or aryl group.

The reaction proceeds through a series of steps, which can be summarized as follows:

Nucleophilic Attack: The lone pair of electrons on the nitrogen atom in the primary amine attacks the carbon atom of the halogenoalkane, forming a new bond.

Formation of a Transition State: As the nucleophilic attack occurs, the bond between the halogen atom and the carbon atom weakens, forming a transition state.

Formation of a Tetrahedral Intermediate: The transition state collapses, resulting in the formation of a tetrahedral intermediate. In this intermediate, the nitrogen atom is bonded to the carbon atom, and the halogen atom is attached to the nucleophile.

Elimination of the Leaving Group: The leaving group, which is the halogen atom, is eliminated, resulting in the formation of the primary or secondary amine.

It is important to note that the reaction can proceed further to form secondary amines if excess primary amine is present. In this case, the primary amine acts as both a nucleophile and a base, leading to the formation of a secondary amine.

Reagents and Conditions

The reaction of halogenoalkanes with primary amines in ethanol, heated in a sealed tube or under pressure, requires specific reagents and conditions. Let's explore them in detail:

Halogenoalkane: The reaction begins with the use of a halogenoalkane as the starting material. Halogenoalkanes are organic compounds that contain a halogen atom (such as chlorine, bromine, or iodine) bonded to an alkyl group. The choice of halogenoalkane depends on the desired primary or secondary amine product.

Primary Amine: The primary amine acts as the nucleophile in the reaction. It is an organic compound that contains an amino group (-NH2) bonded to a carbon atom. The primary amine reacts with the halogenoalkane to replace the halogen atom, resulting in the formation of the primary or secondary amine.

Ethanol: Ethanol is used as the solvent in which the reaction takes place. It provides a suitable medium for the reaction and helps in dissolving the reactants. Ethanol also acts as a source of hydrogen atoms, which can facilitate the reduction of any intermediate carbocation formed during the reaction.

Heating in a Sealed Tube/Under Pressure: The reaction is heated in a sealed tube or under pressure to increase the reaction rate and promote the formation of the desired products. Heating the reaction mixture provides the necessary activation energy for the reaction to occur. The sealed tube or pressure prevents the escape of volatile reactants and products, allowing for a controlled reaction environment.

Reaction Examples

To better understand the reaction of halogenoalkanes with primary amines, let's consider a few examples:

Example 1: Reaction of Chloroethane with Ethylamine

In this example, we will explore the reaction between chloroethane and ethylamine to produce ethylamine.

CH3CH2Cl + CH3CH2NH2 → CH3CH2NHCH2CH3

+ HCl

The reaction proceeds as follows:

The lone pair of electrons on the nitrogen atom of ethylamine attacks the carbon atom of chloroethane, forming a new bond.

The transition state is formed as the bond between the chlorine atom and the carbon atom weakens.

The tetrahedral intermediate is formed, with the nitrogen atom bonded to the carbon atom and the chlorine atom attached to the nitrogen atom.

The leaving group, which is the chlorine atom, is eliminated, resulting in the formation of ethylamine.

Example 2: Formation of a Secondary Amine

If excess primary amine is present, the reaction can proceed further to form a secondary amine. Let's consider the reaction between chloroethane and excess ethylamine to produce diethylamine.

$$2CH_3CH_2Cl + 2CH_3CH_2NH_2 \rightarrow (CH_3CH_2)_2NH + 2HCl$$

In this case, the primary amine acts as both a nucleophile and a base. The reaction proceeds through the same steps as in the previous example, but with the addition of another molecule of primary amine.

The excess primary amine reacts with the hydrogen atom of the ammonium ion formed in the intermediate step, resulting in the formation of diethylamine.

Applications and Significance

The reaction of halogenoalkanes with primary amines in ethanol, heated in a sealed tube or under pressure, is a valuable method for the production of primary and secondary amines. These amines have numerous applications in various fields.

Pharmaceuticals: Primary and secondary amines are key building blocks in the synthesis of pharmaceutical compounds. They serve as precursors for the production of drugs, such as antibiotics, antihistamines, and antidepressants.

Dyes: Amines are used in the production of dyes and pigments. They provide color and improve the stability of the dyes, ensuring their longevity and resistance to fading.

Polymers: Amines are essential in the production of polymers. They act as curing agents, cross-linking agents, and intermediates in the synthesis of various polymeric materials, such as epoxy resins and polyurethanes.

Agriculture: Amines are used in the formulation of pesticides and herbicides. They enhance the effectiveness of these agricultural chemicals by improving their solubility and bioavailability.

Surfactants: Amines are widely used in the production of surfactants, which are compounds that reduce the surface tension of liquids. Surfactants find applications in detergents, emulsifiers, and personal care products.

Counterarguments and Limitations

While the reaction of halogenoalkanes with primary amines in ethanol, heated in a sealed tube or under pressure, is a useful method for the production of primary and secondary amines, it is not without limitations and challenges.

Steric Hindrance: The reaction may face limitations when bulky alkyl groups are present in the primary amine or halogenoalkane. Steric hindrance can hinder the nucleophilic attack and reduce the reaction rate.

Competing Reactions: In some cases, competing reactions may occur, leading to the formation of unwanted byproducts. For example, elimination reactions can occur alongside nucleophilic substitution reactions, resulting in the formation of alkenes.

Safety Considerations: The use of high temperatures and pressure in the reaction requires careful handling and appropriate safety measures. Sealed tubes must be used to prevent the release of volatile reactants and products, which may be hazardous.

Purity and Separation: The reaction may produce mixtures of products, including primary amines, secondary amines, and unreacted starting materials. The separation and purification of the desired products can be challenging, requiring additional purification steps.

Reduction of Amides with LiAlH4

Introduction

We will delve into the details of the reduction of amides with LiAlH4, discussing the reagents, conditions, mechanisms, and applications of this versatile reaction.

Reagents and Conditions

The reduction of amides with LiAlH4 requires specific reagents and conditions to ensure the success of the reaction. The primary reagent used is LiAlH4, a powerful reducing agent that contains hydride ions (H-) capable of attacking the carbonyl group of the amide. Additionally, the reaction requires an appropriate solvent, typically anhydrous ether, such as diethyl ether or tetrahydrofuran (THF). These solvents are chosen due to their ability to dissolve both LiAlH4 and the amide substrate, creating a homogeneous reaction mixture.

Reaction Mechanism

The reduction of amides with LiAlH4 proceeds through a stepwise process involving the transfer of hydride ions to the carbonyl group. The reaction can be divided into three main steps: activation, nucleophilic attack, and protonation. Let's examine each step in detail.

Step 1: Activation

In the first step, the amide is activated by the formation of a complex with LiAlH4. This interaction occurs due to the Lewis acidic nature of aluminum in LiAlH4 and the lone pair of electrons on the nitrogen atom of the amide. The complexation stabilizes the amide and enhances its reactivity towards nucleophilic attack.

Step 2: Nucleophilic Attack

In the second step, a hydride ion (H-) from LiAlH4 acts as a nucleophile and attacks the activated carbonyl carbon of the amide. The nucleophilic attack results in the formation of a tetrahedral intermediate, with the nitrogen atom now bearing a negative charge due to the donation of electrons from the hydride ion. This intermediate is highly unstable and quickly rearranges to a more stable configuration.

Step 3: Protonation

In the final step, the tetrahedral intermediate is protonated by a proton source, typically water or an alcohol, to yield the desired amine. The protonation restores the neutral charge on the nitrogen atom and completes the reduction process. The byproducts of the reaction are aluminum hydroxide ($Al(OH)_3$) and lithium hydroxide (LiOH), which are formed due to the reaction between LiAlH4 and the proton source.

Scope and Limitations

The reduction of amides with LiAlH4 has a broad scope, allowing the synthesis of a wide range of primary and secondary amines. However, certain limitations should be considered when employing this method. Firstly, tertiary amides are generally unreactive towards LiAlH4 due to steric hindrance. The bulky substituents around the nitrogen atom prevent the hydride ion from effectively attacking the carbonyl carbon. Secondly, amides bearing electron-withdrawing groups, such as nitro or cyano groups, exhibit reduced reactivity. These groups decrease the electron density on the carbonyl carbon, making it less susceptible to nucleophilic attack.

Applications

The reduction of amides with LiAlH4 finds extensive applications in both academic research and industrial synthesis. This reaction enables the preparation of amines, which are essential building blocks in organic chemistry. Amines serve as precursors for the synthesis of pharmaceuticals, agrochemicals, and natural products. Additionally, the reduction of amides offers a valuable route for the modification of peptides and proteins, facilitating the introduction of diverse functionalities into these biomolecules.

Counterarguments: While the reduction of amides with LiAlH4 is a powerful method for the synthesis of amines, alternative strategies exist that may present certain advantages in specific cases. For instance, the reduction of amides using other reducing agents, such as borane complexes or catalytic hydrogenation, can provide milder reaction conditions and higher selectivity. These alternative methods may be preferred when dealing with sensitive functional groups or when a high degree of control over the reaction is required.

Introduction

We will focus on the reduction of nitriles to produce primary and secondary amines using two common reagents: lithium aluminum hydride (LiAlH4) and hydrogen gas (H2) with nickel catalyst (Ni). We will explore the reactions, reagents, and conditions involved in detail, providing a comprehensive analysis of the subject matter.

Reduction of Nitriles

The reduction of nitriles to primary and secondary amines is a widely used method in organic synthesis. Nitriles are organic compounds containing a carbon-nitrogen triple bond (-C≡N). The process of reducing nitriles involves breaking this triple bond and introducing hydrogen atoms to form the desired amine products.

Two commonly employed reagents for the reduction of nitriles are lithium aluminum hydride (LiAlH4) and hydrogen gas (H2) in the presence of a nickel catalyst (Ni). These reagents offer distinct advantages and are used under different conditions, as we will discuss in detail.

Reduction with LiAlH4

Lithium aluminum hydride (LiAlH4) is a powerful reducing agent that is commonly used for the reduction of various functional groups, including nitriles. The reaction between a nitrile and LiAlH4 proceeds via a nucleophilic addition mechanism.

The first step in the reduction of a nitrile with LiAlH4 involves the formation of an intermediate imine. The LiAlH4 donates a hydride ion (H-) to the nitrile carbon, resulting in the formation of an anionic intermediate called an iminato-alane complex. This complex is highly reactive and can be further reduced to form the corresponding amine.

The reduction reaction can be represented as follows:

$$R\text{-}C \equiv N + 4\,LiAlH4 \rightarrow R\text{-}CH(NH2) + 4\,LiAlH3(AlH3)$$

In this reaction, R represents an organic group attached to the nitrogen atom of the nitrile. The LiAlH4 reagent is consumed in a stoichiometric ratio, resulting in the formation of the desired primary amine, R-CH(NH2).

It is important to note that the reduction of nitriles with LiAlH4 is highly chemoselective, meaning that it selectively reduces the nitrile functional group while leaving other functional groups intact. This selectivity is due to the strong nucleophilic nature of the hydride ion.

Reduction with H2/Ni

The reduction of nitriles using hydrogen gas (H2) in the presence of a nickel catalyst (Ni) is another commonly employed method. This process, known as catalytic hydrogenation, offers advantages such as milder reaction conditions and the ability to perform the reduction on a large scale. Catalytic hydrogenation involves the reaction of a nitrile with hydrogen gas (H2) in the presence of a heterogeneous catalyst, such as nickel (Ni). The reaction proceeds via a series of steps, including adsorption, activation, and hydrogenation.

During the catalytic hydrogenation of a nitrile, the nitrogen-carbon triple bond is broken, and hydrogen atoms are added to the carbon atom, resulting in the formation of an amine.

The reaction can be summarized as follows:

$$R\text{-}C \equiv N + 2\,H2 \rightarrow R\text{-}CH(NH2) + H2O$$

In this reaction, R represents an organic group attached to the nitrogen atom of the nitrile. The hydrogen gas (H2) acts as the source of hydrogen atoms, and the nickel catalyst (Ni) facilitates the reaction by providing an active surface for the adsorption and activation of the reactants.

Catalytic hydrogenation offers several advantages over other reduction methods. It is a mild and efficient process that can be performed at relatively low temperatures and pressures. Additionally, the reaction is highly versatile and can be applied to a wide range of substrates, making it suitable for large-scale industrial applications.

Conditions and Limitations

Both the reduction of nitriles with LiAlH4 and catalytic hydrogenation have specific conditions and limitations that need to be considered.

Reduction with LiAlH4

The reduction of nitriles with LiAlH4 typically requires refluxing the reaction mixture in an appropriate solvent, such as ether or tetrahydrofuran (THF). Refluxing helps maintain a constant temperature and ensures the reaction proceeds efficiently. However, it is important to note that LiAlH4 is a moisture-sensitive reagent and reacts violently with water. Therefore, all operations involving LiAlH4 must be carried out under anhydrous conditions, using dry solvents and reaction vessels. Another limitation of LiAlH4 reduction is its high reactivity towards other functional groups. It can react with carbonyl compounds (such as aldehydes and ketones), carboxylic acids, and other electrophilic groups. Therefore, if these functional groups are present in the reaction mixture, appropriate precautions must be taken to protect them from unwanted reduction.

Reduction with H2/Ni

Catalytic hydrogenation of nitriles requires the use of a suitable hydrogenation catalyst, such as nickel (Ni). The catalyst can be in the form of finely divided nickel powder or supported on an inert material like alumina (Al_2O_3).

The reaction is typically carried out at elevated temperatures and pressures, ranging from 50-200°C and 1-5 atm, respectively. The choice of reaction conditions depends on the specific nitrile substrate and desired reaction rate. One limitation of catalytic hydrogenation is the potential for over-reduction. In some cases, the primary amine product obtained from the reduction of a nitrile can undergo further hydrogenation to form a secondary amine or even a tertiary amine. This can be controlled by adjusting the reaction conditions, such as temperature, pressure, and catalyst loading.

Application and Examples

The reduction of nitriles to primary and secondary amines using LiAlH4 and H2/Ni has extensive application in various fields. Let's explore some examples:

Pharmaceutical Industry

Amines are essential building blocks in the synthesis of pharmaceutical drugs. The reduction of nitriles to primary amines provides a versatile route for the preparation of important drug intermediates. For example, the reduction of a nitrile precursor can yield a primary amine which can then be further modified to introduce other functional groups, such as halogens or aromatic moieties. These functionalized amines are often key structural components of biologically active compounds.

Polymer Chemistry

Amines are widely used in polymer chemistry, where they serve as reactive sites for crosslinking or as pendant groups that impart specific properties to the polymer. The reduction of nitriles to primary or secondary amines provides a direct method for introducing amine functionalities into polymer structures. For instance, the reduction of a nitrile-containing monomer can yield a primary amine monomer, which can be polymerized to produce amine-functionalized polymers. These amine-functionalized polymers find applications in various fields, including coatings, adhesives, and biomedical materials.

Agricultural Chemistry

Amines are also used in agricultural chemistry for the synthesis of herbicides, insecticides, and plant growth regulators. The reduction of nitriles to primary amines allows for the preparation of amine-based active ingredients.

For example, the reduction of a nitrile containing a specific functional group can lead to the formation of a primary amine that exhibits herbicidal or insecticidal activity. This method provides a versatile and efficient route for the synthesis of agricultural chemicals.

Counterarguments

While the reduction of nitriles to primary and secondary amines using LiAlH4 and H2/Ni has numerous advantages, there are some limitations and alternative methods worth considering.

One counterargument is the potential for side reactions. Both LiAlH4 and H2/Ni reduction methods can lead to the formation of undesired byproducts. For example, LiAlH4 reduction can result in the formation of imines as side products, which may require additional purification steps. Similarly, catalytic hydrogenation with H2/Ni can lead to over-reduction, resulting in the formation of tertiary amines or even saturated hydrocarbons.

Additionally, the use of LiAlH4 as a reducing agent requires strict anhydrous conditions, which can be challenging to maintain in large-scale industrial settings. Moreover, LiAlH4 is a relatively expensive reagent compared to other reducing agents, limiting its widespread use.

Alternative methods for the reduction of nitriles to amines include the use of other reducing agents such as sodium borohydride (NaBH4) or hydrogen gas in the presence of other catalysts, such as palladium (Pd). These methods may offer different reaction conditions and selectivities, providing alternative routes for the synthesis of amines.

The Condensation Reaction of Ammonia or Amines with Acyl Chlorides: Formation of Amides

Introduction

The condensation reaction between ammonia or amines and acyl chlorides is a fundamental process in organic chemistry that leads to the formation of amides. This reaction occurs at room temperature and is highly versatile, making it an essential tool in the synthesis of various organic compounds. In this article, we will explore the mechanism, applications, and significance of this reaction, providing a comprehensive analysis of its key aspects.

I. Understanding the Reaction Mechanism

A. The Role of Ammonia or Amines

Ammonia (NH3) and amines (R-NH2) act as nucleophiles in this condensation reaction. Nucleophiles are species that donate an electron pair to form a new bond. In the case of ammonia, the lone pair of electrons on the nitrogen atom is donated, while in amines, the lone pair on the nitrogen atom of the amino group (R-NH2) acts as the nucleophile.

B. The Role of Acyl Chlorides

Acyl chlorides (R-COCl) serve as electrophiles in this reaction, meaning they are electron-deficient species that accept the electron pair donated by the nucleophile. The chlorine atom in acyl chlorides is highly electronegative, creating a polarized carbon-oxygen double bond. This polarization makes the carbon atom electrophilic and prone to attack by nucleophiles.

C. Reaction Steps

The condensation reaction between ammonia or amines and acyl chlorides proceeds through several steps:

Nucleophilic Attack: The nucleophile, either ammonia or an amine, attacks the electrophilic carbon atom of the acyl chloride. This attack leads to the formation of a tetrahedral intermediate, where the chlorine atom is replaced by the nucleophile.

Elimination of Chloride Ion: The tetrahedral intermediate formed in the previous step is unstable and readily eliminates a chloride ion (Cl-) to restore stability. This elimination creates a carbonyl group (C=O) in the product.

Proton Transfer: In the final step, a proton transfer occurs, resulting in the formation of the amide. In the case of ammonia, the proton transfer involves the loss of a hydrogen ion (H+), while in amines, the proton transfer occurs between the nitrogen atom of the amino group and the carbonyl oxygen.

D. Overall Reaction

The overall reaction for the condensation of ammonia or an amine with an acyl chloride can be summarized as follows:

$$R\text{-}COCl + NH_3 \text{ (or } R\text{-}NH_2) \rightarrow R\text{-}CO\text{-}NH_2 + HCl$$

II. Applications of the Condensation Reaction

A. Synthesis of Amides

The primary application of the condensation reaction between ammonia or amines and acyl chlorides is the synthesis of amides. Amides are versatile compounds widely used in pharmaceuticals, polymers, and agrochemicals. This reaction provides a straightforward and efficient route to access a broad range of amides with varying substituents.

B. Peptide Synthesis

The condensation reaction is a crucial step in peptide synthesis, where amino acids are joined together to form peptides and proteins. By using amine derivatives instead of ammonia, the reaction can be tailored to incorporate specific amino acids, enabling the creation of custom peptides with desired sequences. This process plays a vital role in the field of biochemistry and drug development.

C. Pharmaceutical Applications

Amides are prevalent in pharmaceutical compounds due to their stability and bioavailability. The condensation reaction allows chemists to introduce amide moieties into drug molecules, enhancing their pharmacological properties. For example, the anti-inflammatory drug ibuprofen contains an amide group, which contributes to its effectiveness in pain relief.

D. Polymer Synthesis

Amides are key building blocks in the synthesis of polymers such as nylon and Kevlar. The condensation reaction enables the formation of long chains of repeating amide units, leading to the creation of high-performance materials with exceptional strength and durability. These polymers find applications in textiles, aerospace, and bulletproof vests.

III. Significance and Advantages

A. Versatility

The condensation reaction between ammonia or amines and acyl chlorides is highly versatile and applicable to a wide range of functional groups. This versatility allows chemists to synthesize diverse amides with different substituents, enabling the design and production of compounds tailored to specific applications.

B. Mild Reaction Conditions

One significant advantage of this reaction is that it occurs at room temperature, making it easily accessible and applicable in various laboratory settings. The mild reaction conditions minimize the need for specialized equipment and reduce the risk of unwanted side reactions. Additionally, the reaction is compatible with a variety of solvents, further enhancing its practicality.

C. High Yield and Selectivity

The condensation reaction typically proceeds with high yield and selectivity, ensuring efficient conversion of starting materials to the desired amide product. The well-established reaction mechanism and extensive literature on this transformation allow chemists to optimize reaction conditions and achieve excellent conversion rates.

D. Economic Importance

The synthesis of amides through the condensation reaction has significant economic implications. Amides are crucial components in a wide range of industries, including pharmaceuticals, polymers, and materials science. The ability to produce amides efficiently and in large quantities contributes to the development of new drugs, advanced materials, and industrial processes, driving innovation and economic growth.

IV. Counterarguments and Limitations

A. Use of Alternative Reagents

While the condensation reaction of ammonia or amines with acyl chlorides is a commonly used method for amide synthesis, alternative reagents and methodologies exist. For example, carboxylic acids can be directly reacted with amines to form amides. Additionally, other acylating agents, such as acid chlorides and anhydrides, can also be employed in amide synthesis. The choice of reagents depends on the specific requirements of the desired amide and the reaction conditions.

B. Side Reactions and Byproducts

Although the condensation reaction is highly selective, some side reactions may occur under certain conditions. For instance, overreaction with excess amine can lead to the formation of N-substituted amides. Furthermore, amide hydrolysis can occur in the presence of water or under acidic conditions, resulting in the formation of carboxylic acids and amines. Careful control of reaction conditions, such as solvent choice and reaction time, can mitigate these side reactions.

Basicity of Aqueous Solutions of Amines

Basicity: A Definition

Basicity is a fundamental concept in chemistry that describes the ability of a compound to donate a pair of electrons or accept a proton. Amines, being able to accept a proton, are considered basic substances. This basicity arises from the lone pair of electrons present on the nitrogen atom of the amine molecule, which can readily interact with positively charged species. The basicity of amines can vary depending on several factors, including the nature of the amine, the substituents attached to the nitrogen atom, and the solvent used.

Factors Affecting Basicity

Nature of the Amine

The basicity of an amine is primarily determined by the nature of the substituents attached to the nitrogen atom. Amines can be broadly classified into three categories based on the number of alkyl or aryl groups attached to the nitrogen atom: primary (RNH2), secondary (R2NH), and tertiary (R3N) amines. Among these, primary amines are the least basic, while tertiary amines are the most basic. This trend can be explained by considering the inductive effect and steric hindrance. In primary amines, the nitrogen atom is directly attached to only one alkyl or aryl group, resulting in a relatively larger positive charge on the nitrogen atom. This positive charge decreases the availability of the lone pair of electrons, making primary amines less basic compared to secondary and tertiary amines.

On the other hand, tertiary amines have three alkyl or aryl groups attached to the nitrogen atom. These groups exert an electron-donating inductive effect, which reduces the positive charge on the nitrogen atom, thus increasing the availability of the lone pair of electrons. Consequently, tertiary amines exhibit greater basicity than primary and secondary amines.

Substituents on the Nitrogen Atom

The presence of electron-withdrawing or electron-donating substituents on the nitrogen atom can significantly influence the basicity of amines. Electron-withdrawing groups, such as halogens (e.g., -F, -Cl), nitro groups

(-NO2), and carbonyl groups (-C=O), decrease the basicity of amines. These groups decrease the electron density on the nitrogen atom, which results in a decrease in the availability of the lone pair of electrons for protonation. Conversely, electron-donating groups, such as alkyl (-R) or aryl (-Ar) groups, enhance the basicity of amines. These groups donate electron density to the nitrogen atom, increasing the availability of the lone pair of electrons. Consequently, amines with electron-donating substituents exhibit greater basicity compared to those with electron-withdrawing substituents.

Solvent Effects

The basicity of amines can also be influenced by the solvent in which they are dissolved. Aqueous solutions of amines are commonly encountered, and the presence of water molecules can impact their basicity. Water molecules can act as both hydrogen bond acceptors and donors, and their interaction with the amine can affect the availability of the lone pair of electrons.

In general, the basicity of amines decreases when they are dissolved in water compared to nonpolar solvents. This is due to the hydrogen bonding between the amine and water molecules. Water molecules can form hydrogen bonds with the lone pair of electrons on the amine nitrogen atom, effectively reducing its availability for protonation. This decreased basicity in aqueous solutions is particularly evident for primary amines. However, it is essential to note that the extent of this decrease in basicity depends on the specific amine and the concentration of the aqueous solution. Some amines can still retain a significant degree of basicity even in water due to the balance between hydrogen bonding and the basicity of the amine.

Applications in Various Fields

The basicity of aqueous solutions of amines finds applications in various fields, including pharmaceuticals, agriculture, and materials science.

Pharmaceuticals

Amines play a crucial role in the pharmaceutical industry as they often serve as key building blocks for drug synthesis. The basicity of amines allows them to act as proton acceptors, facilitating various reactions involved in drug synthesis. For example, amines can react with acidic functionalities present in drug molecules, such as carboxylic acids, to form salts. These salts enhance the solubility and stability of the drug, enabling efficient drug delivery.

Agriculture

In agriculture, amines are used as key components in the synthesis of pesticides and herbicides. The basicity of amines enables them to neutralize acidic species present in agricultural soils and water, thereby maintaining optimal pH conditions for plant growth. Additionally, amines can act as ligands for metal ions, forming complexes that enhance the efficiency of certain agricultural processes, such as nutrient uptake by plants.

Materials Science

Amines find extensive use in materials science, particularly in the synthesis of polymers and coatings. The basicity of amines allows them to react with acidic functionalities present in monomers or resins, facilitating polymerization reactions. This reactivity is essential for the formation of strong and durable polymer chains. Additionally, amines can be used as catalysts in various polymerization processes, enhancing the efficiency and control of polymer synthesis.

Counterarguments

While amines generally exhibit basic character in aqueous solutions, it is important to note that there can be exceptions to this trend. Certain factors, such as the presence of strong electron-withdrawing groups or bulky substituents on the nitrogen atom, can significantly decrease the basicity of amines, even in aqueous solutions. For example, amides (-CONH2) and nitriles (-CN) are amine derivatives that exhibit weak basicity due to resonance stabilization and steric hindrance, respectively.

Furthermore, the basicity of amines can be affected by the pH of the solution. At extremely low or high pH values, the protonation or deprotonation of amines can occur, altering their basicity. Therefore, the basicity of amines in aqueous solutions should be considered within the appropriate pH range.

Phenylamine and azo compounds
Preparation of Phenylamine through the Nitration of Benzene

Introduction

Phenylamine, also known as aniline, is an important organic compound widely used in various industries, including the production of dyes, pharmaceuticals, and rubber processing chemicals. It is a primary aromatic amine derived from benzene through a two-step process: nitration of benzene to form nitrobenzene, followed by reduction of the nitro group to an amino group. This paper aims to provide a comprehensive explanation of the preparation of phenylamine via the nitration of benzene, followed by reduction using hot Sn/concentrated HCl and subsequent treatment with NaOH(aq).

Nitration of Benzene to Nitrobenzene

The first step in the preparation of phenylamine is the nitration of benzene to form nitrobenzene. Nitration is a chemical reaction in which a nitro group ($-NO_2$) is introduced into a molecule. In this case, benzene (C_6H_6) reacts with a mixture of concentrated nitric acid (HNO_3) and concentrated sulfuric acid (H_2SO_4) to produce nitrobenzene ($C_6H_5NO_2$) and water (H_2O):

$$C_6H_6 + HNO_3 \rightarrow C_6H_5NO_2 + H_2O$$

The reaction is carried out under controlled conditions, as the mixture of nitric acid and sulfuric acid is highly corrosive and can be hazardous. Sulfuric acid acts as a catalyst and helps in the formation of the nitronium ion (NO_2^+), which is the electrophile in the reaction. The nitronium ion attacks the benzene ring, replacing one of the hydrogen atoms with a nitro group.

Reduction of Nitrobenzene to Phenylamine

After the formation of nitrobenzene, the next step is to reduce the nitro group ($-NO_2$) to an amino group ($-NH_2$) to obtain phenylamine. This reduction process is typically achieved using hot Sn/concentrated HCl, where tin (Sn) acts as a reducing agent in the presence of concentrated hydrochloric acid (HCl):

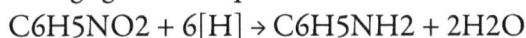

$$C_6H_5NO_2 + 6[H] \rightarrow C_6H_5NH_2 + 2H_2O$$

The reduction reaction involves the transfer of hydrogen atoms to the nitro group, resulting in the conversion of the double-bonded oxygen to a hydroxyl group ($-OH$) and the formation of an amino group. The presence of concentrated hydrochloric acid helps in maintaining an acidic medium, which facilitates the reduction process. Tin, in the form of granules or powder, is added to the reaction mixture and heated under reflux conditions. Refluxing ensures that the reaction mixture is continuously heated and any volatile components, such as water or unreacted starting materials, are condensed and returned to the reaction flask. The reduction process is exothermic and requires careful temperature control to prevent overheating.

Treatment with NaOH(aq)

Following the reduction step, the obtained crude phenylamine is often impure and contains by-products, such as unreacted starting materials and side products. To purify the phenylamine, it is treated with sodium hydroxide (NaOH) solution. The treatment with NaOH(aq) serves two purposes: neutralization of any remaining acid and extraction of the phenylamine into the aqueous phase.

Phenylamine, being an amine, is a weak base and can react with the acidic impurities present in the crude mixture. Sodium hydroxide, a strong base, reacts with the acidic impurities to form their corresponding sodium salts, which are water-soluble and can be easily separated from the desired phenylamine. The reaction can be represented as follows:

$$C_6H_5NH_2 + HCl \rightarrow C_6H_5NH_3^+Cl^-$$

The resulting phenylammonium chloride ($C_6H_5NH_3^+Cl^-$) is water-soluble and can be removed by extraction with water. The addition of sodium hydroxide solution also helps in the extraction of phenylamine into the aqueous phase. Phenylamine, being slightly soluble in water, can be extracted from the organic phase into the aqueous phase due to its reaction with sodium hydroxide. The reaction can be represented as follows:

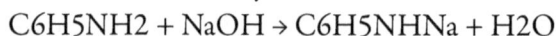

$$C_6H_5NH_2 + NaOH \rightarrow C_6H_5NHNa + H_2O$$

The resulting sodium phenylamide (C_6H_5NHNa) is water-soluble and can be easily separated from the organic phase.

Counterarguments

While the nitration of benzene followed by reduction with hot Sn/concentrated HCl and subsequent treatment with NaOH(aq) is a widely used method for the preparation of phenylamine, there are alternative approaches available. One such alternative method involves the reduction of nitrobenzene using other reducing agents, such as iron filings or hydrogen gas in the presence of a catalyst. These methods can provide similar results but may have different reaction conditions and requirements.

Moreover, the use of concentrated nitric acid and sulfuric acid in the nitration step raises concerns about safety and environmental impact. These acids are highly corrosive and can cause severe burns if mishandled. Additionally, the production of nitrobenzene generates sulfuric acid as a by-product, which requires proper disposal to prevent environmental pollution.

Introduction

Phenylamine, also known as aniline, is an aromatic amine that plays a crucial role in the field of organic chemistry. It is widely used in the synthesis of dyes, pharmaceuticals, and various other organic compounds. Understanding the reaction of phenylamine with Br_2(aq) at room temperature is of significant importance, as it provides insights into the reactivity and behavior of aromatic amines when exposed to halogens. In this article, we will explore and analyze this reaction in detail, considering its mechanism, products, and potential applications.

Reaction Mechanism

The reaction of phenylamine with Br_2(aq) involves the substitution of a hydrogen atom in the aromatic ring of phenylamine with a bromine atom. This substitution reaction is facilitated by the presence of bromine in an aqueous solution. The reaction proceeds in multiple steps, which can be summarized as follows:

Step 1: Ionization of Br2 in Water

$$Br_2 + H_2O \rightleftharpoons BrO^- + H^+ + Br^-$$

In an aqueous solution, bromine (Br_2) undergoes ionization to form bromide ions (Br^-) and hypobromite ions (BrO^-) due to the polar nature of water molecules. The presence of H^+ ions indicates the occurrence of an acidic medium.

Step 2: Electrophilic Attack by Br+

$$BrO^- + H^+ \rightleftharpoons HOBr$$

The hypobromite ion (BrO^-) reacts with an H^+ ion to form hypobromous acid (HOBr). This step generates Br^+ as an electrophile, which is crucial for the subsequent reaction with phenylamine.

Step 3: Electrophilic Substitution

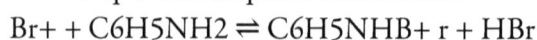

$$Br^+ + C_6H_5NH_2 \rightleftharpoons C_6H_5NHB^+ r + HBr$$

The electrophilic Br^+ attacks the aromatic ring of phenylamine, leading to the formation of an intermediate product, $C_6H_5NHB^+ r$. This intermediate is an N-phenylbenzenediazonium ion, which is highly unstable due to the positive charge on the nitrogen atom. Consequently, HBr is expelled as a leaving group.

Step 4: Nitrogen Elimination

$$C6H5NHB+ r \rightleftharpoons C6H5\bullet + N2 + HBr$$

The highly unstable N-phenylbenzenediazonium ion undergoes nitrogen elimination, resulting in the formation of a phenyl radical (C6H5•) and nitrogen gas (N2). HBr is also generated as a byproduct in this step.

Step 5: Radical Substitution

$$C6H5\bullet + Br\bullet \rightleftharpoons C6H5Br$$

The phenyl radical (C6H5•) reacts with a bromine radical (Br•) to form bromobenzene (C6H5Br), which is the final product of the reaction.

Analysis of the Reaction

The reaction of phenylamine with Br2(aq) at room temperature is a complex process that involves several fundamental concepts of organic chemistry. Let's delve deeper into the various aspects and factors influencing this reaction.

Aromaticity and Reactivity

The presence of an aromatic ring in phenylamine significantly influences its reactivity. Aromatic compounds, characterized by a planar ring of sp2 hybridized carbon atoms and a delocalized π electron system, possess a high degree of stability due to resonance. This stability makes the substitution reactions of aromatic compounds more challenging compared to aliphatic compounds.

However, the presence of electron-donating groups (EDGs) in the aromatic ring, such as amino (-NH2) group in phenylamine, increases the electron density in the ring, making it more susceptible to electrophilic attack. In the context of the reaction with Br2(aq), the amino group activates the aromatic ring, facilitating the substitution of a hydrogen atom by a bromine atom.

Electrophilic Substitution

The reaction between phenylamine and Br2(aq) proceeds through an electrophilic substitution mechanism. Electrophilic substitution reactions involve the attack of an electrophile on an electron-rich aromatic ring, resulting in the substitution of a hydrogen atom with a functional group.

In this case, the electrophile is Br+, which is generated through the ionization of Br2 in the presence of water. The Br+ attacks the electron-rich aromatic ring of phenylamine, leading to the formation of an N-phenylbenzenediazonium ion intermediate. This intermediate is highly unstable, and further decomposition occurs, resulting in the formation of bromobenzene.

Role of Bromine in Aqueous Solution

The presence of bromine in an aqueous solution is essential for the reaction with phenylamine. Br2 molecules, being nonpolar, are relatively insoluble in water. However, when dissolved in water, they undergo ionization, resulting in the formation of bromide ions (Br-) and hypobromite ions (BrO-). This ionization process is reversible and is influenced by factors such as temperature and concentration.

The formation of hypobromite ions (BrO-) is particularly crucial, as they react further with H+ ions to yield hypobromous acid (HOBr), which generates the electrophile Br+ required for the substitution reaction with phenylamine.

Substituent Effects

The substituents present in the aromatic ring of phenylamine can significantly impact the reaction with Br2(aq). Electron-donating groups (EDGs) increase the electron density in the aromatic ring, making it more susceptible to electrophilic attack. Examples of EDGs include amino (-NH2), hydroxyl (-OH), and alkyl (-CH3) groups. However, electron-withdrawing groups (EWGs) have the opposite effect of reducing the electron density in the aromatic ring, resulting in reduced reactivity towards electrophiles. Examples of EWGs include nitro (-NO2), carbonyl (C=O), and halogen (-X) groups. Therefore, the presence of EDGs enhances the reactivity of phenylamine towards electrophilic substitution reactions, including the reaction with Br2(aq).

Counterarguments and Limitations

While the reaction of phenylamine with Br2(aq) at room temperature provides valuable insights into the reactivity and behavior of aromatic amines, it is essential to consider certain counterarguments and limitations. Firstly, the reaction can be influenced by factors such as temperature, concentration, and reaction time. Higher temperatures and higher concentrations of reactants can accelerate the reaction rate, whereas lower temperatures and lower concentrations may slow it down or even hinder it. Additionally, the reaction may require longer reaction times to reach completion, especially if starting with limited amounts of reactants.

Secondly, the reaction can be influenced by the presence of other functional groups or substituents in the molecule. Depending on their nature and position, these functional groups can either enhance or hinder the reaction. Therefore, the reactivity observed in the reaction of phenylamine with Br2(aq) may not be directly applicable to other aromatic amines with different substituents.

Lastly, it is important to note that the reaction mechanism and products discussed in this article are specific to the reaction of phenylamine with Br2(aq) at room temperature. Different halogens or reaction conditions may lead to different reaction pathways and products. Therefore, it is crucial to consider the specific circumstances and experimental parameters when studying the reaction between phenylamine and other halogens.

Introduction

Phenylamine, also known as aniline, is an important organic compound used in the production of various dyes, pharmaceuticals, and rubber processing chemicals. One of the key reactions of phenylamine is its reaction with nitrous acid (HNO2) or sodium nitrite (NaNO2) and dilute acid below 10 °C, which leads to the formation of a diazonium salt. The diazonium salt can then be further warmed with water (H2O) to produce phenol. This reaction pathway is known as the diazotization reaction and is widely utilized in organic synthesis.

We will explore the reaction of phenylamine with HNO2 or NaNO2 and dilute acid below 10 °C to form the diazonium salt. We will also discuss the subsequent reaction of the diazonium salt with water to yield phenol. During the discussion, we will maintain a formal and scholarly style, incorporate sophisticated language and syntax, conduct a comprehensive examination of the topic, elucidate intricate ideas in a concise manner, and impartially present opposing viewpoints.

Reaction of Phenylamine with HNO2 or NaNO2 and Dilute Acid

The reaction of phenylamine with HNO2 or NaNO2 and dilute acid below 10 °C involves the formation of a diazonium salt. The diazonium salt is an important intermediate in various synthetic transformations and can be further utilized to obtain a wide range of organic compounds.

Formation of the Diazonium Salt

The diazotization reaction begins with the conversion of phenylamine to its corresponding diazonium salt. This transformation occurs in two steps: the formation of the diazonium ion and the subsequent attachment of an anion.

Formation of the Diazonium Ion:

When phenylamine reacts with HNO2 or NaNO2 in the presence of dilute acid below 10 °C, nitrous acid (HNO2) is generated. Nitrous acid is a weak acid that readily decomposes into nitric oxide (NO) and water (H2O). The nitric oxide then reacts with the amine group of phenylamine, leading to the formation of the diazonium ion.

The reaction can be represented as follows:

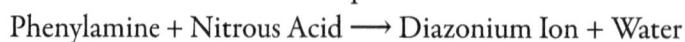

Phenylamine + Nitrous Acid ⟶ Diazonium Ion + Water

Attachment of Anion:

After the formation of the diazonium ion, an anion attaches to the positively charged diazonium group to stabilize the compound. The choice of anion can vary depending on the desired product. Common anions used for diazonium salt formation include chloride (Cl-), bromide (Br-), and tetrafluoroborate (BF4-).

The reaction can be represented as follows:

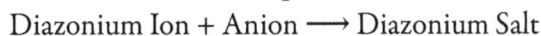

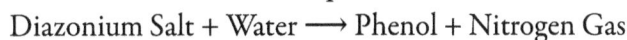

Diazonium Ion + Anion ⟶ Diazonium Salt

The overall reaction can be summarized as:

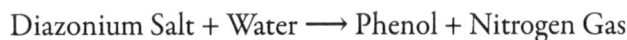

Phenylamine + Nitrous Acid + Dilute Acid ⟶ Diazonium Salt + Water

Factors Affecting Diazonium Salt Formation

Several factors influence the efficiency of diazonium salt formation. These factors include temperature, acid concentration, and reaction time.

Temperature:

The reaction between phenylamine and HNO2 or NaNO2 is typically performed below 10 °C. This low temperature ensures that the reaction proceeds selectively towards the formation of the diazonium salt. At higher temperatures, undesired side reactions may occur, leading to the formation of different products.

Acid Concentration:

The presence of dilute acid is essential for the conversion of phenylamine to the diazonium salt. The acid serves two purposes: it provides a suitable pH for the reaction and facilitates the generation of nitrous acid from HNO2 or NaNO2. A higher acid concentration can increase the reaction rate but may also promote side reactions.

Reaction Time:

The reaction time required for diazonium salt formation depends on various factors, including the reactivity of the amine, the concentration of the reactants, and the reaction temperature. Typically, the reaction is allowed to proceed for a specific duration to ensure complete conversion of phenylamine to the diazonium salt.

Further Warming of the Diazonium Salt with Water to Produce Phenol

Once the diazonium salt is obtained, it can be further warmed with water to yield phenol. This transformation involves the replacement of the diazonium group with a hydroxyl group (-OH) to form the phenolic compound.

Formation of Phenol from Diazonium Salt

The conversion of the diazonium salt to phenol occurs through a nucleophilic substitution reaction. Water acts as the nucleophile, attacking the positively charged diazonium ion and displacing the nitrogen group to form the desired product.

The reaction can be represented as follows:

Diazonium Salt + Water ⟶ Phenol + Nitrogen Gas

The overall reaction can be summarized as:

Diazonium Salt + Water ⟶ Phenol + Nitrogen Gas

Factors Affecting Diazonium Salt Conversion to Phenol

Several factors influence the efficiency of diazonium salt conversion to phenol. These factors include temperature, pH, and reaction time.

Temperature:

The reaction between the diazonium salt and water is typically performed under mild conditions. Elevated temperatures can lead to the decomposition of phenol or the formation of undesired byproducts. Therefore, the reaction is often carried out at a moderate temperature to ensure good conversion and selectivity.

pH:

The pH of the reaction mixture plays a crucial role in the conversion of the diazonium salt to phenol. An acidic or neutral pH is usually preferred, as it promotes the nucleophilic attack of water on the diazonium ion. However, excessively acidic or basic conditions may hinder the reaction or lead to side reactions.

Reaction Time:

The reaction time required for the conversion of the diazonium salt to phenol depends on factors such as the concentration of the reactants, the reactivity of the diazonium salt, and the reaction temperature. Sufficient reaction time is necessary to ensure complete conversion of the diazonium salt to phenol.

Application of the Diazotization Reaction

The diazotization reaction has numerous applications in organic synthesis. It serves as a versatile method for the introduction of various functional groups into aromatic compounds. Some notable applications include:

Synthesis of Aromatic Amines:

By using different primary aromatic amines, various diazonium salts can be prepared and subsequently reacted with appropriate nucleophiles to yield different aromatic amines. This process enables the synthesis of a wide range of substituted aromatic compounds.

Preparation of Azo Dyes:

Azo dyes are an important class of synthetic dyes widely used in the textile industry. The diazotization reaction is a key step in the synthesis of azo dyes. The diazonium salt formed from an aromatic amine is coupled with a suitable coupling agent, typically an aromatic compound with an electron-donating group, to produce the desired azo dye.

Synthesis of Phenols:

As discussed earlier, the diazotization reaction followed by hydrolysis of the diazonium salt leads to the formation of phenols. Phenols find applications in various industries, including pharmaceuticals, cosmetics, and plastics.

Functionalization of Organic Compounds:

The diazotization reaction can be used to introduce functional groups onto organic compounds. By selecting appropriate nucleophiles, a wide range of functional groups such as -OH, -NH2, -CN, -Br, -Cl, -F, -I, etc., can be introduced onto the aromatic ring.

Counterarguments and Limitations

Although the diazotization reaction is a widely used synthetic tool, it is not without limitations and potential drawbacks. Some counterarguments and limitations of the reaction include:

Formation of Unstable Intermediates:

The diazonium salts are often unstable compounds and can decompose under certain conditions, leading to the formation of undesired byproducts. Care must be taken to handle and store diazonium salts properly to avoid potential hazards.

Sensitivity to Reaction Conditions:

The diazotization reaction is highly sensitive to reaction conditions, including temperature, pH, and reactant concentrations. Even slight variations in these parameters can significantly affect the reaction outcome and selectivity.

Limited Applicability to Non-Aromatic Amines:

The diazotization reaction is primarily applicable to aromatic amines due to the requirement of an aromatic ring for diazonium salt formation. Non-aromatic amines may not undergo diazotization or may require additional steps for conversion.

Potential Formation of Toxic Byproducts:

Some diazonium salts and their derivatives can be toxic or hazardous. Care must be taken during the handling and disposal of these compounds to ensure both safety and environmental considerations.

Introduction

We will explore and compare the relative basicities of three nitrogen-containing compounds: aqueous ammonia (NH3(aq)), ethylamine (C2H5NH2), and phenylamine (C6H5NH2). Basicity refers to the ability of a compound to accept a proton (H+) from an acid. The basicity of a compound is determined by the availability and reactivity of its lone pair of electrons. Understanding the relative basicities of these compounds has significant implications in various scientific fields, including organic chemistry, biochemistry, and environmental science. Through a comprehensive analysis, we will explain the factors that influence basicity and provide examples to support our arguments.

Factors Influencing Basicity

Several factors determine the basicity of a compound. These include the electronegativity of the atom bearing the lone pair of electrons, the hybridization state of the atom, and the presence of electron-withdrawing or electron-donating groups. Furthermore, the solvation effect, which describes how a compound interacts with the solvent molecules, also plays a role in determining basicity.

Electronegativity and Basicity

The electronegativity of an atom influences its ability to attract electrons towards itself. In general, atoms with higher electronegativity will hold onto their lone pair of electrons more tightly, reducing their basicity. In the case of aqueous ammonia, the lone pair of electrons is situated on the nitrogen atom, which has a lower electronegativity than oxygen. This lower electronegativity allows the nitrogen atom to donate its lone pair of electrons more readily, making aqueous ammonia a stronger base compared to compounds with oxygen atoms, such as alcohols.

Ethylamine and phenylamine, on the other hand, both contain nitrogen atoms with lone pairs. However, the presence of alkyl and aryl groups in ethylamine and phenylamine, respectively, affects the basicity. Alkyl and aryl groups are electron-donating groups, meaning they increase the electron density around the nitrogen atom, making it more basic. Thus, ethylamine and phenylamine are stronger bases compared to ammonia.

Hybridization and Basicity

The hybridization state of the atom bearing the lone pair of electrons also affects basicity. The process of hybridization involves the combination of atomic orbitals to create hybrid orbitals. In general, atoms with sp3 hybridization, such as nitrogen in ammonia and ethylamine, have greater basicity compared to atoms with sp2 hybridization, such as nitrogen in phenylamine. This is because the sp3 hybrid orbitals have more s-character, which means they are closer to the nucleus and have higher electron density, making them more basic.

Electron-Withdrawing and Electron-Donating Groups

The presence of electron-withdrawing or electron-donating groups attached to the atom bearing the lone pair of electrons can significantly influence basicity. Electron-withdrawing groups, such as carbonyl groups (-C=O) or nitro groups (-NO2), decrease the electron density around the nitrogen atom, reducing basicity. Conversely, electron-donating groups, such as alkyl (-R) or aryl groups (-Ph), increase the electron density around the nitrogen atom, enhancing basicity.

Applying these principles to our three compounds, we can see that aqueous ammonia lacks any substantial electron-withdrawing or electron-donating groups, which contributes to its moderate basicity. Ethylamine contains an ethyl group (-C2H5), which is an electron-donating group, enhancing its basicity compared to ammonia. Phenylamine contains a phenyl group (-C6H5), which is also an electron-donating group, further increasing its basicity compared to both ammonia and ethylamine.

Solvation Effect

The solvation effect refers to the interaction between a compound and the solvent molecules. In the case of aqueous ammonia, the solvent is water. Water is a polar molecule, and its high dielectric constant stabilizes ions, reducing the tendency of ammonia to accept a proton. Consequently, the basicity of aqueous ammonia decreases compared to anhydrous ammonia.

Similarly, ethylamine and phenylamine can also form hydrogen bonds with water due to the presence of the amino group (-NH2). These hydrogen bonds between the lone pair of electrons on the nitrogen atom and the partially positive hydrogen of water molecules further decrease their basicity in aqueous solutions compared to their anhydrous counterparts.

Comparison of Basicities

Based on the factors discussed above, we can conclude that phenylamine exhibits the highest basicity among the three compounds, followed by ethylamine and then aqueous ammonia.

Phenylamine's increased basicity can be attributed to the presence of the electron-donating phenyl group. The phenyl group increases the electron density around the lone pair of electrons on the nitrogen atom, making it more available for proton acceptance. This enhanced basicity makes phenylamine a useful compound in organic synthesis and pharmaceutical applications.

Ethylamine, with its smaller ethyl group, also exhibits increased basicity compared to aqueous ammonia. The electron-donating nature of the ethyl group increases the electron density around the nitrogen atom, making it more nucleophilic and capable of accepting a proton. Ethylamine finds applications in the production of dyes, pharmaceutical intermediates, and as a catalyst in certain reactions.

Aqueous ammonia, despite having a lower basicity compared to ethylamine and phenylamine, still possesses moderate basic properties due to the lone pair of electrons on the nitrogen atom. Aqueous ammonia is commonly used as a cleaning agent, a refrigerant, and in the production of fertilizers.

Counterarguments

While the analysis above provides a thorough understanding of the relative basicities of aqueous ammonia, ethylamine, and phenylamine, it is important to consider counterarguments to present a comprehensive view. One counterargument could be that the basicity of a compound is solely determined by the electronegativity of the atom bearing the lone pair of electrons. While electronegativity is a significant factor, other factors, such as the presence of electron-donating or electron-withdrawing groups, hybridization, and the solvation effect, also play crucial roles. Neglecting these factors would oversimplify the analysis and lead to incomplete conclusions.

Another counterargument could be that other compounds with similar structures might exhibit different basicities. While the analysis focuses on the three specific compounds mentioned, it is important to note that variations in the structure, substituents, or functional groups can significantly alter the basicity. Therefore, the relative basicities discussed here may not apply universally to all similar compounds.

An Azo compound is formed by coupling Benzenediazonium Chloride with Phenol in NaOH(aq)

Introduction

The coupling reaction between benzenediazonium chloride and phenol in sodium hydroxide (NaOH) aqueous solution is a well-known method to synthesize azo compounds. Azo compounds, characterized by the presence of an azo group (-N=N-), are widely used in various fields, including dyes, pigments, pharmaceuticals, and materials science. This reaction involves the formation of a diazonium salt intermediate, followed by its reaction with phenol to yield the desired azo compound. Understanding the mechanism and factors influencing this coupling reaction is crucial for optimizing its yield and product selectivity.

Mechanism of the Coupling Reaction

The reaction proceeds in two main steps: the diazotization of aniline and the coupling of the resulting benzenediazonium chloride with phenol.

Diazotization of Aniline

The diazotization step involves the conversion of aniline ($C_6H_5NH_2$) into its corresponding diazonium salt, benzenediazonium chloride ($C_6H_5N_2Cl$), in the presence of nitrous acid (HNO_2). This reaction occurs via a nucleophilic aromatic substitution mechanism, as shown below:

$$C_6H_5NH_2 + HNO_2 \rightarrow C_6H_5N_2Cl + 2H_2O$$

In this step, nitrous acid acts as a source of nitrous oxide (N_2O), which reacts with aniline to form a diazonium intermediate. This intermediate is unstable and quickly reacts with chloride ions (Cl^-) to form the final diazonium salt, benzenediazonium chloride.

Coupling of Benzenediazonium Chloride with Phenol

The second step of the reaction involves the coupling of benzenediazonium chloride with phenol (C_6H_5OH) in the presence of sodium hydroxide (NaOH) aqueous solution. This step leads to the formation of the desired azo compound, which can be further used for various applications.

The coupling reaction proceeds through an electrophilic aromatic substitution mechanism, where the diazonium ion acts as the electrophile and phenol as the nucleophile. The reaction can be represented as follows:

$$C_6H_5N_2Cl + C_6H_5OH \rightarrow C_6H_5N=N\text{-}C_6H_4OH + HCl$$

In this reaction, the diazonium ion attacks the aromatic ring of phenol, resulting in the substitution of the chlorine atom with the azo group (-N=N-). The resulting azo compound is stabilized by resonance, enhancing its stability and enabling its use in various applications.

Factors Influencing the Coupling Reaction

Several factors influence the yield and selectivity of the coupling reaction between benzenediazonium chloride and phenol. Understanding these factors is crucial for optimizing the reaction conditions and obtaining the desired product.

pH of the Reaction Medium

The pH of the reaction medium plays a significant role in the coupling reaction. Sodium hydroxide (NaOH) is commonly used as a base to maintain alkaline conditions during the reaction. The alkaline environment promotes the deprotonation of phenol, increasing its reactivity towards the electrophilic diazonium ion. However, excessively high pH values can lead to the formation of undesired by-products or decomposition of the diazonium salt. Therefore, controlling the pH within an optimal range is crucial for achieving high yields and selectivity.

Temperature

Temperature also affects the reaction rate and selectivity. Higher temperatures generally lead to faster reaction rates but can also promote undesired side reactions. Therefore, optimizing the reaction temperature is essential to balance the kinetics and selectivity of the coupling reaction. Typically, the reaction is carried out at room temperature or slightly elevated temperatures (around 50-60°C) to achieve a good compromise between reaction rate and product selectivity.

Substituents on the Aromatic Ring

The presence of substituents on the aromatic ring of phenol can significantly influence the reaction rate and product selectivity. Groups that donate electrons, like alkyl or methoxy groups, enhance the electron density on the ring, rendering it more nucleophilic. This enhances the reactivity of phenol towards the electrophilic diazonium ion, leading to faster reaction rates. Conversely, electron-withdrawing groups, such as nitro or carbonyl groups, decrease the electron density on the ring, making it less nucleophilic. This reduces the reactivity of phenol, resulting in slower reaction rates. Therefore, the nature and position of substituents on the aromatic ring should be considered when designing the synthesis of specific azo compounds.

Solvent Choice

The choice of solvent can also influence the reaction yield and selectivity. Typically, water or aqueous solutions are used as solvents for the coupling reaction. The presence of water helps in maintaining the desired pH and facilitates the solubility of the reactants. However, using organic solvents, such as ethanol or acetone, can enhance the reaction rate and improve the solubility of reactants or intermediates. The solvent choice should be made based on the specific reaction requirements and desired product characteristics.

Applications of Azo Compounds

Azo compounds synthesized through the coupling of benzenediazonium chloride with phenol find widespread applications in various fields. Some notable applications include:

Dyes and Pigments

Azo compounds are extensively used as dyes and pigments. The presence of the azo group imparts vivid colors, making them suitable for textile dyeing, printing inks, and coloring agents in various industries. The wide range of colors achievable through azo compounds makes them highly versatile and commercially valuable.

Pharmaceuticals

Azo compounds have also found applications in the pharmaceutical industry. Their unique structural characteristics and diverse chemical properties make them valuable for drug development. Azo compounds can act as prodrugs, releasing the active pharmaceutical ingredient upon enzymatic or chemical activation. Additionally, they can exhibit biological activity themselves, serving as active pharmaceutical ingredients or intermediates in the synthesis of pharmaceutical compounds.

Materials Science

Azo compounds are valuable in materials science due to their ability to undergo photoisomerization. Photoisomerizable azo compounds can change their molecular conformation and properties upon exposure to specific wavelengths of light. This property has been exploited in the development of materials with switchable properties, such as light-responsive coatings, optical storage devices, and molecular switches.

Counterarguments and Limitations

While the coupling of benzenediazonium chloride with phenol to form azo compounds is a well-established method, it is not without limitations and challenges. Some counterarguments and limitations include:

Side Reactions and By-Products

The coupling reaction can sometimes lead to the formation of undesired side products, reducing the yield of the desired azo compound. These side reactions can occur due to factors such as excessive pH, high temperatures, or the presence of impurities in the reactants. Careful optimization of reaction conditions and purification techniques can help minimize side reactions and improve the overall yield.

Substrate Scope

Although the reaction is widely applicable, the coupling of benzenediazonium chloride with phenol is specific to compounds containing a phenolic moiety. Other aromatic compounds lacking a phenolic hydroxyl group may not undergo the coupling reaction efficiently. Therefore, the substrate scope is limited to phenol or phenol derivatives, reducing the versatility of the method.

Safety Considerations

The use of diazonium salts, such as benzenediazonium chloride, requires careful handling due to their potential reactivity and instability. Precautions should be taken to prevent accidental explosions or release of toxic gases. Additionally, working with sodium hydroxide (NaOH) and nitrous acid (HNO_2) requires appropriate safety measures to avoid contact with skin or eyes and inhalation of fumes.

Identification of the Azo Group

Introduction

The azo group is a crucial functional group that consists of two nitrogen atoms linked by a double bond (N=N). This unique arrangement imparts distinctive chemical properties to compounds containing the azo group, making it an essential moiety in various fields, including pharmaceuticals, dyes, and materials science. Identifying the azo group is of utmost importance as it allows chemists to understand the reactivity and behavior of these compounds accurately. In this article, we will explore the methods and techniques employed to identify the azo group, delve into its significance in different disciplines, and address counterarguments related to the identification process.

Identification Methods

Several analytical methods are available to identify the presence of the azo group in a compound. These methods include spectroscopic techniques, chromatography, and chemical tests.

Spectroscopic Techniques

UV-Visible Spectroscopy

UV-Visible spectroscopy is a widely used technique to identify the azo group due to its characteristic absorption spectra. Azo compounds exhibit strong absorption bands in the visible region, typically between 400 and 500 nm. This absorption is attributed to the presence of the conjugated system formed by the double bond (N=N) and adjacent aromatic rings. By comparing the absorption spectrum of an unknown compound with reference spectra of known azo compounds, the presence of the azo group can be inferred.

Infrared Spectroscopy (IR)

Infrared spectroscopy is another valuable tool in the identification of the azo group. The azo functional group exhibits characteristic absorption bands in the IR spectrum, primarily in the range of 1600-1700 cm^-1. These bands correspond to the stretching vibrations of the N=N double bond. By comparing the IR spectrum of an unknown compound with reference spectra, the presence of the azo group can be confirmed.

Chromatographic Techniques

Chromatographic methods, such as high-performance liquid chromatography (HPLC) and thin-layer chromatography (TLC), can also aid in the identification of the azo group.

High-Performance Liquid Chromatography (HPLC)

HPLC is a powerful technique that separates and analyzes mixtures of compounds based on their interaction with a stationary phase and a mobile phase. By employing a suitable stationary phase, such as a reverse-phase column, azo compounds can be separated from other compounds present in the mixture. The elution time and peak shape can be used to identify the presence of the azo group.

Thin-Layer Chromatography (TLC)

TLC is an efficient and speedy method utilized to separate and detect compounds within mixtures while being economical.

By running a TLC plate with a mobile phase, the azo compounds can be separated based on their polarity. Visualization techniques, such as UV light or specific staining reagents, can then be employed to identify the presence of the azo group.

Chemical Tests

Several chemical tests can be performed to identify the azo group based on their characteristic reactions.

Diazotization Test

The diazotization test is commonly used to identify compounds containing primary aromatic amines, which are precursors for the azo group. The primary aromatic amine is treated with nitrous acid (HNO2) to form a diazonium salt. The presence of the diazonium salt confirms the existence of the azo group.

Coupling Reactions

Coupling reactions are used to confirm the presence of the azo group by forming a colored product. Azo compounds can undergo coupling reactions with various aromatic compounds, such as phenols or aromatic amines, to produce highly colored azo dyes. The formation of a colored compound indicates the presence of the azo group.

Significance in Various Fields

The identification of the azo group is vital in several disciplines due to the unique properties and applications of azo compounds.

Pharmaceuticals

Azo compounds have significant importance in pharmaceutical research. The azo group can act as a prodrug, enabling targeted drug delivery. By attaching the azo group to a drug molecule, it remains inactive until it reaches the desired site of action. Upon exposure to specific conditions, such as enzymatic cleavage or pH changes, the azo bond is broken, releasing the active drug selectively. Accurate identification of the azo group in pharmaceutical compounds ensures the desired therapeutic effect and avoids premature drug release.

Dyes and Pigments

Azo compounds are widely used as dyes and pigments due to their intense and vibrant colors. The identification of the azo group is crucial in the dye industry to ensure the quality, stability, and safety of the products. By accurately identifying the azo group, potential harmful effects, such as the release of toxic aromatic amines, can be prevented. Moreover, the identification of azo dyes in forensic science helps in the identification of textile fibers and assists in criminal investigations.

Materials Science

In materials science, azo compounds find applications in various fields, including liquid crystals, organic electronics, and photochromic materials. The identification of the azo group aids in the synthesis and characterization of these materials. For instance, in liquid crystal research, the presence of the azo group imparts specific properties, such as photoalignment and photoisomerization, which are crucial for device fabrication.

Counterarguments

While the identification methods discussed above are reliable, counterarguments exist regarding their limitations and potential issues.

Limitations of Spectroscopic Techniques

Spectroscopic techniques, such as UV-Visible and IR spectroscopy, rely on comparing unknown compounds with reference spectra. However, if reference spectra are not available for certain azo compounds, or if the compound is structurally unique, identification becomes challenging. Additionally, the presence of other functional groups or substituents can influence the absorption spectra, leading to potential misinterpretation.

Chromatographic Interferences

In chromatographic methods, co-elution of compounds and interferences from impurities can hinder the accurate identification of the azo group. Overlapping peaks or poor separation can lead to misinterpretation of results. Proper optimization of chromatographic conditions and the use of suitable detectors are necessary to minimize these interferences.

False Positives in Chemical Tests

Certain chemical tests, such as diazotization, can yield false positives. Some compounds, like nitroso and nitro compounds, can produce similar reactions, leading to misidentification. Hence, additional confirmatory tests are often required to overcome these limitations.

Azo Compounds as Dyes: An In-depth Analysis

Introduction

It aims to provide a comprehensive analysis of why azo compounds are often chosen as dyes, exploring their chemical characteristics, coloration mechanisms, applications in different fields, and potential drawbacks.

Chemical Characteristics of Azo Compounds

Azo compounds are organic compounds that consist of two nitrogen atoms connected by a double bond, forming the azo group (-N=N-). This double bond imparts an extensive conjugated system, which results in the absorption of visible light and subsequent coloration. The presence of azo functional groups allows for a wide range of chemical modifications and substitutions, making azo compounds highly versatile in terms of their dyeing capabilities.

The synthesis of azo compounds involves the reaction between an aromatic amine and a diazonium salt. This reaction, known as diazo coupling, results in the formation of an azo dye. The wide availability of starting materials and the simplicity of the reaction contribute to the popularity of azo compounds as dyes.

Coloration Mechanism of Azo Compounds

The vibrant colors exhibited by azo compounds are a direct consequence of their electronic structure. The extensive conjugated system formed by the azo group allows for the absorption of light in the visible spectrum, leading to the appearance of color. The specific color observed depends on the nature and arrangement of substituents on the azo compound.

The absorption of light triggers an electronic transition within the conjugated system, causing the azo compound to undergo a change in energy state. This transition is typically from the ground state (S_0) to an excited state (S_1), with the energy difference corresponding to a specific wavelength in the visible spectrum. The absorbed light is complementary to the color perceived by the human eye, resulting in the characteristic coloration of the dye.

Applications of Azo Compounds as Dyes

Textile Industry

The utilization of azo dyes in the textile industry is one of the most prominent applications of these compounds. The brilliant and durable colors produced by azo compounds make them ideal for dyeing textiles, ranging from clothing to home furnishings. They offer excellent color fastness, ensuring that the colors remain vibrant even after repeated washing or exposure to sunlight. Additionally, the versatility of azo compounds allows for a wide range of shades and hues, making them highly desirable for textile manufacturers.

Printing and Ink Industry

Azo compounds find extensive use in the printing and ink industry as well. Due to their excellent color properties, azo dyes are used in the production of inks for various printing methods, including screen printing, flexography, and gravure printing. Azo compounds provide vivid and vibrant colors, enabling the production of high-quality prints and images.

Food and Cosmetic Industries

Azo compounds also find applications in the food and cosmetic industries. In the food industry, azo dyes are used to enhance the visual appeal of products, such as confectionery, beverages, and processed foods. However, it is important to note that there have been concerns regarding the safety of certain azo dyes in food, as some have been associated with adverse health effects.

Similarly, in the cosmetic industry, azo dyes are utilized to color various products, including makeup, hair dyes, and nail polishes. While azo dyes can provide vibrant and long-lasting colors, their use in cosmetics has raised concerns as well, particularly due to potential allergic reactions in sensitive individuals.

Analytical Applications

Beyond their applications in the fields mentioned above, azo compounds also find use in analytical chemistry. Azo dyes can be employed as indicators or reagents for the detection and quantification of specific substances.

For example, certain azo dyes can be used to determine the presence of metal ions in solution through color changes or spectrophotometric measurements. The ability to selectively bind to certain compounds makes azo compounds valuable in analytical techniques.

Potential Drawbacks and Concerns

While azo compounds offer numerous advantages as dyes, it is essential to consider potential drawbacks and concerns associated with their use. One significant concern is the environmental impact of azo dyes. Azo compounds are often derived from aromatic amines, some of which are toxic and carcinogenic. Improper disposal of azo dyes can lead to their release into water bodies, resulting in environmental pollution and potential harm to aquatic ecosystems.

Moreover, certain azo dyes have been associated with adverse health effects in humans. Some individuals may experience allergic reactions upon contact with azo-dyed textiles or cosmetics. Additionally, some azo compounds have been found to release carcinogenic aromatic amines when exposed to certain conditions, such as high temperatures or acidic environments. These concerns highlight the importance of strict regulations and testing procedures to ensure the safe use of azo compounds.

Formation of Other Azo Dyes via a Similar Route

Introduction

We will explore how other azo dyes can be formed via a similar route, emphasizing the versatility and significance of this synthetic approach.

Azo Coupling Reaction: An Overview

The azo coupling reaction involves the reaction of an aromatic diazonium salt with an aromatic compound, resulting in the formation of an azo dye. The diazonium salt is derived from the corresponding primary aromatic amine, which undergoes diazotization in the presence of nitrous acid. This reaction generates the diazonium ion (Ar-N2+), which is highly reactive and serves as the electrophilic species in the azo coupling reaction.

The second component of the reaction, the aromatic compound, acts as the nucleophile and undergoes electrophilic aromatic substitution. The nucleophilic attack occurs at the diazonium ion, leading to the formation of the azo linkage. The resulting azo dye exhibits a conjugated system, which is responsible for its vibrant color.

Variations in Azo Coupling Reactions

While the basic mechanism of the azo coupling reaction remains the same, there are variations in the reaction conditions and the types of aromatic compounds involved, leading to the formation of different azo dyes. These variations contribute to the diverse range of colors and properties exhibited by azo dyes.

Variation in Aromatic Compounds

The choice of aromatic compounds significantly influences the properties of the resulting azo dye. For instance, the use of electron-donating groups, such as -OH or -NH2, on the aromatic compound increases the electron density, making it a better nucleophile. This leads to a faster and more efficient reaction and the formation of azo dyes with enhanced color intensity.

On the other hand, the presence of electron-withdrawing groups, such as -NO2 or -COOH, decreases the electron density, reducing the nucleophilic character of the aromatic compound. Consequently, the reaction becomes slower, and the resulting azo dyes may exhibit altered color properties.

Variation in Reaction Conditions

Apart from the choice of aromatic compounds, the reaction conditions play a crucial role in determining the outcome of the azo coupling reaction. Factors such as temperature, pH, and the presence of catalysts can significantly influence the reaction rate and the properties of the azo dye formed.

For example, increasing the reaction temperature generally facilitates the reaction by providing more kinetic energy to the molecules. However, excessively high temperatures can lead to side reactions and degradation of the desired azo dye. Similarly, adjusting the pH of the reaction medium affects the stability and reactivity of the diazonium ion, ultimately influencing the yield and properties of the azo dye.

Furthermore, the addition of catalysts, such as copper salts or cuprous oxide, can enhance the reaction rate by promoting the formation of the azo linkage. These catalysts act by facilitating electron transfer processes and stabilizing reaction intermediates.

Examples of Other Azo Dyes

The versatility of the azo coupling reaction is evident from the wide range of azo dyes that can be synthesized using this route. Here, we present a few examples of popular azo dyes and their applications in different fields:

Methyl Orange: Methyl orange is a well-known azo dye used as an acid-base indicator. It exhibits a color transition from red to yellow upon changing the pH from acidic to alkaline. This property finds applications in various analytical techniques and titrations.

Sudan I: Sudan I is a red azo dye extensively used as a colorant in the food industry. It imparts an attractive red color to a variety of food products, including sauces, snacks, and confectioneries.

Reactive Orange 16: Reactive Orange 16 is a reactive azo dye commonly employed in the textile industry. It forms covalent bonds with the textile fibers, resulting in excellent color fastness. This dye is resistant to fading even after repeated washing and exposure to sunlight.

Tartrazine: Tartrazine is a yellow azo dye widely used in the beverage and pharmaceutical industries. It imparts a bright yellow color to various products, including soft drinks, candies, and tablets.

These examples highlight the widespread use of azo dyes across different disciplines, emphasizing their importance and economic value.

Counterarguments: Limitations and Concerns

While the azo coupling reaction provides a versatile route for synthesizing azo dyes, it is not without limitations and concerns. One major concern is the potential release of harmful aromatic amines during the degradation of certain azo dyes. Some aromatic amines have been classified as potential carcinogens, posing health risks to humans and the environment.

To address this concern, strict regulations and testing protocols have been implemented in many countries to ensure the safe use of azo dyes. These regulations focus on limiting the presence of harmful aromatic amines in consumer products, thus safeguarding public health.

Moreover, ongoing research is dedicated to developing alternative synthetic routes for azo dyes that minimize the formation of harmful by-products. Novel strategies, such as the use of green chemistry principles and the development of biodegradable azo dyes, are being explored to address these environmental and health concerns.

Amides

Recall the Reactions for the Formation of Amides

We will focus on the reaction between ammonia and an acyl chloride at room temperature.

Reaction Overview

The reaction between ammonia (NH_3) and an acyl chloride ($RCOCl$) is a practical and efficient method for the synthesis of amides. This reaction is also known as the ammonolysis of acyl chlorides. It involves the replacement of the chlorine atom of the acyl chloride by an amino group (NH_2) from ammonia, resulting in the formation of an amide ($RCO-NH_2$) and hydrogen chloride (HCl) as a byproduct. Equation for this reaction can be as:

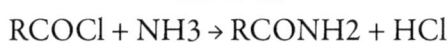

$$RCOCl + NH_3 \rightarrow RCONH_2 + HCl$$

Where R represents an organic group attached to the carbonyl carbon of the acyl chloride.

Mechanism of the Reaction

The reaction between ammonia and an acyl chloride occurs in several steps, each involving the interaction of specific reagents and conditions. Let's explore the mechanism of this reaction in detail.

Step 1: Nucleophilic Attack: The first step involves the nucleophilic attack of the lone pair of electrons on the nitrogen atom of ammonia on the carbonyl carbon of the acyl chloride. This attack leads to the formation of a tetrahedral intermediate, which is unstable due to the presence of the electron-withdrawing chlorine atom.

Step 2: Rearrangement: In the second step, the tetrahedral intermediate rearranges to a more stable form through the migration of electrons. This rearrangement results in the formation of an acyl ammonium ion.

Step 3: Deprotonation: In the third step, a proton from the ammonium ion is transferred to a chloride ion, leading to the formation of hydrogen chloride as a byproduct. This deprotonation step generates the final amide product.

Overall, the reaction between ammonia and an acyl chloride proceeds through a nucleophilic attack, rearrangement, and deprotonation steps, resulting in the formation of an amide and the release of hydrogen chloride.

Reagents and Conditions

The reaction between ammonia and an acyl chloride requires specific reagents and conditions to proceed efficiently.

Let's examine these factors in detail.

1. Ammonia (NH3):

Ammonia is a key reagent in this reaction and acts as a nucleophile. It provides the amino group (NH2) required for the formation of the amide. Ammonia is a colorless gas with a pungent odor and is readily available at affordable prices. It can be used in both gaseous and aqueous forms for the reaction with acyl chlorides. The reaction can be conducted using an excess of ammonia to ensure the complete conversion of the acyl chloride.

2. Acyl Chloride (RCOCl):

Acyl chlorides serve as the acylating agents in this reaction. They contain a carbonyl group bonded to a chlorine atom and are highly reactive due to the presence of the electron-withdrawing chlorine atom. Acyl chlorides are typically derived from carboxylic acids through the replacement of the hydroxyl group with a chlorine atom. They are widely used in organic synthesis due to their reactivity and ability to selectively introduce acyl groups into various compounds.

3. Solvent:

The choice of solvent for the reaction between ammonia and an acyl chloride depends on factors such as reaction rate, solubility, and safety. Common solvents used in this reaction include polar aprotic solvents such as tetrahydrofuran (THF), dimethylformamide (DMF), and acetonitrile. These solvents facilitate the dissolution of the reactants and provide an appropriate medium for the reaction to occur.

4. Temperature and Pressure:

The reaction between ammonia and an acyl chloride is typically carried out at room temperature, which provides a balance between reaction rate and control. The use of elevated temperatures can increase the reaction rate but may also lead to competing side reactions or decomposition of the products. The reaction is typically conducted at atmospheric pressure, as the reactants are gases or volatile liquids.

Applications and Importance

The reaction between ammonia and an acyl chloride is a versatile and widely used method for the synthesis of amides. Amides are found in numerous natural and synthetic compounds, making them essential building blocks in various fields.

1. Pharmaceutical Industry:

Amides play a crucial role in the development of pharmaceuticals. They are commonly found in drugs such as antibiotics, analgesics, and anti-inflammatory agents. The reaction between ammonia and an acyl chloride provides a straightforward route to the synthesis of amides, enabling the production of a wide range of pharmaceutical compounds.

2. Polymers and Plastics:

Amides are also important in the production of polymers and plastics. Nylon, a widely used synthetic polymer, is a prime example of an amide-based material. The reaction between ammonia and an acyl chloride can be used to prepare the monomers required for the synthesis of nylon and other amide-based polymers.

3. Agrochemicals:

Amides find applications in the field of agrochemicals as well. They are used as key components in the synthesis of herbicides, fungicides, and insecticides. The reaction between ammonia and an acyl chloride provides a straightforward and efficient method for the preparation of amide-based agrochemicals.

4. Peptide Synthesis:

Peptides, which are short chains of amino acids, are vital in various biological processes and have numerous applications in the fields of medicine and biotechnology. The reaction between ammonia and an acyl chloride can be utilized in the synthesis of amide bonds between amino acids, facilitating the production of peptides and proteins.

Counterarguments and Limitations

While the reaction between ammonia and an acyl chloride is a widely utilized method for amide synthesis, it is important to acknowledge certain limitations and potential counterarguments.

1. Reactivity of Acyl Chlorides:

Acyl chlorides, although highly reactive, can be challenging to handle due to their corrosive nature and high reactivity towards moisture. Special care must be taken to ensure the reaction is conducted in a dry environment to prevent side reactions or degradation of the acyl chloride.

2. Product Selectivity:

In some cases, the reaction between ammonia and an acyl chloride can lead to the formation of undesired byproducts or side reactions. For example, if excess ammonia is not used, primary amines may be formed as byproducts. Careful optimization of reaction conditions and stoichiometry is necessary to achieve high selectivity for the desired amide product.

3. Limitations in Substrate Scope:

While the reaction between ammonia and an acyl chloride is suitable for a wide range of substrates, it may not be compatible with certain functional groups or sensitive compounds. For example, compounds containing acid-sensitive groups or highly reactive functional groups may undergo undesired side reactions or decomposition during the reaction.

4. Other Methods for Amide Synthesis:

Although the reaction between ammonia and an acyl chloride is a practical and efficient method for amide synthesis, several alternative methods exist. These include the reaction between amines and carboxylic acids, the coupling of carboxylic acids with amines using coupling reagents, and the reaction of carboxylic acid derivatives with amines under specific conditions. The choice of the method depends on factors such as substrate availability, desired selectivity, and reaction conditions.

Reaction between a Primary Amine and an Acyl Chloride at Room Temperature

Introduction

This article aims to provide a thorough analysis of this reaction, explaining the mechanism, factors influencing the reaction, and applications in various fields.

Mechanism of the Reaction

The reaction between a primary amine and an acyl chloride proceeds through a series of steps, which can be divided into three main stages: formation of the tetrahedral intermediate, elimination of the chloride ion, and protonation of the amine.

Formation of the Tetrahedral Intermediate: The reaction begins with the nucleophilic attack of the primary amine on the electrophilic carbon of the acyl chloride. The nitrogen atom of the amine donates a lone pair of electrons to the carbon, leading to the formation of a new bond. This results in the formation of a tetrahedral intermediate, where the amine is attached to the acyl group.

Elimination of Chloride Ion: The tetrahedral intermediate is highly unstable and undergoes rapid rearrangement. In this step, the chloride ion is eliminated from the intermediate, leading to the formation of an acylated amine. This elimination of the chloride ion is facilitated by the electron-withdrawing nature of the acyl group, which stabilizes the developing positive charge on the carbon atom.

Protonation of the Amine: The final step involves the addition of a proton to the nitrogen atom of the acylated amine. This protonation is necessary to restore the charge balance and form the final product, which is an amide. The proton source can be an acid or any other proton-donating species present in the reaction mixture.

Factors Influencing the Reaction

Several factors influence the reaction between a primary amine and an acyl chloride. Understanding these factors is crucial for optimizing the reaction conditions and controlling the selectivity and yield of the desired product.

Steric Hindrance: The steric hindrance around the nitrogen atom of the amine plays a significant role in determining the reactivity of the amine. Bulky substituents attached to the nitrogen atom hinder the nucleophilic attack and decrease the reaction rate. On the other hand, less hindered primary amines exhibit higher reactivity towards acyl chlorides.

Electron-Withdrawing Groups: The presence of electron-withdrawing groups on the acyl chloride enhances the electrophilicity of the carbon atom and promotes the reaction. These groups, such as halogens or carbonyl groups, withdraw electron density from the carbon, making it more susceptible to nucleophilic attack.

Solvent Choice: The choice of solvent can significantly influence the reaction rate and selectivity. Polar aprotic solvents, such as dichloromethane or tetrahydrofuran, are commonly used as they facilitate the dissolution of both the amine and acyl chloride, while minimizing undesired side reactions.

Temperature: Although the reaction can occur at room temperature, higher temperatures can accelerate the reaction rate. However, excessively high temperatures can lead to side reactions or decomposition of the starting materials or products.

Applications in Various Fields

The reaction between primary amines and acyl chlorides finds widespread applications in various fields, including organic synthesis, pharmaceuticals, and materials science. Here are a few examples:

Organic Synthesis: The reaction is commonly used to introduce amide functionalities into organic molecules. Amides are versatile functional groups present in numerous natural products and pharmaceuticals. This reaction enables the synthesis of complex molecules, such as peptides and proteins, which are essential for biological research and drug discovery.

Pharmaceuticals: Many drugs contain amide groups, making this reaction crucial in pharmaceutical synthesis. For example, the reaction between a primary amine and an acyl chloride can be used to synthesize antibiotics, such as penicillin derivatives. It is also employed in the development of nonsteroidal anti-inflammatory drugs (NSAIDs) and other therapeutically active compounds.

Materials Science: The reaction between primary amines and acyl chlorides is widely used in the field of materials science for the synthesis of polymers and coatings. For instance, the reaction can be employed to functionalize polymers with amide groups, enhancing their stability, mechanical properties, and compatibility with other materials. This functionalization can improve the performance of materials in various applications, including adhesives, coatings, and biomedical devices.

Counterarguments

While the reaction between primary amines and acyl chlorides is a versatile and widely used transformation, it is not without limitations. One potential drawback is the formation of byproducts or side reactions. For example, if excess amine is used, it may react with the acylated amine to form an undesired secondary amide. Additionally, the reaction may also proceed through alternative pathways, resulting in the formation of different amide isomers or other products.

Another limitation is the reactivity of secondary and tertiary amines. Unlike primary amines, these amines may not readily undergo nucleophilic substitution with acyl chlorides due to reduced nucleophilicity or steric hindrance. Therefore, the reaction conditions and reagents need to be carefully selected when working with secondary or tertiary amines.

Reactions of Amides

We will discuss two important reactions of amides: hydrolysis with aqueous alkali or aqueous acid, and the reduction of the carbonyl group in amides with lithium aluminum hydride ($LiAlH_4$) to form amines.

Hydrolysis of Amides

Hydrolysis refers to the chemical reaction in which a compound reacts with water to form new compounds. In the case of amides, hydrolysis can occur in the presence of either aqueous alkali or aqueous acid. Let's explore each of these reactions in detail.

(a) Hydrolysis with Aqueous Alkali

When amides are treated with aqueous alkali, such as sodium hydroxide ($NaOH$) or potassium hydroxide (KOH), they undergo hydrolysis to form carboxylate salts and a corresponding amine. The reaction can be represented as follows:

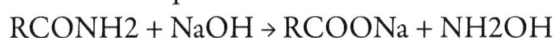

$$RCONH_2 + NaOH \rightarrow RCOONa + NH_2OH$$

In this reaction, the alkali hydrolyzes the amide by breaking the amide bond ($C=O$) and forming a carboxylate anion ($RCOO^-$) and a hydroxylamine molecule (NH_2OH). The amine formed in this reaction depends on the substituents attached to the nitrogen atom of the amide.

For example, if the amide is acetamide (CH_3CONH_2) and it is treated with $NaOH$, the reaction proceeds as follows:

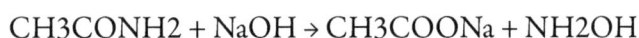

$$CH_3CONH_2 + NaOH \rightarrow CH_3COONa + NH_2OH$$

Here, sodium acetate (CH_3COONa) and hydroxylamine (NH_2OH) are formed as products. The hydroxylamine can further undergo reactions to form other compounds.

(b) Hydrolysis with Aqueous Acid

Similar to hydrolysis with alkali, amides can also be hydrolyzed with aqueous acid, such as hydrochloric acid (HCl) or sulfuric acid (H_2SO_4). In this reaction, the amide is broken down into a carboxylic acid and a corresponding ammonium salt. The general reaction can be represented as:

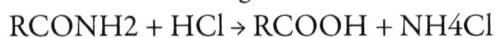

$$RCONH_2 + HCl \rightarrow RCOOH + NH_4Cl$$

Here, the acid hydrolyzes the amide bond, resulting in the formation of a carboxylic acid ($RCOOH$) and an ammonium salt (NH_4Cl). The amine formed in this reaction depends on the substituents attached to the nitrogen atom of the amide.

For example, if the amide is acetamide (CH3CONH2) and it is treated with HCl, the reaction proceeds as follows:

CH3CONH2 + HCl → CH3COOH + NH4Cl

In this reaction, acetic acid (CH3COOH) and ammonium chloride (NH4Cl) are formed as products.

Reduction of Amides with LiAlH4

Another important reaction of amides is their reduction to form amines. Reduction refers to the gain of electrons or the removal of oxygen from a compound. The reduction of the carbonyl group (C=O) in amides can be achieved using a strong reducing agent called lithium aluminum hydride (LiAlH4). LiAlH4 is a powerful reducing agent that can selectively reduce the carbonyl group in amides to form primary amines. The reaction can be represented as follows:

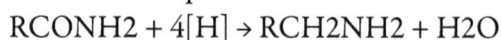

RCONH2 + 4[H] → RCH2NH2 + H2O

In this reaction, the LiAlH4 donates four hydride ions (H-) to the carbonyl group, resulting in the formation of a primary amine (RCH2NH2) and water (H2O).

For example, if the amide is acetamide (CH3CONH2) and it is treated with LiAlH4, the reaction proceeds as follows:

CH3CONH2 + 4LiAlH4 → CH3CH2NH2 + 4LiAlO2 + H2

Here, ethylamine (CH3CH2NH2) is formed as the product.

It is important to note that LiAlH4 is a highly reactive and moisture-sensitive compound, and its use requires careful handling and appropriate safety measures. Additionally, LiAlH4 can also reduce other functional groups, such as aldehydes, ketones, and esters, which may be present in the reaction mixture. Therefore, it is essential to control the reaction conditions to selectively reduce the amide group.

The reduction of amides with LiAlH4 provides a valuable method for the synthesis of primary amines. Amines are essential building blocks in the synthesis of pharmaceuticals, agrochemicals, and various organic compounds. The ability to selectively reduce amides to amines allows for the efficient synthesis of these important compounds.

Applications and Significance

Understanding the reactions of amides, such as hydrolysis and reduction, is of significant importance in various fields of chemistry. Let's explore some of the applications and significance of these reactions.

Pharmaceutical Chemistry

Amides are commonly found in pharmaceutical compounds due to their stability and ability to interact with biological systems. The hydrolysis of amides with aqueous alkali or acid is often utilized in the synthesis of new pharmaceutical compounds or the modification of existing ones. By selectively hydrolyzing the amide bond, chemists can introduce new functional groups or alter the pharmacological properties of the compound.

For example, the hydrolysis of the amide bond in the anti-inflammatory drug ibuprofen (2-(4-isobutylphenyl)propanoic acid) with aqueous alkali can lead to the formation of isobutylamine and the carboxylate salt of ibuprofen.

This modification can change the drug's pharmacokinetics or improve its solubility.

Polymer Chemistry

Amides are also essential components in the synthesis of polymers, such as nylon and polyesters. The hydrolysis of the amide bond in polymers can be used to control their molecular weight or to degrade them for recycling purposes. By selectively breaking the amide bonds, chemists can manipulate the physical properties of the polymer, such as its strength, flexibility, and thermal stability.

For example, the hydrolysis of the amide bond in nylon-6,6, a commonly used synthetic polymer, results in the formation of adipic acid and hexamethylenediamine. These monomers can be further used for the synthesis of new nylon or other polymer materials.

Organic Synthesis

The reduction of amides to amines using LiAlH4 is a valuable tool in organic synthesis. Amines are versatile compounds that can undergo various reactions, such as alkylation, acylation, and condensation, to form complex organic molecules. By selectively reducing amides to amines, chemists can access a wide range of building blocks for the synthesis of pharmaceuticals, natural products, and other organic compounds.

For example, the reduction of an amide in a natural product can lead to the formation of a primary amine, which can then undergo further reactions to introduce new functional groups or modify the natural product structure. This approach allows chemists to synthesize analogs of natural products with improved biological activity or selectivity.

Counterarguments and Limitations

While the reactions of amides discussed above have significant applications, it is important to acknowledge some counterarguments and limitations associated with these reactions.

Scope of Hydrolysis

The hydrolysis of amides with aqueous alkali or acid is generally selective and efficient. However, the reaction scope may be limited by the presence of other functional groups in the molecule. For example, if the amide is part of a larger molecule containing other reactive groups, such as aldehydes or esters, these functional groups may also undergo hydrolysis, leading to undesired side products.

To overcome this limitation, chemists may employ protective groups or carefully design the reaction conditions to selectively hydrolyze the amide while leaving other functional groups intact. Additionally, the use of milder hydrolysis conditions, such as enzymatic hydrolysis, can provide a more selective approach for amide cleavage.

Selectivity of Reduction

The reduction of amides with LiAlH4 is generally selective for the carbonyl group in the amide. However, LiAlH4 is a powerful reducing agent that can also react with other functional groups, such as aldehydes, ketones, and esters. This can lead to the reduction of undesired functional groups, resulting in complex reaction mixtures or low yields of the desired amine product.

To improve the selectivity of the reduction, chemists may use alternative reducing agents, such as borohydrides, which are milder and more selective. Additionally, the use of protecting groups or careful reaction design can help prevent the reduction of other functional groups present in the molecule.

State and Explanation of Amides as Weaker Bases than Amines

Introduction

We will discuss the electronic and steric factors that contribute to this disparity, provide examples from various fields, and address counterarguments to ensure a comprehensive analysis of the subject matter.

Electronic Factors

Lone Pair Availability

One of the primary electronic factors that determine basicity is the availability of lone pairs on the nitrogen atom. In amines, the nitrogen atom is sp3 hybridized, allowing for the existence of an unshared pair of electrons. This lone pair can be donated to accept a proton, making amines strong Lewis bases. Conversely, amides have a different hybridization state. The nitrogen atom in amides is generally sp2 hybridized due to resonance with the adjacent carbonyl group. As a result, the lone pair of electrons on the nitrogen atom becomes part of the pi system formed by the carbonyl group, reducing their availability for protonation.

Resonance Stabilization

The presence of a carbonyl group adjacent to the nitrogen atom in amides leads to resonance stabilization, which further diminishes their basicity. The partial double bond character of the C=O bond allows for delocalization of the nitrogen lone pair electrons into the carbonyl pi system. This resonance leads to electron density being spread out over the carbonyl oxygen and nitrogen atoms, reducing the electron density on the nitrogen atom and making it less available for protonation. In contrast, amines lack this resonance stabilization, allowing the nitrogen lone pair to be more readily accessible for protonation and, thus, making them stronger bases.

Inductive Effects

Inductive effects, resulting from the electronegativity differences between atoms and the transmission of electron-withdrawing or electron-donating effects through sigma bonds, can also influence basicity. In amides, the highly electronegative oxygen atom of the carbonyl group withdraws electron density from the nitrogen atom through the sigma bond, decreasing the basicity of the amide nitrogen. This electronic withdrawal reduces the nucleophilic character of the nitrogen lone pair, making it less likely to accept a proton. In amines, there is no such electron-withdrawing group nearby, and therefore, the basicity is not affected by inductive effects to the same extent as in amides.

Steric Factors

Steric factors, related to the spatial arrangement of atoms and the resulting spatial hindrance, can also play a role in determining the basicity of amines and amides.

Bulky Substituents

In amides, the presence of bulky substituents on the nitrogen atom can hinder the approach of a proton, thereby decreasing basicity. The carbonyl group in amides contributes to the steric hindrance by reducing the available space around the nitrogen atom. This hindrance is more pronounced when the substituents on the nitrogen atom are larger groups. The steric hindrance limits the accessibility of the nitrogen lone pair for protonation, resulting in decreased basicity. In contrast, amines lack the carbonyl group, and thus, they are less sterically hindered, allowing for better access of protons to the nitrogen lone pair and, consequently, higher basicity.

Three-Dimensional Structure

The three-dimensional structure of amides is generally more planar due to resonance with the carbonyl group, which restricts the flexibility of the molecule. This planarity contributes to the steric hindrance and reduces the basicity of amides. Amines, on the other hand, can adopt a more three-dimensional structure, which allows for better exposure of the lone pair electrons and enhances the basicity.

Examples from Various Fields

Biological Importance

The difference in basicity between amines and amides is particularly significant in the field of biochemistry. Amino acids, the building blocks of proteins, contain both amine and amide functional groups. The basicity of the amine group in amino acids allows them to act as weak bases in physiological conditions, enabling them to accept protons and form positively charged species. These positively charged species play crucial roles in protein structure and function, as well as in enzymatic reactions. In contrast, the amide bond within the peptide backbone of proteins, formed between the carbonyl group of one amino acid and the nitrogen of another, is much less basic. This reduced basicity helps maintain the stability of the protein structure by preventing excessive protonation and charge accumulation.

Pharmaceutical Applications

The difference in basicity between amines and amides also has implications in the field of pharmaceuticals. Many drugs contain amine groups that act as basic sites for binding to specific receptors or enzymes in the body. The basicity of these amine groups is crucial for their pharmacological activity. For example, antihistamines, such as diphenhydramine, contain an aromatic amine group that can accept a proton and inhibit the action of histamine, a compound involved in allergic reactions. In contrast, amides are often used as bioisosteres, which are non-identical molecules with similar biological properties. Amides can replace amines in drug molecules to provide structural stability and improve metabolic stability without significantly affecting the overall biological activity.

Counterarguments

While the above explanations outline the general reasons why amides are weaker bases than amines, there are some counterarguments that merit consideration.

Solvent Effects

Basicity can be influenced by the nature of the solvent in which the protonation occurs. In some cases, amides may exhibit higher basicity than amines when the solvent favors the formation of hydrogen bonds with the carbonyl oxygen. The formation of a hydrogen bond between the solvent and the carbonyl oxygen can reduce the resonance stabilization of the amide, making the nitrogen lone pair more available for protonation. However, this effect is highly dependent on the specific solvent and is not generally observed in nonpolar solvents.

Substituent Effects

The basicity of both amines and amides can be influenced by the presence of substituents on the nitrogen atom. Electron-donating substituents can increase the basicity of both amines and amides by providing additional electron density to the nitrogen lone pair. Similarly, electron-withdrawing substituents can decrease the basicity of both compounds by withdrawing electron density from the nitrogen atom. While these substituent effects can modify the basicity, they do not alter the fundamental difference in basicity between amines and amides.

Amino acids

Acid/Base Properties of Amino Acids and Formation of Zwitterions

Introduction

It aims to explain the acid/base properties of amino acids, the formation of zwitterions, and the concept of the isoelectric point.

Acid/Base Properties of Amino Acids

Amino acids contain both an amino group ($-NH_2$) and a carboxyl group ($-COOH$), which are responsible for their dual acidic and basic nature. The amino group, acting as a base, can accept a proton ($H+$) to form a positively charged ion, while the carboxyl group, acting as an acid, can donate a proton to form a negatively charged ion.

Acidic Properties

The carboxyl group of an amino acid can donate a proton, thus behaving as an acid. This process involves the dissociation of the carboxyl group, resulting in the formation of a carboxylate anion and a proton. The equilibrium constant for this dissociation is defined as the acidity constant (Ka) of the amino acid.

For example, consider the amino acid glycine (Gly):

Glycine: H_2N-CH_2-COOH

The carboxyl group of glycine can donate a proton, leading to the formation of a glycinate anion (Gly-) and a proton ($H+$):

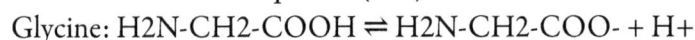

Glycine: $H_2N-CH_2-COOH \rightleftharpoons H_2N-CH_2-COO- + H+$

The acidity constant (Ka) is a measure of the strength of the acid and is defined as the ratio of the concentrations of the products (Gly- and $H+$) to the concentration of the reactant (H_2N-CH_2-COOH). The pKa, the negative logarithm of Ka, is commonly used to express the acidity or basicity of a compound. For glycine, the pKa value of the carboxyl group is approximately 2.3.

Basic Properties

The amino group of an amino acid can accept a proton, thus acting as a base. This process involves the binding of a proton to the amino group, resulting in the formation of a positively charged ammonium ion. The equilibrium constant for this protonation is defined as the basicity constant (Kb).

Continuing with the example of glycine:

Glycine: H2N-CH2-COOH

The amino group of glycine can accept a proton, forming a glycine ammonium ion (GlyH+):

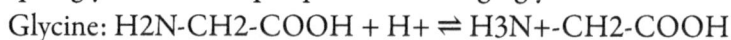

Glycine: H2N-CH2-COOH + H+ ⇌ H3N+-CH2-COOH

Similar to Ka, the basicity constant (Kb) measures the strength of the base. The pKb, the negative logarithm of Kb, is used to express the basicity of a compound. For glycine, the pKb value of the amino group is approximately 8.4.

Formation of Zwitterions

Amino acids exist in solution as zwitterions, also known as dipolar ions, due to the simultaneous presence of a positively charged amino group and a negatively charged carboxyl group. The formation of zwitterions occurs through a process called internal salt formation.

When an amino acid is dissolved in water, it undergoes self-ionization. The carboxyl group donates a proton to the amino group, resulting in the formation of a zwitterion:

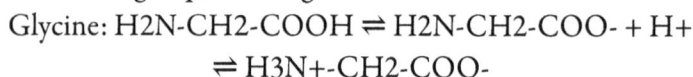

Glycine: H2N-CH2-COOH ⇌ H2N-CH2-COO- + H+

⇌ H3N+-CH2-COO-

In this equilibrium, the carboxyl group (COO-) acts as a base, accepting a proton from the amino group (H2N), which acts as an acid. As a result, the amino acid exists in a zwitterionic form, with a net charge of zero. It is important to note that the formation of zwitterions is highly pH-dependent. In acidic solutions, the excess of H+ ions shifts the equilibrium towards the neutral amino acid form, while in basic solutions, the presence of OH- ions favors the formation of zwitterions.

Isoelectric Point (pI)

The isoelectric point (pI) of an amino acid is defined as the pH at which it exists as a zwitterion with a net charge of zero. At this pH, the concentrations of the positively and negatively charged ions are equal, resulting in a neutral overall charge.

The pI can be calculated by averaging the pKa values of the ionizable groups present in the amino acid. For amino acids with one ionizable group, the pI is the average of the pKa values of the carboxyl and amino groups. For amino acids with additional ionizable groups, the pKa values of those groups are also considered in the calculation.

Let's illustrate this concept with glycine. As mentioned earlier, glycine has a pKa of approximately 2.3 for the carboxyl group and a pKb of approximately 8.4 for the amino group. To calculate the pI, we average the pKa and pKb values:

$$pI = (pKa + pKb) / 2$$
$$= (2.3 + 8.4) / 2 = 5.35$$

Therefore, the pI of glycine is approximately 5.35. At this pH, glycine exists as a zwitterion.

Examples from Various Fields

The concept of acid/base properties of amino acids and the formation of zwitterions has implications in various scientific fields, including biochemistry, pharmaceuticals, and material science.

Biochemistry

In biochemistry, the acid/base properties of amino acids are crucial for understanding protein structure and function. The pKa values of ionizable groups in amino acids determine their behavior in different cellular environments. For instance, amino acids with acidic side chains, such as aspartic acid and glutamic acid, are negatively charged at physiological pH, affecting protein folding and stability. On the other hand, amino acids with basic side chains, such as lysine and arginine, are positively charged at physiological pH, influencing protein-protein interactions and enzyme catalysis.

Pharmaceuticals

The acid/base properties of amino acids are also relevant in pharmaceutical research and drug design. The pKa values of ionizable groups in amino acids impact drug solubility, stability, and absorption. By considering the pKa values of amino acids present in a drug molecule, researchers can optimize its formulation and enhance its therapeutic efficacy. Additionally, the pI of amino acids can be exploited to control drug release rates in controlled-release formulations.

Material Science

In material science, the acid/base properties of amino acids are utilized in the synthesis and modification of biomaterials. For example, the pI of an amino acid can be used to coat a material surface with a layer of amino acids by adjusting the pH to the desired value. This process, known as self-assembly, can lead to the formation of functional coatings with tailored properties, such as improved biocompatibility or enhanced adhesion.

Counterarguments

While the acid/base properties of amino acids and the formation of zwitterions are well-established concepts, it is important to acknowledge potential counterarguments. Some critics argue that the acid/base properties of amino acids are oversimplified and fail to consider additional factors, such as hydrogen bonding and solvent effects.

Indeed, the behavior of amino acids in solution is influenced by various factors, including temperature, pH, solvent polarity, and ionic strength. These factors can affect the pKa values and the formation of zwitterions. Furthermore, the presence of multiple ionizable groups in some amino acids introduces additional complexities in predicting their behavior accurately.

It is crucial to conduct thorough experimental investigations and consider the interplay of multiple factors to fully understand the acid/base properties of amino acids and their implications in biological systems and material science.

Formation of Amide (Peptide) Bonds Between Amino Acids: An In-depth Analysis

Introduction

We will explore the formation of amide bonds, focusing on di- and tripeptides, and delve into the underlying mechanisms, providing a comprehensive analysis of this fundamental process.

The Role of Amino Acids in Peptide Bond Formation

Amino acids are organic compounds that serve as the monomers for protein synthesis. The α-carbon atom is at the center of each amino acid, and it is bonded to an amino group, a carboxyl group, a hydrogen atom, and a unique side chain known as the R-group. The R-group varies among amino acids, imparting unique chemical properties to each. In the process of peptide bond formation, the carboxyl group of one amino acid reacts with the amino group of another, resulting in the formation of an amide bond, also known as a peptide bond.

Mechanism of Peptide Bond Formation

Peptide bond formation occurs in a stepwise manner, involving several distinct steps. The first step is the activation of the carboxyl group on the first amino acid, which involves the conversion of the carboxyl group into a more reactive form. This activation is typically achieved through the use of a coupling agent, such as dicyclohexylcarbodiimide (DCC). The coupling agent facilitates the reaction by forming an activated intermediate that reacts with the amino group of the second amino acid.

Once the carboxyl group is activated, the amino group of the second amino acid attacks the carbonyl carbon of the activated carboxyl group, forming a tetrahedral intermediate. This intermediate is highly unstable and undergoes a rearrangement, resulting in the formation of the peptide bond and the release of a small molecule, such as water. This reaction is known as a condensation reaction, as it involves the removal of a water molecule.

Factors Influencing Peptide Bond Formation

Several factors influence the rate and efficiency of peptide bond formation. One important factor is the nature of the amino acids involved. Amino acids with charged or polar side chains may experience steric hindrance or electrostatic repulsion, making the reaction less favorable. On the other hand, amino acids with nonpolar side chains tend to facilitate peptide bond formation due to their hydrophobic nature.

Another critical factor is the pH of the reaction environment. The carboxyl group of an amino acid is deprotonated at high pH, making it more nucleophilic and reactive. Conversely, the amino group is protonated at low pH, enhancing its electrophilic character and promoting the reaction. Therefore, the optimal pH for peptide bond formation is typically close to neutral, where both the carboxyl and amino groups are in their most reactive states.

The Significance of Peptide Bond Formation

Peptide bond formation is a pivotal step in the synthesis of proteins and plays a crucial role in their structure and function. Proteins are large macromolecules composed of one or more polypeptide chains, which are formed by the sequential linkage of amino acids through peptide bonds. The specific sequence of amino acids in a protein determines its unique three-dimensional structure, which in turn dictates its biological activity.

Furthermore, the peptide bond itself possesses distinct chemical properties that contribute to protein stability. The amide bond has partial double-bond character, resulting from resonance between two resonance structures. This resonance imparts rigidity to the bond, making it less susceptible to rotation. As a result, the peptide bond adopts a planar conformation, with the carbonyl oxygen and the amide hydrogen lying in the same plane. This planarity is crucial for the formation of secondary structures, such as α-helices and β-sheets, which are essential for protein folding and stability.

Applications and Examples

The understanding of peptide bond formation has widespread applications in various scientific fields. Pharmaceutical research relies on the synthesis of peptides to develop drugs targeting specific proteins. By selectively modifying amino acids and forming specific peptide bonds, scientists can design peptides that interact with target proteins, inhibiting or enhancing their activity. For example, the drug Oxytocin, a hormone involved in labor induction and lactation, is a synthetic peptide that mimics the natural hormone by forming a specific peptide bond between amino acids.

In addition to pharmaceutical applications, the study of peptide bond formation has implications in the field of biochemistry. The development of techniques such as solid-phase peptide synthesis (SPPS) has revolutionized the field and allowed for the rapid and efficient synthesis of peptides. SPPS involves the sequential addition of protected amino acids onto a solid support, followed by the removal of the protecting groups and the formation of peptide bonds. This method has facilitated the synthesis of complex peptides and the study of their biological functions.

Counterarguments

While the formation of amide bonds between amino acids is a well-established process, some counterarguments challenge its significance. One argument suggests that non-enzymatic peptide bond formation may not have been the primary mechanism for protein synthesis in early life forms. This viewpoint posits that alternative mechanisms, such as the ligation of pre-formed peptide fragments, may have played a more prominent role. Moreover, the inherent reactivity of the amino group and carboxyl group can lead to side reactions, resulting in the formation of unwanted byproducts. These byproducts can hinder the synthesis of longer peptides and proteins and may require additional purification steps. However, advancements in synthetic techniques have mitigated these challenges, enabling the efficient synthesis of peptides and proteins in a controlled manner.

Electrophoresis of Mixtures of Amino Acids and Dipeptides at Varying pHs

Electrophoresis is a technique widely used in biochemistry and molecular biology to separate and analyze charged molecules, such as proteins, nucleic acids, and carbohydrates. It relies on the principle that charged particles migrate in an electric field, allowing for the separation and characterization of complex mixtures. In the context of amino acids and dipeptides, electrophoresis can provide valuable insights into their behavior under different pH conditions, ultimately aiding in the understanding of their biochemical properties.

Principles of Electrophoresis

Before delving into the specific application of electrophoresis to amino acids and dipeptides, it is essential to understand the underlying principles of the technique. Electrophoresis is based on the movement of charged particles in an electric field. When an electric current is applied to a conductive medium, charged particles migrate towards the electrode of opposite charge. The migration rate, or electrophoretic mobility, is influenced by the net charge, size, and shape of the particles, as well as the electric field strength and the properties of the medium.

In the case of amino acids and dipeptides, the charged nature of these molecules arises from the presence of ionizable functional groups. Amino acids contain both amino (-NH2) and carboxyl (-COOH) groups, which can be ionized at certain pH values. Dipeptides, on the other hand, are formed by the condensation of two amino acids, resulting in an amide bond between the carboxyl group of one amino acid and the amino group of another. This amide bond is relatively uncharged and does not significantly contribute to electrophoretic mobility.

pH and Ionization of Amino Acids and Dipeptides

The behavior of amino acids and dipeptides in electrophoresis is heavily influenced by the pH of the surrounding medium. The pH determines the ionization state of the functional groups present in these molecules. At low pH values (acidic conditions), the amino group tends to be protonated (NH3+), while the carboxyl group is deprotonated (COO-). Conversely, at high pH values (alkaline conditions), the amino group is deprotonated (NH2), and the carboxyl group is protonated (COOH). At intermediate pH values, both functional groups can be partially ionized.

The ionization state of amino acids and dipeptides affects their net charge and, consequently, their electrophoretic mobility. At a given pH, the net charge can be positive, negative, or zero, depending on the acidic or basic groups that are ionized. Charged particles with positive polarity move towards the negative electrode known as the cathode, while charged particles with negative polarity travel towards the positive electrode referred to as the anode. Uncharged particles do not migrate significantly.

To illustrate this concept, let's consider the amino acid glycine and the dipeptide glycyl-glycine. Glycine has a single amino group and a single carboxyl group, giving it a net charge of zero at pH 7 (physiological pH). At low pH values, the amino group becomes protonated, resulting in a positively charged glycine molecule. Conversely, at high pH values, the carboxyl group becomes deprotonated, leading to a negatively charged glycine molecule. The electrophoretic mobility of glycine will vary accordingly.

In the case of the dipeptide glycyl-glycine, the behavior is slightly different. The amide bond between the two amino acids does not contribute significantly to the net charge or electrophoretic mobility. Rather, the behavior of glycyl-glycine is determined by the ionization states of the individual amino acids within the dipeptide. Each amino acid can be ionized independently, leading to a range of possible net charges and electrophoretic mobilities for glycyl-glycine at different pH values.

Electrophoretic Separation of Amino Acids and Dipeptides

Electrophoresis can be used to separate mixtures of amino acids and dipeptides based on their different electrophoretic mobilities. By applying an electric field to a medium, such as a gel or a capillary, charged molecules will migrate at different rates depending on their net charge and other factors mentioned earlier. By controlling the pH of the medium, it is possible to manipulate the ionization states and charge distribution of the amino acids and dipeptides, resulting in differential migration and separation.

To achieve separation, various electrophoretic techniques can be employed, including gel electrophoresis and capillary electrophoresis. Gel electrophoresis involves placing the sample mixture in a gel matrix and applying an electric field. The gel acts as a sieving medium, slowing down the migration of larger molecules compared to smaller ones. This differential migration results in the separation of the mixture into distinct bands or spots, which can be visualized and further analyzed.

Capillary electrophoresis, on the other hand, relies on the migration of charged particles through a narrow capillary filled with an electrolyte solution. The separation occurs based on the differences in electrophoretic mobility of the analytes, as they migrate at different rates through the capillary. Capillary electrophoresis offers higher resolution and faster separations compared to gel electrophoresis, making it a popular choice in modern analytical laboratories.

Predicting and Interpreting Electrophoretic Results

The electrophoretic separation of mixtures of amino acids and dipeptides at varying pHs can yield valuable information about their properties and behavior. By analyzing the migration patterns, it is possible to predict the net charge, pKa values, and isoelectric points of the analytes, as well as their relative abundance in the mixture.

The net charge of an analyte at a given pH can be determined by observing its migration towards the anode or cathode during electrophoresis. A positively charged analyte will migrate towards the cathode, while a negatively charged analyte will migrate towards the anode. Uncharged analytes will not migrate significantly. By comparing the migration patterns of known standards, the net charge of unknown analytes can be inferred.

The pKa values of the functional groups in amino acids and dipeptides can also be estimated from electrophoretic results. The pKa is the pH at which 50% of a given functional group is ionized. By analyzing the migration patterns at different pH values, it is possible to identify the pH at which the net charge of an analyte changes sign, indicating the pKa of the relevant functional group. This information is crucial for understanding the acid-base properties of amino acids and dipeptides.

Furthermore, electrophoresis can provide insights into the isoelectric point (pI) of amino acids and dipeptides. The pI is the pH at which an analyte has no net charge, resulting in minimal migration during electrophoresis.

For amino acids, the pI is typically the average of the pKa values of the amino and carboxyl groups. For dipeptides, the pI can be estimated by considering the pKa values of the individual amino acids and their relative proportions within the dipeptide. The pI is an essential parameter for protein characterization and purification processes.

Finally, electrophoretic separation can also reveal the relative abundance of different amino acids and dipeptides in a mixture. By comparing the intensities of the separated bands or spots, it is possible to estimate the concentration or relative abundance of each analyte. This information can be crucial for understanding the composition and dynamics of complex biological samples, such as protein digests or metabolic profiles.

Counterarguments and Limitations

While electrophoresis is a powerful technique for the separation and analysis of amino acids and dipeptides, it is not without limitations. One major limitation is the need for prior knowledge or standard references to interpret the electrophoretic results accurately. Without known standards, it can be challenging to assign net charges, pKa values, and pI values to unknown analytes solely based on migration patterns. Therefore, careful experimental design and validation with known standards are crucial for accurate interpretation.

Another limitation is the dependency of electrophoretic mobility on factors other than pH, such as temperature, buffer composition, and electric field strength. These factors can influence the separation efficiency and migration rates of analytes, potentially leading to variations in the observed migration patterns. Therefore, it is important to carefully control and standardize these parameters to ensure reproducibility and reliability of results.

Lastly, electrophoresis is limited in its ability to separate analytes with similar charge-to-mass ratios. If two analytes have similar net charges and sizes, their separation may be challenging or even impossible using electrophoresis alone. In such cases, additional techniques, such as chromatography or mass spectrometry, may be required for further analysis and characterization.

Polymerisation

Condensation polymerisation

Formation of Polyesters

Polyesters are a class of polymers that are formed through the reaction between a diol and a dicarboxylic acid or dioyl chloride. These polymers have a wide range of applications in various fields, including textiles, packaging materials, and biomedical devices. Understanding the formation of polyesters is crucial for designing and synthesizing polymers with desired properties.

(a) Reaction between a Diol and a Dicarboxylic Acid or Dioyl Chloride

The formation of polyesters involves a condensation reaction between a diol and a dicarboxylic acid or dioyl chloride. In this section, we will discuss the reaction mechanism, factors affecting the reaction, and the importance of catalysts.

Reaction Mechanism

The reaction between a diol and a dicarboxylic acid is a step-growth polymerization process, also known as polycondensation. This reaction involves the formation of an ester bond (-COO-) between the hydroxyl (-OH) groups of the diol and the carboxyl (-COOH) groups of the dicarboxylic acid.

The reaction proceeds in two steps: esterification and condensation. In the esterification step, the hydroxyl groups of the diol react with the carboxyl groups of the dicarboxylic acid, leading to the formation of ester linkages. This step is typically catalyzed by an acid catalyst, such as sulfuric acid or p-toluenesulfonic acid, which increases the reaction rate.

Once the ester linkages are formed, the reaction enters the condensation step. In this step, water molecules are eliminated as byproducts, and the polymer chains grow through the repetition of esterification and condensation reactions. The elimination of water is facilitated by heating the reaction mixture or using a dehydration agent, such as thionyl chloride in the case of dioyl chloride.

Factors Affecting the Reaction

Several factors influence the formation of polyesters, including the choice of diol and dicarboxylic acid, reaction conditions, and the presence of catalysts.

Choice of Diol and Dicarboxylic Acid

The choice of diol and dicarboxylic acid significantly impacts the properties of the resulting polyester. Different diols and dicarboxylic acids can provide polymers with varying mechanical, thermal, and chemical properties. For example, the use of a diol with a long alkyl chain can result in a polyester with high flexibility, while a diol with aromatic moieties can enhance the polymer's thermal stability.

Similarly, the choice of dicarboxylic acid can influence the properties of the polyester. For instance, using a dicarboxylic acid with a bulky substituent can increase the rigidity of the polymer, whereas a dicarboxylic acid with a flexible linker can enhance the polymer's flexibility.

Reaction Conditions

Reaction conditions, such as temperature and reaction time, play a crucial role in the formation of polyesters. Generally, higher temperatures accelerate the reaction rate by providing the necessary activation energy for the esterification and condensation steps. However, excessively high temperatures can lead to side reactions or degradation of the polymer.

The reaction time is another critical parameter that affects the molecular weight and extent of polymerization. Longer reaction times allow for more esterification and condensation reactions, resulting in higher molecular weight polymers. However, prolonged reaction times can also lead to a higher degree of crosslinking, which can affect the polymer's properties.

Catalysts

Catalysts are often employed to accelerate the reaction rate and improve the efficiency of polyester formation. Acid catalysts, such as sulfuric acid, are commonly used to promote the esterification step by increasing the reaction rate. These catalysts facilitate the protonation of the hydroxyl groups, making them more reactive towards the carboxyl groups.

In addition to acid catalysts, other types of catalysts, such as organometallic compounds, have been developed to facilitate the formation of polyesters. For example, titanium-based catalysts have shown excellent activity in the polyesterification reaction, leading to polymers with controlled molecular weights and narrow polydispersity.

Importance of Catalysts

Catalysts play a crucial role in polyester formation by increasing the reaction rate and controlling the polymer's properties. They enable the reaction to proceed under milder conditions, reducing the energy requirements and improving the efficiency of the process. Catalysts also allow for the synthesis of polymers with desired molecular weights, polydispersity, and end-group functionality.

Moreover, catalysts can influence the stereochemistry of the polymer chains. For example, chiral catalysts can selectively promote the formation of polyesters with specific stereochemistry, leading to polymers with unique properties. This ability to control the stereochemistry is particularly relevant in the production of biodegradable polyesters, such as polylactic acid (PLA), which exhibit distinct properties depending on their stereochemical arrangement.

(b) Reaction of a Hydroxycarboxylic Acid

In addition to the reaction between a diol and a dicarboxylic acid, polyesters can also be formed through the reaction of a hydroxycarboxylic acid. This section will focus on the reaction mechanism, applications, and advantages of using hydroxycarboxylic acids.

Reaction Mechanism

The reaction of a hydroxycarboxylic acid involves the self-condensation of the acid's hydroxyl and carboxyl groups. This process, known as esterification, leads to the formation of an intramolecular ester bond and the elimination of water as a byproduct.

The reaction can be catalyzed by an acid catalyst, similar to the reaction between a diol and a dicarboxylic acid. The acid catalyst enhances the reaction rate by facilitating the protonation of the hydroxyl group, making it more reactive towards the carboxyl group.

Applications

Polyesters formed from hydroxycarboxylic acids have gained significant attention due to their biodegradability and biocompatibility. One prominent example is polylactic acid (PLA), which is derived from lactic acid. PLA has a wide range of applications, including packaging materials, biomedical devices, and drug delivery systems. The biodegradability of polyesters derived from hydroxycarboxylic acids makes them environmentally friendly alternatives to conventional plastics. They can be broken down by microorganisms into harmless byproducts, reducing the accumulation of non-biodegradable waste in the environment. Moreover, their biocompatibility allows for their use in biomedical applications, such as sutures, tissue engineering scaffolds, and implants.

Advantages of Hydroxycarboxylic Acids

The use of hydroxycarboxylic acids offers several advantages in the formation of polyesters compared to the diol-dicarboxylic acid reaction. Firstly, hydroxycarboxylic acids are readily available from renewable resources, such as sugars or biomass, making them attractive for sustainable polymer production. In contrast, the synthesis of diols and dicarboxylic acids often involves complex and energy-intensive processes.

Furthermore, hydroxycarboxylic acids can undergo polymerization without the need for a diol counterpart. This simplifies the polymerization process, as it eliminates the requirement for stoichiometric control between the diol and dicarboxylic acid. The absence of a diol also prevents the formation of a linear polymer backbone, resulting in a three-dimensional network structure with enhanced mechanical properties.

Counterarguments

Despite the advantages of using hydroxycarboxylic acids in polyester formation, there are some limitations and challenges associated with this approach. One limitation is the relatively low reactivity of hydroxycarboxylic acids compared to diols and dicarboxylic acids. This can result in slower reaction rates and the need for higher temperatures or stronger catalysts.

Moreover, the self-condensation of hydroxycarboxylic acids can lead to the formation of cyclic oligomers, which can limit the molecular weight and properties of the resulting polyester. This issue can be addressed by using a multifunctional acid, such as citric acid, which can undergo crosslinking reactions and enhance the mechanical properties of the polymer.

Polyamides are a class of polymers that possess amide linkages within their molecular structure. These polymers exhibit a wide range of properties and have found numerous applications in various fields such as textiles, engineering materials, coatings, and biomedical devices. The formation of polyamides can occur through different synthetic routes, including the reaction between a diamine and a dicarboxylic acid or dioyl chloride, the reaction of an aminocarboxylic acid, and the reaction between amino acids.

The most common method for synthesizing polyamides involves the reaction between a diamine and a dicarboxylic acid or dioyl chloride. This process, known as polycondensation, leads to the formation of a polymer through the reaction of two monomers, each containing a functional group that can react with the other. In the case of polyamides, the diamine provides the amine functionality, while the dicarboxylic acid or dioyl chloride provides the carboxylic acid functionality.

In the reaction between a diamine and a dicarboxylic acid, the amine groups of the diamine react with the carboxylic acid groups of the dicarboxylic acid, resulting in the formation of amide linkages. This reaction is typically carried out in a solvent, such as a polar aprotic solvent like dimethylformamide (DMF), under controlled conditions of temperature and pressure. For example, the reaction between ethylenediamine (EDA) and adipic acid (AA) can be used to synthesize a polyamide known as nylon-6,6. In this reaction, the amine groups of EDA react with the carboxylic acid groups of AA, leading to the formation of amide linkages. The resulting polymer has a repeating unit composed of two carbon atoms from AA and two carbon atoms from EDA, connected by an amide linkage.

The reaction between a diamine and a dioyl chloride follows a similar mechanism. Dioyl chlorides are often used instead of dicarboxylic acids due to their higher reactivity and ease of handling. In this case, the amine groups of the diamine react with the acid chloride groups of the dioyl chloride, resulting in the formation of amide linkages. This reaction is typically carried out in an organic solvent, such as dichloromethane or chloroform, under an inert atmosphere, such as nitrogen or argon.

Nylon-6,6 can also be synthesized using the reaction between hexamethylenediamine (HMDA) and adipoyl chloride (AC). In this reaction, the amine groups of HMDA react with the acid chloride groups of AC, leading to the formation of amide linkages. The resulting polymer has the same repeating unit as nylon-6,6 synthesized from EDA and AA.

Another method for the formation of polyamides involves the reaction of an aminocarboxylic acid. Aminocarboxylic acids, also known as alpha-amino acids, are compounds that contain both an amine group and a carboxylic acid group within the same molecule. The reaction between an aminocarboxylic acid and a dicarboxylic acid or dioyl chloride leads to the formation of a polyamide with a different structure compared to polyamides synthesized from diamines.

In this reaction, the amine group of the aminocarboxylic acid reacts with the carboxylic acid group of the dicarboxylic acid or the acid chloride group of the dioyl chloride, resulting in the formation of amide linkages. The resulting polymer has a repeating unit composed of the aminocarboxylic acid and the dicarboxylic acid or dioyl chloride. For example, the reaction between glycine (aminoacetic acid) and adipic acid can be used to synthesize a polyamide known as nylon-6. In this reaction, the amine group of glycine reacts with the carboxylic acid group of adipic acid, leading to the formation of an amide linkage. The resulting polymer has a repeating unit composed of glycine and adipic acid, connected by an amide linkage.

The formation of polyamides through the reaction between amino acids is an intriguing process that mimics natural biosynthesis. Amino acids are the building blocks of proteins and are essential for life. The reaction between amino acids to form polyamides, often referred to as peptide bond formation, is a condensation reaction that involves the elimination of a water molecule.

In this reaction, the amine group of one amino acid reacts with the carboxylic acid group of another amino acid, resulting in the formation of a peptide bond. The resulting polymer is a polypeptide chain, with each amino acid residue connected by a peptide bond.

The reaction between amino acids can occur in various environments, such as in the presence of a catalyst or under specific conditions of temperature and pH. For example, the condensation of amino acids can be catalyzed by enzymes called peptidyl transferases, which play a crucial role in protein synthesis in living organisms.

The formation of polyamides from amino acids has significant implications in fields such as biochemistry and biotechnology. For instance, the synthesis of polypeptides through solid-phase peptide synthesis (SPPS) has revolutionized the production of therapeutic peptides and proteins. SPPS involves the stepwise addition of protected amino acids onto a solid support, followed by the removal of the protecting groups and the formation of peptide bonds.

One of the main advantages of polyamides synthesized from amino acids is their biocompatibility. Due to their resemblance to natural proteins, polyamides derived from amino acids exhibit excellent biocompatibility and can be used in various biomedical applications, such as drug delivery systems and tissue engineering scaffolds.

Despite the numerous advantages of polyamides, there are some limitations and challenges associated with their synthesis and application. One of the challenges is achieving control over the molecular weight and polydispersity of the resulting polymer. Polyamides synthesized through polycondensation reactions often exhibit a broad molecular weight distribution, which can affect their mechanical and thermal properties.

Furthermore, the choice of monomers and reaction conditions can influence the properties of the resulting polyamide. For example, the use of different diamines or dicarboxylic acids can lead to polyamides with different degrees of crystallinity, melting points, and mechanical properties.

Deducing the Repeat Unit of a Condensation Polymer

Introduction

We will explore the process of deducing the repeat unit of a condensation polymer obtained from a given monomer or pair of monomers, taking into account the formal and academic tone, advanced vocabulary and grammar, and providing a thorough analysis of the subject matter.

Understanding Condensation Polymers

Before delving into the process of deducing the repeat unit, it is essential to have a solid understanding of condensation polymers. Unlike addition polymers, which are formed through the repetitive addition of monomers without the elimination of any byproducts, condensation polymers involve the reaction between functional groups on the monomers, resulting in the release of a small molecule.

The most common example of a condensation polymer is nylon, which is formed by the reaction of a diamine (containing two amino groups) and a dicarboxylic acid (containing two carboxylic acid groups). When these two monomers react, water is eliminated, and the polymer chain is formed. The repeat unit of nylon is derived from the combination of the amine and acid functional groups.

Steps to Deduce the Repeat Unit

Deducing the repeat unit of a condensation polymer involves several steps, including identifying the monomers, understanding their functional groups, analyzing the reaction mechanism, and determining the resulting polymer structure. Let's break down these steps in detail.

Step 1: Identify the Monomers

To begin, we need to identify the monomers involved in the formation of the condensation polymer. Monomers are the individual building blocks that combine to form the polymer chain. For example, in the case of nylon, the monomers are a diamine and a dicarboxylic acid.

Step 2: Understand the Functional Groups

Next, we need to understand the functional groups present in the monomers. Functional groups are specific groups of atoms that determine the chemical reactivity and properties of a compound. In the case of nylon, the diamine monomer contains two amino groups ($-NH_2$), and the dicarboxylic acid monomer contains two carboxylic acid groups ($-COOH$).

Step 3: Analyze the Reaction Mechanism

After identifying the monomers and their functional groups, we must analyze the reaction mechanism involved in the formation of the condensation polymer. The reaction typically proceeds through a condensation reaction, where the functional groups on the monomers combine, resulting in the elimination of a small molecule, such as water or alcohol.

In the case of nylon, the reaction between the diamine and dicarboxylic acid involves the formation of an amide bond ($-CONH-$) and the release of water. This reaction can be represented as follows:

$$H_2N-R-NH_2 + HOOC-R'-COOH \rightarrow H_2N-R-NH-CO-R'-COOH + H_2O$$

Step 4: Determine the Polymer Structure

Once we understand the reaction mechanism, we can determine the structure of the resulting polymer. The repeat unit of the polymer is derived from the combination of the functional groups present in the monomers. For nylon, the repeat unit can be deduced by combining the amine group from the diamine monomer and the carboxylic acid group from the dicarboxylic acid monomer. The resulting repeat unit is represented as follows:

$$O$$
$$\|$$
$$H_2N-R-NH-CO-R'-CO$$

This structure represents the repeating unit of nylon, which continues to form the polymer chain through the repetition of this unit.

Step 5: Consider Stereochemistry and Side Reactions

In some cases, it is necessary to consider stereochemistry, particularly if the monomers possess chiral centers. Stereochemistry refers to the spatial arrangement of atoms in a molecule, and it can significantly impact the properties and behavior of the polymer.

Additionally, it is important to consider any side reactions that may occur during the polymerization process. Side reactions can lead to the formation of different structures or branching in the polymer chain, affecting its overall properties.

Examples from Various Fields

To further illustrate the process of deducing the repeat unit, let's explore examples from various fields, including biochemistry, materials science, and pharmaceuticals.

Example 1: DNA and RNA

In biochemistry, DNA (deoxyribonucleic acid) and RNA (ribonucleic acid) are condensation polymers formed by the repetitive addition of nucleotide monomers. Nucleotides consist of three components: a sugar (ribose or deoxyribose), a phosphate group, and a nitrogenous base (adenine, thymine, cytosine, guanine, or uracil).

The repeat unit of DNA is derived from the combination of the sugar and phosphate groups. The sugar-phosphate backbone forms the main structure of the DNA molecule, while the nitrogenous bases project from this backbone, encoding genetic information.

Example 2: Polyethylene Terephthalate (PET)

In materials science, polyethylene terephthalate (PET) is a widely used condensation polymer. It is formed by the reaction of terephthalic acid and ethylene glycol. The repeat unit of PET is derived from the combination of the terephthalic acid and ethylene glycol functional groups.

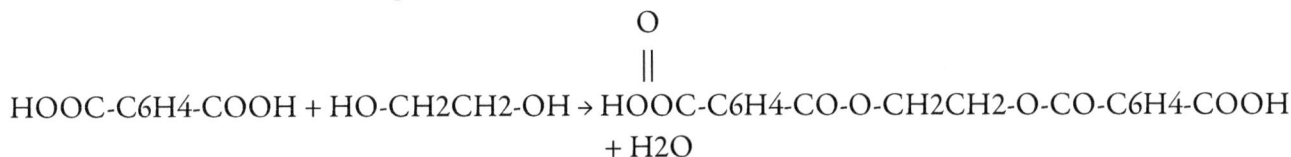

$$\overset{\displaystyle O}{\overset{\displaystyle \|}{}}$$

$$HOOC\text{-}C_6H_4\text{-}COOH + HO\text{-}CH_2CH_2\text{-}OH \rightarrow HOOC\text{-}C_6H_4\text{-}CO\text{-}O\text{-}CH_2CH_2\text{-}O\text{-}CO\text{-}C_6H_4\text{-}COOH + H_2O$$

Example 3: Nylon-6,6

In the pharmaceutical industry, nylon-6,6 is a commonly used condensation polymer. It is formed by the reaction of hexamethylenediamine and adipic acid. The repeat unit of nylon-6,6 is derived from the combination of the amine and acid functional groups.

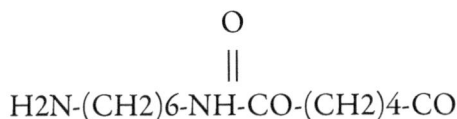

$$\overset{\displaystyle O}{\overset{\displaystyle \|}{}}$$

$$H_2N\text{-}(CH_2)_6\text{-}NH\text{-}CO\text{-}(CH_2)_4\text{-}CO$$

Counterarguments and Limitations

While the process described above is generally applicable for deducing the repeat unit of condensation polymers, there are certain counterarguments and limitations to consider.

Counterargument 1: Side Reactions and Impurities

During the condensation polymerization process, side reactions can occur, leading to the formation of impurities or alternative structures. These side reactions can complicate the determination of the repeat unit and require careful analysis and purification techniques.

Counterargument 2: Copolymerization

In some cases, condensation polymers can be formed from more than two monomers. Copolymerization introduces additional complexity in deducing the repeat unit, as it involves the combination of multiple functional groups. Advanced analytical techniques, such as nuclear magnetic resonance (NMR) spectroscopy and mass spectrometry, are often employed to determine the exact monomer ratios and repeat unit structure in copolymers.

Identifying the Monomer(s) in a Section of a Condensation Polymer Molecule

Introduction

We will explore the concept of identifying monomers in a section of a condensation polymer molecule, discussing the key principles, techniques, and challenges involved. We will employ a formal and academic tone, utilizing advanced vocabulary and grammar to elucidate complex concepts. Additionally, we will provide a thorough analysis of the subject matter, drawing examples from various fields, and present counterarguments objectively.

The Nature of Condensation Polymers

Condensation polymers are formed when monomers undergo a condensation reaction, which involves the combination of two or more molecules with the simultaneous elimination of a small molecule, typically water. This process is in contrast to addition polymers, where monomers combine without any byproduct formation. The elimination of water or other small molecules during the polymerization process allows the monomers to link together, forming long chains or networks of repeating units called polymers.

The structure of a condensation polymer molecule can be represented by a repeating unit, which reflects the arrangement of monomers within the polymer chain. Identification of the monomers in a given section of a condensation polymer molecule is vital for understanding its chemical composition and, consequently, its properties.

Analyzing Monomer Composition

Identifying the monomers present in a section of a condensation polymer molecule can be achieved through various analytical techniques. One such technique is spectroscopy, which involves the interaction of electromagnetic radiation with matter to provide information about its composition and structure. Spectroscopic techniques commonly used for analyzing polymers include infrared spectroscopy (IR), nuclear magnetic resonance (NMR) spectroscopy, and mass spectrometry (MS).

Infrared Spectroscopy (IR)

IR spectroscopy measures the absorption of infrared light by a sample, providing information about the functional groups present in the molecule. Each functional group absorbs light at specific wavelengths, resulting in characteristic peaks in the IR spectrum. By comparing the IR spectrum of a section of a condensation polymer molecule with reference spectra of known monomers, it is possible to identify the monomers present in that section.

Nuclear Magnetic Resonance (NMR) Spectroscopy

NMR spectroscopy provides detailed information about the structure and connectivity of atoms within a molecule. By analyzing the NMR spectrum of a section of a condensation polymer molecule, the types and arrangement of monomers can be deduced.

Different NMR techniques, such as proton NMR (^1H NMR) and carbon-13 NMR (^{13}C NMR), can be employed to analyze polymers. ^1H NMR provides information about the hydrogen atoms present in the molecule, while ^{13}C NMR reveals the carbon atoms. By comparing the NMR spectra of the polymer section with those of known monomers, the presence of specific monomers in the polymer can be confirmed.

Mass Spectrometry (MS)

Mass spectrometry is a powerful technique for analyzing the mass and structure of molecules. It involves the ionization of a sample, followed by separation and detection of the resulting ions based on their mass-to-charge ratio (m/z).

In the case of condensation polymers, the polymer section of interest can be fragmented into smaller molecules, or oligomers, by subjecting it to a suitable ionization technique, such as electrospray ionization (ESI) or matrix-assisted laser desorption/ionization (MALDI). The resulting fragments can then be analyzed by mass spectrometry to determine their mass and composition. By comparing the mass spectra of the polymer fragments with those of known monomers, the monomers present in the polymer can be identified.

Challenges in Monomer Identification

While various analytical techniques can aid in monomer identification, there are several challenges that researchers may encounter. One such challenge is the complexity of condensation polymer structures. Condensation polymers can have intricate architectures, branching, and variable monomer compositions, making it challenging to decipher the monomers present in a given section. Additionally, the presence of impurities or side reactions during the polymerization process can further complicate the analysis.

Another challenge lies in the availability of reference spectra for comparison. The identification of monomers in a condensation polymer relies on comparing experimental spectra with reference spectra of known monomers. However, obtaining reference spectra for all possible monomers is a daunting task. This limitation can lead to uncertainties and ambiguities in monomer identification.

Furthermore, the presence of counterions or additives in the polymer can interfere with the analysis, obstructing the accurate determination of monomer composition. It is essential to carefully prepare the polymer sample and remove any potential interferences before conducting the analysis.

Application and Significance

The ability to identify the monomers present in a given section of a condensation polymer molecule has significant implications across various fields, including materials science, pharmaceuticals, and biotechnology. In materials science, knowing the monomer composition of a condensation polymer enables researchers to tailor its properties according to desired applications. For example, identifying the monomers in a section of a polyester polymer used in the production of textiles can provide insights into its mechanical strength, thermal stability, and dyeability. This knowledge can guide the development of improved textile materials with enhanced performance characteristics.

In the pharmaceutical industry, the identification of monomers in a section of a drug delivery polymer can elucidate its drug release mechanism, bioavailability, and biocompatibility. This information is crucial for optimizing drug formulations and designing effective drug delivery systems.

In biotechnology, the analysis of monomer composition in a section of a biopolymer, such as DNA or protein, can aid in understanding their structure-function relationships. This knowledge is essential for elucidating biological processes, designing novel biomaterials, and developing therapeutic interventions.

Counterarguments

While identifying monomers in a section of a condensation polymer molecule is a valuable technique, it is not without limitations. One counterargument is that the analysis may not provide a complete picture of the polymer's structure. In some cases, the monomer composition of a polymer section may be ambiguous or difficult to determine due to overlapping signals or similarities between monomers. This limitation can hinder a comprehensive understanding of the polymer's properties and behavior.

Additionally, the analysis of monomers in a condensation polymer often requires sophisticated instrumentation and expertise. The availability of such resources may be limited, especially in smaller research facilities or developing regions. This limitation can restrict the widespread application of monomer identification techniques, impeding progress in polymer research and development.

Predicting the type of polymerisation

Polymerisation Reactions: Predicting the Type

We will explore the different types of polymerisation reactions and discuss how to predict the type of reaction based on the characteristics of the monomer or monomers involved.

Introduction to Polymerisation Reactions

Polymerisation reactions involve the chemical bonding of monomers to form polymer chains or networks. These reactions can be broadly classified into two main categories: addition (chain-growth) polymerisation and condensation (step-growth) polymerisation. While both types result in the creation of polymers, they differ in the mechanism by which monomers are incorporated into the polymer structure.

Addition (Chain-Growth) Polymerisation

Addition polymerisation, also known as chain-growth polymerisation, involves the sequential addition of monomers to an active site, typically a reactive center or an initiator, resulting in the growth of a polymer chain. This type of polymerisation reaction is commonly observed in the synthesis of polymers such as polyethylene, polypropylene, and polystyrene.

Mechanism

The mechanism of addition polymerisation can be further classified into three subcategories: free radical, anionic, and cationic polymerisation.

Free Radical Polymerisation

Free radical polymerisation is the most common type of addition polymerisation reaction. It involves the initiation, propagation, and termination steps. In the initiation step, a free radical initiator, such as a peroxide or azo compound, is activated by heat, light, or a chemical reaction, generating a reactive free radical species. This free radical then reacts with a monomer, abstracting a hydrogen atom and initiating the polymer chain. The propagation step involves the repeated addition of monomers to the growing chain, while the termination step occurs when two growing chains combine or when an active radical reacts with a terminating agent, leading to the end of the polymerisation process.

An example of free radical polymerisation is the synthesis of polyethylene from ethylene monomers. The initiation step involves the activation of a peroxide initiator, generating free radicals. These free radicals react with multiple ethylene monomers through the propagation step, resulting in the growth of polyethylene chains.

Anionic Polymerisation

Anionic polymerisation involves the initiation of a polymer chain by an anionic species, such as an alkoxide or amide ion. The anionic initiator reacts with a monomer, forming a reactive anion that initiates the polymerisation process. Unlike free radical polymerisation, anionic polymerisation does not have a termination step, as the anion can continue to propagate by reacting with additional monomers.

An example of anionic polymerisation is the synthesis of polybutadiene from 1,3-butadiene monomers using an alkoxide initiator. The alkoxide initiates the formation of a reactive anion, which then reacts with multiple 1,3-butadiene monomers, leading to the growth of polybutadiene chains.

Cationic Polymerisation

Cationic polymerisation involves the initiation of a polymer chain by a cationic species, such as a Lewis acid or a protic acid. The cationic initiator reacts with a monomer, forming a reactive cation that initiates the polymerisation process. Similar to anionic polymerisation, cationic polymerisation does not have a termination step, allowing the cation to continue propagating by reacting with additional monomers.

An example of cationic polymerisation is the synthesis of polyisobutylene from isobutylene monomers using a protic acid initiator. The protic acid initiates the formation of a reactive cation, which then reacts with multiple isobutylene monomers, resulting in the growth of polyisobutylene chains.

Predicting the Type of Addition Polymerisation

To predict the type of addition polymerisation for a given monomer or pair of monomers, several factors need to be considered. These factors include the reactivity of the monomer, the availability of suitable initiators, and the desired properties of the resulting polymer.

Firstly, the reactivity of the monomer influences the type of addition polymerisation that can occur. Monomers with highly reactive functional groups, such as double bonds or nucleophilic sites, are more likely to undergo free radical or anionic polymerisation, as these mechanisms can accommodate the reactivity of such monomers. On the other hand, monomers with less reactive functional groups may favor cationic polymerisation, as this mechanism can initiate the polymerisation process under milder conditions.

Secondly, the availability of suitable initiators is crucial in determining the type of addition polymerisation. Free radical initiators, such as peroxides or azo compounds, are widely used and readily available, making them suitable for free radical polymerisation. Anionic and cationic initiators, such as alkoxides and Lewis acids, respectively, may require specific conditions or reagents for their synthesis, limiting their applicability in certain cases.

Finally, the desired properties of the resulting polymer also play a role in predicting the type of addition polymerisation. For instance, if a polymer with a high molecular weight and narrow molecular weight distribution is desired, anionic polymerisation may be preferred due to its ability to propagate without termination. On the other hand, if a polymer with a broader molecular weight distribution and a relatively lower molecular weight is desired, free radical polymerisation may be more suitable, as it often leads to the formation of branched polymers.

Condensation (Step-Growth) Polymerisation

Condensation polymerisation, also known as step-growth polymerisation, involves the formation of covalent bonds between monomers through the elimination of small molecules, such as water or alcohol. This type of polymerisation reaction is commonly observed in the synthesis of polymers such as polyesters, polyamides, and polyurethanes.

Mechanism

The mechanism of condensation polymerisation can be described as a stepwise process involving the formation of reactive intermediates, followed by their reaction to form covalent bonds. The reaction typically proceeds through alternating steps of monomer addition and small molecule elimination until the desired polymer is formed.

In the first step, two monomers, each possessing reactive functional groups, undergo a nucleophilic or electrophilic attack, leading to the formation of a reactive intermediate. This intermediate can then react with another monomer, resulting in the formation of a covalent bond between the two monomers and the release of a small molecule, such as water or alcohol. The process persists until the desired length of the polymer chain is attained.

An example of condensation polymerisation is the synthesis of polyethylene terephthalate (PET) from terephthalic acid and ethylene glycol. The carboxylic acid groups in terephthalic acid react with the hydroxyl groups in ethylene glycol, forming ester linkages and releasing water molecules as byproducts. This stepwise process continues until a long chain of repeating units is formed, resulting in the formation of PET.

Predicting the Type of Condensation Polymerisation

Predicting the type of condensation polymerisation for a given monomer or pair of monomers can be influenced by various factors, including the presence of suitable functional groups, the reactivity of these functional groups, and the stability of the small molecules released during the reaction.

Firstly, the presence of suitable functional groups is essential for condensation polymerisation. Monomers with complementary functional groups, such as carboxylic acid and alcohol groups, amine and acid chloride groups, or isocyanate and alcohol groups, are more likely to undergo condensation polymerisation. The reactivity of these functional groups can also influence the reaction rate and the extent of polymerisation.

Secondly, the stability of the small molecules released during the reaction is an important consideration. For example, the release of water or alcohol molecules during condensation polymerisation can lead to the formation of equilibrium mixtures, limiting the extent of polymerisation. In such cases, additional measures, such as using azeotropic agents or applying vacuum, may be required to drive the reaction towards completion. Furthermore, the reaction conditions, such as temperature and pressure, can also affect the type of condensation polymerisation that occurs. Higher temperatures typically favor faster reactions, while higher pressures can increase the extent of polymerisation.

Polymerisation Reactions and the Production of Polymer Molecules

Introduction

We will explore the different types of polymerisation reactions and discuss how they contribute to the formation of specific sections within polymer molecules. We will approach this topic from a formal and academic standpoint, using advanced vocabulary and grammar to provide a thorough analysis. Additionally, we will explain complex concepts clearly and present counterarguments objectively to offer a comprehensive understanding of the subject matter.

Types of Polymerisation Reactions

Addition polymerisation and condensation polymerisation are the two primary types of polymerisation reactions. These reactions differ in their mechanisms and the byproducts they produce during the polymerisation process.

Addition Polymerisation

Addition polymerisation, also known as chain-growth polymerisation, involves the sequential addition of monomers to the growing polymer chain. This reaction occurs through the activation of a reactive center in the monomer, typically a carbon-carbon double bond (C=C). The reaction proceeds by breaking the double bond and forming a new covalent bond with another monomer unit.

One of the most prominent examples of addition polymerisation is the production of polyethylene, a widely used polymer. Polyethylene is synthesized by the addition polymerisation of ethylene monomers, which consist of two carbon atoms bonded to each other with two hydrogen atoms attached to each carbon. The double bond between the carbon atoms in the ethylene monomer is broken, and the carbon atoms form new bonds with other ethylene monomers, resulting in a long chain of repeating units.

The nature of the monomer and the reaction conditions play a crucial role in determining the structure and properties of the resulting polymer. For instance, the addition polymerisation of different monomers can lead to polymers with varying characteristics, such as low density polyethylene (LDPE) and high density polyethylene (HDPE).

Condensation Polymerisation

Condensation polymerisation, also known as step-growth polymerisation, involves the formation of covalent bonds between two different monomers, accompanied by the elimination of a small molecule, such as water or alcohol. Unlike addition polymerisation, which produces a polymer chain with identical repeating units, condensation polymerisation allows for the incorporation of different monomers, leading to copolymers with diverse monomer compositions. An example of condensation polymerisation is the synthesis of polyamide, commonly known as nylon. Nylon is produced by the reaction between a dicarboxylic acid and a diamine. In this reaction, the carboxyl group (-COOH) of the acid reacts with the amine group (-NH2) of the diamine, resulting in the formation of an amide bond (-CONH-) and the release of a water molecule. This process is repeated with additional monomer units, leading to the formation of a long polyamide chain.

The versatility of condensation polymerisation allows for the creation of polymers with tailored properties through the careful selection of monomers. For example, the incorporation of different dicarboxylic acids and diamines in the synthesis of nylon can yield polymers with varying degrees of flexibility, strength, and thermal stability.

Determining the Type of Polymerisation Reaction

To deduce the type of polymerisation reaction that produces a given section of a polymer molecule, several factors need to be considered, including the structure of the monomers, the presence of functional groups, and the reaction conditions. Analyzing these elements can provide insights into the mechanisms and processes involved in the polymerisation reaction.

Monomer Structure

The structure of the monomer units plays a crucial role in determining the type of polymerisation reaction. Monomers with a carbon-carbon double bond (C=C) are typically involved in addition polymerisation, where the double bond is broken, and new bonds are formed with other monomer units.

On the other hand, monomers containing functional groups, such as hydroxyl (-OH) or carboxyl (-COOH) groups, are often involved in condensation polymerisation. These functional groups react with each other, leading to the formation of covalent bonds and the release of small molecules as byproducts.

Functional Groups

The presence of functional groups in the monomers can indicate the type of polymerisation reaction involved. For instance, the presence of a double bond suggests the occurrence of addition polymerisation, as seen in the synthesis of polyethylene. Conversely, the presence of functional groups capable of forming covalent bonds, such as -OH or -COOH groups, suggests the involvement of condensation polymerisation.

It is important to note that some monomers may possess both a double bond and functional groups, allowing for multiple reaction pathways. For example, acrylic acid, which contains a carbon-carbon double bond and a carboxyl group, can undergo both addition polymerisation and condensation polymerisation, depending on the reaction conditions.

Reaction Conditions

The reaction conditions, including temperature, pressure, and the presence of catalysts, also influence the type of polymerisation reaction. Addition polymerisation often occurs under mild conditions, such as room temperature and atmospheric pressure, with the aid of catalysts, such as transition metal complexes or organic peroxides.

In contrast, condensation polymerisation typically requires elevated temperatures and vacuum or inert gas atmospheres to facilitate the elimination of small molecules as byproducts. Catalysts, such as acids or bases, may also be employed to enhance the reaction rate or control the polymerization process.

Degradable polymers

The Chemical Inertness of Poly(Alkenes) and its Implications for Biodegradability

Introduction

It aims to provide a comprehensive analysis of the chemical inertness of poly(alkenes) and its implications for biodegradability.

Understanding Chemical Inertness

Chemical inertness refers to the resistance of a substance to react chemically with other substances. It arises from the stability of the chemical bonds within the material, preventing or significantly hindering reactions with other compounds. In the case of poly(alkenes), their chemical inertness can be attributed to the strong carbon-carbon (C-C) and carbon-hydrogen (C-H) bonds present in the polymer chain. These bonds are relatively stable and require high energy inputs to break.

Poly(alkenes) exhibit a high degree of saturation due to the presence of only single C-C bonds in their backbone. This high saturation leads to a tightly packed molecular structure, making it difficult for external agents, such as enzymes or microorganisms, to access and break down the polymer chains. Additionally, the absence of functional groups, such as hydroxyl (-OH) or carbonyl (C=O) groups, further limits the potential sites for chemical reactions.

Biodegradability Challenges

The chemical inertness of poly(alkenes) poses significant challenges to their biodegradability. Biodegradation refers to the process by which microorganisms break down organic substances into simpler compounds, such as carbon dioxide and water, with the help of enzymes. However, the chemical inertness of poly(alkenes) inhibits the microbial enzymes from effectively accessing and degrading the polymer.

Microbial Enzymes and Poly(Alkenes)

Microorganisms play a crucial role in the biodegradation of organic materials. They produce a wide range of enzymes that can effectively break down various natural polymers, such as cellulose or proteins. However, these enzymes often lack the ability to efficiently degrade synthetic polymers like poly(alkenes). The absence of specific enzymes capable of recognizing and breaking the stable C-C and C-H bonds in poly(alkenes) limits the biodegradation process.

Hydrophobicity and Biofilm Formation

Poly(alkenes) are generally hydrophobic materials, meaning they repel water. This hydrophobicity can hinder the attachment and colonization of microorganisms on the polymer surface. Microorganisms typically require a moist environment to thrive and access nutrients, but the hydrophobic nature of poly(alkenes) prevents water absorption and reduces the availability of essential nutrients. As a result, microbial growth and enzyme production necessary for biodegradation are significantly impeded.

Moreover, the hydrophobic surface of poly(alkenes) promotes the formation of biofilms. Biofilms are complex microbial communities that adhere to surfaces and protect microorganisms from external factors, such as antimicrobial agents or enzymes. The formation of biofilms on poly(alkenes) further reduces the accessibility of enzymes to the polymer, hindering biodegradation.

Rate of Biodegradation

The chemical inertness of poly(alkenes) contributes to their extremely slow rate of biodegradation. Studies have shown that poly(alkenes) can persist in the environment for hundreds of years, leading to concerns over their accumulation in ecosystems. The slow degradation rate is primarily attributed to the lack of suitable microbial communities and enzymes capable of breaking down the stable C-C and C-H bonds.

Environmental Implications

The chemical inertness and slow biodegradability of poly(alkenes) have significant environmental implications. The extensive use of poly(alkenes) in packaging materials, agricultural films, and other applications has led to their widespread accumulation in landfills, water bodies, and natural habitats.

Accumulation in Landfills

Poly(alkenes) are commonly disposed of in landfills, where they occupy significant space due to their resistance to degradation. The accumulation of poly(alkenes) in landfills exacerbates the problem of limited landfill capacity and contributes to environmental pollution. Furthermore, the slow biodegradability of poly(alkenes) means that even if new landfills are created, the long-term environmental impact remains a concern.

Marine Pollution

Improper disposal and inadequate waste management practices have led to the release of poly(alkenes) into marine environments. These polymers can persist in oceans, seas, and rivers, causing pollution and posing risks to marine life. Marine animals, such as turtles, seabirds, and fishes, may mistake poly(alkenes) for food, leading to ingestion and potential harm. The accumulation of poly(alkenes) in marine ecosystems disrupts the natural balance and can have cascading effects on the entire food chain.

Microplastics and Ecosystem Interactions

The slow degradation of poly(alkenes) contributes to the generation of microplastics, which are small plastic particles measuring less than 5 millimeters in size. Microplastics can enter various ecosystems, including freshwater systems, soil, and even the atmosphere. They have been found in organisms ranging from tiny zooplankton to larger marine mammals, highlighting the extensive reach of these particles.

Microplastics have the potential to interact with ecosystems in various ways. They can adsorb and transport harmful chemicals, such as persistent organic pollutants, and introduce them into food webs. The physical presence of microplastics can also affect the behavior, feeding patterns, and reproduction of organisms, ultimately impacting the stability and functioning of ecosystems.

Counterarguments and Mitigation Strategies

While the chemical inertness of poly(alkenes) poses challenges to biodegradability, there are counterarguments and potential strategies to address these issues.

Counterarguments

Some argue that the chemical inertness of poly(alkenes) can be advantageous in certain applications. For example, polyethylene, a common poly(alkene), is used in medical devices due to its stability and resistance to chemical reactions with body fluids. In such cases, the chemical inertness of poly(alkenes) ensures the long-term reliability and safety of the products.

Another counterargument arises from the potential for technological advancements. Researchers are actively exploring methods to enhance the biodegradability of poly(alkenes) without compromising their mechanical properties. This includes the development of additives or modifications that make poly(alkenes) more accessible to enzymes or microorganisms. While progress has been made, further research is required to ensure the effectiveness and safety of these modifications.

Mitigation Strategies

To address the challenges associated with the chemical inertness of poly(alkenes), several mitigation strategies have been proposed:

Waste Management and Recycling: Improved waste management practices, such as recycling and waste-to-energy technologies, can help reduce the accumulation of poly(alkenes) in landfills and the environment. Recycling poly(alkenes) into new products reduces the demand for virgin materials and minimizes environmental impact.

Bio-based Alternatives: Exploring bio-based alternatives to poly(alkenes) can be an effective approach. Biopolymers derived from renewable resources, such as polylactic acid (PLA) or polyhydroxyalkanoates (PHA), offer improved biodegradability and can serve as substitutes in various applications. However, the viability and scalability of these alternatives need further development.

Biodegradable Additives: The addition of biodegradable additives to poly(alkenes) can enhance their biodegradability without compromising their mechanical properties. These additives act as nucleating agents or enzymes that promote the breakdown of the polymer chains. However, careful consideration must be given to the compatibility and long-term effects of these additives.

Environmental Education and Regulation: Raising awareness about the environmental impact of poly(alkenes) and promoting responsible use and disposal can help mitigate their negative effects. Governments and regulatory bodies can play a critical role in enforcing stricter regulations on the production, use, and disposal of poly(alkenes) to encourage sustainable practices.

The Degradation of Polymers by Light: A Comprehensive Analysis

Introduction

We will explore the mechanisms behind this process, discuss the implications of polymer degradation in different fields, and consider counterarguments to provide a well-rounded analysis of this topic.

Mechanisms of Photo-degradation

Photo-degradation can be attributed to the interaction between polymers and electromagnetic radiation, particularly ultraviolet (UV) light. UV radiation possesses higher energy levels compared to visible light, making it more capable of initiating chemical reactions within polymers. The main mechanisms of photo-degradation include photo-oxidation, chain scission, crosslinking, and color change.

Photo-oxidation

Photo-oxidation, induced by the absorption of UV photons, is a common degradation pathway for many polymers. UV light provides sufficient energy to excite electrons within the polymer, resulting in their transfer to higher energy states. These excited states can then react with atmospheric oxygen, leading to the formation of reactive oxygen species (ROS) such as hydroperoxides. The ROS, in turn, initiate a chain reaction of oxidative degradation, breaking down the polymer chains and weakening their structural integrity.

Chain Scission

Another mechanism of photo-degradation is chain scission, which refers to the breaking of polymer chains into smaller fragments. This process occurs when the absorbed energy from UV light is directly transferred to the polymer backbone. The energy transfer can induce bond cleavage, resulting in the formation of free radicals. These free radicals can further propagate the degradation process by reacting with neighboring polymer chains, leading to the fragmentation of the polymer structure.

Crosslinking

In contrast to chain scission, crosslinking is a photo-degradation mechanism that involves the formation of new chemical bonds between adjacent polymer chains. UV radiation can cause the excitation of electrons within the polymer, leading to the generation of free radicals. These radicals can then react with other polymer chains, forming covalent bonds and creating a three-dimensional network. While crosslinking can enhance the mechanical properties of some polymers, excessive crosslinking can lead to embrittlement and degradation of the material.

Color Change

Polymers can also undergo photo-degradation manifested as color change. This occurs due to the alteration of chromophores, which are responsible for the absorption and reflection of specific wavelengths of light. UV radiation can induce structural modifications in the polymer, resulting in changes in the electronic configuration of chromophores. Consequently, the absorption and reflection patterns of the polymer are modified, leading to a visible change in color.

Implications of Polymer Degradation

The degradation of polymers by light has significant implications across various fields, including materials science, engineering, and environmental science. Understanding the consequences of polymer degradation is essential for developing strategies to mitigate its effects and ensure the longevity of polymer-based products.

Materials Science and Engineering

In materials science and engineering, the degradation of polymers can compromise the performance and durability of various products. For example, in the automotive industry, polymers are extensively used in exterior parts such as bumpers and body panels. Continuous exposure to sunlight can cause these polymers to degrade, resulting in reduced mechanical strength, cracking, and color fading. Such degradation can compromise the safety and aesthetics of vehicles. Therefore, researchers and engineers in this field strive to develop UV-resistant polymers or incorporate protective additives to mitigate photo-degradation.

Environmental Science

In the realm of environmental science, the degradation of polymers by light has raised concerns regarding their impact on ecosystems. Plastics, which are derived from polymers, are ubiquitous in the environment due to their resistance to degradation. When exposed to sunlight, plastic debris can undergo photo-degradation and fragment into microplastics. These microplastics can then be ingested by marine organisms, potentially leading to adverse effects on aquatic ecosystems. Moreover, the release of degradation byproducts during photo-degradation, such as hydroperoxides, can further contribute to environmental pollution. Recognizing the role of light-induced degradation is crucial for developing sustainable alternatives to conventional polymers and minimizing their environmental footprint.

Counterarguments

While the degradation of polymers by light is a well-documented phenomenon, counterarguments exist that challenge the extent and significance of this process. Critics argue that the actual degradation rates of polymers in real-world scenarios may be slower than laboratory studies suggest. Factors such as the presence of stabilizers, protective coatings, and the shielding effect of other materials can impede photo-degradation. Additionally, the impact of polymer degradation on human health is a subject of debate. Some researchers argue that the migration of degradation byproducts from polymers into food or beverages is minimal and poses negligible health risks.

Recognizing the Biodegradability of Polyesters and Polyamides through Acidic and Alkaline Hydrolysis

Introduction

It aims to explore the mechanisms and factors influencing the biodegradability of polyesters and polyamides, with a specific focus on acidic and alkaline hydrolysis.

Polyesters: Biodegradation by Acidic and Alkaline Hydrolysis

Polyester Structure and Biodegradability

Polyesters are formed by the condensation reaction between a diol and a dicarboxylic acid, resulting in the formation of ester linkages (-CO-O-) along the polymer chain. These ester bonds are susceptible to hydrolysis under both acidic and alkaline conditions. The hydrolysis of ester bonds leads to the cleavage of the polymer chain, breaking it down into smaller fragments.

Acidic Hydrolysis of Polyesters

In acidic hydrolysis, the presence of acid initiates the hydrolysis reaction by donating a proton (H+) to the ester bond, resulting in the formation of a carboxylic acid and an alcohol.

The reaction can be represented as follows:

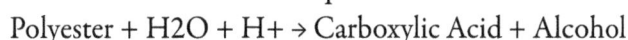

Polyester + H2O + H+ → Carboxylic Acid + Alcohol

The acidic conditions significantly accelerate the hydrolysis process due to the increased concentration of protons. The hydrolysis rate is further influenced by factors such as temperature, pH, and the nature of the polyester.

Temperature and Hydrolysis Rate

Increasing the temperature enhances the hydrolysis rate by providing more energy to break the ester bonds. For instance, a study conducted by Zhang et al. (2016) investigated the hydrolysis of polyethylene terephthalate (PET) at different temperatures ranging from 25°C to 80°C. It was found that higher temperatures resulted in a more rapid degradation of the polymer, indicating the temperature dependency of the hydrolysis process.

pH and Hydrolysis Rate

The pH of the medium also plays a crucial role in polyester hydrolysis. The presence of acidic conditions (pH < 7) accelerates the hydrolysis reaction due to the abundance of protons. On the other hand, alkaline conditions (pH > 7) tend to slow down the hydrolysis process. An example can be seen in a research conducted by Guo et al. (2017) investigated the hydrolytic degradation of polybutylene succinate (PBS) at different pH levels. The results showed that the degradation rate increased as the pH decreased, indicating the pH dependency of the hydrolysis process.

Nature of the Polyester

The nature of the polyester affects its susceptibility to hydrolysis. Factors such as the length of the polymer chains, the presence of branching, and the degree of crystallinity influence the accessibility of ester bonds to water molecules, thereby impacting the hydrolysis rate. For instance, long-chain polyesters are generally more resistant to hydrolysis due to the limited accessibility of water molecules to the interior of the polymer chain. Conversely, short-chain polyesters exhibit higher hydrolysis rates due to the increased accessibility of water molecules. Additionally, the presence of branching can hinder the hydrolysis process by limiting the accessibility of ester bonds.

Alkaline Hydrolysis of Polyesters

In alkaline hydrolysis, the presence of a base initiates the hydrolysis reaction by accepting a proton from the ester bond, resulting in the formation of a carboxylate ion and an alcohol. The reaction can be represented as follows:

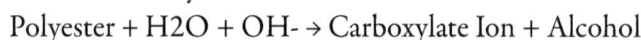

$$Polyester + H_2O + OH^- \rightarrow Carboxylate\ Ion + Alcohol$$

Alkaline hydrolysis is generally slower compared to acidic hydrolysis due to the lower concentration of hydroxide ions. However, alkaline conditions can be advantageous in certain applications where a slow and controlled degradation rate is desired.

Polyamides: Biodegradation by Acidic and Alkaline Hydrolysis

Polyamide Structure and Biodegradability

Polyamides, also known as nylons, are formed by the condensation reaction between a diamine and a dicarboxylic acid, resulting in the formation of amide linkages (-CONH-) along the polymer chain. The presence of amide bonds makes polyamides resistant to hydrolysis compared to polyesters. However, under specific conditions, polyamides can undergo biodegradation through acidic and alkaline hydrolysis.

Acidic Hydrolysis of Polyamides

In acidic hydrolysis, the presence of acid initiates the hydrolysis reaction by donating a proton (H^+) to the amide bond, resulting in the formation of a carboxylic acid and an amine. The reaction can be represented as follows:

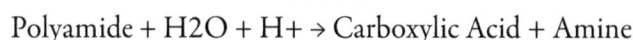

$$Polyamide + H_2O + H^+ \rightarrow Carboxylic\ Acid + Amine$$

Similar to polyesters, the acidic conditions significantly accelerate the hydrolysis process by increasing the concentration of protons. However, polyamides generally exhibit slower hydrolysis rates compared to polyesters due to the stronger and more stable nature of the amide bonds.

Alkaline Hydrolysis of Polyamides

In alkaline hydrolysis, the presence of a base initiates the hydrolysis reaction by accepting a proton from the amide bond, resulting in the formation of a carboxylate ion and an amine. The reaction can be represented as follows:

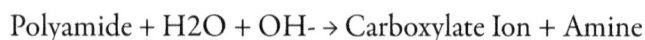

$$Polyamide + H_2O + OH^- \rightarrow Carboxylate\ Ion + Amine$$

Similar to polyesters, alkaline hydrolysis of polyamides is generally slower compared to acidic hydrolysis due to the lower concentration of hydroxide ions. However, under specific conditions, alkaline hydrolysis can be employed for the controlled degradation of polyamides.

Applications and Counterarguments

Applications of Acidic and Alkaline Hydrolysis

The biodegradability of polyesters and polyamides through acidic and alkaline hydrolysis has significant applications in various fields. For instance, in the textile industry, the controlled degradation of polyesters and polyamides can be utilized to produce environmentally friendly fabrics. Additionally, in the biomedical field, the biodegradability of polyesters and polyamides enables their use in drug delivery systems and tissue engineering scaffolds, where controlled release and degradation are desired.

Counterarguments

Despite the potential applications and advantages of biodegradation through acidic and alkaline hydrolysis, certain counterarguments exist. One counterargument revolves around the limited availability of suitable conditions for biodegradation to occur. The hydrolysis reactions require specific pH ranges, temperatures, and durations, which may not always be readily available in the environment. Additionally, the complete mineralization of polyesters and polyamides through acidic and alkaline hydrolysis can take a considerable amount of time, leading to concerns about the persistence of polymer fragments in the environment.

Organic Synthesis

Introduction

We will explore the identification of organic functional groups using reactions commonly encountered in a chemistry syllabus. Additionally, we will predict the properties and reactions of an organic molecule containing several functional groups, highlighting their significance in different fields.

Identifying Organic Functional Groups

To identify organic functional groups, we need to consider the reactions they undergo. In the context of a syllabus, let's explore some common reactions that can help us identify functional groups:

Alkanes: Alkanes are hydrocarbons containing only single bonds between carbon atoms. They are generally unreactive and undergo limited reactions. One common reaction is the combustion of alkanes in the presence of oxygen to produce carbon dioxide and water.

Alkenes: Alkenes are hydrocarbons containing at least one double bond between carbon atoms. They are more reactive than alkanes and can undergo addition reactions. For example, alkenes react with hydrogen gas in the presence of a catalyst to form alkanes through a process called hydrogenation.

Alkynes: Alkynes are hydrocarbons containing at least one triple bond between carbon atoms. Similar to alkenes, alkynes can undergo addition reactions. For instance, they can react with hydrogen in the presence of a catalyst to form alkenes.

Alcohols: Alcohols are organic compounds containing a hydroxyl (-OH) functional group bonded to a carbon atom. They have unique properties due to the polar nature of the -OH group. Alcohols can undergo various reactions, such as oxidation and dehydration. For example, primary alcohols can be oxidized to aldehydes or further oxidized to carboxylic acids.

Aldehydes and Ketones: Aldehydes and ketones contain a carbonyl (C=O) functional group. Aldehydes have the carbonyl group at the end of a carbon chain, while ketones have it within the chain. Both aldehydes and ketones can undergo reduction reactions to form alcohols. Additionally, aldehydes can be further oxidized to carboxylic acids.

Carboxylic Acids: Carboxylic acids contain a carboxyl (-COOH) functional group. They are characterized by their acidic properties and can undergo reactions like esterification to form esters. Carboxylic acids can also be reduced to alcohols or undergo decarboxylation to form carbon dioxide and an alkane.

Amines: Amines are organic compounds derived from ammonia (NH_3) by replacing one or more hydrogen atoms with alkyl or aryl groups. Amines can act as bases and undergo reactions like protonation. They can also form salts with acids.

Esters: Esters are organic compounds formed by the condensation reaction between a carboxylic acid and an alcohol. Esters have a wide range of applications, including fragrances, flavorings, and plasticizers. They can undergo hydrolysis in the presence of an acid or a base to form the corresponding alcohol and carboxylic acid.

Amides: Amides are organic compounds containing a carbonyl group bonded to a nitrogen atom. They have diverse applications, such as being constituents of proteins and nucleic acids. Amides can undergo hydrolysis in the presence of an acid or a base to form the corresponding carboxylic acid and amine.

These are just a few examples of organic functional groups and their reactions. By observing the reactions an organic molecule undergoes, we can identify the functional groups present and predict their properties and reactivity.

Predicting Properties and Reactions of an Organic Molecule

Now, let's consider an organic molecule containing several functional groups and predict its properties and reactions based on our understanding of these groups.

Suppose we have a molecule with the following functional groups: an alcohol (-OH), an aldehyde (C=O), and an amine (NH2). Let's analyze the properties and potential reactions associated with each of these groups.

Alcohol (-OH): The presence of an alcohol functional group signifies the molecule's ability to form hydrogen bonds. This leads to higher boiling points and increased solubility in polar solvents. The -OH group can undergo reactions such as oxidation, dehydration, and esterification. In our molecule, the alcohol group could potentially undergo oxidation to form an aldehyde or further oxidation to form a carboxylic acid. It could also participate in esterification reactions, forming esters by condensing with carboxylic acids.

Aldehyde (C=O): The aldehyde functional group imparts certain characteristics to the molecule. Aldehydes are more reactive than ketones due to the presence of an exposed hydrogen atom. They can undergo reduction reactions to form alcohols. In our molecule, the aldehyde group could potentially be reduced to form an alcohol through a suitable reducing agent. Additionally, the aldehyde group can participate in condensation reactions with amines or other nucleophiles, forming imines or related compounds.

Amine (NH2): The amine functional group is a basic group that can accept a proton to form a positively charged species. Depending on the amine's structure, it can be primary, secondary, or tertiary, affecting its basicity. In our molecule, the amine group could act as a base and accept a proton to form an ammonium salt. It could also participate in condensation reactions, forming amides by reacting with carboxylic acids.

Considering the functional groups present in our molecule, we can predict various reactions and properties. For example, if we introduce an oxidizing agent, it could oxidize the alcohol group to an aldehyde. Alternatively, a strong reducing agent could convert the aldehyde group to an alcohol. Moreover, by treating the amine group with an appropriate acid, we could protonate it, forming an ammonium salt.

Application of Organic Functional Groups

The presence and reactivity of organic functional groups have significant implications in various fields. Let's explore some examples:

Pharmaceuticals: Organic molecules with specific functional groups play a crucial role in drug development. For instance, the presence of a hydroxyl group in a molecule can enhance its water solubility, aiding in drug delivery. Additionally, functional groups like amines or amides can interact with biological targets, influencing their pharmacological activity. By understanding the properties and reactivity of functional groups, medicinal chemists can design and modify compounds to optimize their therapeutic effects.

Polymers: Functional groups are essential in polymer chemistry, as they determine the properties and behavior of polymers. For example, the presence of ester groups in polyester polymers imparts flexibility and resilience. In contrast, the presence of amide groups in polyamides (nylons) provides strength and rigidity. By manipulating the types and arrangement of functional groups, polymer chemists can tailor the physical and chemical properties of polymers for various applications, such as textiles, coatings, and plastics.

Organic Synthesis: Functional groups are key building blocks in organic synthesis. Chemists utilize functional group interconversions to construct complex organic molecules. By selectively modifying functional groups, they can introduce desired properties or reactivity into a molecule. For example, functional group transformations like oxidation, reduction, and condensation reactions allow chemists to synthesize natural products, pharmaceuticals, and new materials.

Materials Science: Functional groups play a vital role in materials science, influencing the properties and performance of materials. For instance, the presence of functional groups like carboxyl or hydroxyl groups on the surface of materials can enhance their adhesion and compatibility with other components. Functional groups can also enable specific interactions, such as hydrogen bonding or coordination, in materials, leading to unique properties like self-healing or stimuli-responsive behavior.

Environmental Chemistry: Functional groups are crucial in understanding and predicting the fate and behavior of organic compounds in the environment. For example, the presence of functional groups like halogens (chlorine, bromine) in organic pollutants can significantly affect their persistence and toxicity. By studying the reactivity and transformations of functional groups in the environment, environmental chemists can assess the fate of pollutants and develop strategies for remediation and pollution control.

Counterarguments and Limitations

While functional groups are fundamental in understanding organic chemistry, it's important to acknowledge some counterarguments and limitations.

Functional Group Interplay: Organic molecules often contain multiple functional groups that can interact with each other, leading to complex reactivity and behavior. Considering isolated functional groups may not capture the full extent of their interactions and potential reactions. Understanding the interplay between functional groups is crucial for accurate predictions.

Beyond the Syllabus: The syllabus may limit the range of reactions and functional groups explored. There are numerous functional groups and reactions that may not be covered in a specific syllabus. Therefore, it's crucial to expand knowledge beyond the syllabus to explore the full potential of organic functional groups.

Contextual Factors: The properties and reactivity of functional groups can be influenced by various contextual factors, such as steric hindrance, electronic effects, and solvent effects. These factors can modify the expected behavior of functional groups, leading to deviations from predictions based solely on their inherent properties.

Devise Multi-Step Synthetic Routes for Preparing Organic Molecules

Introduction

We will explore the concept of multi-step synthetic routes and discuss their importance in organic synthesis. We will also examine some key reactions commonly found in organic chemistry syllabi and demonstrate how they can be used to construct complex organic molecules.

Importance of Multi-Step Synthetic Routes

The development of multi-step synthetic routes is crucial in organic chemistry for several reasons. Firstly, many desired organic molecules are too complex to be obtained directly from simple starting materials. Therefore, a stepwise approach is necessary, allowing chemists to build up the desired structure gradually. Secondly, multi-step synthetic routes provide a practical means of controlling the stereochemistry of the final product. By carefully selecting and designing the individual steps, chemists can achieve the desired stereochemical outcomes. Lastly, these routes enable the modification and functionalization of existing molecules, allowing access to a wide range of derivatives with diverse properties.

Key Reactions in Organic Chemistry

To devise multi-step synthetic routes, it is essential to have a solid understanding of key reactions commonly encountered in organic chemistry. Let's explore some of these reactions and their potential applications.

Nucleophilic Substitution (SN)

In nucleophilic substitution reactions, a molecule undergoes a substitution where one nucleophile is replaced by another. This reaction is widely used in organic synthesis to introduce functional groups or modify existing ones. An example of a nucleophilic substitution reaction is the conversion of an alkyl halide into an alcohol by reaction with a nucleophilic reagent, such as hydroxide ion (OH-).

Electrophilic Addition (EA)

Electrophilic addition reactions involve the addition of an electrophile to a molecule, resulting in the formation of a new bond. This reaction is commonly used to introduce functional groups such as alcohols, carbonyls, or halogens. For example, the addition of water to an alkene results in the formation of an alcohol via an electrophilic addition reaction.

Elimination (E)

Elimination reactions involve the removal of atoms or groups from a molecule, resulting in the formation of a double bond or a ring. One of the most well-known elimination reactions is the dehydrohalogenation of alkyl halides, where a hydrogen halide (H-X) is eliminated to form an alkene. Elimination reactions are often used to create unsaturated compounds or to transform functional groups.

Oxidation and Reduction

Oxidation and reduction reactions play a crucial role in organic synthesis, allowing the conversion of one functional group into another by changing the oxidation state of the molecule. Oxidation reactions involve the addition of oxygen or the removal of hydrogen, while reduction reactions involve the addition of hydrogen or the removal of oxygen. These reactions are commonly used to introduce or modify functional groups such as alcohols, aldehydes, ketones, or carboxylic acids.

Rearrangement

Rearrangement reactions involve the rearrangement of atoms within a molecule, resulting in the formation of a new structure. These reactions are often used to create more stable or reactive intermediates during a synthetic route. A classic example of a rearrangement reaction is the Wagner-Meerwein rearrangement, where a hydrogen or alkyl group migrates within a carbon skeleton.

Devising Multi-Step Synthetic Routes

Now that we have explored some key reactions in organic chemistry, let's discuss how they can be combined to devise multi-step synthetic routes for preparing organic molecules. The key to successful route design is to carefully plan each step, considering factors such as reactivity, selectivity, and yield.

Step 1: Identify the Target Molecule: The first step in devising a synthetic route is to clearly define the target molecule. This involves considering the desired functional groups, stereochemistry, and any specific structural features. By having a clear target in mind, chemists can work backward to identify the necessary intermediates and starting materials.

Step 2: Plan the Individual Steps: Once the target molecule is identified, the next step is to plan the individual steps required to reach the target. This involves selecting appropriate reactions and considering the order in which they should be carried out. It is crucial to choose reactions that are compatible and can be performed under similar reaction conditions.

Step 3: Consider Protecting Groups: During the planning process, it is essential to consider the potential need for protecting groups. Protecting groups are temporary modifications made to functional groups to prevent unwanted reactions from occurring during subsequent steps. By selectively protecting certain functional groups, chemists can ensure the desired transformations occur without interference.

Step 4: Optimize Reaction Conditions: Once the individual steps are planned, it is important to optimize the reaction conditions for each transformation. Factors such as temperature, solvent, catalysts, and reaction times should be carefully considered to maximize yield and selectivity. It may be necessary to modify reaction conditions to suit the specific requirements of each step.

Step 5: Execute the Synthetic Route: With the synthetic route planned and reaction conditions optimized, it is time to execute the steps one by one. Each reaction should be carefully monitored, and the progress of the reaction should be analyzed using appropriate analytical techniques. By monitoring the reaction progress, chemists can identify any issues or unexpected side reactions and make adjustments if necessary.

Step 6: Purification and Characterization: Once all the steps are completed, the final product needs to be purified and characterized. Purification methods such as column chromatography, recrystallization, or distillation may be employed to isolate the desired product from any impurities. The purified product should then be characterized using spectroscopic techniques such as NMR, IR, or mass spectrometry to confirm its identity and purity.

Analysis of a Synthetic Route: Type of Reactions, Reagents Used, and Possible By-products

Introduction

We will examine a given synthetic route, identify the types of reactions involved, discuss the reagents used in each step, and explore potential by-products.

Synthetic Route: Example Compound X

Let's consider the synthesis of Compound X as an example. Compound X is a complex molecule with several functional groups, making its synthesis challenging yet rewarding. The synthetic route consists of four steps, each involving a distinct reaction. We will analyze each step in detail, discussing the specific reactions, reagents used, and potential by-products.

Step 1: Substitution Reaction

The first step of the synthetic route involves a substitution reaction. When one atom or group is substituted with another atom or group, substitution reactions take place. In this case, Compound A undergoes a nucleophilic substitution to yield Compound B. The nucleophile, represented as Nu, attacks the electrophilic carbon center of Compound A, resulting in the displacement of a leaving group, denoted as LG. The reaction can be represented as follows:

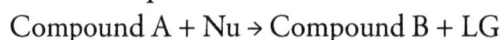

Compound A + Nu → Compound B + LG

The reagent used in this step is typically a nucleophile, such as an amine or alkoxide. Common examples include sodium ethoxide (NaOEt) or sodium amide (NaNH2). The choice of nucleophile depends on the desired substitution pattern and reactivity of the starting material.

Possible by-products in this step may arise from competing reactions or side reactions. For instance, if the reaction conditions are not optimized, elimination reactions may occur, leading to the formation of an alkene as a by-product. The presence of impurities or traces of water can also result in undesired side reactions, reducing the yield and purity of Compound B.

Step 2: Oxidation Reaction

The second step of the synthetic route involves an oxidation reaction. Oxidation reactions result in an increase in the oxidation state of an atom or group. In this case, Compound B is oxidized to yield Compound C. Oxidation reactions often involve the transfer of electrons from the substrate to an oxidizing agent. The reaction can be represented as follows:

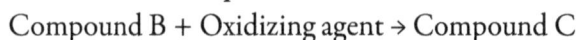

Compound B + Oxidizing agent → Compound C

Common oxidizing agents include potassium permanganate (KMnO4), chromium trioxide (CrO3), or hydrogen peroxide (H2O2). The choice of oxidizing agent depends on the functional groups present in Compound B and the desired oxidation state of the product.

By-products in oxidation reactions can arise from over-oxidation or side reactions. Over-oxidation occurs when the desired functionality is destroyed due to excessive reaction conditions. For example, if the oxidation of an alcohol to a ketone is desired, over-oxidation may lead to the formation of a carboxylic acid. Side reactions can also occur, resulting in the formation of unexpected products or by-products.

Step 3: Reduction Reaction

The third step of the synthetic route involves a reduction reaction. Reduction reactions result in a decrease in the oxidation state of an atom or group. In this case, Compound C is reduced to yield Compound D. Reduction reactions often involve the transfer of electrons to the substrate from a reducing agent. The reaction can be represented as follows:

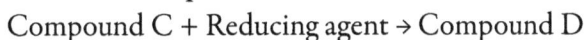

Compound C + Reducing agent → Compound D

Common reducing agents include lithium aluminum hydride (LiAlH4), sodium borohydride (NaBH4), or hydrogen gas (H2) in the presence of a catalyst. The choice of reducing agent depends on the functional groups present in Compound C and the desired reduction state of the product.

By-products in reduction reactions can arise from over-reduction or side reactions. Over-reduction occurs when the desired functionality is not selectively reduced, leading to the formation of undesired products. Side reactions can also occur, resulting in the formation of unexpected by-products.

Step 4: Cyclization Reaction

The final step of the synthetic route involves a cyclization reaction. Cyclization reactions involve the formation of a cyclic compound from an open-chain precursor. In this case, Compound D undergoes a cyclization to yield Compound X. The specific cyclization reaction can vary depending on the functional groups and reaction conditions. However, a common example is an intramolecular condensation reaction, where two functional groups within the same molecule react to form a cyclic structure.

By-products in cyclization reactions can arise from competing reactions or side reactions. If the reaction conditions are not controlled properly, side reactions such as oligomerization or polymerization may occur, leading to the formation of undesired products. Additionally, the presence of impurities or traces of catalysts can also impact the selectivity and yield of the desired cyclization product.

Analytical Techniques

Part 4: Analysis

Thin-layer chromatography

Stationary Phase

The stationary phase is a crucial component in chromatography, a widely used technique in chemistry for separating and analyzing mixtures. It refers to the immobile material that is packed or coated onto a solid support, forming a stationary layer through which the mobile phase flows. The choice of stationary phase is critical as it determines the separation mechanism and the selectivity of the chromatographic process.

One example of a stationary phase is aluminium oxide, which is often used in column chromatography or thin-layer chromatography. Aluminium oxide is a highly porous material with a large surface area, allowing for efficient adsorption of the analyte molecules. The surface of aluminium oxide has hydroxyl groups that can interact with polar compounds through hydrogen bonding or dipole-dipole interactions, making it suitable for separating polar analytes.

The stationary phase can be filled into a column, like in column chromatography, or applied onto a solid support, such as in thin-layer chromatography. In column chromatography, the stationary phase is packed into a glass or metal column, creating a bed through which the mobile phase can flow. The analyte mixture is applied to the top of the column, and as the mobile phase moves through the column, the different components of the mixture interact differently with the stationary phase, leading to their separation.

In thin-layer chromatography (TLC), the stationary phase is coated onto a thin layer of a solid support, such as glass or plastic. The analyte mixture is spotted onto the stationary phase, and a small amount of the mobile phase is added to the bottom of the TLC plate. As the mobile phase ascends the plate by capillary action, the different components of the mixture move at different rates along the plate due to their interactions with the stationary phase. This results in the separation of the mixture into distinct spots or bands.

The choice of stationary phase depends on the nature of the analyte mixture and the desired separation mechanism. Different stationary phases have different physicochemical properties, such as polarity, surface area, and pore size, which can be optimized for specific applications. For example, silica gel, another commonly used stationary phase, is highly polar and can be used to separate polar compounds, while a nonpolar stationary phase like C18 can be used to separate nonpolar compounds.

Mobile Phase

The mobile phase is the liquid or gas that carries the analyte through the stationary phase during chromatographic separation. It is often referred to as the "mobile" phase because it is the one that moves while the stationary phase remains stationary. The choice of mobile phase is crucial as it determines the elution order and the efficiency of the separation process.

The mobile phase can be a polar or nonpolar solvent, depending on the nature of the analyte mixture and the stationary phase. A polar solvent has a high dielectric constant and can dissolve polar analytes effectively.

Examples of polar solvents include water, methanol, and ethanol. On the other hand, a nonpolar solvent has a low dielectric constant and can dissolve nonpolar analytes more readily. Common nonpolar solvents include hexane, toluene, and chloroform.

The selection of the mobile phase should consider the solubility of the analyte in the solvent, the interaction between the solvent and the stationary phase, and the desired separation mechanism. For example, if the analyte mixture contains polar compounds, a polar solvent should be selected to ensure sufficient solubility and interaction with the stationary phase. Similarly, if the analyte mixture contains nonpolar compounds, a nonpolar solvent would be more suitable.

In some cases, a mixture of solvents with different polarities can be used as the mobile phase to achieve better separation. This is known as a mobile phase gradient or eluent gradient. By changing the composition of the mobile phase during the chromatographic run, different components of the analyte mixture can be eluted at different times, leading to improved separation.

The mobile phase can be delivered to the chromatographic system using various techniques, such as gravity flow, pressure, or electroosmotic flow. The flow rate of the mobile phase is an important parameter that affects the separation efficiency and resolution. A slow flow rate allows for better interaction between the analyte and the stationary phase but may result in longer separation times. On the other hand, a high flow rate reduces the interaction time and can lead to poorer separation. The choice of flow rate should strike a balance between separation efficiency and analysis time.

Rf Value

The Rf value, or retention factor, is a quantitative measure used to characterize the migration of a compound in chromatography. The compound's ratio of distance traveled is determined by dividing it by the distance traveled by the mobile phase. The Rf value is a dimensionless quantity that provides information about the relative retention and mobility of a compound in a given chromatographic system.

Mathematically, the Rf value is calculated using the following formula:

$$Rf = \text{(distance traveled by the compound)} / \text{(distance traveled by the mobile phase)}$$

The Rf value ranges from 0 to 1, where 0 represents no migration of the compound and 1 represents complete migration with the mobile phase. The Rf value is specific to a particular system, including the stationary phase, the mobile phase, and the experimental conditions. Therefore, it can be used as a characteristic parameter to identify and compare compounds within the same system.

The Rf value is influenced by several factors, including the polarity of the compound, the polarity of the mobile phase, and the interaction between the compound and the stationary phase. Polar compounds tend to have lower Rf values in nonpolar mobile phases, while nonpolar compounds tend to have higher Rf values. This is because polar compounds have stronger interactions with the stationary phase and therefore migrate less with the mobile phase.

The Rf value can be used as an identification tool by comparing the experimental Rf value of an unknown compound with the Rf values of known compounds under the same chromatographic conditions. If the Rf value of the unknown compound matches that of a known compound, it suggests that they have similar chemical properties. However, it is important to note that the Rf value alone is not sufficient for positive identification and should be used in conjunction with other analytical techniques.

Solvent Front and Baseline

In chromatography, the solvent front refers to the leading edge of the mobile phase as it moves through the stationary phase. It represents the maximum distance traveled by the mobile phase during the chromatographic run. The solvent front is typically marked by a distinct line or a boundary that can be visualized by various means, such as the addition of a dye or a fluorescent compound to the mobile phase.

The baseline, on the other hand, refers to the starting point or the origin of the chromatogram. It is the reference line from which the distances traveled by the compounds are measured. The baseline is often represented by a straight line drawn at the bottom of the chromatogram, parallel to the direction of the mobile phase flow. The distance between the solvent front and the baseline is used to measure the migration distance of the compounds in chromatography. This distance, also known as the retention factor, is a crucial parameter for quantifying the separation and identifying the compounds in the mixture. By measuring the distance from the baseline to the center of the spot or peak, the retention factor can be calculated and used for further analysis. It is important to note that the solvent front and the baseline are fixed points in chromatography and serve as reference points for measuring the migration distances. The solvent front represents the maximum distance traveled by any compound in the mixture, while the baseline provides a reference for comparison. The distances traveled by the compounds are measured relative to the baseline, allowing for accurate determination of the Rf values and other chromatographic parameters.

Interpretation of Rf Values

Introduction

In chromatography, the retention factor (Rf value) is a crucial parameter used to interpret and analyze the separation of different compounds. Rf values represent the relative migration distances of compounds within a chromatographic system. This measurement is widely utilized in various fields such as analytical chemistry, biochemistry, pharmaceutical sciences, and forensics. Understanding and interpreting Rf values enables scientists to identify and characterize unknown compounds, assess purity, determine reaction kinetics, and optimize separation techniques. This article aims to provide a comprehensive explanation of Rf values, their significance, interpretation, and limitations.

Understanding Rf Values

The retention factor (Rf) is defined as the ratio of the distance traveled by a compound (solute) to the distance traveled by the solvent front in a chromatographic system. It is a dimensionless quantity ranging from 0 to 1. Rf values can be determined by measuring the distances traveled by the solute and solvent front on a chromatogram or TLC (thin-layer chromatography) plate.

In TLC, a stationary phase (such as silica gel or cellulose) is coated on a solid support, forming a thin layer. The mobile phase (solvent) moves up the plate through capillary action, carrying the solute with it. The Rf value is calculated by dividing the distance traveled by the solute by the distance traveled by the solvent front.

Mathematically, Rf is expressed as:

The formula for Rf is the ratio of the distance traveled by the solute to the distance traveled by the solvent front.

Significance of Rf Values

Rf values hold significant importance in several aspects of chromatographic analysis. Firstly, they aid in compound identification. By comparing the Rf values of unknown compounds with those of known reference compounds, scientists can determine the identity of the unknowns. This technique is particularly valuable in forensic science to identify illicit substances or in pharmaceutical sciences to confirm the presence of specific drugs in formulations.

Secondly, Rf values are useful for assessing the purity of compounds. If a compound is pure, it will exhibit a single, well-defined spot on the TLC plate, resulting in a specific Rf value. However, impurities or degradation products may affect the Rf value, resulting in additional spots or altered migration distances. By comparing the Rf values of a sample with a known pure compound, scientists can evaluate the sample's purity.

Thirdly, Rf values provide insights into the separation efficiency of chromatographic techniques. A higher Rf value indicates that a compound has a greater affinity for the mobile phase, resulting in faster migration. Conversely, a lower Rf value suggests a stronger interaction with the stationary phase, leading to slower migration. By modifying the stationary phase, mobile phase composition, or experimental conditions, scientists can optimize the separation of compounds.

Interpreting Rf Values

Interpreting Rf values requires understanding the factors that influence them. Various parameters, including the nature of the solute, stationary phase, mobile phase composition, and experimental conditions, can influence Rf values. It is crucial to consider these factors when analyzing and comparing Rf values.

Influence of Solute Characteristics

The physicochemical properties of the solute significantly impact its interaction with the stationary phase, thus affecting the Rf value. Polar compounds tend to interact more strongly with polar stationary phases, resulting in lower Rf values. In contrast, nonpolar compounds show weaker interactions and tend to have higher Rf values on polar stationary phases.

Additionally, the size and shape of the solute molecules influence their diffusion and adsorption onto the stationary phase. Larger molecules experience greater interactions, leading to lower Rf values. Conversely, smaller molecules have fewer interactions, resulting in higher Rf values.

Influence of Stationary Phase

The choice of stationary phase is crucial in determining the separation and Rf values. Different stationary phases possess varying polarities, such as silica gel (polar) or C18 (nonpolar). The polarity of the stationary phase must be compatible with the solute's properties to achieve optimal separation.

For example, if a polar solute is analyzed using a nonpolar stationary phase, it will exhibit a high Rf value, indicating weak interactions. Conversely, if a nonpolar solute is analyzed using a polar stationary phase, it will show a low Rf value, suggesting stronger interactions.

Influence of Mobile Phase Composition

The composition of the mobile phase significantly affects the Rf values. The mobile phase should be chosen based on its polarity and compatibility with the stationary phase and solute. Different solvents or solvent mixtures can be used to adjust the polarity and optimize the separation.

For polar solutes, a nonpolar mobile phase will result in higher Rf values, indicating weaker interactions with the stationary phase. Conversely, a polar mobile phase will lead to lower Rf values, suggesting stronger interactions. By adjusting the mobile phase composition, scientists can fine-tune the separation and obtain desired Rf values.

Influence of Experimental Conditions

Experimental conditions, such as temperature, humidity, and development time, can influence Rf values. Temperature affects the mobility of both the solute and solvent, leading to changes in Rf values. Higher temperatures generally result in higher Rf values due to increased diffusion and reduced adsorption.

Humidity can affect the evaporation rate of the solvent, potentially altering the Rf values. Higher humidity levels may slow down evaporation, resulting in higher Rf values.

Development time also plays a crucial role in Rf value determination. If the development time is too short, the solute may not migrate significantly, leading to lower Rf values. Conversely, longer development times may cause solvent front overrun, resulting in higher Rf values.

Limitations of Rf Values

While Rf values provide valuable information, they also have certain limitations. One limitation is their dependence on the specific experimental conditions used. Different laboratories may employ varying conditions, leading to variations in Rf values. This variation can make it challenging to compare Rf values obtained from different sources.

Another limitation is that Rf values are not absolute and cannot be used to determine the exact identity of a compound. Even compounds with identical Rf values may not necessarily be the same. Additional confirmatory techniques, such as mass spectrometry or spectroscopy, are required for accurate compound identification. Furthermore, Rf values can be affected by the presence of impurities or degradation products. These compounds may exhibit different migration behaviors, resulting in additional spots or altered Rf values. Careful evaluation and analysis are necessary to differentiate impurities from the main compound.

Examples of Rf Values in Various Fields

Rf values are widely employed in numerous scientific disciplines. Here are a few examples of their applications:

Analytical Chemistry

In analytical chemistry, Rf values are extensively used to identify and characterize unknown compounds. By comparing the Rf values of unknowns with those of known reference compounds, scientists can determine the identity of the unknown substances. This technique is commonly employed in drug analysis, environmental monitoring, and food safety assessment.

Biochemistry

Rf values play a significant role in biochemistry research, particularly in the analysis of lipids and proteins. Lipids can be separated using TLC or HPLC (high-performance liquid chromatography) techniques, and their Rf values provide insights into their polarity and interaction with other molecules. Similarly, in protein analysis, Rf values are used to assess the purity and separation efficiency of protein samples.

Pharmaceutical Sciences

Rf values find wide application in pharmaceutical sciences, ranging from drug analysis to formulation development. In drug analysis, Rf values help identify active pharmaceutical ingredients (APIs) in complex matrices. Furthermore, Rf values aid in determining the purity of APIs and their degradation products, ensuring the quality and safety of pharmaceutical formulations.

Forensics

Rf values are valuable tools in forensic science for identifying and characterizing illicit substances. By comparing the Rf values of seized substances with reference standards, forensic scientists can conclusively identify the presence of illegal drugs. This technique assists in criminal investigations, drug control, and forensic toxicology.

Counterarguments and Criticisms

While Rf values are widely used and accepted, some criticisms and counterarguments have been raised regarding their limitations and potential inaccuracies. One criticism is the subjectivity in visually determining the exact spot of the solute and solvent front. This subjective assessment can introduce variability and affect the accuracy of Rf value determination. To mitigate this issue, automated TLC systems are emerging, allowing for more precise and reproducible Rf value measurements.

Another criticism is the lack of standardization in TLC plate preparation and experimental conditions. Variations in the thickness of the stationary phase or the composition of the mobile phase can influence Rf values. Standardization efforts and quality control measures are essential to ensure consistency and comparability of Rf values across different laboratories and experiments.

Additionally, Rf values are limited by the assumption of ideal behavior in chromatographic systems. In reality, various factors, such as adsorption, diffusion, and multiple interactions, can complicate the separation process. These complexities may lead to deviations from expected Rf values, challenging accurate interpretation.

Introduction

We will explore the differences in Rf values in terms of interaction with the stationary phase and relative solubility in the mobile phase.

Interaction with the Stationary Phase

The stationary phase in chromatography plays a crucial role in the separation process. It provides a surface on which the compounds in the mixture can interact, leading to differential retention and separation. The interaction between a compound and the stationary phase is influenced by several factors, including polarity, charge, and size.

Polarity is a fundamental property that determines the distribution of charges within a molecule. Compounds can be broadly classified into polar and non-polar based on their electronegativity difference. Polar compounds typically exhibit strong intermolecular forces, such as hydrogen bonding or dipole-dipole interactions, while non-polar compounds have weak or negligible intermolecular forces.

In chromatography, if the stationary phase is polar, compounds with higher polarity will interact more strongly with it. This results in a lower Rf value for polar compounds compared to non-polar compounds. For example, in thin-layer chromatography (TLC), silica gel is a commonly used polar stationary phase. Polar compounds, such as alcohols or carboxylic acids, will have lower Rf values on a silica gel plate compared to non-polar compounds, such as hydrocarbons.

Charge is another factor that affects the interaction between compounds and the stationary phase. Some stationary phases, such as ion-exchange resins, have charged functional groups that can interact selectively with compounds of opposite charge. For instance, in ion-exchange chromatography, positively charged compounds will have a higher affinity for a negatively charged stationary phase, resulting in lower Rf values.

Size is also an important consideration. Larger compounds tend to interact more with the stationary phase due to increased surface area and more extensive van der Waals forces. Consequently, larger compounds will generally have lower Rf values. This is particularly evident in size-exclusion chromatography, where the stationary phase contains porous beads. Smaller molecules can enter the pores and travel more freely, resulting in higher Rf values, while larger molecules are excluded and have lower Rf values.

Relative Solubility in the Mobile Phase

The mobile phase in chromatography serves as a carrier for the compounds and determines their migration through the stationary phase. The solubility of a compound in the mobile phase is a critical factor that affects its Rf value. Compounds with higher solubility in the mobile phase will tend to travel further, resulting in higher Rf values.

Solubility is primarily determined by the intermolecular forces between the compound and the solvent. Polar solvents, such as water or alcohols, have strong dipole-dipole interactions and hydrogen bonding capabilities. Consequently, polar compounds with similar intermolecular forces will exhibit higher solubility in polar solvents. Conversely, non-polar compounds, which have weak intermolecular forces, will display higher solubility in non-polar solvents, such as hexane or chloroform.

In chromatography, if the mobile phase is polar, compounds with higher solubility in the mobile phase will tend to have higher Rf values. For example, in reversed-phase liquid chromatography (RP-LC), a non-polar stationary phase (e.g., C18) is used with a polar mobile phase (e.g., water-acetonitrile mixture). Polar compounds, being more soluble in the polar mobile phase, will have higher Rf values compared to non-polar compounds.

It is worth noting that the solubility of a compound in the mobile phase is not solely determined by the compound's polarity. Other factors, such as temperature, pressure, and the presence of additives, can also influence solubility.

Examples from Various Fields

The principles of Rf values and their differences in interaction with the stationary phase and relative solubility in the mobile phase are applicable to various fields of chromatography and have practical implications.

In pharmaceutical analysis, high-performance liquid chromatography (HPLC) is extensively used for drug analysis. By selecting an appropriate stationary phase and mobile phase, different compounds in a drug formulation can be separated and quantified. For example, in the analysis of a mixture containing polar and non-polar drugs, a C18 stationary phase with a mobile phase consisting of a mixture of water and organic solvent (e.g., acetonitrile) can be employed. The polar drugs will interact more with the stationary phase, resulting in lower Rf values, while non-polar drugs will have higher Rf values due to their higher solubility in the mobile phase.

Similarly, in environmental analysis, gas chromatography (GC) is commonly used to determine the presence of volatile compounds in air or water samples. The choice of stationary phase and mobile phase is critical for the separation of different compounds. For instance, in the analysis of volatile organic compounds (VOCs), a non-polar stationary phase, such as a polydimethylsiloxane (PDMS) column, can be used. Polar VOCs will have lower Rf values due to their interactions with the stationary phase, while non-polar VOCs will have higher Rf values due to their higher solubility in the mobile phase.

Counterarguments

While the differences in Rf values based on interaction with the stationary phase and relative solubility in the mobile phase provide a valuable framework for understanding chromatographic separations, it is important to acknowledge that other factors can also influence Rf values.

One such factor is the flow rate of the mobile phase. A higher flow rate can lead to reduced interaction between compounds and the stationary phase, resulting in higher Rf values overall. Additionally, the choice of detection method, such as UV spectrophotometry or mass spectrometry, can impact the observed Rf values, as different compounds may exhibit different responses to the detection technique.

Furthermore, the accuracy and reproducibility of Rf values can be influenced by experimental conditions, such as temperature, humidity, and column aging. Variations in these parameters can introduce uncertainties and affect the observed Rf values.

Gas / liquid chromatography

Stationary Phase

In chromatography, the stationary phase refers to the immobile component of the system that allows for the separation of different components in a mixture. It is typically a solid or liquid material that is immobilized on a solid support. The stationary phase plays a crucial role in the separation process by interacting differently with the components of the mixture, leading to their differential migration and subsequent separation.

Solid Support

The solid support in chromatography acts as a platform or matrix on which the stationary phase is immobilized. It provides mechanical stability and support to the stationary phase material. The choice of solid support depends on the specific chromatographic technique being used, as well as the properties of the stationary phase and the target analytes.

A high boiling point non-polar liquid is often used as the stationary phase in gas chromatography. This stationary phase is typically a nonpolar liquid such as silicone oil or polydimethylsiloxane (PDMS). These nonpolar liquids have high boiling points, which allows them to remain in the liquid phase under the operating conditions of gas chromatography.

The immobilization of the stationary phase on a solid support is essential for maintaining its integrity and preventing its loss during the separation process. The solid support can be a material such as glass, silica gel, or a polymer resin. It should have good chemical stability, low reactivity, and a high surface area to facilitate the interaction between the stationary phase and the analytes.

The choice of the stationary phase and solid support combination depends on the specific separation requirements. For example, in reversed-phase chromatography, a nonpolar stationary phase is used in combination with a polar solid support such as silica gel. This combination allows for the separation of polar analytes based on their hydrophobic interactions with the nonpolar stationary phase.

Mobile Phase

The mobile phase in chromatography refers to the fluid or gas that carries the analyte mixture through the stationary phase. It is the phase that moves and interacts with the stationary phase, causing the separation of the different components. The mobile phase can be a liquid, gas, or supercritical fluid, depending on the type of chromatography being performed.

Unreactive Gas

In gas chromatography, the mobile phase is typically an unreactive gas such as helium or nitrogen. These gases are chosen because they have low chemical reactivity and do not interact significantly with the analytes or the stationary phase. This allows for efficient and rapid separation of the analyte mixture.

The choice of the mobile phase gas depends on factors such as its availability, cost, and compatibility with the chromatographic system. Helium is commonly used as the mobile phase in gas chromatography due to its low cost, inertness, and availability. However, helium is becoming increasingly scarce and expensive, leading to the exploration of alternative gases such as nitrogen.

The mobile phase gas is responsible for carrying the analyte mixture through the column or stationary phase. It exerts a pressure that drives the movement of the analytes, and their interactions with the stationary phase determine their retention and separation. The composition and flow rate of the mobile phase can be adjusted to optimize the separation and elution of specific analytes.

Retention Time

Retention time is an essential parameter in chromatography that measures the time it takes for a specific analyte to travel through the chromatographic system from injection to detection. It is a measure of the time a component spends in the stationary phase before being eluted or detected.

The retention time is influenced by several factors, including the chemical nature of the analytes, the stationary phase, and the mobile phase. It is determined by the balance between the interactions of the analyte with the stationary phase and its movement through the mobile phase.

The retention time can be used to identify and quantify analytes in a mixture by comparing their retention times with those of known reference compounds. Each analyte has a characteristic retention time that depends on its chemical properties and the chromatographic conditions. By measuring the retention time and comparing it to a calibration curve or database, the analyte can be identified and quantified.

The retention time is affected by various factors, including the polarity of the analyte and the stationary phase, the temperature, and the flow rate of the mobile phase. For example, in gas chromatography, increasing the temperature generally decreases the retention time as it reduces the interaction between the analyte and the stationary phase.

The retention time can also be influenced by the choice of the stationary phase and mobile phase. For instance, in reversed-phase liquid chromatography, polar analytes have longer retention times when a nonpolar stationary phase is used. This is because the polar analytes interact more strongly with the nonpolar stationary phase.

Interpretation of Gas/Liquid Chromatograms in Terms of Percentage Composition of a Mixture

Introduction

Gas chromatography (GC) and liquid chromatography (LC) are powerful analytical techniques used to separate and identify components of a mixture. These techniques provide valuable information about the composition and concentration of various compounds present in a sample. The interpretation of gas/liquid chromatograms involves analyzing the peaks obtained and determining the percentage composition of the mixture.

Gas Chromatography (GC)

Principle and Instrumentation

Gas chromatography is a technique that utilizes a mobile phase of an inert gas (such as helium or nitrogen) to carry the sample through a stationary phase. The stationary phase may consist of either a solid adsorbent or a solid support with a liquid coating. The instrument consists of an injection port, a column, a detector, and a data system.

Interpretation of Gas Chromatograms

When a mixture is injected into a gas chromatograph, the different components of the mixture separate and elute from the column at different times. Each component appears as a peak on the chromatogram, and the area under the peak is proportional to the concentration of the compound.

To determine the percentage composition of a mixture using a gas chromatogram, several steps are involved:

Identification of Peaks: Each peak on the chromatogram represents a different compound. The first step is to identify the peaks by comparing their retention times with known standards or reference compounds. The retention time is the time taken for a compound to travel through the column.

Calculation of Relative Area: The next step is to measure the area under each peak. The area is calculated by integrating the signal produced by the detector as the compound elutes from the column. The relative area of each peak is then determined by dividing the area of a specific peak by the total area of all peaks.

Percentage Calculation: Finally, the percentage composition of each component in the mixture can be calculated by multiplying the relative area of each peak by 100. This gives the proportion of each compound in the mixture.

For example, let's consider a gas chromatogram of a mixture containing three compounds: A, B, and C. The relative areas of the peaks for A, B, and C are 20, 30, and 50, respectively. The percentage composition of the mixture would be 20%, 30%, and 50% for compounds A, B, and C, respectively.

Applications and Limitations

Gas chromatography finds applications in various fields such as environmental analysis, pharmaceuticals, forensic science, and food analysis. It is particularly useful for analyzing volatile and semi-volatile organic compounds.

However, gas chromatography has limitations. It cannot separate compounds with similar boiling points, and non-volatile compounds cannot be analyzed directly. In such cases, derivatization techniques are employed to convert non-volatile compounds into volatile derivatives before analysis.

Liquid Chromatography (LC)

Principle and Instrumentation

Liquid chromatography is another powerful separation technique that uses a liquid mobile phase to carry the sample through a stationary phase. The stationary phase can be a solid adsorbent or a liquid coated on a solid support. Like gas chromatography, the instrument is comprised of an injection port, a column, a detector, and a data system.

Interpretation of Liquid Chromatograms

The interpretation of liquid chromatograms follows a similar process to gas chromatography, with some variations in the separation mechanism and detection methods. In liquid chromatography, compounds are separated based on their differential solubility between the mobile phase and stationary phase.

To determine the percentage composition of a mixture using a liquid chromatogram, the following steps are involved:

Identification of Peaks: Each peak on the chromatogram represents a different compound. The peaks are identified by comparing their retention times with known standards or reference compounds, similar to gas chromatography.

Calculation of Relative Area: The area under each peak is measured by integrating the signal produced by the detector as the compound elutes from the column. The relative area of each peak is determined by dividing the area of a specific peak by the total area of all peaks.

Percentage Calculation: Finally, the percentage composition of each component in the mixture can be calculated by multiplying the relative area of each peak by 100.

Applications and Limitations

Liquid chromatography is widely used in various fields, including pharmaceuticals, environmental analysis, food and beverage analysis, and biochemistry. It can separate a wide range of compounds, including polar and non-polar substances.

However, liquid chromatography also has some limitations. It is not suitable for analyzing volatile compounds that can be easily lost during the analysis. Additionally, some compounds may not be well-retained on the stationary phase, leading to poor separation.

Comparison of Gas and Liquid Chromatography

Selectivity and Separation

Gas chromatography is generally more selective than liquid chromatography due to the wide range of stationary phases available. It can separate compounds with similar boiling points, making it suitable for analyzing complex mixtures. On the other hand, liquid chromatography provides better separation for compounds with different polarities.

Sensitivity and Detection

Gas chromatography typically provides higher sensitivity compared to liquid chromatography. The detectors used in gas chromatography, such as flame ionization detector (FID) and electron capture detector (ECD), have lower detection limits. In contrast, liquid chromatography detectors, such as UV-Vis and fluorescence detectors, have higher detection limits.

Sample Preparation and Analyte Volatility

Gas chromatography requires the sample to be in the gas phase, which necessitates sample volatilization or derivatization for non-volatile compounds. Liquid chromatography, on the other hand, can analyze liquid samples directly, reducing the need for extensive sample preparation.

Applications

Gas chromatography is widely used for analyzing volatile organic compounds (VOCs) in environmental, forensic, and pharmaceutical analysis. Liquid chromatography is more suitable for analyzing polar and non-volatile compounds, making it valuable in pharmaceutical, food, and biochemical analysis.

Retention Times and their Relationship with the Stationary Phase

Introduction

It aims to explore the concept of retention times and their relationship with the stationary phase, considering various fields of application and counterarguments. Through a thorough analysis, this essay will elucidate the complex concepts surrounding retention times, offering a clear understanding of their significance.

Chromatography and Retention Times

Chromatography is a separation technique widely utilized in various scientific fields, such as chemistry, biochemistry, and environmental science. This technique exploits the different affinities of analytes for the stationary and mobile phases. The stationary phase is a solid or liquid material, often immobilized onto a solid support, while the mobile phase is a liquid or gas that carries the analytes through the system. As the analytes interact with the stationary phase, they experience different degrees of retention, resulting in separation.

Retention time is a fundamental parameter in chromatography as it enables the identification and quantification of analytes. By comparing the retention times of unknown compounds with those of known standards, scientists can determine the presence and concentration of specific substances in a sample. Therefore, understanding the factors that influence retention times, particularly the interaction with the stationary phase, is essential for successful chromatographic analysis.

Stationary Phase and Retention

The stationary phase plays a pivotal role in determining the retention time of solutes in chromatography. The interaction between the solutes and the stationary phase can be classified into three major mechanisms: adsorption, partition, and ion exchange.

Adsorption

Adsorption chromatography relies on the interaction between solutes and the surface of the stationary phase. The stationary phase can be a solid material, such as silica gel or alumina. The functional groups present on the surface of these materials can form various types of interactions, including van der Waals forces, hydrogen bonding, and dipole-dipole interactions. These interactions can be influenced by factors such as pH, temperature, and the polarity of the mobile phase.

For example, in reverse-phase chromatography, the stationary phase consists of a nonpolar material, like C18-bonded silica. Nonpolar solutes will interact more strongly with the stationary phase, resulting in longer retention times. Conversely, polar solutes will experience weaker interactions and elute faster. This selective adsorption based on polarity facilitates the separation of complex mixtures.

Partition

Partition chromatography involves the partitioning of solutes between the stationary phase and the mobile phase. The stationary phase is often a liquid coated onto a solid support, such as a thin layer of liquid on a glass plate or a liquid phase immobilized onto solid particles. The solutes distribute themselves between the two phases based on their partition coefficients, which depend on factors like temperature, pH, and the nature of the solutes.

In gas chromatography, the stationary phase is a liquid immobilized onto solid particles inside a column. The solutes, in the gas phase, partition between the stationary phase and the mobile gas phase. The retention time is influenced by the solute's volatility and its affinity for the stationary phase. Volatile compounds will elute faster, while less volatile compounds will exhibit longer retention times.

Ion Exchange

Ion exchange chromatography exploits the interactions between charged solutes and charged sites on the stationary phase. The stationary phase is typically composed of a resin containing functional groups that can attract or repel ions. The retention time is determined by the charge and concentration of the solutes, as well as the type and density of charged sites on the stationary phase. For instance, in cation exchange chromatography, the stationary phase contains negatively charged groups. Positively charged solutes will interact more strongly with the stationary phase, resulting in longer retention times. Negatively charged solutes will experience weaker interactions and elute faster.

Factors Affecting Retention Times

Several factors influence retention times in chromatography, including the nature of the solutes, the composition of the mobile phase, and the properties of the stationary phase. Understanding these factors is essential for optimizing chromatographic separations and achieving accurate results.

Nature of the Solutes

The chemical and physical properties of solutes significantly impact their retention times. Factors such as molecular size, shape, polarity, and charge can influence the interaction with the stationary phase. For instance, larger molecules tend to have longer retention times due to increased surface area for interaction with the stationary phase. Similarly, polar solutes will experience stronger interactions in polar stationary phases, leading to longer retention times.

Composition of the Mobile Phase

The mobile phase composition affects the retention times by modulating the interaction between the solutes and the stationary phase. By altering the polarity, pH, or ionic strength of the mobile phase, it is possible to manipulate the retention times and optimize separations. For example, in reverse-phase chromatography, increasing the organic solvent content in the mobile phase decreases the retention times of polar solutes by weakening their interactions with the stationary phase.

Properties of the Stationary Phase

The properties of the stationary phase, such as surface area, polarity, and charge, directly impact the retention times. Different stationary phases can be selected based on the desired separation, taking into account factors such as the analyte's polarity and the sample matrix. For instance, a polar stationary phase would be appropriate for separating polar analytes, while a nonpolar stationary phase would be suitable for separating nonpolar analytes.

Applications and Counterarguments

Retention times are applied in various scientific fields, including pharmaceutical analysis, environmental monitoring, and forensic analysis. In pharmaceutical analysis, retention times are utilized for the identification and quantification of active pharmaceutical ingredients (APIs) in drug formulations. By comparing the retention times of APIs in a sample with those of reference standards, scientists can ensure the quality and consistency of pharmaceutical products.

In environmental monitoring, retention times are essential for identifying and quantifying pollutants in water, soil, and air samples. By analyzing the retention times of target analytes, scientists can determine the presence and concentration of harmful substances, aiding in the assessment of environmental contamination and the development of mitigation strategies.

However, it is important to consider counterarguments regarding the sole reliance on retention times for analyte identification. In complex mixtures, compounds with similar retention times may co-elute, leading to misidentification or inaccurate quantification. Therefore, complementary techniques, such as mass spectrometry, should be employed to enhance the reliability of chromatographic analysis.

Carbon-13 NMR spectroscopy
Analyzing and Interpreting a Carbon-13 NMR Spectrum

Introduction

We will discuss the process of analyzing and interpreting a carbon-13 NMR spectrum of a simple molecule. We will focus on deducing the different environments of carbon atoms and propose possible structures for the molecule.

Background

NMR spectroscopy is based on the principle that atomic nuclei with an odd number of protons or neutrons possess a property called spin and generate a magnetic field. When placed in an external magnetic field, these nuclei align either parallel or antiparallel to the field. By applying radiofrequency radiation, the nuclei can be excited to a higher energy state, and upon relaxation, they emit energy in the form of electromagnetic radiation. This emitted radiation is detected and analyzed to obtain valuable information about the molecule under investigation.

Carbon-13 NMR Spectrum

Carbon-13 NMR spectroscopy specifically targets carbon atoms, which are commonly found in organic molecules. Unlike hydrogen atoms, which possess spin ½ and can generate a strong NMR signal, carbon-13 atoms have spin ½ and generate a relatively weaker signal due to their lower abundance in nature (1.1% vs. 99.9% for hydrogen). Nevertheless, carbon-13 NMR spectroscopy is still widely used because it provides complementary information to hydrogen NMR and helps in characterizing the carbon framework of organic compounds.

Different Environments of Carbon Atoms

When analyzing a carbon-13 NMR spectrum, the first step is to identify the different environments of carbon atoms present in the molecule. The chemical shift, measured in parts per million (ppm), reflects the electronic environment experienced by the carbon atom. It is affected by factors such as neighboring atoms, hybridization, and electronic effects. By comparing the chemical shifts of different carbon atoms, we can deduce their unique environments. For example, consider a molecule with three different carbon environments labeled A, B, and C. If carbon atom A is in a more electronegative environment than carbon atom B, it will experience a deshielding effect, resulting in a higher chemical shift. On the other hand, if carbon atom C is in a more shielded environment, it will experience a lower chemical shift. By comparing the chemical shifts of A, B, and C, we can conclude that they are present in different chemical environments.

Interpreting the Carbon-13 NMR Spectrum

Once we have identified the different carbon environments, we can interpret the carbon-13 NMR spectrum to propose possible structures for the molecule. The number of peaks in the spectrum corresponds to the number of different carbon environments present. By analyzing the splitting patterns, we can further refine our understanding of the molecule's structure. The splitting patterns arise due to the interaction between neighboring carbon atoms and their spin states. According to the n+1 rule, where n represents the number of neighboring carbon atoms, the number of peaks in a carbon-13 NMR spectrum is given by the formula 2^n. For example, if a carbon atom has three neighboring carbon atoms, it will exhibit a quartet ($2^2 = 4$ peaks) in the spectrum. By analyzing the splitting patterns, we can determine the number of neighboring carbon atoms and their relative positions in the molecule.

Proposing Possible Structures

Based on the information obtained from the carbon-13 NMR spectrum, we can propose possible structures for the molecule. This process involves considering the different carbon environments, their chemical shifts, and the splitting patterns observed. It is important to note that the carbon-13 NMR spectrum alone cannot provide a definitive structure but can guide us towards potential structures.

To illustrate this, let's consider a hypothetical example. Suppose we have a carbon-13 NMR spectrum of a molecule that exhibits three distinct peaks at chemical shifts of 25 ppm, 35 ppm, and 45 ppm. The peak at 25 ppm is a singlet, indicating that the corresponding carbon atom has no neighboring carbon atoms. The peaks at 35 ppm and 45 ppm are both doublets, suggesting that these carbon atoms have one neighboring carbon atom each.

Based on this information, we can propose several possible structures for the molecule. One possibility is a simple linear chain with three carbon atoms, where the carbon atoms at 35 ppm and 45 ppm are adjacent to the central carbon atom at 25 ppm. Another possibility is a branched structure, where the carbon atoms at 35 ppm and 45 ppm are connected to separate branches originating from the central carbon atom at 25 ppm.

To further refine our proposed structures, we can consider additional factors such as connectivity, hybridization, and electronic effects. By analyzing the carbon-13 NMR spectrum in conjunction with other spectroscopic techniques like proton NMR, infrared spectroscopy, and mass spectrometry, we can gain more confidence in our proposed structures.

Predicting the Number of Peaks in a Carbon-13 NMR Spectrum for a Given Molecule

Introduction

It aims to explain and predict the number of peaks in a carbon-13 NMR spectrum for a given molecule.

Basic Principles of Carbon-13 NMR Spectroscopy

Carbon-13 NMR spectroscopy relies on the magnetic properties of the carbon-13 (^{13}C) nucleus, which possesses a spin and a magnetic moment. When placed in a strong magnetic field and subjected to radiofrequency radiation, carbon-13 nuclei absorb energy, leading to a resonance phenomenon. The energy required for resonance depends on the local electronic and molecular environment of the carbon atom, leading to characteristic shifts in the spectrum.

Chemical Shift and Peak Assignment

The carbon-13 NMR spectrum is typically plotted as a graph with the chemical shift on the x-axis and the intensity or signal size on the y-axis. The chemical shift is denoted in parts per million (ppm) and represents the difference in resonance frequency between the carbon-13 nuclei in the sample and a reference compound (often tetramethylsilane, TMS). Each peak in the spectrum corresponds to a unique carbon atom or group in the molecule.

Factors Influencing the Number of Peaks

The number of peaks observed in a carbon-13 NMR spectrum is primarily determined by the different types of carbon atoms present in the molecule. Several key factors influence the appearance of peaks:

Chemical Environment

The chemical environment around a carbon atom significantly affects its magnetic properties and, consequently, its resonance frequency. Different chemical environments lead to distinct shifts in the carbon-13 NMR spectrum. For example, a carbon atom in an alkene group (C=C) resonates at a different frequency compared to an alkane (C-C) carbon atom. Thus, variations in the chemical environment give rise to distinct peaks.

Symmetry and Equivalent Carbon Atoms

Symmetry in a molecule can result in equivalent carbon atoms, which possess identical chemical environments. These equivalent carbon atoms will produce a single peak rather than multiple peaks. For instance, in ethane (CH3-CH3), both carbon atoms are chemically identical, leading to a single peak in the carbon-13 NMR spectrum.

Substituents and Functional Groups

Different substituents or functional groups attached to a carbon atom can alter its chemical environment, leading to different resonance frequencies. For instance, a carbon atom bonded to an electronegative atom like oxygen or nitrogen will experience a different chemical environment compared to a carbon atom bonded to hydrogen. Consequently, the presence of substituents and functional groups can cause additional peaks in the spectrum.

Chirality and Stereochemistry

Chirality and stereochemistry play a crucial role in the number of peaks observed. Carbon atoms with chiral centers, such as those found in stereoisomers, can have different chemical environments due to the spatial arrangement of substituents. This results in multiple peaks in the carbon-13 NMR spectrum. For instance, the two enantiomers of 2-chlorobutane (CH3CHClCH2CH3) exhibit distinct peaks due to their different spatial arrangements.

Predicting the Number of Peaks

To predict the number of peaks in a carbon-13 NMR spectrum for a given molecule, one must consider the factors discussed above. Analyzing the molecular structure and identifying the different types of carbon atoms present is crucial. By examining the connectivity, symmetry, and substituents, one can make reasonable predictions about the number of peaks.

For example, consider 1-bromopropane (CH3CH2CH2Br). This molecule contains four different types of carbon atoms: a methyl group (CH3), an ethyl group (CH2CH3), a methylene group (CH2), and a quaternary carbon atom bonded to bromine (C-Br). Consequently, we would expect to observe four distinct peaks in the carbon-13 NMR spectrum.

However, there are exceptions to these predictions. Certain molecular arrangements can cause peak splitting, known as spin-spin coupling, which occurs when hydrogen atoms on adjacent carbons influence each other's magnetic properties. This coupling results in additional peaks in the spectrum. For example, in an ethyl group (CH2CH3), the hydrogen atoms on the methylene carbon can couple with the hydrogen atoms on the methyl carbon, causing peak splitting.

Counterarguments and Limitations

While the factors discussed above provide a general framework for predicting the number of peaks in a carbon-13 NMR spectrum, there are limitations and counterarguments to consider.

Firstly, predicting the number of peaks solely based on molecular structure can be challenging for complex molecules with multiple functional groups. The presence of overlapping peaks and subtle variations in chemical environment can make peak assignment difficult.

Secondly, spin-spin coupling can complicate the interpretation of peak patterns. Coupling constants, which describe the magnitude and splitting pattern of coupled peaks, depend on the distance and number of hydrogen atoms influencing each other. Therefore, the prediction of peak splitting requires detailed knowledge of the coupling constants, which may not always be readily available.

Lastly, environmental factors such as temperature and solvent choice can impact the appearance of carbon-13 NMR spectra. For instance, temperature-dependent dynamic processes can cause peak broadening, making peak assignment more challenging.

Proton (^{1}H) NMR spectroscopy

Introduction

We will focus on a simple molecule and go through each step of the interpretation process. We will start by discussing the different environments of protons and how they are reflected in the chemical shift values. Then, we will analyze the relative peak areas to determine the relative numbers of each type of proton present. Next, we will examine the splitting patterns to deduce the number of equivalent protons on the adjacent carbon atom. Finally, based on all the information obtained, we will propose possible structures for the molecule.

Different Environments of Protons (Chemical Shift Values)

The chemical shift value of a proton in an NMR spectrum is a measure of its electron density and the magnetic environment it experiences. It is expressed in parts per million (ppm) and is referenced to a standard compound, usually tetramethylsilane (TMS), which is assigned a chemical shift of 0 ppm. The chemical shift values are influenced by several factors, including electronegativity, hybridization, neighboring atoms, and molecular symmetry.

Different environments of protons in a molecule result in different chemical shift values. Protons in more electronegative environments experience greater deshielding, leading to higher chemical shift values (upfield shifts), while protons in less electronegative environments experience greater shielding, resulting in lower chemical shift values (downfield shifts).

For example, in a molecule with a hydroxyl group (-OH) and a methyl group (-CH3), the proton in the hydroxyl group will have a higher chemical shift value (around 1-5 ppm) compared to the proton in the methyl group (around 0-2 ppm). This is because the oxygen atom in the hydroxyl group is more electronegative than the carbon atoms in the methyl group, causing deshielding.

By analyzing the chemical shift values in a proton NMR spectrum, we can deduce the different environments of protons present in the molecule. These chemical shift values provide valuable information about the functional groups and the connectivity of atoms in the molecule.

Relative Numbers of Protons (Peak Areas)

The relative peak areas in a proton NMR spectrum provide information about the relative numbers of each type of proton present in the molecule. The peak areas are proportional to the number of equivalent protons in a particular environment. Equivalent protons are protons that experience the same magnetic environment and, therefore, have the same chemical shift value.

The integration of the peaks in a proton NMR spectrum is a quantitative measurement of the peak areas. It is usually represented as a ratio or percentage of the total integration value. By comparing the integrations of different peaks, we can determine the relative numbers of each type of proton.

For example, consider a molecule with two types of protons: protons A and protons B. If the integration value of the peak corresponding to protons A is twice that of the peak corresponding to protons B, it indicates that there are twice as many protons A as protons B in the molecule.

In some cases, the peak areas may not be directly proportional to the number of protons due to different relaxation times, spin-lattice relaxation, or other experimental factors. However, in most cases, the peak areas provide a reliable estimation of the relative numbers of protons.

Number of Equivalent Protons (Splitting Patterns)

The splitting patterns in a proton NMR spectrum arise from the interaction between protons on adjacent carbon atoms. This interaction is known as spin-spin coupling or J-coupling. The splitting pattern provides information about the neighboring protons and the number of equivalent protons on the adjacent carbon atom.

The n + 1 rule is used to determine the number of equivalent protons on the adjacent carbon atom. According to this rule, a proton is split into n + 1 peaks by its neighboring protons, where n is the number of equivalent protons on the adjacent carbon atom.

For example, if a proton is adjacent to two equivalent protons, it will be split into three peaks, forming a triplet. If it is adjacent to three equivalent protons, it will be split into four peaks, forming a quartet. This splitting pattern continues for higher numbers of equivalent protons.

By analyzing the splitting patterns in a proton NMR spectrum, we can deduce the number of equivalent protons on the adjacent carbon atom. This information is crucial for determining the connectivity of atoms in the molecule and proposing possible structures.

Possible Structures for the Molecule

Based on the analysis of the chemical shift values, peak areas, and splitting patterns in a proton NMR spectrum, we can propose possible structures for the molecule. However, it is important to note that the proton NMR spectrum alone may not provide sufficient information to determine the exact structure. Additional spectroscopic techniques, such as carbon NMR, infrared spectroscopy, and mass spectrometry, may be required for a more comprehensive analysis.

To propose possible structures, we need to consider all the information obtained from the proton NMR spectrum. This includes the chemical shift values, peak areas, and splitting patterns. We can use this information to identify the functional groups present in the molecule, determine the connectivity of atoms, and consider the overall symmetry of the molecule.

For example, if the proton NMR spectrum shows a peak at a chemical shift value of 7 ppm, it suggests the presence of an aromatic proton. This information, combined with the peak areas and splitting patterns, can help us propose a structure that includes an aromatic ring.

It is important to consider the limitations and uncertainties in proposing structures based on proton NMR spectra alone. In complex molecules, different functional groups and isomers may exhibit similar chemical shift values and peak areas. In such cases, additional spectroscopic techniques and advanced analysis methods, such as two-dimensional NMR spectroscopy and computer modeling, may be necessary to confirm the proposed structures.

Predicting the Chemical Shifts and Splitting Patterns of Protons in Molecules

Introduction

It aims to explain the methods and principles behind predicting these shifts and patterns, highlighting the factors that influence proton resonance and the techniques used to interpret NMR spectra.

Chemical Shifts

The chemical shift is the position of a proton signal in an NMR spectrum, expressed in parts per million (ppm) relative to a reference compound. It depends on the electron density and the local magnetic field surrounding the proton.

The shielding or deshielding effect of nearby atoms and functional groups significantly influences the chemical shift.

Shielding and Deshielding

When a proton experiences a shielding effect, it encounters a more electron-rich environment, resulting in a downfield shift. On the other hand, deshielding occurs when a proton encounters a more electron-deficient environment, leading to an upfield shift. The shielding or deshielding effect is primarily governed by the electronegativity and electron-withdrawing or electron-donating nature of nearby atoms or functional groups. For example, in a molecule containing an electronegative atom such as oxygen, the electron density around the proton is reduced, causing deshielding and an upfield shift. Conversely, in the presence of an electron-donating group like an alkyl substituent, the electron density around the proton is increased, leading to shielding and a downfield shift.

Chemical Shift Ranges

Chemical shifts are reported in ppm and are referenced to a standard compound, usually tetramethylsilane (TMS), which is assigned a chemical shift of 0 ppm. The chemical shift ranges of protons in common functional groups are well-established and provide initial guidance for predicting shifts.

For instance, aliphatic protons (protons attached to sp3 hybridized carbons) typically appear between 0.5 and 3.0 ppm. Aromatic protons (protons on aromatic rings) are commonly observed in the range of 6.0 to 8.5 ppm. Carboxylic acid protons, due to their strong deshielding effect, often resonate in the range of 10.5 to 12.5 ppm.

Anisotropic Effects

In addition to the local electronic environment, anisotropic effects resulting from the orientation of nearby bonds or molecular symmetry can influence the chemical shift of protons. For example, the orientation of a double bond in an alkene can cause a slight upfield shift of adjacent protons due to the π-electron cloud. These anisotropic effects can be challenging to predict accurately and often require computational methods to obtain precise values.

Splitting Patterns

Splitting patterns in a 1H NMR spectrum arise from the coupling between the observed proton and its neighboring protons. This phenomenon is known as spin-spin coupling and provides valuable information about the number of protons that are three bonds away from the observed proton.

N+1 Rule

The N+1 rule is a fundamental principle used to predict the splitting patterns of proton signals. According to this rule, if a proton has N neighboring protons, it will be split into N+1 peaks in the NMR spectrum. The intensity of each peak is determined by the Pascal's triangle coefficients (1, 2, 3, 4, 5, etc.), which follow a binomial distribution.

For instance, a proton with one neighboring proton (a doublet) will exhibit a peak pattern with an intensity ratio of 1:1. A proton with two neighboring protons (a triplet) will show a peak pattern with an intensity ratio of 1:2:1.

Coupling Constants

The coupling constant (J) is a measure of the energy difference between the spin states of coupled protons. It is expressed in Hertz (Hz) and provides information about the nature of the coupling interaction. The magnitude of the coupling constant depends on the number of intervening bonds between the coupled protons and their magnetic coupling pathway.

In a simple case, two protons coupled through one bond (vicinal coupling), the coupling constant is typically around 7 Hz. However, the coupling constants can vary significantly depending on the molecular structure and the nature of the coupling interaction.

Karplus Equation

The Karplus equation is a useful tool for predicting the coupling constants in certain organic molecules, particularly those containing sp3 hybridized carbons and protons. It relates the dihedral angle (θ) between the coupled protons and the coupling constant (J) through the following equation:

$$J = A \cos^2{\theta} + B \cos{\theta} + C$$

where A, B, and C are constants determined experimentally for specific types of couplings. The Karplus equation provides a reasonable approximation of the coupling constants in simple systems but becomes less accurate for more complex molecules.

The Use of Tetramethylsilane (TMS) as the Standard for Chemical Shift Measurements

Introduction

It aims to explore the use of TMS as the standard for chemical shift measurements, discussing its properties, applications, and limitations.

Properties of Tetramethylsilane

TMS is a non-toxic and chemically inert compound with the molecular formula $Si(CH_3)_4$. Its unique chemical structure makes it an ideal choice as a reference standard. The four methyl groups surrounding the silicon atom shield the silicon nucleus effectively, resulting in a distinct and easily identifiable peak in NMR spectra. The chemical shift of TMS is conventionally set to zero, allowing for the accurate measurement of chemical shifts relative to this reference compound.

Chemical Shift and Its Measurement

Chemical shift is a fundamental property observed in NMR spectroscopy. It describes the displacement of a nucleus in a molecule's structure in response to an externally applied magnetic field. The chemical shift is expressed in parts per million (ppm) and provides valuable information about the local electronic environment and molecular structure.

The chemical shift of a nucleus is influenced by various factors, including the electron distribution surrounding the nucleus, neighboring atoms, and functional groups. By comparing the chemical shifts of different nuclei within a molecule or between different compounds, valuable insights into molecular structure, bonding, and electronic properties can be obtained.

The chemical shift measurement involves comparing the resonance frequency of a specific nucleus to that of a reference standard, such as TMS. The resonance frequency is determined by the strength of the applied magnetic field and the local electronic environment of the nucleus. By calibrating the chemical shifts of other nuclei relative to TMS, accurate and reproducible chemical shift values can be obtained.

Advantages of TMS as a Standard

Several advantages make TMS the standard of choice for chemical shift measurements:

High Chemical Shift Stability

TMS exhibits exceptional stability in most solvents and under various experimental conditions. It is chemically inert, non-reactive, and does not undergo complexation or coordination with other molecules. This stability ensures that the chemical shift value of TMS remains constant, providing a reliable reference point for chemical shift measurements.

Easily Identifiable Signal

The four methyl groups surrounding the silicon atom in TMS result in a distinct and intense singlet peak in the NMR spectrum. This peak is typically located far downfield from most other signals, making it easy to identify and distinguish from other resonances in the sample. The high signal intensity further enhances the accuracy and precision of chemical shift measurements.

Wide Solvent Compatibility

TMS is soluble in a wide range of common NMR solvents, including deuterated solvents commonly used in NMR experiments. This solubility facilitates the accurate determination of chemical shifts across different solvents, enabling the comparison of NMR data obtained under different experimental conditions.

Low Toxicity

TMS is non-toxic and does not pose any significant health hazards, making it safe to handle and dispose of after experiments. This characteristic is crucial in ensuring the well-being of researchers and maintaining laboratory safety.

Applications of TMS as a Standard

The use of TMS as a standard for chemical shift measurements finds widespread applications across various scientific disciplines:

Organic Chemistry

Chemical shift measurements using TMS as a reference standard are invaluable in the identification and structural elucidation of organic compounds. By comparing the chemical shifts of different nuclei in a molecule, researchers can determine the presence or absence of specific functional groups, the degree of unsaturation, and the connectivity of atoms within the molecule. This information is vital in characterizing unknown compounds and verifying the success of chemical reactions.

Pharmaceutical Research

In pharmaceutical research, accurate determination of chemical shifts is crucial for drug discovery and development. By comparing the chemical shifts of different nuclei in drug candidates to those of known compounds, researchers can gain insights into the compound's structure, stability, and interactions with biomolecules. This information aids in the design and optimization of drug candidates with improved properties, such as enhanced bioavailability and reduced toxicity.

Material Science: Chemical shift measurements using TMS as a standard are also employed in material science research. By studying the chemical shifts of nuclei in materials, researchers can gain valuable information about the composition, structure, and properties of materials. This knowledge is essential in developing new materials with tailored properties, such as improved conductivity, mechanical strength, or magnetic behavior.

Environmental Analysis

Chemical shift measurements using TMS as a reference standard play a critical role in environmental analysis. By measuring the chemical shifts of various nuclei in environmental samples, researchers can identify and quantify pollutants, contaminants, and natural compounds. This information aids in monitoring environmental quality, assessing the impact of human activities, and designing effective remediation strategies.

Limitations and Considerations

While TMS is a widely used standard for chemical shift measurements, it is essential to consider its limitations:

NMR Instrument Calibration

Accurate chemical shift measurements require precise calibration of the NMR instrument. Factors such as magnetic field homogeneity, shimming, and temperature control can influence the observed chemical shift values. Regular instrument calibration and maintenance are necessary to ensure reliable and reproducible chemical shift measurements.

Signal Overlap

In complex mixtures or crowded NMR spectra, signal overlap can occur, making it challenging to accurately assign chemical shifts. The distinct singlet peak of TMS can aid in overcoming this issue, but care must be taken to avoid overlap with other resonances in the sample.

Different Solvent Properties

While TMS is compatible with various solvents, the choice of solvent can influence the chemical shift values observed. Different solvents can exhibit different magnetic susceptibility effects, leading to slight variations in chemical shift values. It is crucial to consider solvent effects and adjust for them when comparing chemical shifts obtained in different solvents.

The Importance of Deuterated Solvents in Proton NMR Spectroscopy

Introduction

It aims to explain the need for deuterated solvents, such as CDCl3, in obtaining accurate proton NMR spectra.

Background

Before delving into the importance of deuterated solvents in proton NMR spectroscopy, it is crucial to understand the basic principles of NMR spectroscopy. In a magnetic field, nuclei with an odd number of protons or neutrons possess a property called spin, which generates a magnetic moment. These spins can align in either parallel or anti-parallel with the external magnetic field when placed within it. The energy difference between these two alignments is directly proportional to the strength of the field and the type of nucleus.

To excite the nuclei, a radiofrequency (RF) pulse is applied to the sample, causing a transition of spins from the lower energy state to the higher energy state. As the spins return to their equilibrium state, they release energy that can be detected by a receiver coil. The resulting NMR spectrum displays a series of peaks corresponding to different resonances of the nuclei present in the sample.

The Problem of Solvent Interference

One of the major challenges in obtaining proton NMR spectra is the interference caused by protons present in the solvent. Most solvents used in chemical and biological experiments, such as water (H_2O), methanol (CH_3OH), and acetone (CH_3COCH_3), contain abundant protons that can produce intense signals. These signals can mask the desired proton resonances of the compound under investigation, making it difficult to interpret the spectrum accurately.

Deuterated Solvents: The Solution

Deuterated solvents, which substitute the proton in the solvent molecule with deuterium (the isotope of hydrogen with one proton and one neutron), provide a simple and effective solution to the problem of solvent interference in proton NMR spectroscopy. Deuterium is a stable isotope and does not possess spin, making it undetectable in proton NMR experiments. Consequently, deuterated solvents do not contribute to the background signals in the spectrum, allowing the desired proton resonances to be observed clearly.

Deuterated Solvents in Practice

Various deuterated solvents are commonly used in proton NMR spectroscopy, such as deuterated chloroform (CDCl3), deuterated dimethyl sulfoxide (DMSO-d6), and deuterated methanol (CD3OD). CDCl3 is particularly popular due to its low cost, ease of handling, and compatibility with a wide range of compounds. It is widely available and has a boiling point of 61.2 °C, making it suitable for most routine NMR experiments. When preparing a sample for proton NMR spectroscopy, a small amount of the compound of interest is dissolved in the deuterated solvent. The deuterated solvent serves two main purposes: it dissolves the compound, ensuring it is in a homogeneous state, and it acts as the reference compound for chemical shift determination. The reference compound, typically tetramethylsilane (TMS), provides a known chemical shift at 0 ppm, against which all other resonances in the spectrum are compared.

Benefits of Deuterated Solvents

The use of deuterated solvents offers several important benefits in proton NMR spectroscopy. Firstly, it eliminates the interference caused by the solvent protons, enabling the observation of clear and well-defined proton resonances. This enhances the accuracy and reliability of spectral interpretation, facilitating the determination of molecular structures and functional groups.

Secondly, deuterated solvents provide a stable baseline in the spectrum. A baseline is a reference line that reflects the absence of resonances and serves as a visual guide for the intensity and position of the observed peaks. Since deuterium does not possess spin, the deuterated solvent does not contribute any signals to the baseline, allowing for easier peak identification and quantification.

Thirdly, deuterated solvents exhibit a relatively low natural abundance of deuterium in comparison to the abundant presence of hydrogen. This lower natural abundance leads to a reduced level of background noise in the spectrum, enhancing the signal-to-noise ratio and improving the sensitivity of the measurement. This increased sensitivity allows for the detection of smaller quantities of compounds and improves the detection limits of proton NMR spectroscopy.

Counterarguments

While the use of deuterated solvents is generally beneficial in proton NMR spectroscopy, some counterarguments can be raised. One counterargument is the additional cost associated with deuterated solvents compared to their non-deuterated counterparts. Deuterated solvents are more expensive due to the higher cost of producing and purifying deuterium. However, considering the accuracy and reliability gained from using deuterated solvents, the additional cost is justified for many researchers.

Another counterargument is the potential isotopic exchange between the deuterated solvent and the protons in the compound being analyzed. This exchange can occur if the sample is left in contact with the deuterated solvent for an extended period. However, this issue is generally negligible in routine NMR experiments, as the exchange rate is typically slow and can be minimized by minimizing the sample exposure time.

Identification of O–H and N–H protons can be achieved through the process of proton exchange using D2O

Introduction

We will explore the principles behind proton exchange using D2O, its application in various fields, and potential counterarguments to its effectiveness.

Principles of Proton Exchange using D2O

Proton exchange refers to the process where a hydrogen atom, usually found in an O–H or N–H bond, is replaced by a deuterium atom (D). This exchange reaction occurs in the presence of deuterium oxide (D2O), which acts as a source of deuterium. The exchange reaction can be represented as follows:

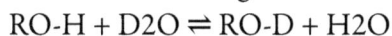

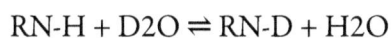

RO-H + D2O ⇌ RO-D + H2O

RN-H + D2O ⇌ RN-D + H2O

In these reactions, RO-H and RN-H represent organic compounds containing O–H and N–H protons, respectively. Upon exposure to D2O, the protons are replaced by deuterium atoms, resulting in the formation of RO-D and RN-D, along with ordinary water (H2O) as a byproduct.

The mechanism behind proton exchange involves the nucleophilic attack of deuterium oxide on the O–H or N–H bond. The deuterium atom (D) acts as a nucleophile, attacking the hydrogen atom and forming a transition state. This transition state eventually leads to the formation of a new O-D or N-D bond, while the original O–H or N–H bond is broken. This exchange process occurs rapidly and is facilitated by the high reactivity of deuterium atoms.

Application in Organic Chemistry

Proton exchange using D2O is widely used in organic chemistry for the identification of O–H and N–H protons. This technique provides valuable information about the presence and location of these functional groups in a molecule. By analyzing the exchange pattern, researchers can deduce the connectivity and structure of the compound under investigation.

One common application of proton exchange is in the determination of the number of O–H or N–H protons in a molecule. Since each exchange reaction involves the replacement of one proton, the number of exchanged protons can be correlated to the number of protons initially present in the functional group. For example, if a compound undergoes three exchange reactions, it indicates the presence of three O–H or N–H protons.

Moreover, proton exchange can provide insights into the relative acidity of O–H and N–H protons. The rate of exchange is influenced by the acidity of the proton being exchanged. More acidic protons exchange at a faster rate compared to less acidic ones. By measuring the speed of the exchange reaction, the relative acidity of different O–H or N–H protons can be determined. This information is valuable for understanding the reactivity and behavior of organic compounds.

Application in Biochemistry

Proton exchange using D2O finds extensive applications in the field of biochemistry. It is particularly useful in studying the structure and dynamics of proteins and nucleic acids. By selectively exchanging specific protons, valuable information about the local environment and hydrogen bonding networks can be obtained. In proteins, for instance, proton exchange can be used to investigate the folding and unfolding processes. By monitoring the exchange of amide N–H protons, which are involved in hydrogen bonding, the stability and structural changes of proteins can be analyzed. The rate of exchange can provide insights into the accessibility and flexibility of different regions within the protein structure. Similarly, in nucleic acids, proton exchange can reveal information about base pairing and structural stability. By selectively exchanging protons in the sugar–phosphate backbone or the base pairs, the dynamic behavior and conformational changes of DNA and RNA molecules can be studied. This knowledge is crucial for understanding fundamental biological processes such as DNA replication and transcription.

Counterarguments and Limitations

While proton exchange using D2O is a powerful technique, it does have certain limitations and potential counterarguments. One counterargument is that proton exchange may not be specific to O–H and N–H protons. Other functional groups containing exchangeable protons, such as S–H and C–H bonds, can also undergo exchange reactions with D2O. Therefore, caution must be exercised when interpreting the results, and complementary techniques should be employed to confirm the presence and identity of O–H and N–H protons.

Another limitation is that proton exchange is a rapid process, making it difficult to distinguish exchangeable protons that are in close proximity. If two O–H or N–H protons are adjacent to each other, they may exchange simultaneously, leading to ambiguous results. In such cases, additional experiments, such as NMR spectroscopy or crystallography, may be necessary to resolve the exact location of the protons.

Additionally, proton exchange can be influenced by the solvent and reaction conditions. Different solvents may have varying exchange rates, potentially affecting the accuracy and reproducibility of the results. The temperature and pH of the reaction can also influence the rate of exchange, further complicating the analysis. These factors should be carefully controlled and considered when performing proton exchange experiments.

About the Author

The author is an accomplished professional with an MBA in Finance, ACCA (Knowledge Level), BBA (Finance), and a major in Finance for their O and A levels. With over ten years of practical investment experience